Vertical-Cavity Surface-Emitting Lasers

One of the key advances in photonic technology in recent years is the development of vertical-cavity surface-emitting lasers, or VCSELs. These devices have a huge range of potential applications, in areas such as communications, printing, and optical switching. This book provides a clear insight into the physics of VCSELs and describes details of their fabrication and applications.

All of the book's contributors are at the forefront of VCSEL research and development. Together they provide complete and coherent coverage of the current state of the art. The opening chapters cover VCSEL design, emission from microcavities, growth, fabrication, and characterization. These are followed by chapters on long- and short-wavelength VCSELs, optical data links, and free-space optical processing.

The book will be of great interest to graduate students and researchers in electrical engineering, applied physics, and materials science. It will also be an excellent reference volume for practicing engineers in the photonics industry.

Carl Wilmsen is Professor of Electrical Engineering at Colorado State University, where he has served as Head of the Department and as Associate Director of the Optoelectronic Computing System Center.

Henryk Temkin holds the Jack Maddox Chair in Electrical Engineering at Texas Tech University. He is a Fellow of the IEEE.

Larry Coldren is Professor of Electrical and Computer Engineering at the University of California, Santa Barbara. He is a Fellow of the IEEE and the OSA. He has previously published a book entitled *Diode Lasers and Photonic Integrated Circuits*.

CAMBRIDGE STUDIES IN MODERN OPTICS

Series Editors

P. L. KNIGHT

Department of Physics, Imperial College of Science, Technology and Medicine

A. MILLER

Department of Physics and Astronomy, University of St. Andrews

Vertical-Cavity Surface-Emitting Lasers

Cambridge Studies in Modern Optics

TITLES IN PRINT IN THIS SERIES

Vertical-Cavity Surface-Emitting Lasers

Design, Fabrication, Characterization, and Applications

Edited by

CARL W. WILMSEN
Colorado State University

HENRYK TEMKIN
Texas Tech University

and

LARRY A. COLDREN
University of California, Santa Barbara

CAMBRIDGE
UNIVERSITY PRESS

PUBLISHED BY THE PRESS SYNDICATE OF THE UNIVERSITY OF CAMBRIDGE
The Pitt Building, Trumpington Street, Cambridge, United Kingdom

CAMBRIDGE UNIVERSITY PRESS
The Edinburgh Building, Cambridge CB2 2RU, UK   http://www.cup.cam.ac.uk
40 West 20th Street, New York, NY 10011-4211, USA   http://www.cup.org
10 Stamford Road, Oakleigh, Melbourne 3166, Australia

First published 1999

Transferred to digital printing 2001

Printed in Great Britain by Biddles Short Run Books, King's Lynn

Typeset in Times Roman 10/12.5 pt. in LATEX 2_ε [TB]

A catalog record for this book is available from the British Library.

Library of Congress Cataloging-in-Publication Data

Vertical-cavity surface-emitting lasers : design, fabrication,
characterization, and applications / edited by Carl Wilmsen, Henryk
Temkin, Larry A. Coldren.
    p.   cm. – (Cambridge studies in modern optics)
                ISBN 0-521-59022-1
1. Semiconductor lasers.   I. Wilmsen, Carl W.   II. Temkin. H.
III. Coldren, L. A. (Larry A.)   IV. Series: Cambridge studies in
                modern optics (Unnumbered)
TA1700.V474   1999
                                                                98-31984
621.36′6 – dc21   CIP
                ISBN 0 521 59022 1 hardback

To our wives
Ann, Bharti, and Donna

Contents

Contributors

Dr. Dubravko I. Babić
Hewlett-Packard Laboratories
Palo Alto, CA 94304

Professor John E. Bowers
Department of Electrical & Computer
Engineering
University of California, Santa Barbara
Santa Barbara, CA 93106

Dr. Kent D. Choquette
Sandia National Laboratories
Albuquerque, NM 87185

Professor Larry A. Coldren
Department of Electrical & Computer
Engineering
University of California, Santa Barbara
Santa Barbara, CA 93106

Kent M. Geib
Sandia National Laboratories
Albuquerque, NM 87185

Dr. Kirk S. Giboney
Hewlett-Packard Laboratories
Palo Alto, CA 94304

Dr. Kenneth Hahn
Hewlett-Packard Laboratories
Palo Alto, CA 94304

Eric R. Hegblom
Department of Electrical & Computer
Engineering

University of California, Santa Barbara
Santa Barbara, CA 93106

Y. H. Houng
Hewlett-Packard Laboratories
Palo Alto, CA 94304

Dr. N. E. J. Hunt
Mitra Imaging
Waterloo, Ontario,
Canada

Dr. Kenichi Kasahara
NEC Corporation
Tsukuba,
Japan

Dr. Dmitri V. Kuksenkov
Polaroid Corp.
Cambridge, MA

Dr. Joachim Piprek
Materials Science Program
University of Delaware
Newark, DE 19716

Dr. Rick Schneider
Hewlett-Packard Laboratories
Palo Alto, CA 94304

Professor E. Fred Schubert
Department of Electrical and Computer
Engineering
Boston University
Boston, MA 02215

Professor Henryk Temkin
Department of Electrical Engineering
Texas Tech University
Lubbock, TX 79409

Dr. Robert L. Thornton
Xerox Palo Alto Research Center
Palo Alto, CA 94304

Dr. Richard C. Williamson
MIT Lincoln Laboratory
Lexington, MA 02173

Professor Carl W. Wilmsen
Department of Electrical Engineering
Colorado State University
Fort Collins, CO 80523

1

Introduction to VCSELs

L. A. Coldren, H. Temkin, and C. W. Wilmsen

1.1 Introduction

In this book we give up-to-date details on a new type of diode laser that promises to find widespread use in many future applications. In fact, these vertical-cavity surface-emitting lasers (VCSELs) are already being used in some prototype products. This is a remarkable turn of events because these devices were considered little more than laboratory novelties at the beginning of this decade. Their rise in credibility has largely been forced by the rapid evolution of their performance as well as the more widespread recognition of their compatibility with low-cost wafer-scale fabrication and characterization technologies. They are especially interesting for array applications, since specific arrays can be formed with no change in the fabrication procedure. In this chapter we give a general introduction to what VCSELs are, how they are made, and for what applications they are being considered.

Most characteristics of GaAs-based VCSELs in the 0.8 to 1.0 μm wavelength range are now comparable to or better than those of edge-emitters in the lower power ($\sim$1 mW) regime. Figure 1.1 gives plots of the lowest threshold current versus the year of publication for 980 nm diode lasers of any type. It indicates the rapid progress in VCSELs that has taken place over the past few years. As can be seen, VCSELs have had lower threshold currents than uncoated edge-emitters for several years, and since 1995 they have had the overall lowest thresholds. Figure 1.2(a) shows the improvement in the electrical to optical powerconversion (or "wall-plug") efficiency versus year. Oxide-apertured devices have yielded efficiencies as high as 57%. While the peak efficiency is an important figure of merit, a more practical characteristic is the efficiency at a desired operating power, which is generally around 0.5 mW for transmission over a few hundred meters of multimode fiber and considerabely lower for free-space interconnects over a few centimeters. Figure 1.2(b) plots the wall-plug efficiency versus output power for several leading results. Figure 1.3 gives the 3 dB modulation bandwidths versus year, along with the required bias current. Because of their small volume, VCSELs are able to achieve very high bandwidths with lower bias currents and dissipated powers than edge-emitters.

Efficiency and high speed at low powers are of paramount importance for such applications as short-haul data communications. For these applications, fiber loss and dispersion are generally not significant factors, and so the 830 to 980 nm wavelength is appropriate. VCSELs are also proposed for many other applications, ranging from printing to optical switching. As we shall see in the following chapters, their improved characteristics are a natural consequence of scaling the active and modal volumes of these diode lasers. The appeal of the VCSEL structure is that it enables this scaling in a simpler way than with

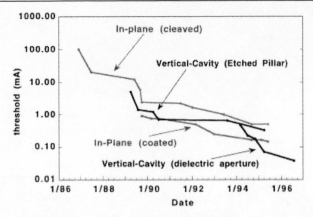

Figure 1.1. Laser threshold current as a function of publication date (only strained-layer InGaAs/GaAs devices are included).

edge-emitters. Besides the manufacturability feature, additional attractive characteristics of VCSELs include their circularly shaped, low-numerical-aperture output beams for easy coupling to fibers or free-space optics, their single-axial-mode spectra (although lateral modes still need to be dealt with) for potential wavelength-division multiplexing (WDM) or wavelength addressing schemes, their high power conversion efficiency in the low power range for reduced heating in highly integrated circuits, and their natural vertical emission for array applications. The chapters to follow detail the state of the art in VCSEL performance as well as their use in many new applications.

Efforts in InP-based longer-wavelength VCSELs, favored for long-haul fiber-based telecommunications (1.3–1.6 μm), have met with slower progress because of inherent difficulties in constructing highly reflecting mirrors as well as high-gain active regions. Nevertheless, recent work with wafer-bonded GaAs mirrors and strained-quantum-well active regions suggests that viable devices are forthcoming. Chapter 8 specifically deals with recent progress in these devices.

Similarly, shorter-wavelength, visible VCSELs have been somewhat more difficult to engineer, but again, recent progress suggests that viable devices will shortly emerge. These are desired for plastic fiber data links (∼650 nm) and storage and display applications. Chapter 7 will review recent progress in red VCSELs. With the advent of gallium-nitride-based light emitters, VCSELs with wavelengths extending even to the UV seem possible.

1.2 Structures

Figure 1.4 illustrates four possible VCSEL structures. The first three have emerged as viable candidates in the GaAs-based devices, and the fourth is seen as the logical evolution in the quest for ever smaller and more efficient devices. Many variations on these general structures have been explored, including combinations of two or more. The primary purpose of these structures is to provide a VCSEL of limited lateral extent without introducing unwanted losses, so that relatively small devices can be made with practical continuous wave (CW) operating characteristics. Ultimately, one might hope to have operating current densities only slightly larger than those of large-area devices. We now discuss some of the features

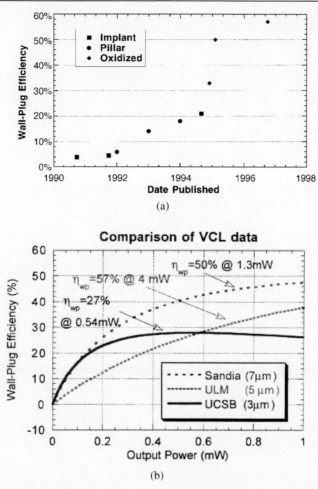

Figure 1.2. (a) Plot of the peak wall-plug efficiency of 850 nm and 980 nm VCSELs versus date of publication. Initial VCSELs had low efficiency due to high mirror resistance and an excessive number of mirror periods. Work of Peters [9] to reduce the mirror resistance improved the efficiency just above 17% in etched pillar VCSELs. Later oxide apertures enabled low-loss confinement of small modes with a large contact area, and these improvements lead to efficiencies as high as 50% [24] and 57% [110]. (b) Plot of the wall-plug efficiency versus output power for three of the leading results [24, 111, 112, and 113]. The UCSB result used a lenslike tapered aperture.

of each structure shown in Figure 1.4 in hopes of giving the reader a feel for the general trade-offs involved.

1.2.1 Etched Mesa

The first structure, Figure 1.4(a), is termed the etched-mesa structure [1–4], and as might be obvious, it is very analogous to the edge-emitting ridge structure. Only here, a two-dimensional round or square postlike mesa is formed. As in the edge-emitter case, the etch

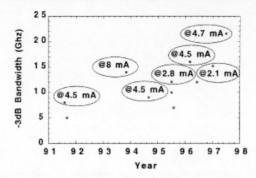

Figure 1.3. Small-signal 3 dB modulation bandwidth versus publication date with bias currents indicated.

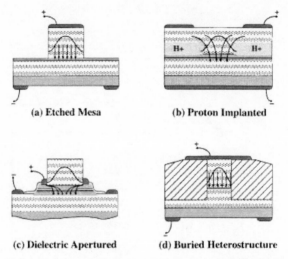

(a) Etched Mesa   **(b) Proton Implanted**

(c) Dielectric Apertured   **(d) Buried Heterostructure**

Figure 1.4. Schematics of some common VCSEL structures. (a) Etched mesa (bottom-emitting): Drive current confined to mesa width; current conducted through the mirrors; optical index guide formed by mesa walls. (b) Proton implanted (top or bottom emitting): Current confined by semi-insulating region of implant; current conducted through the mirrors; optical guiding by gain and thermal profiles. (c) Dielectric apertured with intracavity drive layers (top or bottom emitting): Current apertured by dielectric layer; current conducted laterally along contact layers; optical index guiding by lensing action of dielectric aperture and/or mesa walls.

is usually stopped just above the active layer to avoid surface recombination of carriers and reliability problems. Thus, current is confined or apertured to the lateral dimensions of the mesa, but carriers are free to diffuse laterally in the active region. Thus, there is a *lateral leakage current* that becomes important for dimensions below about 10 μm in diameter. In typical multiple quantum well (MQW) InGaAs 980 nm VCSELs, this leakage current has been estimated to account for about half of the threshold current for a 6 μm diameter mesa with a value of about 70–90 μA per quantum well [5].

The etched-mesa structure also provides a lateral index of refraction step over the upper etched portion of the cavity. Since lateral diffraction is small for diameters greater than about 5 μm in typical length vertical-cavity structures [6], the lateral mode can be approximated as the solution to a uniform step-index (vertical) waveguide with an effective index step equal to the fraction of the mode occupying the upper portion times the actual index step in the upper portion [7, 8]. Because this index step tends to be large ($\Delta n \sim$ 1.5-2 for nitride or oxide coatings), these devices will support multilateral modes for all but submicron diameter mesas. However, since the etched surface is not perfect, the higher order modes tend to suffer higher losses than the fundamental mode. Consequently, there is less scattering for larger diameters, and single lateral mode operation is typically observed for diameters up to about 8 μm [9, 10]. For smaller sizes, where finite diffraction does occur, there also tends to be a scattering of energy at the boundary between the upper and lower portions of the axial cavity, and this is larger for the higher order modes as well. Of course, this mode filtering action is accompanied by unwanted loss for the fundamental mode, and this increases the threshold currents and decreases the differential quantum efficiency. Figure 1.5 gives results from etched-mesa devices illustrating typical results for this structure [9].

Because of the relatively small area of the etched-mesa top, it is most convenient to construct bottom-emitting etched-mesa structures. These allow the entire top surface to be used for contact formation for relatively low contact resistance. Also, this bottom-emitting structure is very compatible with flip-chip bonding technology. In this case, the etched-mesa is typically plated over with some metal such as gold, and this is then used to solder bond to a matching pad on the host substrate. Such a configuration provides a lower thermal impedance, and most parasitic capacitance is eliminated, since the bonding pads can be small [10]. Bottom emission implies that the substrate must be transparent to the emitted light. Thus, InGaAs/GaAs structures with emission wavelengths in the 0.9 to 1.0 μm range have been chosen with GaAs substrates. For relatively low data rates silicon detectors can be used in the 0.9 to 0.95 μm range. In many cases emission in the 0.83–0.85 μm range is desired. For the etched-mesa structure, this implies using a transparent AlGaAs substrate, removing the GaAs substrate material in the vicinity of the beam, or going to a top-emission structure.

A problem with the simple etched-mesa structure, as well as the other VCSELs that employ conduction through the mirrors, is the series voltage that develops across the mirrors due to the potential barriers at the numerous heterointerfaces [2, 3]. This is especially problematic in the p-type mirror since small barriers will inhibit thermionic emission for the heavy holes. To solve this problem bandgap engineering techniques have been employed [3]. These involve matching a doping gradient to the compositional gradient within the transition between the different mirror layers to flatten the valence band. Using these techniques excess p-mirror voltage drops have been reduced to as low as 0.2 V over 18 mirror periods [11]. The lowest voltages occur in proton-implanted or dielectric-apertured structures, which can have considerably larger contact area.

An alternative to solving the problem with conduction through the mirrors is to use intracavity contact layers. Figure 1.4(c) uses these in combination with dielectric aperturing to avoid shunt current paths outside the cavity [12–14]. However, with the simple etched-mesa structure that employs no dielectric aperture, only a bottom intracavity contact is practical [15]. In this case, the etch must cut through the active region, but it is possible to use a semi-insulating substrate. The use of semi-insulating substrates also reduces electrical parasitics for reduced electrical crosstalk and increased modulation bandwidth. Because of the small modal volume of VCSELs, they should have relatively high modulation bandwidths at low current levels [13].

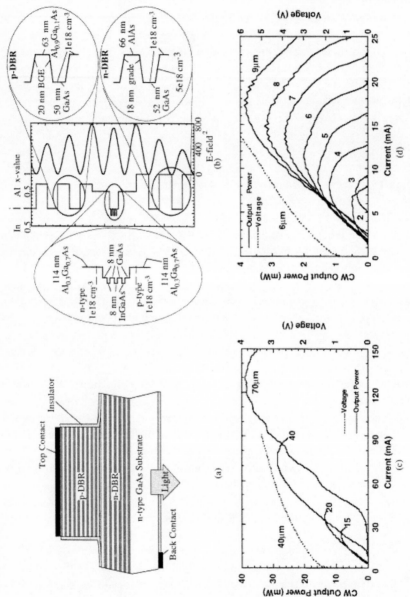

Figure 1.5. Example results from etched mesa VCSEL at 980 nm grown by MBE [3]. (a) Schematic of bottom-emitting structure. (b) Material composition and electric-field-squared standing wave profiles in vicinity of three quantum-well InGaAs active region: Interfaces in p-DBR are bandgap engineered (BGE) with parabolic compositional gradings and a dipole doping profile; linear digital alloy grading is used in the n-DBR. (c) Light-out versus drive current for larger devices (numbers give diameters); voltage-current characteristic given for 40 μm diameter device. (d) Light-current and voltage–current characteristics for smaller devices.

1.2.2 Proton Implanted

Figure 1.4(b) illustrates the planar, proton-implanted, gain-guided structure [11, 16–20]. This structure is very simple to manufacture, and it has therefore been touted by many to be the most important of the VCSEL structures. The proton implant apertures the current to provide a desired current confinement. The planar structure also allows relatively large area electrodes for low contact resistance. Both top- and bottom-emitting structures are possible, although most work has been with a ring-contacted top-emitting configuration to allow 0.85 μm emission with GaAs substrates. The thermal impedance of this planar structure can also be lower than the etched mesa, if the mesa is not plated in metal, since heat can now flow laterally from the upper mirror.

The proton-implanted structure shares the problems associated with electrical conduction through the mirrors pointed out above. Also, it has additional problems in properly aperturing the current as compared to the etched-mesa (or dielectric apertured) case. One can consider two cases: one in which the implant stops above the active region and one in which it penetrates the active region. To ensure that the implant damage does not penetrate the active region, the deep (>2 μm) implant must necessarily stop some distance above it. Thus, *lateral current spreading* tends to occur above the active region, and this adds a significant unproductive shunt current. If the implant proceeds through the active, the current is effectively apertured, but since the carrier lifetime now goes to zero at the edge of the implant in the active layer, the effective interface recombination velocity is infinite, and this leads to much larger carrier losses than in the etched-mesa (or dielectric-apertured) case, where the carriers must diffuse away. Reliability problems are also anticipated in such deeply implanted structures.

For many applications the main problem with the proton-implanted structure is the lack of any deliberate index guiding. The primary lateral index guiding is due to thermal lensing effects [16]. This can provide a stable mode under CW conditions, but under dynamic conditions the lateral mode tends to be dependent upon the data pattern. Thus, data-coding schemes that guarantee nearly constant average power have been employed [21].

Modulation experiment [18, 19] and bit error rate (BER) measurements with both nominally single-mode and multimode VCSELs have been investigated with step and graded-index multimode fibers [21]. The effects of mode selective loss and feedback were studied. Mode selective loss at the transmitter seemed to be most critical, and only high levels of feedback seems to create error floors. Error-free operation at 1 Gb/s with a nonreturn to zero (NRZ) data format has been observed for fiber lengths up to about 500 m. This was obtainable with both multimode and single-mode VCSELs with significant levels of mode selective loss.

Despite the listed problems, the proton-implanted structure has provided attractive terminal characteristics, and it remains a viable candidate for many potential communications applications. Again, its producibility and reproducibility are very attractive in a manufacturing environment. Figure 1.6 gives examples of the characteristics of such structures [20].

1.2.3 Dielectric Apertured

Figure 1.4(c) illustrates a dielectric-apertured VCSEL structure with two intracavity contacts [12]. Dielectric aperturing has been found to be desirable even without the intracavity contacts, but this combined structure addresses several of the problem areas introduced above. Both underetching [12–14] as well as oxidation [22–25] of a high aluminum content

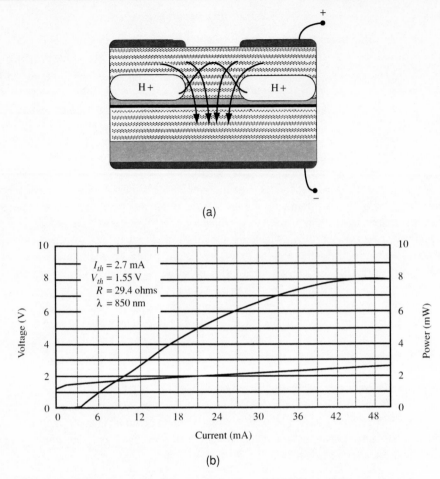

(a)

(b)

Figure 1.6. Example results from proton-implanted VCSELs at 850 nm grown by MOCVD [20]. Device schematic similar to Figure 1.2(b). (a) Schematic of top-emitting proton implanted structure: Current confined by semi-insulating region of implant; current conducted through the mirrors; optical guiding by gain and thermal profiles. (b) Low-voltage result active region and mirror compositions, but with a special doping and grading profile at the mirror interfaces; proton implant window $g = 20$ μm and metalization window $w = 15$ μm.

AlGaAs layer have been used. The first purpose of the dielectric aperture is to block the shunt current that would otherwise flow between the p- and n-regions of the device. This current aperturing is very similar to that in the etched mesa, since the aperture can exist just above the active layer. In either case, the aperture is superior to that resulting from proton implantation, since good aperturing is not accompanied by reduced carrier lifetime in the active region (unless an etch with high ion damage is used). The dielectric aperture in principle also may allow injection of the current into a region smaller than the optical mode width. In fact, this latter feature was the primary reason that dielectric aperturing was first investigated [26]. More recently it has become apparent that the *optical lensing* or waveguiding properties of the dielectric aperture also needs to be considered [7, 8, 27, 28]. As discussed in Chapter 2, this index-guiding technique provides a lower loss waveguide than other approaches such as the etched mesa.

Perhaps the most attractive feature of the dielectric apertured VCSEL is that it seems to combine several desirable features into one structure without removing flexibility of design. As illustrated in Figure 1.4(c) it enables both electrical contact layers to be between the mirrors so that conduction through the mirrors is avoided. Of course, either or both may be placed beyond the mirrors if desired. By placing a narrow aperture at a null of the electric-field-squared standing wave, the current aperture will have little effect on the optical mode [27]. It can thus be moved inside the effective lateral dielectric waveguide, which may be formed by another aperture placed away from the standing-wave null. By grading the aluminum content both functions can be performed in one layer. Thus, both current and photon confinement can both be provided, but with independent control [27, 29]. Of course, lateral carrier diffusion will still occur, unless some lateral potential barrier is provided in the active region.

Figure 1.7 gives experimental results from dielectric-apertured devices using lateral oxidation. Earlier devices used lateral underetching to form the dielectric aperture [12, 13]. These also used a double intracavity contacted structure to avoid mirror resistance. Figure 1.7(a) uses a thin oxide aperture and a single intracavity contact [30]. The thin aperture reduces optical scattering losses in small sizes as indicated by the small reduction in differential efficiency for $D < 3$ μm. Details will be given in Chapter 2.

Figure 1.7(b) shows a through-the-mirror pumped oxidized structure [24, 31]. This structure had optimized mirror doping and very low series resistance. Subsequent experiments with etched and oxidized apertures on similar material show very similar results (see Chapter 2). Modulation experiments on structures with intracavity contacts on a semi-insulating substrate have given some of the best examples of high bandwidths at low drive currents. The dielectric-apertured devices have provided the lowest threshold currents and highest power conversion efficiencies of any VCSEL structure. Figure 1.7(b) illustrates overall power efficiencies of about 50% at output powers $\sim$1 mW [24]. This performance has never been achieved in edge-emitting lasers. It is a good example of the ease in scaling VCSELs. The improved performance of the dielectric-apertured VCSELs as compared to the etched mesas seems to be directly correlated to a reduction in the size-dependent optical scattering loss [30, 32]. This will be discussed in Chapter 2.

1.2.4 Buried Heterostructure

Figure 1.4(d) illustrates one embodiment of a buried-heterostructure (BH) VCSEL. As in the edge-emitter case, the BH-VCSEL should be able to overcome the shortcomings discussed above. It is the only structure of the group in Figure 1.4 that includes lateral carrier confinement. If the regrown material surrounding the etched pillar has a larger bandgap and lower index than the mirror layers, then this structure would include all of the desired lateral confinement features desired in optimal lasers (i.e., optical, current, and carrier confinement). Although this has been well known for some time, the fabrication of such structures has met with only partial success at this writing [33]. Much of the problem results from the difficulties with direct regrowth over a deeply etched mesa containing high-Al content layers. Surface passivation with sulfide treatments has also been attempted with encouraging results [34], but no technique of stabilizing the surface has been found. Impurity-induced disordering may be more promising in this situation, but efforts here also have met with only marginal success [35].

Figure 1.8 gives results from a VCSEL that uses impurity-induced disordering (IID) (intermixing) to create buried heterostructures without regrowth on high-aluminum content

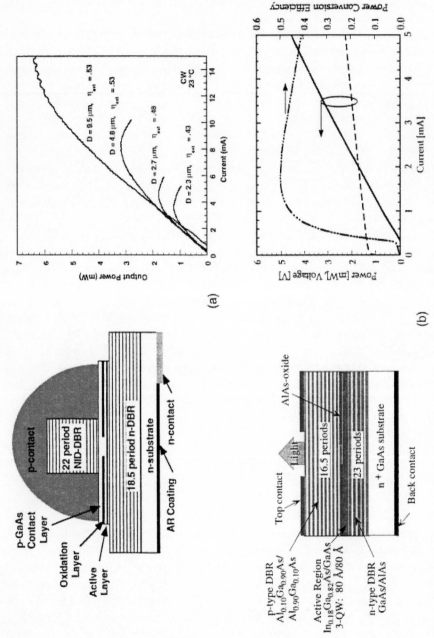

Figure 1.7. Examples of top-emitting dielectric apertured VCSELs. (a) Results from dielectric-apertured, back-emitting 980 nm VCSEL with lateral oxidation grown by MBE [30]; left: schematic; right: light-out versus current for various indicated diameters. A thin aperture leads to lower scattering losses at small diameters D as indicated by high differential efficiencies, η_{ext}. (b) VCSEL grown by MOCVD [24]. Power, voltage, and power conversion efficiency is plotted versus current for a 7 μm square device. Due to the short output coupler and low voltages, 50% power conversion efficiency at 1.5 mA is obtained.

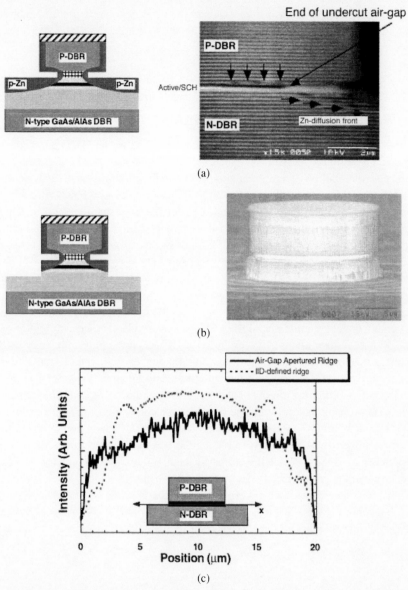

Figure 1.8. (a) Illustration of underetched VCSEL after diffusion step and cross-section SEM image of ridge on the same sample after 25 min/600°C Zn diffusion. The Zn diffusion front is delineated by the black arrows. The enhanced Zn diffusion in the region around the air gap is revealed by stain etching resulting in a dark band much wider than the gap left from the selective undercut etching. (b) Illustration of VCL structure after second RIE step to limit the extent of the parasitic junction and SEM image of the structure after the etch. (c) Intensity versus position plot across the active region luminescence from near-field images before and after IID. The luminescence intensity in the air-gap sample is roughly constant and appears to extend across most of the ridge width, dropping off gradually at the edges. In contrast, the luminescence in the IID-defined ridges drops off substantially over the same distance [35].

layers. These initial results show clear carrier confinement, but little improvement in the $P–I$ characteristics owing to additional leakage paths [35]. Very good edge-emitting lasers have been formed with impurity-induced disordering and therefore this seems like a promising course to pursue.

1.3 Long- and Short-Wavelength VCSELs

1.3.1 Visible

To move to shorter wavelengths aluminum can be added to the GaAs quantum wells to increase their bandgap (see Chapter 7). Although AlGaAs has a direct bandgap for wavelengths down to about 640 nm, population of the indirect minima as well as nonradiative recombination associated with the Al vastly inhibit performance below 780 nm. For viable visible wavelengths (<700 nm), the AlGaInP/GaAs system tends to be superior to AlGaAs/GaAs. This system allows the use of ternary GaInP quantum wells, completely avoiding Al in these carrier storage areas. The most successful work in this area at this writing has used the structure indicated in Figure 1.9 or some close variant on it [36, 37]. The latest results are given in Chapter 8. As can be seen, a simple dielectric-apertured structure has been adopted [37] to reduce the required threshold currents. Prior work used a proton-implanted structure [36], but as seems to be generally the case, the threshold currents were considerably higher, resulting in additional device heating and reduced output powers.

The primary issue with these red-emitting VCSELs is providing sufficiently high transverse carrier confinement barriers as well as eliminating nonradiative recombination in the active region [36–38]. Minimizing nonradiative recombination appears to be associated with eliminating or minimizing Al in and near the quantum wells. Hole confinement is not a significant problem, but the available barrier in the conduction band between GaInP and $Al_x In_y Ga_{1-y}P$ is limited to about 170 meV for $x = 0.67$. This is not so small, but with quantum-confinement effects used to reduce the wavelength, the lowest electron energy level will rise, thus reducing the net barrier to about 100 meV at 610 nm. Also, possible alloy ordering during growth tends to lower the effective barrier height. With larger Al content in the barriers, the conduction band offset actually is reduced, since the indirect bandgap is offset in favor of the valence band. To further improve confinement, one possible approach is to use multiple quantum barriers, as has been applied with some success in edge-emitters.

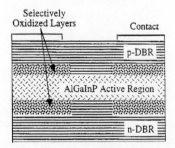

Figure 1.9. Top-emitting red (<680 nm) VCSEL grown by low-pressure MOCVD on (311) GaAs substrate using lateral oxidation [36]. Device schematic showing oxidized layers. The active region uses four compressively strained InGaP quantum wells.

To obtain still shorter wavelengths (blue-green) the II–VI ZnSe-related compounds as well as the group-III nitrides (e.g., GaN) are being investigated [39–45]. At this writing intensive work on II–VI edge-emitters has been carried out for a number of years in several laboratories. This has resulted in room-temperature CW operating devices in the blue-green spectral regions [41]. Threshold currents are relatively low, but significant series resistance has led to relatively inefficient operation. GaAs substrates are used for much of this work, and so the lattice-mismatched heteroepitaxy has led to a large number of defects that limit radiative efficiency as well as reliability. CW lifetimes are still limited to hours rather than many years as is desired.

The group-III nitride technology is very new at this writing. Work has expanded dramatically since the demonstration of practical, high-efficiency light-emitting diodes (LEDs) [42]. Several Japanese groups and two U.S. groups have recently demonstrated edge-emitting diode lasers [43–48], but only the Nichia results show viable room-temperature CW characteristics with good lifetime. These InGaAlN materials are typically grown on sapphire or SiC, and very high dislocation densities are reported. Surprisingly, this high defect count does not appear to severely limit radiative efficiency or device lifetime in LEDs [46]. Thus, although alternative substrates are being studied, it is not clear that the defects must be eliminated as is clearly the case in the II–VIs. VCSELs with epitaxial mirrors appear viable since the index difference between GaN and AlN is large. Dielectric mirrors may also find use.

1.3.2 InP-Based

VCSELs that operate in the 1.3 to 1.55 μm wavelength range [49–52] have not evolved as rapidly as the shorter wavelength devices. As discussed in Chapter 8, this is because both highly reflecting mirrors as well as high-gain active regions are more difficult to obtain. The common InGaAsP/InP and InGaAlAs/InP lattice-matched materials systems both suffer from a relatively small range of achievable optical indices. Thus, very long epitaxial mirror stacks are necessary to theoretically develop the required reflectivity, and because of inherent optical losses, the net reflectivity saturates at about 98% in typical cases [53]. As a result, the epitaxial mirrors have not been used successfully in CW devices. Nevertheless, work continues to try to develop viable epitaxial mirrors in these systems as well as the AlGaAsSb/InP system [54–56]. Most effort, however, has turned to either deposited dielectric [49] or wafer-bonded AlGaAs/GaAs mirror stacks [50–52], both of which can provide >99% of net reflectivity.

The achievable gain in these materials is limited because of significant nonradiative carrier recombination due to Auger processes. These processes tend to vary as the cube of the carrier density and exponentially with temperature [57], resulting in runaway nonradiative recombination for carrier densities greater than about 4×10^{18} cm^{-3}, especially above room temperature. Relatively thick ($\sim$1 μm) bulk InGaAsP active layers have been explored in order to obtain reasonable per-pass gain at relatively low carrier densities [49]. However, the high required current densities and the high thermal impedance of the VCSEL structure have made this approach very difficult. From analytical work it has become clear that one must maximize the gain per unit current density for success in these structures [58]. This suggests that strained multiple quantum wells, which have been found to provide increased gain at lower current densities, should be used. The InGaAlAs/InP system may be the best choice [59].

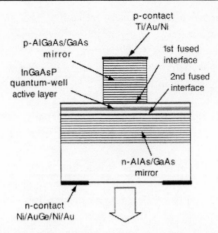

Figure 1.10. Schematic and results from a 1.55 μm wavelength, etched-mesa VCSEL with fused AlGaAs/GaAs mirrors incorporating bandgap engineering [51]. Active region consists of seven strain-compensated quantum wells. Room-temperature light–current results given for several mesa diameters as indicated.

Figure 1.10 gives a schematic of the wafer-fused structure that is being pursued for VCSELs at 1.55 μm [51]. It is very analogous to the GaAs-based etched-mesa configuration shown in Figure 1.4(a). In fact, except for the thin InGaAs/InP strained-MQW active region, the device is basically the same. It is formed by a two-step wafer-bonding procedure in which epitaxial mirror stacks grown on GaAs are fused to the active region grown on InP. After fusing epitaxial faces of the InP wafer to the first GaAs wafer (which contains a p-type AlAs/GaAs mirror), the InP substrate is removed and this is fused to a second GaAs wafer with an n-type epitaxial mirror. The first GaAs substrate is then removed and the etched mesa formed. As in the case of Figure 1.4(a) electrical conduction to the active region is through the GaAs-based mirror stacks. Further details are given in Chapter 8. Optical losses have been shown to be even more problematic at 1.3 and 1.55 μm due to increased absorption in p-doped AlGaAs at longer wavelengths [60]. Of course, with this technology any of the other structures introduced above for GaAs could also be employed. Moreover, with the wafer-bonding procedure it is also possible to prepattern the substrates before bonding to provide desired lateral current, carrier, and optical confinement. For the wafer-fused etched-mesa device of Figure 1.10, a CW threshold current as low as 0.9 mA has obtained for a 4 μm diameter etched mesa VCSEL, and pulsed operation up to 60°C has been observed with a 6 μm structure [52].

Figure 1.11 gives plots of I_{th} versus year in analogy to Figure 1.1 for long-wavelength devices. As can be seen, there has been very rapid progress in recent years.

1.4 Growth and Fabrication Issues

All epitaxially grown GaAs-based VCSELs are now recognized as being truly practical and producible in the 0.8 to 1.0 μm wavelength range. Molecular-beam-epitaxy (MBE), gas-source MBE, and organo-metallic vapor-phase epitaxy (OMVPE) have all been used with good success as discussed in Chapter 4 and refs. 61–63. However, this has not been a trivial accomplishment. In fact, standard growth technology, as developed for electronic

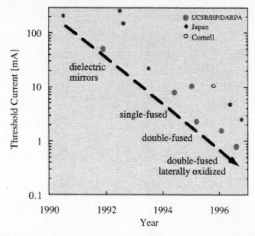

Figure 1.11. Threshold current of long wavelength (1.3–1.6 μm) VCSELs versus year of first report [53].

devices or edge-emitting lasers, has not provided sufficient material thickness or uniformity control for VCSELs. This is why many of the major breakthroughs in this field occurred in laboratories where efforts to significantly improve the control of epitaxial growth were taking place. To obtain good wafer yield, some form of in-situ monitoring is now employed in most facilities.

1.4.1 In Situ Monitoring

The most common form of in situ monitoring during growth is optical reflectometry [64]. As discussed in Chapter 4, this can involve a relatively simple apparatus. For example, a broadband light source can be reflected off the wafer surface while in the growth system and the result analyzed by a simple optical spectrum analyzer. The apparatus is typically used only after the bottom mirror and active region are grown to calculate what additional trimming layer must be added to center the Fabry–Perot mode in the mirror reflection band before the upper mirror is grown. This approach can be used with either MBE or OMVPE [64, 65].

Other monitoring techniques that are used involve either optical pyrometry [66] or molecular beam flux monitoring by optical absorption [67, 68]. The first makes use of the natural blackbody radiation from the sample. Optical pyrometry is commonly used to measure the substrate temperature. However, if a multilayer structure is being grown, this relatively broadband signal is filtered by the transfer function of that structure. Thus, the received optical spectrum can be used to determine the properties of the layer structure by analyzing this spectrum.

In situ beam flux monitoring has been accomplished with light sources that emit the atomic spectra of the atoms to be monitored. In this case, a relatively crude spectrometer can be used to isolate the lines of interest [68]. Feedback can be used to set the bandgap of quantum wells, and integration of the fluxes can be used to set the layer thickness. Unfortunately, the technique can not be used for OMVPE, but in that case the limitation on layer composition and thickness is usually controlled more by reactor flow and chemistry rather than the source fluxes, which are generally well controlled.

1.4.2 Processing for Lateral Definition

Although the initial growth of the VCSEL wafers is undoubtedly the most crucial step in the device's formation, the subsequent processing is also critical to obtain a high-performance, high-yield result as discussed in Chapter 5. The growth provides the vertical (axial) structure, and the processing defines the lateral structure. Since a number of lateral structures are being considered, a similar number of fabrication procedures are involved.

For the etched-mesa structure, reactive-ion-etching with Cl_2 has been the principle technique of forming the mesas [69–71]. This provides highly anisotropic etching with vertical and relatively smooth sidewalls. Photoresist, Ni, or dielectric layers can be used for masking. To stop the etch at the desired point, in situ optical monitoring is necessary [72]. Use of etchable masks in combination with nonetchable masks has also been used to create accurate double step height structures as required for some of the intracavity contacted structures [12]. Figure 1.12 gives an example result from this process.

As already indicated, the proton-implanted structures are perhaps the easiest to produce, since they only require implant masking, the proton implant, and contact formation. The main issues here are the energy of the implant and the relative diameters of the implant and the iris in the top electrode. A relatively wide variation in device performance has been observed as these parameters are varied [73].

Laterally underetched or oxidized dielectric-apertured VCSELs require accurate control of the etching or oxidation process. In the underetching case it is common to include some smaller control structures that will literally fall off when the etching has proceeded a desired amount [74]. Following underetching the structure is coated with a chemical vapor deposition (CVD) dielectric to provide step coverage and increased mechanical stability. The laterally oxidized structures are formed in flowing wet nitrogen gas, as will be discussed in detail in Chapter 5. Oxidation has been shown to be highly reproducible in AlAs layers with small amounts of Ga added [75, 76]. Tapered oxidation fronts, desirable for lower loss as discussed in Chapter 2, can also be created with a variable Al composition. Figure 1.12 shows a transmission electron micrograph (TEM) cross section of an oxidized VCSEL and a scanning electron micrograph (SEM) profile of an underetched device with intracavity contacts.

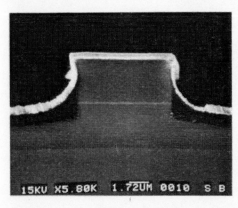 

                         (a)   (b)

Figure 1.12. (a) Cross-sectional transmission electron micrograph of an oxidized VCSEL, and (b) a low-angle scanning electron micrograph view of an underetched VCSEL with step-etching for intracavity contacts.

1.4.3 Nonepitaxial Mirrors and Integrated Microlenses

As discussed above, the all-epitaxial mirror structures are not desired or even practical in some cases. The long-wavelength InP-based VCSELs are important examples. Both conventional dielectric mirrors as well as wafer-bonding technology have been employed as shown in Figure 1.10. The wafer-bonding approach involves atomically fusing two substrates together by placing them in contact under pressure at sufficiently high temperatures for a vapor phase transport to take place, as will be discussed in Chapter 8. Although there is a lattice mismatch, the necessary defects are well confined to the interface. The process is already used in the production of high-efficiency LEDs, where AlGaInP epilayers grown on GaAs are transferred to a transparent GaP substrate [77].

In the GaAs-based system, dielectric mirrors have also been explored. Successful devices have been obtained with TiO_2/SiO_2, $CaF_2/ZnSe$, or $MgF_2/ZnSe$ [18, 22, 78]. The deposition is generally done by electron-beam evaporation or reactive sputtering. In both cases reflectivities above 99% can be achieved. Because of the large refractive index difference between MgF_2 and ZnSe, the high reflectivities can be achieved with only four or five mirror periods. Also, by making the dielectric mirror in the last step, all processing can be done on a planar substrate. VCSEL threshold currents below 100 μA have been achieved with the $CaF_2/ZnSe$ and $MgF_2/ZnSe$ mirrors [22].

Although not a necessary part of VCSEL fabrication, integrated microlenses provide very attractive embellishments to these devices. They can be added at the wafer level using standard lithographic and etching techniques, and they can provide all of the required focusing for a given application. Thus, fiber coupling and free-space interconnects can be accomplished with no additional external optics. Chapters 9, 10, and 12 discuss the desirability of free-space interconnects. Figure 1.13 illustrates one technique of making the integrated microlenses and lists some results [79]. This so-called photoresist melting technique involves spinning on a specific thickness of a special photoresist, patterning it to some desired diameter, performing an initial slight etch to define a small confining step at its periphery, heating the resist until it melts and reflows, and finally dry etching with a process to transfer the lens shape into the semiconductor substrate. For the case of Figure 1.13 the resist is Microelectronics SF15, a polymethylgutylimide (PMGI), the dry etch was a pure Cl_2 RIE @ 350 V, and the substrate was GaAs. Figure 1.14 shows the effect of such integrated backside microlenses on the output beams of 0.98 μm etched-mesa VCSELs [79]. An SiO AR coating was also added to prevent unwanted feedback into the laser. As indicated, a nearly collimated beam is possible with a lens focal length equal to the substrate thickness. Far fields indicate that lens aberrations and scattered energy are both quite small.

1.5 Integration: Photonic and Optoelectronic

1.5.1 Photonic

The photonic integration of numerous identical VCSELs into linear and two-dimensional arrays requires no additional growth effort, since the required layer structure fills the entire wafer surface. Standard processing technology also produces full wafers of devices naturally arrayed to fill the available space. The primary difference in forming different kinds of VCSEL arrays is in the final electrical contacting steps. Chapters 10–12 contain detailed discussions of the use of VCSEL arrays.

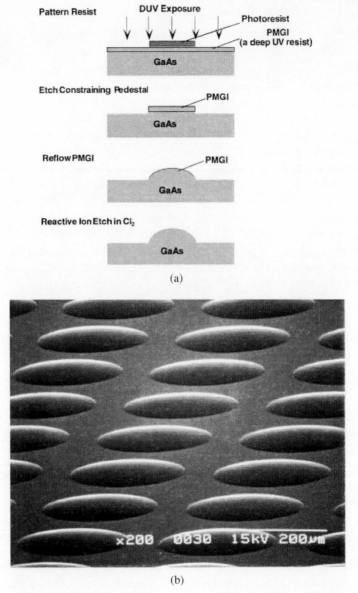

(a)

(b)

Figure 1.13. (a) Fabrication process for and (b) SEM of integrated microlenses on GaAs [79].

Several different contacting schemes have been demonstrated. Arrays of devices have been connected in parallel with a pair of common contacts [80–82]. This has been done to create high-power arrays. Obtaining high power from VCSEL arrays actually makes quite a bit of sense, because of the three-dimensional heat spreading away from small dots. In this sense, such arrays may be viewed as a further extension on the successful philosophy employed in one-dimensional arrays of edge-emitters, which enjoy a two-dimensional heat spreading. Attempts to coherently phase lock arrays of single-mode devices has met with modest success, but high-power phase-locked arrays with single-lobed far fields have not

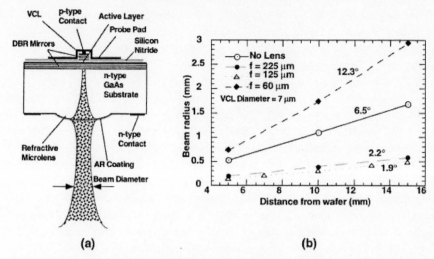

Figure 1.14. (a) Schematic of VCSEL with integrated microlens, and (b) beam radius versus distance from such a structure [79].

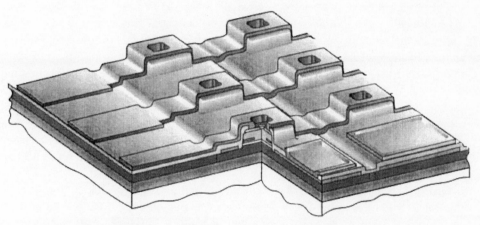

Figure 1.15. A schematic section of a matrix-addressed VCSEL array on a semi-insulating substrate [86].

been demonstrated. Separately contacted, individually addressable devices have been studied most widely [83, 84]. If grown on a doped substrate, all devices will have a common contact (usually the cathode), and the other can be attached to individual bond pads.

Matrix addressing is possible with a semi-insulating substrates [85, 86]. Structures as shown in Figure 1.4(c) would provide candidate devices. In this case, the p-contacts of devices along a row and the n-contacts of devices down a column are connected together. Then by applying a positive voltage to the desired row and an equal negative voltage to the column, the VCSEL at the intersection would be turned on. Other devices along the row and down the column would see only half the voltage of this device. Thus, because it is easy to design devices whose threshold voltage is less than half of the desired operating voltage, excellent on–off contrast is possible. Figure 1.15 gives a schematic of such a structure [86].

Fabricating arrays of VCSELs with multiple wavelengths does require some additional growth and fabrication effort in order to define the necessarily different cavity lengths. Several approaches have been investigated, including the efforts discussed in Chapter 10. A thickness gradient can be created across the wafer by stopping rotation during MBE growth. This has been used to make experimental wavelength chirped arrays [87], but the approach is not viable for manufacturing many arrays per wafer due to the nonlinearity of the growth variation across the wafer. To create a wafer with periodic thickness variations during MBE growth, a special grooved wafer holder has been employed to provide periodic surface temperature variations and thus different sticking coefficients for the In and/or Ga species [88, 89]. VCSEL arrays formed across this thickness gradient have different wavelengths. Reproducibility may be a problem with this approach. Perhaps a more reproducible approach uses conventional multiple step lithography to adjust the reflection phase of one mirror. This has been done on the top surface of a short semiconductor distributed Bragg reflector (DBR) mirror segment prior to completing the mirror either with dielectric/metal layers [90] or with a regrowth of additional semiconductor DBR periods [91]. Another novel approach to change the wavelength of a WDM array element involves a postgrowth adjustment of the thickness of a laterally oxidized layer by varying the distance it is spaced from the edge where the oxidation began [92].

One of the key issues with arrays is crosstalk. It can derive from thermal, electrical, or optical parasitic paths as discussed in later chapters. It has been shown that thermal crosstalk can be minimized with a larger element-to-element spacing or with vertical heat sinking and etched grooves between devices [93]. Electrical crosstalk may be capacitive (radiation) or resistive (leakage). Both tend to be aggravated by closely spaced devices or contacting electrodes. In a well-designed array most of the lateral current confinement techniques should eliminate the resistive paths between devices. Of course, in matrix addressing there is always the half-voltage issue. Radiative or capacitive coupling tends to be the larger problem for high-frequency modulation of elements in an array. Good microwave engineering using thick dielectric spacer layers, thick conductors, and shielding ground planes can address this problem in most cases.

Optical crosstalk may be the most difficult problem to solve, especially in free-space interconnected arrays or in cases where the optical coupling loss to the desired medium (e.g., fiber) is large. In both cases stray light tends to be generated, and unless care is exercised, this may find its way to adjacent receiver elements. It is generally desirable to eliminate large emission angle ray paths from the VCSELs by avoiding significant scattering elements or the generation of very high order lateral modes. Side lobes generated in the far field by the VCSEL mode or nonideal coupling optics can also create problems. For free-space interconnects, Gaussian modes are desirable. Finally, the entire optical link generally needs to be considered in analyzing optical crosstalk. That is, the use of limited numerical aperture fibers or free-space receiver elements may make up for some problems in the emitter. However, it is generally best to avoid generating the unwanted signals in the first place.

Vertical photonic integration is possible by using the backside of the VCSEL wafers, as already shown with the microlenses above, or by employing vertically stacked wafers that might be self-aligned with solder bumps, alignment notches, etc. More complex heterogeneous schemes continue to be studied. Figure 1.16 shows a two-dimensional (2-D) VCSEL array flip-chip bonded to the edge of a stack of Si processing chips [94]. This concept can provide a very dense three-dimensional (3-D) processor/interconnect geometry.

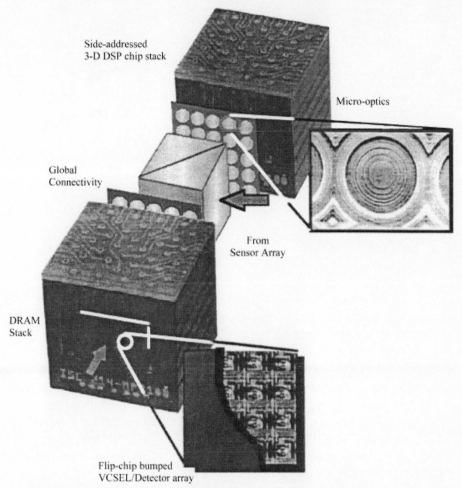

Side-addressed
3-D DSP chip stack

Micro-optics

Global
Connectivity

From
Sensor Array

DRAM
Stack

Flip-chip bumped
VCSEL/Detector array

Figure 1.16. Two-dimensional VCSEL/detector arrays flip-chip bonded to processor blocks together with intermediate optics [94].

1.5.2 Optoelectronic

Because VCSELs require a relatively small footprint on a semiconductor wafer, they appear to be more compatible with electronic devices than edge-emitters, and this has generated some interest in integrating VCSELs with electronic driver or logic circuits. The growth and processing required for heterojunction bipolar transistors (HBTs) appear to be fairly similar to VCSELs. In fact, VCSELs have been directly integrated in the collector circuits of HBTs by simply growing the layers for both sequentially [95]. The HBT provides current gain so that very small base currents can drive several milliamperes through the VCSEL. Although the vertical dimension is much larger, the processing is very similar. Direct illumination of the HBT base can also be used to generate the required minority carriers, and thus, optical-to-optical gain is possible in a single vertical stack.

Direct vertical integration with pnpn latching diodes has also been explored for logic applications. This so-called V-STEP device is discussed more completely in Chapter 10 [96]. Once a pulse of incident light turns on the pnpn diode, the voltage across it collapses, and

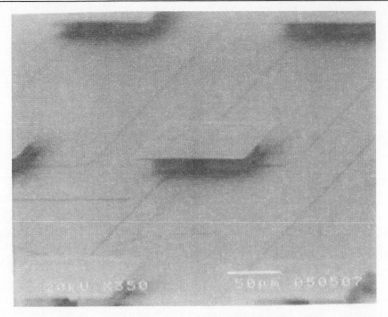

Figure 1.17. Scanning electron photomicrograph of VCSELs flip-chip bonded to GaAs circuits after substrate removal [99].

the VCSEL turns on. It stays on until the terminal voltage is reduced to reset the device. This is a very simple structure that may find use in optoelectronic logic or image processing.

To enhance flexibility in processing of integrated optoelectronic circuits composed of VCSELs, transistors, and photodetectors, discrete optimized devices have also been monolithically integrated. VCSEL-MESFET [97] and VCSEL-HBT [98] circuits have been explored. This provides a technology in which the transistors can be interconnected in various ways to form complex logic circuits.

Hybrid integration using flip-chip bonding and/or multichip module (MCM) technology provides the ultimate in flexibility without sacrificing device and circuit performance. Of course, additional handling and bonding are necessary after the wafers are diced into chips, but the electronic and photonic wafers can be designed and processed without compromise, and chips from both types of wafers can be pretested to reject any defective parts before being integrated. As a result, it has generally been found that unless very large scale production is anticipated, the hybrid form of optoelectronic integration is usually more cost effective. Figure 1.17 illustrates VCSELs flip-chip bonded to a 1.0 μm GaAs circuit after substrate removal [99]. This coplanar contacting technique will be discussed further in Chapter 12.

1.6 Applications

1.6.1 Discrete Devices

Although the advantages of VCSELs for array applications have been widely touted, their potential for low-cost LED-like production coupled to their superiority to LEDs for advanced applications may provide a large LED-upgrade market for discrete VCSELs. VCSELs can be modulated at multigigahertz rates, their output beams are well collimated, and their overall power efficiency can be much higher than an analogous LED. Thus, for higher-capacity data links, efficient coupling to fibers, low-crosstalk free-space links, or minimizing drive power requirements, VCSELs may offer an attractive evolutionary path to systems designers who wish to improve the performance of existing LED-based links. Chapters 9 and 12 discusses a number of VCSEL applications.

The other class of applications comprises those in which lasers are required, but edge-emitting devices appear too costly. Thus, the focus here would be mainly on cost reduction with VCSELs versus edge-emitters. This would derive both from the initial manufacture of the chips as well as in their packaging with fibers or other desired media. In fact, their low numeral aperture (NA) outputs, either directly or in combination with integrated microlenses, may be an even larger attraction than the wafer-scale fabrication and testing feature of VCSELs. Potential applications in this instance include sources for new low-cost optical sensors, plastic fiber–based data links, optical disk storage, and fiber-to-the-home (FTTH). For some of these applications, improvements in either short-wavelength (e.g., blue for optical recording) or longer wavelength (e.g., for FTTH) VCSELs will need to be developed. However, it seems clear that VCSELs could have a major impact in both cases.

1.6.2 Linear Arrays

One-dimensional arrays of VCSELs coupled to multimode fiber ribbon cables have been used in prototype parallel data link applications [100–102]. Chapter 11 will discuss these in some detail. These links take advantage of the fact that entire linear arrays of VCSELs can be simultaneously aligned to the elements of a fiber ribbon cable by only aligning the end two lasers and fibers. Thus, these linear arrays significantly leverage the relatively easy alignment of VCSELs to fibers by carrying along the extra elements. Because the uniformity of the arrays proved to be quite good, it was also decided to only monitor the end VCSELs. This avoided a complex backside monitor or a beamsplitter at the front.

Prototype fiber-based parallel data links are presently generating some market interest. Predictions suggest that they may reach many tens of millions of dollars in sales by the end of the decade [103]. Currently, efforts are underway to reduce the cost, but it seems the connectorized fiber ribbon cables and the general packaging pose more of a problem than the VCSELs themselves.

Linear arrays are also proposed for parallel board-to-board interconnection using free-space paths [79]. In this case the devices can be flip-chip mounted, and integrated backside microlenses will collimate the emission to an analogous receiver array on the second board, which also contains integrated backside microlenses. Since nearly collimated beams are emitted, this system is very tolerant of lateral displacements for movements of up to about half of the spacing between array elements.

1.6.3 Two-Dimensional Arrays

Two-dimensional arrays of devices generally are coupled to free-space optical paths. Massively parallel interconnection between processing planes is possible using the same integrated microlenses introduced above. These refractive microlenses can be integrated off-axis for beam steering, or diffractive optics can also be integrated. For simultaneous fan out and collimation, computer-generated holograms implemented in diffractive optics would be useful [104]. In some cases simple field lenses external to the chips may be desired, but these will result in much lower tolerances in the placement of the chips to be interconnected.

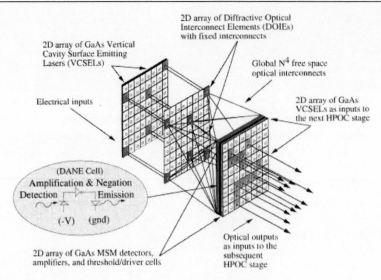

2D array of Diffractive Optical
Interconnect Elements (DOIEs)
with fixed interconnects

2D array of GaAs Vertical
Cavity Surface Emitting
Lasers (VCSELs)

Global N^4 free space
optical interconnects

Electrical inputs

2D array of GaAs
VCSELs as inputs to
the next HPOC stage

(DANE Cell)
Amplification & Negation
Detection   Emission

(-V)   (gnd)

Optical outputs
as inputs to the
subsequent
HPOC stage

2D array of GaAs MSM detectors,
amplifiers, and threshold/driver cells

Figure 1.18. High-performance-optical computing (HPOC) stage using 2-D arrays of VCSELs, diffractive optical interconnect elements, and detector/amplifier/threshold/driver cells [106].

Two important applications for such large numbers of interconnections are in signal processing and in switching as discussed in Chapters 10–12 [105, 106]. For switching, fixed interconnection paths between multiple boards can generally be used to accomplish the desired routing of a signal on one of the inputs to any of the outputs. Thus, all of the switching is done with electronically controlled optical switches at each plane. This architecture would appear to be possible with the current state of the art in VCSELs using flip-chip bonding.

In the signal processing or computing case, a fixed interconnect architecture as shown in Figure 1.18 may still be used for some functions [106]. But in other cases, it may also be desirable to alter the interconnection paths between two boards during the processing. A spatial light modulator (SLM) placed in the interconnect path allows selection of one of several possible paths [107, 108]. Alternatively, all of the desired interconnection paths from a given node can be set with fixed optics, and the switching can be done in the electronics. However, this greatly increases the number of possible beams that must be spatially separated. A variation is to use wavelength encoding of the source and detector arrays to reduce the spatial crosstalk tolerance. Other proposals involve using tunable sources and diffractive optics to route a particular wavelength to a particular node. Generally, this programmable interconnect architecture will require further research.

Displays are another possible application of two-dimensional arrays of VCSELs. Just as LEDs are currently finding their way into displays [109], the VCSEL could be viewed as an efficient high-power LED for such applications.

References

[1] J. L. Jewell, A. Scherer, S. L. McCall, Y. H. Lee, S. Walker, J. P. Harbison, and L. T. Florez, "Low-Threshold Electrically Pumped Vertical-Cavity Surface-Emitting Microlasers," *Electronics Letters*, **25**, 1123–1124, 1989.

[2] R. S. Geels, S. W. Corzine, and L. A. Coldren, "InGaAs Vertical-Cavity Surface-Emitting Lasers," *IEEE Journal of Quantum Electronics*, **27**(6), 1359–1367, 1991.

[3] M. G. Peters, B. J. Thibeault, D. B. Young, J. W. Scott, F. H. Peters, A. C. Gossard, and L. A. Coldren, "Band-Gap Engineered Digital Alloy Interfaces for Lower Resistance Vertical-Cavity Surface-Emitting Lasers," *Applied Physics Letters*, **63**(25), 170–171, 1993.

[4] K. D. Choquette, G. Hasnain, Y. H. Wang, J. D. Wynn, R. S. Freund, A. Y. Cho, and R. E. Leibenguth, "GaAs Vertical-Cavity Surface Emitting Lasers Fabricated by Reactive Ion Etching," *IEEE Photonics Technology Letters*, **3**(10), 859–862, 1991.

[5] B. J. Thibeault, T. A. Strand, T. Wipiejewski, M. G. Peters, D. B. Young, S. W. Corzine, L. A. Coldren, and J. W. Scott, "Evaluating the Effects of Optical and Carrier Losses in Etched Post Vertical Cavity Lasers," *Journal of Applied Physics*, **78**(10), 5871–5875, 1995.

[6] D. I. Babic, Y. Chung, N. Dagli, and J. E. Bowers, "Modal Reflection of Quarter-Wave Mirrors in Vertical-Cavity Lasers," *IEEE Journal of Quantum Electronics*, **29**(6), 1950–1962, 1993.

[7] G. R. Hadley, K. L. Lear, M. E. Warren, K. D. Choquette, J. W. Scott, and S. W. Corzine, "Comprehensive Numerical Modeling of Vertical-Cavity Surface-Emitting Lasers," *IEEE Journal of Quantum Electronics*, **32**(4), 607–616, 1996.

[8] L. A. Coldren, J. W. Scott, B. Thibeault, T. Wipiejewski, D. B. Young, M. G. Peters, G. Thompson, E. Strzelecka, and G. Robinson, "Vertical-Cavity Semiconductor Lasers: Advances and Opportunities," *Integrated Optical Fiber Communication Conference IOOC-95*, Hong Kong, Optical Society of America, 26–27, 1995.

[9] M. G. Peters, D. B. Young, F. H. Peters, J. W. Scott, B. J. Thibeault, and L. A. Coldren, "17.3% Peak Wall Plug Efficiency Vertical-Cavity Surface-Emitting Lasers Using Lower Barrier Mirrors," *IEEE Photonics Technology Letters*, **6**(1), 31–33, 1994.

[10] F. H. Peters, M. L. Majewski, M. G. Peters, J. W. Scott, B. J. Thibeault, D. B. Young, and L. A. Coldren, "Vertical Cavity Surface Emitting Laser Technology," *SPIE Proceedings*, **1851**, 122–127, 1993.

[11] K. L. Lear, R. P. Schneider, K. D. Choquette, S. P. Kilcoyne, J. J. Figiel, and J. C. Zolper, "Vertical Cavity Surface Emitting Lasers with 21% Efficiency by Metalorganic Vapor Phase Epitaxy," *IEEE Photonics Technology Letters*, **6**(9), 1053–1055, 1994.

[12] J. W. Scott, B. J. Thibeault, D. B. Young, L. A. Coldren, and F. H. Peters, "High Efficiency Sub-Milliamp Vertical Cavity Lasers with Intra-Cavity Contacts," *IEEE Photonics Technology Letters*, **6**(6), 678–680, 1994.

[13] J. W. Scott, B. J. Thibeault, C. J. Mahon, L. A. Coldren, and F. H. Peters, "High Modulation Efficiency of Intracavity Contacted Vertical Cavity Lasers," *Applied Physics Letters*, **65**(12), 1483–1485, 1994.

[14] J. W. Scott, B. J. Thibeault, M. G. Peters, D. B. Young, and L. A. Coldren, "Low-Power High-Speed Vertical-Cavity Lasers for Dense Array Applications," *SPIE Proceedings*, **2382**, 280–287, 1995.

[15] B. J. Thibeault, J. W. Scott, M. G. Peters, F. H. Peters, D. B. Young, and L. A. Coldren, "Integrable InGaAs/GaAs Vertical-Cavity Surface-Emitting Lasers," *Electronics Letters*, **29**(25), 2197–2198, 1993.

[16] B. Tell, Y. H. Lee, K. F. Brown-Goebeler, J. L. Jewell, R. E. Leibenguth, M. T. Asom, G. Livescu, L. Luther, and V. D. Mattera, "High-Power CW Vertical-Cavity Top Surface-Emitting GaAs Quantum Well Lasers," *Applied Physics Letters*, **57**(18), 1855–1857, 1990.

[17] R. A. Morgan, L. M. F. Chirovsky, M. W. Focht, G. Guth, M. T. Asom, R. E. Leibenguth, K. C. Robinson, Y. H. Lee, and J. L. Jewell, "Progress in Planarized Vertical Cavity Surface Emitting Laser Devices and Arrays," *SPIE Proceedings*, **1562**, 149–159, 1991.

[18] J. A. Lehman, R. A. Morgan, M. K. Hibbs-Brenner, D. Carlson, "High-Frequency Modulation Characteristics of Hybrid Dielectric/AlGaAs Mirror Singlemode VCSELS," *Electronics Letters*, **31**(15), 1251–1252, 1995.

[19] G. Shtengel, H. Temkin, P. Brusenbach, T. Uchida, M. Kim, C. Parsons, W. E. Quinn, and

S. E. Swirhun, "High-Speed Vertical-Cavity Surface Emitting Laser," *IEEE Photonics Technology Letters*, **5**(12), 1359–1361, 1995.

[20] R. A. Morgan, M. K. Hibbs-Brenner, R. A. Walterson, J. A. Lehman, T. M. Marta, S. Bounnak, E. L. Kalweit, T. Akinwande, and J. C. Nohava, "Producible GaAs-Based Top-Surface Emitting Lasers with Record Performance," *Electronics Letters*, **31**(6), 462–463, 1995.

[21] D. Kupta and C. Mahon, "Mode Selective Loss Penalties in VCSEL Optical Fiber Transmission Links," *IEEE Photonics Technology Letters*, **6**(2), 288–290, 1994.

[22] D. G. Deppe, D. L. Huffaker, J. Shin, and Q. Deng, "Very-Low-Threshold Index-Confined Planar Microcavity Lasers," *IEEE Photonics Technology Letters*, **7**(9), 965–967, 1995.

[23] Y. Hayashi, T. Mukaihara, N. Hatori, N. Ohnoki, A. Matsutani, F. Koyama, and K. Iga, "Record Low-Threshold Index-Guided InGaAs/GaAlAs Vertical-Cavity Surface-Emitting Laser with a Native Oxide Confinement Structure," *Electronics Letters*, **31**(7), 560–561, 1995.

[24] K. L. Lear, K. D. Choquette, R. P. Schneider, Jr., S. P. Kilcoyne, and K. M. Geib, "Selectively Oxidised Vertical Cavity Surface Emitting Lasers with 50% Power Conversion Efficiency," *Electronics Letters*, **31**(3), 208–209, 1995.

[25] M. H. MacDougal, P. D. Dapkus, V. Pudikov, H. Zhao, and G. M. Yang, "Ultralow Threshold Current Vertical-Cavity Surface-Emitting Lasers with AlAs Oxide-GaAs Distributed Bragg Reflectors," *IEEE Photonics Technology Letters*, **7**(3), 229–231, 1995.

[26] J. W. Scott, R. S. Geels, S. W. Corzine, and L. A. Coldren, "Modeling Temperature Effects and Spatial Hole Burning to Optimize Vertical-Cavity Surface-Emitting Laser Performance," *IEEE Journal of Quantum Electronics*, **29**(5), 1295–1308, 1993.

[27] L. A. Coldren, B. J. Thibeault, E. R. Hegblom, G. B. Thompson, and J. W. Scott, "Dielectric Apertures as Intracavity Lenses in Vertical-Cavity Lasers," *Applied Physics Letters*, **68**(3), 313–315, 1996.

[28] K. L. Lear, K. D. Choquette, R. P. Schneider, Jr., and S. P. Kilcoyne, "Modal Analysis of a Small Surface Emitting Laser with a Selectively Oxidized Waveguide," *Applied Physics Letters*, **66**(20), 2616–2618, 1995.

[29] E. R. Hegblom, D. I. Babic, B. J. Thibeault, J. Ko, R. Naone, and L. A. Coldren, "Estimation of Optical Scattering Losses in Dielectric Apertured Vertical Cavity Lasers," *Applied Physics Letters*, **68**(13), 1757–1759, 1996.

[30] B. J. Thibeault, E. R. Hegblom, P. D. Floyd, R. Naone, Y. Akulova, and L. A. Coldren, "Reduced Optical Scattering Loss in Vertical-Cavity Lasers Using a Thin (300 Å) Oxide Aperture," *IEEE Photonics Technology Letters*, **8**(5), 593–595, 1996.

[31] P. D. Floyd, B. J. Thibeault, L. A. Coldren, and J. L. Merz, "Scalable AlAs-Oxide Vertical Cavity Lasers," *Electronics Letters*, **32**(2), 114–116, 1996.

[32] P. D. Floyd, B. J. Thibeault, L. A. Coldren, and J. L. Merz, "Reduced Threshold Bottom Emitting Vertical Cavity Lasers by AlAs Oxidation," *LEOS '95*, Paper SCL14.2, 414–415, 1995.

[33] K. D. Choquette, M. Hong, R. S. Freund, J. P. Mannaerts, R. C. Wetzel, and R. E. Leibenguth, "Vertical-Cavity Surface-Emitting Laser Diodes Fabricated by *In Situ* Dry Etching and Molecular Beam Epitaxial Regrowth," *IEEE Photonics Technology Letters*, **5**(3), 284, 1993.

[34] D. B. Young, A. Kapila, J. W. Scott, V. Malhotra, and L. A. Coldren, "Reduced Threshold Vertical-Cavity Surface-Emitting Lasers," *Electroics Letters*, **30**(3), 233–234, 1994.

[35] P. D. Floyd, B. J. Thibeault, J. Ko, D. B. Young, L. A. Coldren, "Vertical Cavity Lasers with Zn Impurity-Induced Disordering (IID) Defined Active Regions," *Proceedings LEOS '96*, Paper TuJ3, Boston, 1996.

[36] J. A. Lott, R. P. Schneider, Jr., K. J. Malloy, S. P. Kilcoyne, and K. D. Choquette, "Partial Top Dielectric Stack Distributed Bragg Reflectors for Red Vertical Cavity Surface Emitting Lasers," *IEEE Photonics Technology Letters*, **6**(12), 1397–1399, 1994.

[37] K. D. Choquette, R. P. Schneider, M. Hagerott Crawford, K. M. Geib, and J. J. Figiel,

"Continuous Wave Operation of 640–660 nm Selectively Oxidised AlGaInP Vertical-Cavity Lasers," *Electronics Letters*, **31**(14), 1145–1146, 1995.

[38] T. Takagi and F. Koyama, "Design and Photoluminescence Study on a Multiquantum Barrier," *IEEE Journal of Quantum Electronics*, **27**(6), 1511, 1991.

[39] M. A. Haase, J. Qiu, J. M. DePuydt, and H. Cheng, "Blue-Green Laser Diodes," *Applied Physics Letters*, **59**(11), 1272–1274, 1991.

[40] H. Okuyama, E. Kato, S. Itoh, N. Nakayama, T. Ohata, and A. Ishibashi, "Operation and Dynamics of ZnSe/ZnMgSSe Double Heterostructure Blue Laser Diode at Room Temperature," *Applied Physics Letters*, **66**(6), 656–658, 1995.

[41] A. Ishibashi and S. Itoh, "One-Hour-Long Room Temperature CW Operation of ZnMgSSe-Based Blue-Green Laser Diodes," *15th International Semiconductor Laser Conference*, Paper PD1.1, 1995.

[42] S. Nakamura, "Zn-Doped InGaN Growth and InGaN/AlGaN Double-Heterostructure Blue-Light-Emitting Diodes," *Journal of Crystal Growth*, **145**, 911–917, 1994.

[43] S. Nakamura, "InGaN Multiquantum-Well-Structure Laser Diodes with GaN-AlGaN Modulation Dopes Strained-Layer Superlattice," *IEEE Journal of Selected Topics in Quantum Electronics*, **4**(3), 483–489, 1998.

[44] K. Itaya, M. Onomura, J. Nishio, L. Sugiura, S. Saito, M. Suzuki, J. Rennie, S. Nunoue, M. Yamamoto, H. Fujimoto, Y. Kokubun, Y. Ohba, G. Hatakoshi, and M. Ishikawa, "Room Temperature Pulsed Operation of Nitride Based Laser Diodes with Cleaved Facets on C-Face Sapphire Substrates," *15th IEEE International Semiconductor Laser Conf.*, Paper PDP 1, Haifa, 1996.

[45] G. E. Bulman, K. Doverspike, S. T. Sheppard, T. W. Weeks, H. S. Kong, H. M. Dieringer, J. A. Edmond, J. D. Brown, J. T. Swindell, and J. F. Schetzina, "Pulsed Operation Lasing in a Cleaved-Facet InGaN/GaN MQW SCH Laser Grown on 6H-SiC," *Electronics Letters*, **33**(18), 1556–1557, 1997.

[46] K. Domen et al., unpublished. Oral presentation at *IEEE/LEOS Summer Topicals*, Montreal, Canada, Aug. 1997.

[47] M. P. Mack, A. Abare, M. Aizcorbe, P. Kozodoy, S. Keller, U. K. Mishra, L. A. Coldren, and S. P. DenBaars, "Characteristics of Indium-Gallium-Nitride Multiple-Quantum-Well Blue Laser Diodes Grown by MOCVD," http://nsr.mij.mrs.org/2/41/, *MRS Internet J. Nitride Semicond. Res.*, **2**(41), 1997.

[48] S. D. Lester, F. A. Ponce, M. G. Craford, and D. A. Steigerwald, "High Dislocation Densities in High Efficiency GaN-Based Light-Emitting Diodes," *Applied Physics Letters*, **66**(10), 1249–1251, 1995.

[49] T. Baba, Y. Yogo, K. Suzuki, F. Koyama, and K. Iga, "Near Room Temperature Continuous Wave Lasing Characteristics of GaInAsP/InP Surface-Emitting Laser," *Electronics Letters*, **29**(10), 913–914, 1993.

[50] J. J. Dudley, D. I. Babic, R. Mirin, L. Yang, B. I. Miller, R. J. Ram, T. Reynolds, E. L. Hu, and J. E. Bowers, "Low Threshold, Wafer Fused Long Wavelength Vertical Cavity Lasers," *Applied Physics Letters*, **64**(12), 1463–1465, 1994.

[51] D. I. Babic, K. Streubel, R. P. Mirin, N. M. Margalit, E. L. Hu, J. E. Bowers, D. E. Mars, L. Yang, and K. Carey, "Room-Temperature Continuous-Wave Operation of 1.54 μm Vertical-Cavity Lasers," *IEEE Photonics Technology Letters*, **7**(11), 1225–1227, 1995.

[52] N. M. Margalit, R. P. Mirin, J. E. Bowers, E. L. Hu, D. I. Babic, and K. Streubel, "Submilliamp Long-Wavelength Vertical-Cavity Lasers," *Proc. 15th IEEE International Semiconductor Laser Conf.*, Paper M3.5, Haifa, 1996.

[53] D. I. Babic and S. W. Corzine, "Analytic Expressions for the Reflecting Delay, Penetration Depth, and Absorptance of Quarter-Wave Dielectric Mirrors," *IEEE Journal of Quantum Electronics*, **28**(2), 514–524, 1992.

[54] O. Blum, I. J. Fritz, L. R. Dawson, A. J. Howard, T. J. Headly, J. F. Klem, and T. J. Drummond, "Highly Reflective, Long Wavelength AlAsSb/GaAsSb Distributed Bragg

Reflector Grown by Molecular Beam Epitaxy on InP Substrates," *Applied Physics Letters*, **66**(3), 329–331, 1995.

[55] B. Lambert, Y. Toudic, Y. Rouillard, M. Gauneau, M. Baudet, F. Alard, I. Valiente, and J. C. Simon, "High Reflectivity 1.55 μm (Al)GaAsSb/AlAsSb Bragg Reflector Lattice Matched on InP Substrates," *Applied Physics Letters*, **66**(4), 442–444, 1995.

[56] O. Blum, J. F. Klem, K. L. Lear, G. A. Vawter, and S. R. Kurtz, "Photopumped 1.56 μm Vertical Cavity Surface Emitting Laser with AlGaAsSb/AlAsSb Distributed Bragg Reflectors," *LEOS Summer Topical Meetings*, Montreal, Canada, 1997.

[57] G. P. Agrawal and N. K. Dutta, *Semiconductor Lasers*, 2nd ed., Van Nostrad Reinhold, New York, 98–118, 1993.

[58] L. A. Coldren and S. W. Corzine, *Diode Lasers and Photonic Integrated Circuits*, Wiley, New York, Appendix 17, 1995.

[59] M. J. Mondry, E. J. Tarsa, and L. A. Coldren, "Molecular Beam Epitaxial Growth of Strained AlGaInAs Multi-Quantum Well Lasers on InP," *Journal of Electronic Materials*, **25**(6), 948–954, 1996.

[60] D. I. Babic, *Double-Fused Long-Wavelength Vertical-Cavity Lasers*, Ph.D. Dissertation, ECE Technical Report #95-20, University of California at Santa Barbara, p. 95, 1995.

[61] D. B. Young, J. W. Scott, F. H. Peters, B. J. Thibeault, S. W. Corzine, M. G. Peters, S.-L. Lee, and L. A. Coldren, "High-Power Temperature-Insensetive Gain-Offset InGaAs/GaAs Vertical-Cavity Surface-Emitting Lasers," *IEEE Journal of Quantum Electronics*, **5**(2), 129–132, 1993.

[62] Y. M. Houng, M. R. T. Tan, B. W. Liang, S. Y. Wang, et al., "InGaAs (0.98 μm)/GaAs Vertical Cavity Surface Emitting Laser Grown by Gas-Source Molecular Beam Epitaxy," *Journal of Crystal Growth*, **136**(1–4), 216–220, 1994.

[63] R. P. Schneider, Jr., J. A. Lott, K. L. Lear, K. D. Choquette, et al., "Metalorganic Vapor Phase Epitaxial Growth of Red and Infrared Vertical-Cavity Surface-Emitting Laser Diodes," *Journal of Crystal Growth*, **145**(1–4), 838–845, 1994.

[64] S. A. Chalmers, and K. P. Killeen, "Method for Accurate Growth of Vertical-Cavity Surface-Emitting Lasers," *Applied Physics Letters*, **6**(6), 678–680, 1994.

[65] H. Q. Hou, H. C. Chui, K. D. Choquette, B. E. Hammons, W. G. Breiland, and K. M. Geib, "Highly Uniform and Reproducible Vertical-Cavity Surface-Emitting Lasers Grown by Metalorganic Vapor Phase Epitaxy with *In Situ* Reflectometry," *IEEE Photonics Technology Letters*, **8**(10), 1285–1287, 1996.

[66] F. G. Bobel, H. Moller, A. Wowchak, B. Hertl, J. Van Hove, L. A. Chow, and P. P. Chow, "Pyrometric Interferometry for Real Time Molecular Beam Epitaxy Process Monitoring," *Journal of Vacuum Science and Technology B*, **12**(2), 1207–1210, 1994.

[67] S. A. Chalmers, K. P. Killeen, and E. D. Jones, "Accurate Multiple-Quantum-Well Growth Using Real-Time Optical Flux Monitoring," *Applied Physics Letters*, **65**(1), 4–6, 1994.

[68] A. W. Jackson, P. R. Pinsulsanjana, A. C. Gossard, and L. A. Coldren, "In Situ Monitoring and Control for MBE Growth of Optoelectronic Devices," *IEEE Journal of Selected Topics in Quantum Electronics*, **3**(3), 836–844, 1997.

[69] L. A. Coldren, "Reactive-Ion-Etching of III–V Compounds," U.S. Patent No. 4,285,763, 1981.

[70] E. L. Hu and R. E. Howard, "Reactive Ion Etching of GaAs in a Chlorine Plasma," *Journal of Vacuum Science and Technology B*, **2**, 85, 1984.

[71] A. Scherer, J. L. Jewell, Y. H. Lee, J. P. Harbison, and L. T. Florez, "Fabrication of Microlasers and Microresonator Optical Switches," *Applied Physics Letters*, **55**(26), 2724–2726, 1989.

[72] J. A. Skidmore, D. L. Green, D. B. Young, J. A. Olsen, E. L. Hu, L. A. Coldren, and P. M. Petroff, "Investigation of Radical-Beam Etching-Induced Damage in GaAs/AlGaAs Quantum-Well Structures," *Journal of Vacuum Science and Technology B* **9**(6), 3516–3520, 1991.

[73] R. A. Morgan, G. D. Guth, M. W. Focht, M. T. Asom, K. Kojima, L. E. Rogers, and S. E.

Callis, "Transverse Mode Control of Vertical-Cavity Top-Surface-Emitting Lasers," *IEEE Photonics Technology Letters*, **4**(4), 374–376, 1993.

[74] J. W. Scott, D. B. Young, B. J. Thibeault, M. G. Peters, and L. A. Coldren, "Design of Index-Guided Vertical-Cavity Lasers for Low Temperature-Sensitivity, Sub-Milliamp Thresholds, and Single-Mode Operation," *IEEE Journal of Selected Topics in Quantum Electronics*, **1**(2), 638–648, 1995.

[75] J. M. Dallesasse, N. Holonyak, Jr., A. R. Snugg, T. A. Richard, and N. El-Zein, "Hydrolization Oxidation of AlxGa1-xAs-AlAs-GaAs Quantum Well Heterostructures and Superlattices," *Applied Physics Letters*, **57**(26), 2844–2846, 1990.

[76] K. D. Choquette, R. P. Schneider, Jr., K. L. Lear, and K. M. Geib, "Low Threshold Voltage Vertical-Cavity Lasers Fabricated by Selective Oxidation," *Electronics Letters*, **30**(24), 2043–2044, 1994.

[77] F. A. Kish, F. M. Steranka, D. C. DeFevere, D. A. Vanderwater, K. G. Park, C. P. Kuo, T. D. Osentowski, M. J. Peanasky, J. G. Yu, R. M. Fletcher, D. A. Steigerwald, M. G. Craford, and V. M. Robbins, "Very High-Efficiency Semiconductor Wafer-Bonded Transparent-Substrate $(Al_xGa_{1-x})_{0.5}In_{0.5}P$/GaP Light-Emitting Diodes," *Applied Physics Letters*, **64**(21), 2839–2841, 1994.

[78] C. Lei, T. J. Rogers, D. G. Deppe, and B. G. Streetman, "ZnSe/CaF2 Quarter-Wave Bragg Reflector for the Vertical-Cavity Surface-Emitting Laser," *Journal of Applied Physics*, **69**(11), 7430–7434, 1991.

[79] E. M. Strzelecka, G. D. Robinson, L. A. Coldren, and E. L. Hu, "Fabrication of Refractive Microlenses in Semiconductors by Mask Shape Transfer in Reactive Ion Etching," *Microelectronic Engineering*, **35**, 385–388, 1997.

[80] R. A. Morgan, K. Kojima, L. E. Rogers, G. D. Guth, R. E. Leibenguth, M. W. Focht, M. T. Asom, T. Mullally, and W. A. Gault, "Progress and Properties of High-Power Coherent Vertical Cavity Surface Emitting Laser Arrays," *SPIE Laser Diode Technology and Applications V*, **1850**, 100–108, 1993.

[81] P. L. Gourley, M. E. Warren, G. R. Hadley, G. A. Vawter, et al., "Coherent Beams from High Efficiency Two-Dimensional Surface-Emitting Semiconductor Laser Arrays," *Applied Physics Letters*, **58**(9), 890–892, 1991.

[82] M. Orenstein and T. Fishman, "Coupling Mechanism of Two Dimensional Reflectivity Modulated Vertical Cavity Semiconductor Laser Arrays," *1995 Laser Conference Proceedings*, pp. 70–71, 1995.

[83] A. Von Lehman, C. Chang-Hasnain, J. Wullert, L. Carrion, N. Stoffel, L. Florez, and J. Harbison, "Independently Addressable InGaAs/GaAs Vertical-Cavity Surface-Emitting Laser Arrays," *Electronics Letters*, **27**(7), 583–584, 1991.

[84] B. Moller, E. Zeeb, T. Hackbarth, and K. J. Ebeling, "High Speed Performance of 2-D Vertical-Cavity Laser Diode Arrays," *IEEE Photonics Technology Letters*, **6**(9), 1056–1058, 1994.

[85] M. Orenstein, A. C. Von Lehman, C. Chang-Hasnain, N. G. Stoffel, J. P. Harbison, and L. T. Florez, "Matrix Addressable Vertical Cavity Surface Emitting Laser Array," *Electronics Letters*, **27**(5), 437–438, 1991.

[86] R. A. Morgan, G. D. Guth, C. Zimmer, R. E. Leibenguth, M. W. Focht, J. M. Freund, K. G. Glogovsky, T. Mullally, F. F. Judd, and M. T. Asom, "Two-Dimensional Matrix Addressed Vertical Cavity Top-Surface Emitting Laser Array Display," *IEEE Photonics Technology Letters*, **6**(8), 913–915, 1994.

[87] C. J. Chang-Hasnain, M. W. Maeda, J. P. Harbison, L. T. Florez, et al., "Monolithic Multiple Wavelength Surface Emitting Laser Arrays," *Journal of Lightwave Technology*, **9**(12), 1665–1673, 1991.

[88] W. Goodhue, J. Donnelly, and J. Zayhowski, "Technique for Monolithically Integrating GaAs/AlGaAs Lasers of Different Wavelengths," *Journal of Vacuum Science and Technology B*, **7**(2), 409–411, 1989.

[89] L. E. Eng, K. Bacher, Y. Wupen, J. S. Harris, Jr., et al., "Multiple-Wavelength Vertical Cavity Laser Arrays on Patterned Substrates," *IEEE Journal on Selected Topics in Quantum Electronics*, **1**(2), 624–628, 1995.

[90] T. Wipiejewski, M. G. Peters, E. R. Hegblom, and L. A. Coldren, "Vertical-Cavity Surface-Emitting Laser Diodes with Post-Growth Wavelength Adjustment," *IEEE Photonics Technology Letters*, **7**(7), 727–729, 1995.

[91] T. Wipiejewski, J. Ko, B. J. Thibeault, D. B. Young, and L. A. Coldren, "Multiple Wavelength Vertical-Cavity Laser Array Employing Molecular Beam Epitaxy Regrowth," *46th Electron. Components and Tech. Conf.*, Orlando, FL, 1996.

[92] A. Fiore, Y. A. Akulova, J. Ko, E. R. Hegblom, and L. A. Coldren, "Multiple-Wavelength Vertical-Cavity Laser Arrays Based on Postgrowth Lateral-Vertical Oxidation of AlGaAs," *Applied Physics Letters*, **73**(3), 282–284, 1998.

[93] W. Nakwaski and M. Osinski, "Thermal Resistance of Top-Surface-Emitting Vertical-Cavity Semiconductor Lasers and Monolithic Two-Dimensional Arrays," *Electronics Letters*, **28**(6), 572–574, 1992.

[94] S. Esener and P. Marchand, "3D Optoelectronic Stacked Processors: Design and Analysis," *Optics in Computing '98, SPIE*, 3490, Brugge, Belgium, 541–545, 1998.

[95] W. K. Chan, J. P. Harbison, A. C. Von Lehman, L. T. Florez, C. K. Nguyen, and S. A. Schwarz, "Optically Controlled Surface-Emitting Lasers," *Applied Physics Letters*, **58**(21), 2342–2344, 1991.

[96] H. Kasahara, I. Ogura, H. Saito, M. Sugimoto, K. Kurihara, T. Numai, and K. Kasahra, "Pixels Consisting of a Single Vertical-Cavity Laser Thyristor and a Double Vertical-Cavity Phototransistor," *IEEE Photonics Tecnology Letters*, **5**(12), 1409–1411, 1993.

[97] Y. J. Yang, T. G. Dziura, T. Bardin, S. C. Wang, R. Fernandez, and A. S. H. Liao, "Monolithic Integration of a Vertical Cavity Surface Emitting Laser and a Metal Semiconductor Field Effect Transistor," *Applied Physics Letters*, **62**(6), 600–602, 1993.

[98] B. Lu, P. Zhou, J. Cheng, R. E. Leibenguth, A. C. Adams, J. L. Zilco, J. C. Zolper, K. L. Lear, S. A. Chalmers, and G. A. Vawter, "Reconfigurable Binary Optical Routing Switches with Fan-Out Based on the Integration of GaAs/AlGaAs Surface-Emitting Lasers and Heterojunction Phototransistors," *IEEE Photonics Technology Letters*, **6**(2), 222–226, 1994.

[99] R. Pu, E. M. Hayes, R. Jurrat, C. W. Wilmsen, K. D. Choquette, H. Q. Hou, and K. M. Geib, "VCSEL's Bonded Directly To Foundry Fabricated GaAs Smart Pixel Arrays," *IEEE Photonics Technology Letters*, **9**, 1622–1624, 1997.

[100] D. B. Schwartz, C. K. Y. Chun, B. M. Foley, D. H. Hartman, M. Lebby, H. C. Lee, C. L. Shieh, S. M. Kuo, S. G. Shook, and B. Webb, "A Low Cost, High Performance Optical Interconnect," *45th Electronic Components and Technology Conference*, Las Vegas, NV, pp. 376–379, 1995.

[101] D. K. Lewis, P. J. Anthony, J. D. Crow, and M. Hibbs-Brenner, "The Optoelectronics Technology Consortium (OETC)-Program Update," *LEOS '93 Conference Proceedings*, pp. 7–8, 1993.

[102] K. H. Hahn, "POLO – Parallel Optical Links for Gigabyte Data Communications," *45th Electronic Components and Technology Conference*, Las Vegas, NV, 368–375, 1995.

[103] *Optoelectronic Technology Roadmap – Conclusions and Recommendations*, Optoelectronics Industry Development Association, Washington, DC, 1995.

[104] K. S. Urquhart, P. Marchand, Y. Fainman, and S. H. Lee, "Diffractive Optics Applied to Free-Space Optical Interconnects," *Applied Optics*, **33**(17), 3670–3682, 1994.

[105] D. V. Plant, B. Robertson, H. S. Hinton, M. H. Ayliffe, G. C. Boisset, W. Hsiao, D. Kabal, N. H. Kim, Y. S. Liu, M. R. Otazo, D. Pavlasek, A. Z. Shang, J. Simmons, and W. M. Robertson, "A 4 × 4 VCSEL/MSM Optical Backplane Demonstrator System," *IEEE LEOS '95*, San Francisco, CA Paper PD2.4, 1995.

[106] P. S. Guilfoyle, F. F. Zeise, and J. M. Hessenbruch, "'Smart' Optical Interconnects for High Speed Photonic Computing," *Optical Computing 1993 Technical Series Digest*, **7**, 78–81, 1993.

[107] T. Cloonan, G. Richards, A. Lentine, F. McCormick, Jr, H. S. Hinton, and S. J. Hinterlong, "A Complexity Analysis of Smart Pixel Switching Nodes for Photonic Extended Generalized Shuffle Switching Networks," *IEEE Journal of Quantum Electronics*, **29**(2), 619–34, 1993.

[108] F. B. McCormick, "Free-Space Interconnection Techniques," in *Photonics in Switching, Volume II: Systems*, J. E. Midwinter, ed., Academic, San Diego, 1993.

[109] R. A. Metzger, "Turning Blue to Green," *Compound Semiconductor*, **1**(1), 26–28, 1995.

[110] B. Weigl, M. Grabherr, R. Jager, G. Reiner, and K. J. Ebling, "57% Wallplug Efficiency Wide Temperature Range 840 nm Wavelength Oxide Confined GaAs VCSELs," *Proc. 15th IEEE International Semiconductor Laser Conference*, Paper no. PDP2, 1996.

[111] B. Weigl, M. Grabherr, C. Jung, R. Jager, G. Reiner, R. Michalzik, D. Sowada, and K. J. Ebeling, "High Performance Oxide-Confined GaAs VCSELs," *IEEE J. Selected Topics in Quantum Electron.*, **3**, 409–415, 1997.

[112] E. R. Hegblom, N. M. Margalit, A. Fiore, and L. A. Coldren, "Small Efficient Vertical-Cavity Lasers with Tapered Oxide Apertures," *Electron. Lett.*, **34**(9), 895–896, 1998.

[113] L. A. Coldren, E. R. Hegblom, Y. A. Akulova, J. Ko, E. M. Strzelecka, and S. Y. Hu, "VCSELs in '98: What We Have and What We Can Expect," *SPIE Photonics West '98*, Paper 3286-100, San Jose, CA, 1998.

2

Fundamental Issues in VCSEL Design

Larry A. Coldren and Eric R. Hegblom

2.1 Analysis of Light–Current Characteristics and the Parameters Involved

The basic physics governing the operation of VCSELs is no different from that of other types of diode lasers. Current is injected into an active region where stimulated emission provides gain for an optical mode that is confined by an optical cavity, which includes at least one optical output port. The active region generally incorporates quantum wells within a pin-diode as in the edge-emitter case. The photon lifetime also tends to be only slightly shorter than typical edge-emitters. Of course, in the VCSEL case this is obtained with much shorter cavities and much higher mirror reflectivities.

Nevertheless, some of the approximations generally used for edge-emitters are not valid here, although the basic approach to analyzing VCSELs follows a similar course. In fact, some of the more interesting features of VCSELs have been uncovered by avoiding these classic approximations. For example, in edge-emitters the axial confinement factor is generally approximated as the fraction of the cavity length occupied by gain material. But, when the appropriate weighted averages are calculated for very short gain lengths, it is found that the axial confinement factor may vary considerably from this fill factor due to standing-wave effects. For a thin active region at an axial standing-wave peak, the confinement factor is nearly doubled! Also, for common edge-emitters the gain and spontaneous emission are always calculated assuming a bulklike, continuous optical mode density. But, if the optical cavity becomes small in some dimension, then the discrete nature of the optical mode density must be considered, just as we do for the electronic state density in active layers with some small dimension, such as quantum wells. Taking proper account of this discrete optical mode density as well as the standing-wave effects leads to the so-called microcavity effects, which are the subject of Chapter 3.

2.1.1 Rate Equations

The operation of a VCSEL, like any other diode laser, can be understood by observing the flow of carriers into its active region, the generation of photons due to the recombination of some of these carriers, and the transmission of some of these photons out of the optical cavity. These dynamics can be described by a set of rate equations, one for the carriers and one for the photons in each of the optical modes. In fact, the construction of such rate equations provides a clear definition of the basic laser parameters that we will need in describing the terminal characteristics of the VCSEL. Consider the generic VCSEL illustrated in

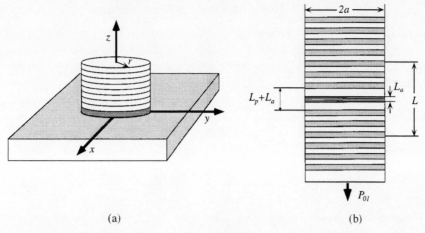

Figure 2.1. (a) General VCSEL schematic indicating coordinate system. (b) Cross-sectional schematic indicating active region thickness, L_a, and VCSEL effective cavity length, L.

Figure 2.1 with an active layer radius of a and active and effective cavity lengths of L_a and L, respectively. We will assume a lightly doped active region and relatively high injection, so that the hole density is equal to the electron density, and that the carriers all share a common reservoir, so that only one carrier density (N) rate equation is needed. If multiple carrier reservoirs exist with weak coupling between them, then additional equations for the separate carrier reservoirs must be used. For simplicity we will also assume that just one optical mode is important in the desired output, so that only one photon density (N_p) rate equation is needed. In both cases, the rate of increase in density is given by generation rates minus recombination rates. In the photon density case, the generation terms derive from both spontaneous and stimulated carrier recombination terms in the carrier density rate equation.

Thus, in the active region we have [1]

$$\frac{dN}{dt} = \eta_i \frac{1}{qV} - R_{sp} - R_{nr} - g v_g N_p \tag{2.1}$$

and

$$\frac{dN_p}{dt} = \Gamma g v_g N_p + \Gamma \beta_{sp} R_{sp} - \frac{N_p}{\tau_p}, \tag{2.2}$$

where the injection efficiency η_i is the fraction of terminal current that provides carriers that recombine in the active region; I is the terminal current; q is the electronic charge; $V = \pi a^2 L_a$ is the active region volume; R_{sp} is the spontaneous recombination rate of carriers (generation rate of photons); R_{nr} is the nonradiative recombination rate; $g v_g N_p$ is the stimulated recombination rate of carriers (generation rate of photons), in which g is the incremental optical gain in the active material and v_g is the group velocity in the axial direction of the mode in question; Γ is the three-dimensional mode confinement factor; β_{sp} is the spontaneous emission factor; and τ_p is the photon lifetime in the cavity. The injection efficiency accounts for current that might be shunted around the active region as well as recombination in the diode depletion region outside of the active region.

2.1.2 Modal Gain and Confinement Factor

The confinement factor is the overlap between the optical mode and the distribution of material gain created by the carriers. It is needed in the photon density generation terms to account for the fact that photons generated in the active volume occupy a larger cavity volume. Mathematically, it is obtained by calculating the net gain given to the optical mode – the so-called modal gain $\langle g \rangle$, and then dividing by g – the plane-wave incremental gain in the active material that would exist if the carriers were evenly distributed. For the general case in which both the optical electric field $\mathcal{E}(r, \theta, z)$ and the gain $g(r, \theta, z)$ may vary across the device, we have

$$\langle g \rangle = \frac{\int \mathcal{E}^*(r, \theta, z)\tilde{g}(r, \theta, z)\mathcal{E}(r, \theta, z)\, dV}{\int |\mathcal{E}(r, \theta, z)|^2\, dV}. \tag{2.3}$$

If we assume the gain is constant at $\tilde{g} = gn/\tilde{n}$ over the active volume of radius a, it can be factored out, and if we also assume a large axial standing-wave ratio as in VCSELs, we obtain [2]

$$\langle g \rangle = g\Gamma = g\Gamma_{xy}\Gamma_z = g\left[\frac{n}{\tilde{n}}\frac{\int_0^{2x}\int_0^a |U(r, \theta)|^2 r\, dr\, d\theta}{\int |U(r, \theta)|^2 r\, dr\, d\theta}\right]\frac{L_a}{L}\xi, \tag{2.4}$$

where n is index of refraction in the active region, $\tilde{n}$ is the effective index of the guided mode, and $U(r, \theta)$ is the normalized transverse electric-field mode shape. The last factor, ξ, is referred to as the axial enhancement factor, because it enhances the normally expected fill factor L_a/L in the axial confinement factor. When the field is approximated as a sinusoid enveloped by a decaying exponential in the DBRs, ξ reduces to [3, 4]

$$\xi = e^{-z_{\text{DBR}}/L_{\text{eff}}}\left[1 + \cos 2\beta z_s \frac{\sin \beta L_a}{\beta L_a}\right], \tag{2.5}$$

where $\beta = 2\pi\tilde{n}/\lambda$ is the axial propagation constant, z_s is the shift between the active layer center and the standing-wave peak, and the exponential prefactor accounts for placement of the active region within the DBR mirror. In this, z_{DBR} is the distance to the active material measured from the cavity–DBR interface and L_{eff} is the penetration depth of optical energy into the DBR. If the active region is placed between the mirrors, as is typical, $z_{\text{DBR}} \equiv 0$, and the prefactor is unity. As stated above, ξ can be as large as 2, if a thin active layer is centered on the standing-wave peak in the cavity. If placed at a null ($2\beta z_s = \pi$), ξ can be near zero.

The spontaneous emission factor, β_{sp}, is the fraction of spontaneous emission generated in the active region that is coupled into the desired mode. The spontaneous emission can of course couple into all of the optical modes of the system, including radiation modes of the cavity being considered. Thus, the desired lasing mode generally receives only a small fraction of the generated spontaneous emission. For common VCSEL designs with diameters of $\sim$3–6 μm, $\beta_{\text{sp}} \sim 0.01$ to 0.001 [5]. As will be discussed in Chapter 3, however, it is possible to make it much larger in certain microcavity structures. It should be realized that β_{sp} and R_{sp} are not independent parameters, since $\beta_{\text{sp}}R_{\text{sp}}V = \Gamma g v_g n_{\text{sp}}$ [6], where n_{sp} is a "population inversion factor" dependent only upon the quasi-Fermi level separation due to pumping. The population inversion factor $n_{\text{sp}} = f_2(1 - f_1)/(f_2 - f_1)$ is defined in terms of the Fermi functions f_1 and f_2 for the upper (conduction band) and lower (valence band) states, respectively. Below transparency n_{sp} and g are negative. Once the quasi-Fermi level separation exceeds the transition energy, n_{sp} becomes positive (typically <2 at threshold) and tends toward unity at complete inversion.

The cavity lifetime τ_p is given by the optical losses in the cavity,

$$\frac{1}{\tau_p} = v_g(\langle\alpha_i\rangle + \alpha_m), \tag{2.6}$$

where $\langle\alpha_i\rangle$ is the incremental internal modal power loss that is calculated from a modal average similar to that for gain given by Eq. (2.3), and $\alpha_m = (1/L)\ln(1/R)$ is the distributed mirror loss. For this, the mean mirror power reflection coefficient is $R = |r_1 r_2|$, where r_1 and r_2 are the amplitude reflection coefficients for the two mirrors.

We can define a threshold modal gain as the value that equals the modal losses in the steady state,

$$\Gamma g_{th} = \frac{1}{v_g \tau_p} = \langle\alpha_i\rangle + \alpha_m. \tag{2.7}$$

For finite β_{sp}, Eq. (2.2) indicates that the actual gain will always be slightly lower than the losses for any finite photon density; nevertheless, Eq. (2.7) is taken as the definition of the threshold gain, g_{th}, which is really the loss level. As β_{sp} approaches unity, the apparent threshold can approach zero, but the character of the optical emission (at low pumping) approaches that of an LED rather than a laser. At higher pumping, the linewidth narrows as n_{sp} approaches unity.

2.1.3 Power Out Versus Current

We can use the rate equations to obtain a steady-state solution for the power out versus the current in by setting the left-hand side derivatives to zero. If we assume that the carrier recombination rates and the gain are monotonically related to the carrier density and that the gain clamps at its threshold value, given in Eq. (2.7), then as the current is increased above I_{th} (the value to obtain g_{th}), we can set $R_{sp} + R_{nr} = \eta_i I_{th}/qV$ for $I > I_{th}$ and solve Eq. (2.1) for $N_p(I > I_{th})$. For the power out of one mirror above threshold we multiply this by the volume of the optical mode, V/Γ, the photon energy $h\nu$, the loss rate (coupling) out of the mirrors, $v_g\alpha_m$, and the fraction of power out of mirror 1, F_1. This yields [1]

$$P_{01} = F_1 \frac{h\nu}{q} \eta_d (I - I_{th}), \tag{2.8}$$

where the differential efficiency is

$$\eta_d = \eta_i \frac{\alpha_m}{\langle\alpha_i\rangle + \alpha_m} = \eta_i \frac{T_m}{A_i + T_m}, \tag{2.9}$$

where $T_m = \ln(1/R)$ and $A_i = \langle\alpha_i\rangle L$. This latter form is sometimes more convenient in VCSELs, because the mirror reflectivity must be large ($\sim$99%) to accommodate the small gain that can be achieved with $L_a < 30$ nm. In this case T_m is a very good approximation to the mean mirror transmission in the absence of losses, and A_i is a very good approximation to the one-pass losses, including those in the mirrors.

Similarly, the one-pass modal power gain, G, is also introduced for convenience. If we explicitly consider a multiple-quantum-well active region with N_w wells of thickness L_w

and a material gain g in the wells, such that $\Gamma = \Gamma_{xy}\xi N_w L_w/L$, we have

$$G = \Gamma_{xy} g \xi N_w L_w = \Gamma g L = T_m + A_i \quad \text{(at threshold)}, \tag{2.10}$$

where it is important to use the proper enhancement factor, ξ, for the N_w wells as positioned relative to the standing-wave peak.

The basic P–I characteristic given by Eq. (2.8) is valid even in cases where there is nonideal alignment of the gain peak and the mode wavelength, lateral leakage currents, size-dependent losses, or dynamic heating. However, to properly use it, we must determine what the relevant constants are for the case at hand. Moreover, we must also realize that some of the constants and relationships may become direct or indirect functions of the applied current. Thus, the threshold current, injection efficiency, gain constants, and losses may shift dynamically as the terminal current is increased. The most obvious example of this is the commonly observed P–I "roll-over" caused by device heating and the subsequent changes in some of these constants. Again, Eq. (2.8) generally holds, but the parameters are changing dynamically. The result is an apparent nonlinear P–I characteristic. These are critical effects, but before exploring how the important parameters of the device might change, we first need to obtain baseline values for them.

2.2 Modeling the Gain, Mirror Reflection, and Effective Cavity Dimensions

2.2.1 Gain Versus Current Relationships

The gain versus current density for bulk, quantum-well, and strained quantum-well active regions has been carefully analyzed over the past several years by a number of contributors. In all cases it is necessary to carefully calculate the conduction and valence bands, including the effects of possible valence band mixing, which generally leads to nonparabolic band shapes. Experimentally measured effective masses and bandgaps are used to obtain the necessary fitting parameters. The lineshape function used is another important aspect of these calculations, especially in quantum-well structures. Subsequent experimental work with edge-emitting lasers has shown very good agreement with these gain calculations. Thus, we can feel confident in proceeding with the existing results.

Generally, the gain at some photon energy E and temperature T can be expressed as [6]

$$g(E, T) = K(E, E_g(T))\rho_r(E, L_w, E_g(T))[f_2(E, E_{F1}(T), T) - f_1(E, E_{F2}(T), T)], \tag{2.11}$$

where K is a slowly varying factor proportional to the transition matrix element, ρ_r is the joint density of states, and f_2 and f_1 are the Fermi functions for the upper (conduction band) and lower (valence band) states, respectively, involved in the stimulated emission. The energies E_g and E_{Fi} are the bandgap and quasi-Fermi level energies, respectively. Equation (2.11) must also be convolved with a lineshape function to account for finite state occupation lifetime. In the case of quantum wells, the density of states is a steplike function with initial step height $\rho_{r1} = m_r^*/(L_w\pi\hbar^2)$, in which m_r^* is a reduced effective carrier mass. The Fermi functions for the upper ($i = 2$) and lower ($i = 1$) states are given by $f_i = [\exp((E_i - E_{Fi})/kT) + 1]^{-1}$. The quasi-Fermi levels are determined by the carrier densities in the valence and conduction bands, and the carrier densities can be determined from the current density $J = \eta_i I/\pi a^2$ that creates carriers in the active region from Eq. (2.1).

Thus, for some moderate pumping level, we find from Eq. (2.11) that the gain versus photon energy rises sharply according to the density of states function (as rounded by the lineshape function) from an energy slightly below (again due to the lineshape function) the energy separating the first energy levels in the conduction and valence bands to a peak gain point, and then it decreases somewhat more slowly due to the roll-off in the Fermi-function difference ($f_2 - f_1$). As the pumping current is increased, the gain peak increases and the high-energy roll-off is somewhat slower.

At some wavelength (photon energy) and temperature a simple two-parameter logarithmic function can fit the gain versus current density characteristic for positive gains. A more complex three-parameter fit does a better job over a wider range of gains, but for most lasers, the range of gains considered is not sufficiently wide to merit the added complexity. Thus, we will model the gain by [6],

$$g = g_0 \ln \frac{J}{J_0}, \qquad (2.12)$$

where g_0 is a gain parameter and J_0 is the transparency current density. The effects of

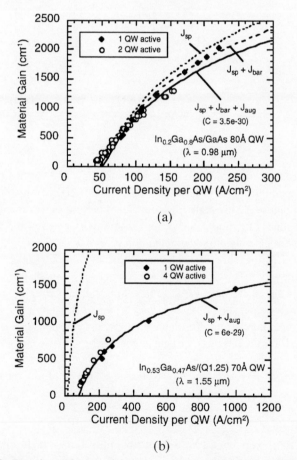

(a)

(b)

Figure 2.2. Experimental material gain versus injected current curves for $In_{0.2}GA_{0.8}As/GaAs$ 1 and 2 80 Å QW active region 0.98 μm lasers (upper), and $In_{0.53}Ga_{0.47}As/(Q1.25)$ 1 and 4 70 Å QW active region 1.55 μm lasers (lower). Theoretical curves are superimposed on the plots. The gain in the lower plot is well represented by the two-parameter expressions, $g = 583 \ln(J/81) \text{ cm}^{-1}$, where J is given in A/cm^2 [6].

Table 2.1. *Two-parameter fits to theoretical gain versus current density relationships for various material systems [6].*

Active Material	$J_{tr}{}^a$	$g_0{}^b$
$J_{sp} + J_{bar} + J_{Aug}$:		
Bulk GaAs	80	700
GaAs/Al$_{0.2}$Ga$_{0.8}$As 80 Å QW	110	1300
In$_{0.2}$Ga$_{0.8}$As/GaAs 80 Å QW	50	1200
J_{sp}:		
Bulk GaAs	75	800
GaAs/Al$_{0.2}$Ga$_{0.8}$As 80 Å QW	105	1500
In$_{0.2}$Ga$_{0.8}$As/GaAs 80 Å QW	50	1440
J_{sp}:		
Bulk In$_{0.53}$Ga$_{0.47}$As	11	500
InGaAs 30 Å QW (+1%)	13	2600
InGaAs 60 Å QW (0%)	17	1200
InGaAs 120 Å QW (−0.37%)	32	1100
InGaAs 150 Å QW (−1%)	35	1500
$J_{sp} + J_{Aug}$:		
In$_{0.53}$Ga$_{0.47}$As/(Q1.25) 70 Å QW	81	583

$^a[J] = $ A/cm^2. $^b[g] = $ cm^{-1}. $g = g_0 \ln[J/J_{tr}]$

nonradiative recombination can also be included in Eq. (2.12) by using the appropriate parameters. For example, in the case of InGaAs/InP, the effects of Auger recombination are very significant. Figure 2.2 illustrates the gain versus current for several example cases [6]. Table 2.1 gives values of g_0 and J_0 for a number of important cases at 300 K for the wavelength of maximum gain under moderate pumping.

For the short VCSEL cavities, the axial mode spacing $\delta\lambda = \lambda^2/(2\tilde{n}_g L)$ is typically larger than the gain bandwidth. This results in inherent single-axial-mode operation, but it is also very common for the mode wavelength to be different from the gain peak. Also, the gain changes with temperature, and this is important because self-heating effects can change the active-region temperature even if the ambient temperature remains fixed. Thus, it is important to understand how the active region gain might change with wavelength and temperature. Figure 2.3 illustrates the changes in gain spectra and logarithmic fit parameters g_0 and J_0 as the temperature or the measurement wavelength is changed for two cases of interest.

2.2.2 Reflection from VCSEL Mirrors

Because of the short gain length, VCSEL mirrors must have a high reflection coefficient. Using the results of Table 2.1 and Eq. (2.12) in Eq. (2.10), we see that the gain per pass in a VCSEL with three properly placed quantum wells can be about 1% for reasonable pumping levels. Thus, if we desire a differential efficiency of 50% (i.e., 50% of the generated light provided as useful output), the output mirror transmission must be about 0.5% and the other

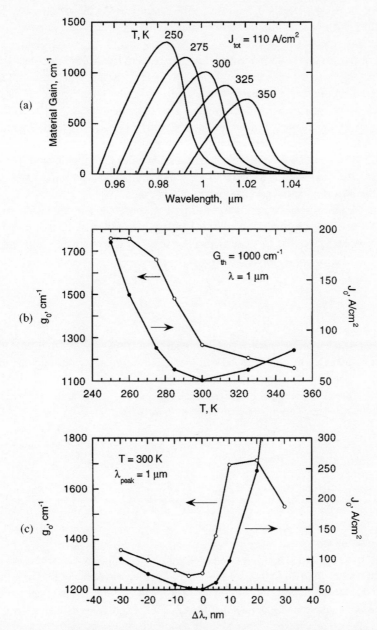

Figure 2.3. Temperature and wavelength dependence of TE gain calculated for 3-$In_{0.2}Ga_{0.8}As$-QW configuration with 80 Å GaAs barriers and 100 Å GaAs claddings on each side: (a) Gain spectrum versus temperature at fixed current density ($J_{tot} = 110A/cm^{-2}$.) (b) and (c) Dependence of the parameters for two-parameter gain versus current density logarithmic curve fit on temperature for $\lambda = 1\mu m$ and wavelength misalignment for 300 K. The fit parameters shown in (b) and (c) are valid over the limited gain range (800–1200 cm^{-1}).

cavity losses, including the back mirror transmission, must be about 0.5%. Mirrors with power reflection coefficients of less than 99% would clearly be of little use. Therefore, we should focus our attention on mirror designs that are capable of providing such high reflection.

The only mirrors that can provide these levels of reflectivity are dielectric stacks composed of multiple layers with alternating high and low index. These are generally referred to as distributed Bragg reflectors (DBRs), because the maximum reflection occurs at the Bragg condition. Metal mirrors are limited to power reflectivities $\sim$95%, although composites consisting of several dielectric mirror periods backed by metal can have power reflectivities >99%. Multilayer dielectric mirrors and the other reflecting layers in a VCSEL are generally analyzed by a transmission matrix approach that includes the individual properties of the various layers [7]. However, we here provide simple analytical formulas that can provide most of the important answers regarding VCSEL design.

The maximum possible amplitude reflectivity is obtained with quarter-wave thick dielectric layers. For a mirror stack with m identical periods, each consisting of a pair of quarter-wave layers ($2m$ layers), one with a low index n_1 and one with a high index n_2, the reflectivity at the Bragg frequency is given by [8]

$$r_{\text{gm}} = \frac{1 - \left(\frac{n_1}{n_2}\right)^{2m}}{1 + \left(\frac{n_1}{n_2}\right)^{2m}} \approx \tanh\left(\frac{m\,\Delta n}{\bar{n}}\right). \tag{2.13}$$

In the rightmost expression, $\Delta n = (n_2 - n_1)$, and $\bar{n} = (n_1 + n_2)/2$. To derive Eq. (2.13), one must assume that the incident and exit media have the same index and that their index is between n_1 and n_2. More generally, if the quarter-wave stack has $2m + 1$ interfaces with the ith interface between n_{Li}, a low index material, and n_{Hi}, a high index material, then the reflectivity at the Bragg frequency can be written as [9]

$$r_{\text{gm}} = \frac{1 - b}{1 + b}, \qquad \text{where } b = \prod_0^{2m} \frac{n_{\text{Li}}}{n_{\text{Hi}}}. \tag{2.14}$$

Here $i = 0$ denotes the interface between the DBR stack and the incident medium and $i = 2m$ denotes the interface between the DBR and the exit medium. For high-reflectivity mirrors, the transmitted power can be approximated by $T_{\text{m}} \approx 4b$. Loss can also be included by using complex indices.

Equations (2.13) and (2.14) can be easily modified to apply to cases where the interfaces are not abrupt or the two layers are not each a quarter-wave thick. This can be seen by realizing that the fundamental reflection is really due to the fundamental Fourier component of the variable index pattern across the mirror. Thus, expressing n_2 and n_1 as $\bar{n} \pm \Delta n/2$, where $\bar{n}$ is the average index, we see that Δn should be multiplied by the ratio of the Fourier coefficient of the actual pattern to that of a square wave. For example, if the index pattern is sinusoidal, then Δn should be multiplied by $\pi/4$. For linearly graded interfaces, the same Bragg frequency can be maintained by keeping the overall period the same. To a good approximation this can be achieved by reducing the thickness of the uniform regions by the width of the grading region. Table 2.2 gives index values for commonly used materials. Figure 2.4 plots the reflection and transmission of some typical mirrors as a function of the number of periods [8].

Table 2.2. *Refractive index of various semiconductors.*

III–V Compounds	$n@E_g$	$n@(\lambda \mu m)$
GaAs	3.62	3.52(0.98)
AlGaAs (0.2)	3.64	3.46(0.87)
		3.39(0.98)
AlAs[a]	3.2	2.98(0.87)
		2.95(0.98)
InGaAs (0.2) comp. strained on GaAs	3.6	
InP	3.41	3.21(1.3);
		3.17(1.55)
InGaAsP (1.3 μm)	3.52	3.40(1.55)
InGaAsP (1.55 μm)	3.55	
InGaAsP (1.65 μm)	3.56	
InAs	3.52	
GaP[a]	3.5	
AlP[a]	2.97	
AlSb[a]	3.5	
GaSb	3.92	
InSb	3.5	
GaN (Hexagonal)	2.67	2.33(1 eV)
AlN (Hexagonal)		2.15(3 eV)

[a] Indirect gap.

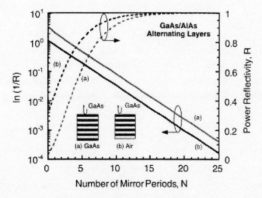

Figure 2.4. Reflectivity of typical surface-emitting laser (SEL) GaAs/AlAs alternating quarter-wavelength mirror stacks as a function of the number of mirror pairs. Case (a) corresponds to a stack terminated on GaAs (the bottom mirror of an SEL). Case (b) corresponds to a stack terminated on air (the top mirror of some types of SELs). The solid curves, corresponding to the left axis, are proportional to the mirror loss experienced by the cavity. The dashed curves, corresponding to the right axis indicate power reflectivity (field reflectivity squared). In the inserts, black corresponds to AlAs and white corresponds to GaAs. Note that for zero number of mirror pairs, case (a) is left with a residual layer of AlAs (for proper matching to the GaAs termination). As a result, $N = 15$ (for example) should be interpreted as $N = 15.5$ for case (a).

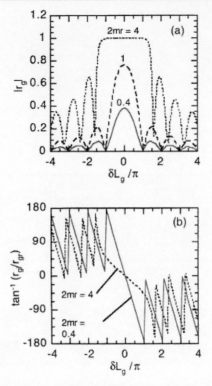

Figure 2.5. (a) Magnitude and (b) phase of the grating reflection coefficient versus the normalized frequency deviation from the Bragg condition for different values of the reflection parameter, $kL_g \equiv 2mr$ [7]. For small r and large m, the reflection spectrum plotted above is only dependent on the product $2mr$. However, for larger values of r and/or smaller values of m, there is some dependence on the individual values of r and m. For example, $r = 0.01$ and $m = 100$ would be roughly equivalent to $r = 0.1$ and $m = 10$. However, some changes in the spectrum would occur with $r = 0.2$ and $m = 5$, even though the $2mr$ product is equivalent in all three cases. The above plots use $r = 0.1, 0.025, 0.01$ with $m = 20$ for all cases. In the phase plot, the $2mr = 1$ case is very similar to the 0.4 case and hence is not plotted. Here $\delta \equiv \beta - \beta_0$, where β is the average propagation constant of the grating.

As shown by the magnitude response in Figure 2.5 [8], the entire wavelength response of a DBR is given by $r_g \approx r_{gm} \sin \delta L_g / \delta L_g$ for amplitude reflectivities less than about 0.4. As indicated in the figure, $\delta = \beta - \beta_0$ is the deviation in propagation constant from the Bragg condition, and L_g is the total length of the grating. For amplitude reflectivities up to 0.8 the nulls still fall in the same place, even though the peak becomes somewhat broadened. For values >0.8 the top becomes relatively flat and the initial nulls begin to spread in wavelength as the effective penetration of light into the DBR is reduced. The effective optical bandwidth over which the reflection is relatively high is given by $\Delta \lambda / \lambda \sim 1/m$.

2.2.3 Effective Mirror Model: Effective Cavity Length and Loss

The phase response in Figure 2.5 shows a linear deviation about the Bragg condition. This allows us to construct an effective mirror model for the DBR as well as the entire VCSEL. That is, a reflection with a linear phase deviation versus frequency can be modeled by a discrete (hard) mirror a length L_{eff} away [8, 10]. Then, the net effective cavity length discussed earlier is $L = L_{\text{eff1}} + L_a + L_p + L_{\text{eff2}}$. By noting that L_{eff} is given by half of the slope of the phase versus the propagation constant [8], we find that

$$L_{\text{eff}} = \frac{\lambda}{4\Delta n} r_{\text{gm}} \approx \frac{\lambda}{4\Delta n} \tanh\left(\frac{m\Delta n}{\tilde{n}}\right). \tag{2.15}$$

For highly reflecting mirrors as in VCSELs, $r_{\text{gm}} \to 1$, and $L_{\text{eff}} \to \lambda/(4\Delta n)$. When high contrast (e.g., oxide/semiconductor) mirrors are used, then it is necessary to account for the difference between the low index material and the material used in front of the hard mirror, n_{m}, by incorporating a correction factor [9], $L_{\text{eff}} = (n_{\text{L0}}n_2/n_{\text{m}}n_{\text{H0}})\lambda/(4\Delta n)$. Again, if the mirror interfaces are graded, or the layer duty ratio altered from $1:1$, we must multiply the peak Δn by the appropriate Fourier coefficient ratio as discussed above.

From the effective mirror model, it is possible to estimate the axial mode spacing as long as the modes fall within the reflection band of the mirror. That is,

$$\delta\lambda = \frac{\lambda^2}{2(\tilde{n}_{\text{g1}}L_{\text{eff1}} + \tilde{n}_{\text{ga}}L_a + \tilde{n}_{\text{gp}}L_p + \tilde{n}_{\text{g2}}L_{\text{eff2}})}, \tag{2.16}$$

where the $\tilde{n}_{\text{gi}}$ refer to the group effective indices. In typical VCSELs, $L_a \sim 20$ nm, $L_p \sim 200$ nm, and $L_{\text{effi}} \sim 500$ nm. So, we see that the mirror penetration depth dominates the cavity length, and it is the primary factor in determining the axial mode spacing. However, in this case it is also found that the mode spacing is larger than the mirror high-reflectivity bandwidth. Once the reflectivity begins to roll off, the L_{eff} lengths begin to get larger, and they can no longer be treated as constants. Thus, the calculation is more useful in determining the rate at which modes will translate in response to a change in cavity thickness, and this is useful in properly aligning the mode in the mirror reflection band during growth. The shift in a particular mode with respect to a change in some cavity length, say L_p, is

$$\Delta\lambda|_p = \lambda \frac{\tilde{n}_p \Delta L_p}{\tilde{n}_{\text{g1}}L_{\text{eff1}} + \tilde{n}_{\text{ga}}L_a + \tilde{n}_{\text{gp}}L_p + \tilde{n}_{\text{g2}}L_{\text{eff2}}}. \tag{2.17}$$

It is also notable that this same penetration length is approximately equal to the energy penetration depth in the mirror [9]. Thus, L_{eff} can also be used in an effective energy storage model in which it is assumed that the optical energy density is uniform at its peak value throughout a cavity of length $L = L_{\text{eff1}} + L_a + L_p + L_{\text{eff2}}$. This further suggests that small mirror losses can be modeled by simply using a complex propagation constant in the effective mirror model. That is, the actual amplitude reflection coefficient including loss can be found from its lossless value calculated from real indices [8, 10],

$$r_{\text{gm}} = r_{\text{gm}}(\text{lossless})\, e^{-\alpha_{\text{ig}}L_{\text{eff}}}, \tag{2.18}$$

where α_{ig} is the incremental loss in the grating. Thus, we see that it is desirable to minimize L_{eff} by using high-contrast materials to reduce cavity losses.

Another issue with DBR mirrors that must be addressed for many VCSEL designs is their net electrical conductance. For many simple VCSEL designs, it is desired to conduct current bound for the active region through one or both mirrors, and it is desired to have minimal series resistance in the device. Although maximum reflectivity is obtained by using very abrupt interfaces, this results in the worst electrical conductivity. Fortunately, a modest amount of compositional grading together with appropriate doping can reduce the effective potential barriers at these interfaces, while only reducing the effective Δn used in the mirror reflection analysis by a small amount [10]. (Recall for a sine wave – the maximum amount of grading – that the reduction factor is only $\pi/4$.) Of course, a slight reduction in Δn can easily be compensated by a similar increase in the number of periods, m. The penalty of decreasing Δn is a slight increase in mirror penetration depth. The increased resistance due to the $2m$ heterointerfaces may be estimated by modeling the DBR as a resistor in series with two diodes – one each for the potential up and down steps [11]. The resistance is calculated from the bulk values in the series of separate high- and low-index layers; the diode models include the effective potential at the up or down steps and an ideality factor of m, the number of up or down interfaces. Experiments have shown rough agreement with this model, although it is difficult to separate the diodes from each other, and since they really do not turn on, they are even difficult to distinguish from the resistance experimentally. Thus, the DBR is generally well modeled by a single diode and resistor.

The other major issue is the increased loss that accompanies the relatively high doping that must be used in the DBR. This has been addressed by using bandgap engineering to grade the composition as well as the doping at the mirror interfaces, so that the doping necessary to reduce the potential barrier is minimized. Figure 2.6 illustrates the general concept and gives a couple of examples of how such bandgap engineering might be accomplished [12]. This will be discussed further in Chapter 4. Neglecting standing-wave effects, the local incremental loss due to doping is approximately 11 cm^{-1} per 10^{18} cm^{-3} of p-doping and 5 cm^{-1} per 10^{18} cm^{-3} of n-doping in the 850 to 980 nm range [11], and it is significantly higher in the 1.3 to 1.55 μm range [13]. Unfortunately, due to the heavy hole mass, it is more important to raise the doping in the p-mirror to reduce series resistance. Thus, loss in the p-mirror is the biggest problem. Also, due to the high-contrast standing wave in VCSELs, this loss can be increased by up to a factor of 2, if the doping is placed at the standing-wave peak, or decreased by a larger factor if placed at the standing-wave null, analogous to the gain. The standing wave is also much stronger nearer the active region. Thus, in the first few mirror periods, it is worthwhile accepting some increased resistance for much lower optical losses. Of course, if intracavity contacts are employed as introduced in Chapter 1, the problem of series resistance is shifted from vertical conduction issues in the DBR mirrors to spreading resistance issues along the intracavity contact layers.

Mirrors grown by MOCVD use carbon p-type doping and continuous compositional gradings [14]. Mirrors grown by MBE have generally used beryllium and graded superlattices [12]. The MBE mirrors tend to be inferior both because the beryllium is not well activated in high-aluminum areas and because it tends to diffuse away from desired locations at the common growth temperatures. MBE efforts using carbon doping in the p-type DBRs have resulted in much lower voltage [12, 15].

2.3 Lateral Effects: Optical Modes and Optical, Current, and Carrier Losses

Except for the inclusion of a lateral confinement factor, Γ_{xy}, the above analysis has been one dimensional. Although this provides the baseline performance of any VCSEL, we must consider lateral effects to be able to design and understand devices with ever shrinking diameters. Broad area characteristics tend to scale down to diameters of about ten microns. At this point the threshold current densities begin to increase due to nonideal current, carrier, and photon confinement. At diameters of less than three microns, the threshold current density in standard VCSEL designs begins to skyrocket, and increased optical losses begin to significantly pull down the differential efficiency as well [16, 17].

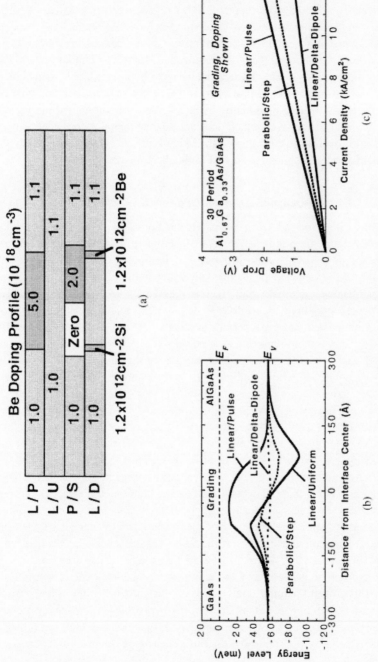

Figure 2.6. Band-gap engineering |12|. (a) Schematic of interface designs. Numbers indicate doping levels in units of 10^{18} cm^{-3}. L/P is Linear grading with Pulse doping over the entire graded region. L/U is Linear grading with Uniform doping. P/S is Parabolic grading with Step doping in the graded region. L/D is Linear grading with Delta-doping at the interface edges. (b) Results from a self-consistent one-dimensional Poisson equation solver of the interface barriers produced by the doping/grading profiles. (c) Voltage drop over 30 period DBR using the various techniques simulated. The experiment verifies the simulated results for lowering barrier voltages.

As introduced in Chapter 1, these problems can be addressed by incorporating some form of optical, current, and carrier confinement, just as in high-performance edge-emitting lasers. In principle, all of this could be provided by a buried-heterostructure design (Figure 1.2(d)), in which a lower-index, higher-bandgap material was laterally regrown around an etched mesa as introduced in Chapter 1. This may eventually be possible, although initial attempts have given marginal results [18]. (See Chapter 5.) As outlined in Chapter 1 surface passivation with sulfide treatments and impurity-induced disordering (intermixing) have shown clear carrier confinement, but little improvement in the $P-I$ characteristics [19].

At this writing most VCSELs only partially address the important lateral confinement issues. In fact, it may be more important to address one or two of the lateral confinement issues well, rather than all three marginally. In this section we summarize the various losses assuming a planar active layer, which does not provide lateral carrier confinement. Nevertheless, it should be realized that future VCSELs could benefit greatly by incorporating some form of lateral carrier confinement such as with a buried heterostructure.

2.3.1 Size-Dependent Current and Carrier Leakage

Confining the current injection radius is perhaps the most obvious requirement in making VCSELs of some limited lateral dimension. The etched-mesa design [Figure 1.2(a)], in which the etch proceeds to just above the active region, has solved this problem quite well for some time. If the etch extends to the undoped diode depletion region, the current has no incentive to spread laterally, and nearly all goes straight down into the active region. The dielectrically apertured structures [Figure 1.2(c)] share this characteristic, although the current may be concentrated near the aperture if the intracavity or large-diameter ring contacts are used [20]. The planar proton-implanted structures [Figure 1.2(b)] address this problem by creating a high-resistivity region to block current above the active layer. However, to avoid carrier losses in the active layer and problems with reliability the implant is generally kept somewhat above the active layer so that current can spread laterally beneath the implant. Also, the current may be crowded near the implant edges if ring contacts are used.

Once the current gets to the diode depletion region, carriers are injected into it, and if the device is properly designed, most carriers are collected by the active layer. (Those that recombine in transit are accounted for in η_i.) Carrier losses may occur by lateral diffusion if no lateral carrier confinement exists [21]. Vertical carrier leakage over the active layer heterobarriers can also occur, but this is well understood from one-dimensional analysis, and it can be addressed by using sufficiently high barriers as in the case of edge-emitting lasers [22]. In this discussion we only address the specific problem of lateral carrier leakage due to diffusion, or surface recombination if the active region is etched. This leakage is especially important at the smaller device sizes where the ratio of lateral surface area (at the current injection aperture) to the volume of the active region becomes large. In all cases, the exact analysis of this problem requires a self-consistent finite element approach to account for nonuniform current injection as well as nonuniform carrier and photon density profiles. Finite element modeling like that shown in Figure 2.7 has quantified several consequences of carrier diffusion. Below or near threshold the carrier density profile rolls off at the edges of the desired active region as determined by the lateral leakage due to diffusion or surface recombination. For device radii much larger than the lateral carrier diffusion length this is an edge effect; for radii similar to or smaller than the diffusion length, the carrier density profile starts rolling off near the center of the device [16, 20] (see Figure 2.7). At higher

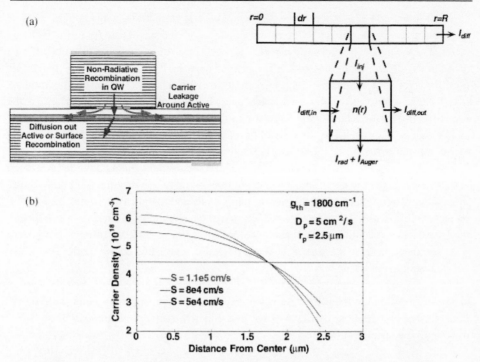

Figure 2.7. (a) Schematic of a finite-element technique used to determine the carrier density profile at threshold and the contribution to the threshold current. (b) Carrier density profile obtained with various levels of surface recombination [16].

optical powers, the interaction of the carriers with the mode can lead to spatial hole burning. The overlap of the carrier (gain) profile with the optical mode determines the spatial profile of the current sunk by stimulated emission. If that profile does not match the profile of the injected current, then the mode will deplete carriers where the field is strongest. This spatial hole burning results in more carrier leakage because the carrier density at the edge now must increase to maintain the same modal gain [20]. Thus, the effective threshold current increases at the higher powers, and the $P-I$ characteristic becomes sublinear due to this shift even with no assumed heating. Moreover, higher order modes now receive more gain, and so lasing in such modes tends to occur.

For either carrier leakage or current spreading the increase of threshold can be modeled as a leakage current proportional to the active region perimeter. When the injected current spreads over a distance larger than the carrier diffusion length, then the gradient for carrier diffusion is significantly reduced, and current leakage will dominate carrier leakage. One can model the current spreading through a resistive sheet on top of a junction diode. Below threshold, one must raise the applied diode voltage to sink more current, which will increase the spreading, but above threshold the active region diode will act as an ideal current sink (since the voltage is clamped). Consequently, current spreading has virtually no impact on the slope efficiency but significant impact on the threshold current. The excess current at threshold is given by [23]

$$I_{ls}(N_{th}, a) = I_0/2 + \sqrt{I_0 \pi a^2 J'_{th}/\eta_i}, \qquad (2.19)$$

in which $I_0 = 8\pi s V_T/\rho$, where $V_T = \eta k T/q$, s is the spacing between the current aperture and the active region, η is the diode ideality factor, and ρ is the resistivity of the included material. As Figure 2.8 shows the current leakage model fits well to the size dependence of the threshold current in dielectrically apertured devices. Even for an aperture in the first p-mirror layer with moderate doping beneath it, the current spreading can easily dominate over leakage due to diffusion. A large leakage current above the active layer also is the primary cause for the relatively high thresholds in proton-implanted devices, and Eq. (2.19) can be used to estimate it. In the case of two apertures surrounding the active region, the resistive paths between the apertures will add in series and so ρ/s can be replaced with $\rho/s \rightarrow \rho_A/s_A + \rho_B/s_B$, where the subscripts A and B denote the regions above and below the active region. Since the hole mobility is about twenty times less than the electron mobility in AlGaAs, we conclude that for similar doping levels and thicknesses, adding an aperture on the n-side provides less than a 5% reduction in the lateral leakage current. In either case, the lateral leakage is reduced by shrinking s or increasing ρ. However, dropping ρ to zero will essentially increase the size of the diode depletion region and permit carriers to diffuse before reaching the wells.

For etched-mesa or properly designed dielectric-apertured devices, the leakage current can be very low and in the ideal case when there is no current spreading, one can estimate the contribution of lateral carrier leakage due either to surface recombination or lateral diffusion. The leakage current is [16]

$$\eta_i I_{ld}(N_{th}, a) = 2\pi aq N_{th}(a) S_{eff}(N), \qquad (2.20)$$

where S_{eff} is the effective surface recombination velocity at $r = a$. (The η_i is included to refer the leakage current to the terminal.) For lateral diffusion, $S_{eff} = \sqrt{D/\tau(N)}$, in which D is the ambipolar diffusion constant and $\tau(N)$ is the carrier lifetime. If primarily radiative recombination is assumed, $\tau(N) \propto 1/N$, and $N \propto \sqrt{J_{th}'}$ and thus from Eq. (2.20), $I_{ld} \propto a J_{th}'^{3/4}$, where J_{th}' is the threshold current density with no leakage, which also depends upon the optical losses. If the recombination becomes dominated by surface or defect recombination, as it might when we care about such recombination, $I_{ld} \propto a J_{th}'$. So, in either case, to a good approximation, we let

$$\eta_i I_{ld}(N_{th}, a) = K_{ld} a J_{th}', \qquad (2.21)$$

where K_{ld} includes the accumulated constants for the case in question. The leakage would decrease linearly with radius for a constant current density. However, for smaller device radii the current density increases rapidly, primarily due to increasing optical losses in typical devices [16, 17].

2.3.2 Size-Dependent Optical Losses

Until recently, optical losses provided the principle obstacle to scaling the size of VCSELs to small sizes. Figure 2.9 gives example data from etched-mesa and dielectrically apertured VCSELs. Figure 2.10 plots de-embedded excess per-pass optical losses, A_{is} versus device radius for the etched-mesa, air-apertured, and oxide-apertured structures [17]. The excess (size-dependent) loss is extracted from the decrease in differential efficiency and is defined by

$$A_i(a_m) = A_{i0} + A_{is}(a_m), \qquad (2.22)$$

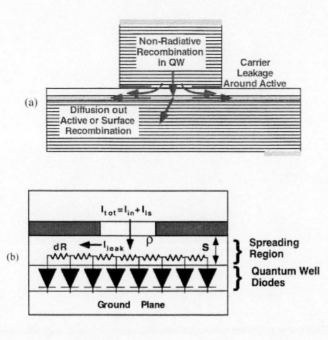

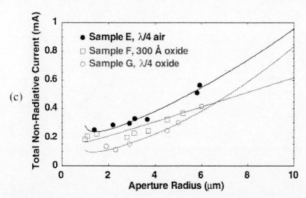

Figure 2.8. (a) Schematic of leakage mechanisms in a VCL. Contributions to the threshold current arise from current spreading between the aperture and the active region, from diffusion or surface recombination of carriers in the intrinsic region, and from nonradiative recombination in the quantum well [17]. (b) Electrical model for the current leakage. Current spreads laterally in a resistive sheet above an active region with a diode characteristic. A higher current density in the center, a lower resistance spreading region, or bad diodes will increase leakage. (c) Plot of the nonradiative current at threshold versus device size for various apertured VCLs. The nonradiative current is primarily due to current spreading between the aperture and the active region and fits well to the current-spreading model. The variation in the amount of spreading is primarily due to variations in doping.

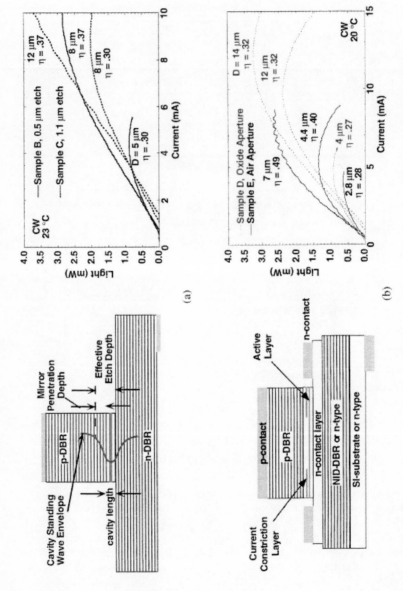

Figure 2.9. Schematics and L–I curves for (a) etched-post and (b) dielectrically apertured VCSELs [16].

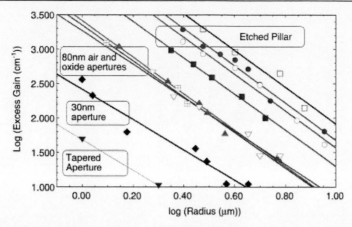

Figure 2.10. Comparison of the size-dependent optical losses for etched pillar devices and for air or oxide apertured ones taken from Figure 2.9 and elsewhere [17, 43]. The losses due to roughness of the pillar are much larger than the losses due to scattering from the thin apertures. The tapered aperture has the least excess loss.

where $A_{i0} = \langle \alpha_{i0} \rangle L$ is the broad area per pass loss. The linearity of the plots in the log–log format shows that optical losses vary as $1/a_m^{\gamma}$, with $\gamma \sim 2$–3, for a wide variety of device configurations. The $1/e$ mode radius, a_m, should be used here because the optical losses will tend to depend upon this rather than the active region radius or the optical aperture radius. Since the plot uses the optical aperture radius, we need to correct the abscissa to give the proper power law. Given these data, we express the size-dependent per pass optical losses for the mode by $A_{is}(a_m) \approx A_{i1}/a_m^{\gamma}$, with $\gamma \sim 2.6$.

Figure 2.10 shows that the dominant loss in most etched-mesa VCSELs is due to surface roughness of the mesa walls. As the mesa is etched deeper the differential efficiency goes down. If diffraction were the primary loss mechanism, the differential efficiency should go up. Also, analytical studies of diffraction loss show that it should be relatively small in typical devices for diameters >5 μm [24].

To address the excess loss in etched-mesa devices, as well as improve current leakage in proton-implanted VCSELs, dielectric-apertured devices have gained favor. The aperture confines the current as effectively as an etched mesa, but it also acts as an intracavity optical lens to at least partially compensate the effects of diffraction [25, 26]. A nonideal lens will give rise to scattering losses as shown for the unfolded optical cavity in Figure 2.11 [4]. The size-dependent scattering loss in a dielectrically apertured VCSEL has been quantified by propagating light through an unfolded cavity that uses DBR mirrors (instead of hard mirrors) and contains a region of gain. As the aperture radius is decreased, the threshold modal gain (equivalent to modal losses) increases. This increase is plotted in Figure 2.12(a) for a variety of different apertures (shown schematically in Figure 2.12(b)) for a particular cavity length at a wavelength of 980 nm [4]. The curves are generated by varying the radius of the aperture (or index gradient of the parabolic taper) and then calculating the mode profile and modal gain of the resonator. The mode radius, indicated along the horizontal axis in Figure 2.12(a), is found from the mode profile as indicated in Figure 2.12(b). As might be expected, the excess loss is nearly zero for an aperture with a parabolic phase variation (i.e., an ideal lens). The rapid increase in loss for very small mode radii in this case is due to the roll-off in DBR reflection at the large angular spectrum contained in such small modes. Mirrors with higher index

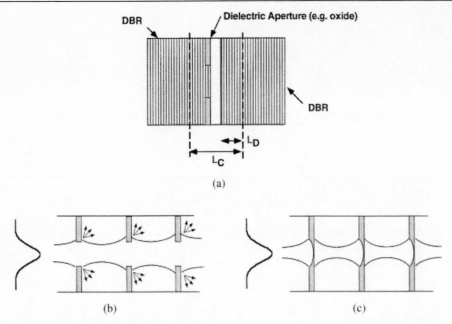

Figure 2.11. Schematic of dielectrically apertured VCSEL and equivalent unfolded cavity models. (a) The mirrors can be approximated as hard mirror planes at a diffraction equivalent distance, L_D, from the incident plane. (b) An unfolded cavity with an abrupt dielectric aperture. Because the aperture cannot exactly compensate for the diffraction of the mode, it creates scattering loss. (c) An unfolded cavity with a parabolically tapered aperture (ideal lens) can exactly compensate for diffraction of the mode and eliminate the scattering loss [4].

contrast layers would have a wider angular reflection spectrum. Alternatively, this limit on mode size could be overcome by using multiple apertures to continually refocus the mode as it travels deeper into the DBR. For the abrupt or linearly tapered apertures, the parameter Δn_d denotes the effective index step between the apertured and unapertured regions of the cavity (and will be discussed further in Section 2.3.3). Quarter-wave thick abrupt apertures (with $\Delta n_d \sim 0.06$) formed by etching or oxidizing one layer of the DBR show considerable loss for small diameters. Thin (or displaced) apertures or ones with tapered tips have significantly lower loss. In the thin abrupt-aperture case illustrated, $\Delta n_d = 0.02$, we observe that the optical mode reaches a minimum radius and then again expands slightly as the aperture radius is decreased. This is similar to what happens in a weak uniform dielectric waveguide.

As might be surmised from the illustration of Figure 2.11, the scattering loss is caused by the amount of diffraction between apertures, if the apertures are not ideal lenses. Since this diffraction is related to the Fresnel number, $F = n_0 a^2 / (\lambda L)$, we might expect that the calculated losses can also be plotted in a normalized fashion versus F. This is shown by the universal curves in Figure 2.13 for abrupt apertures [4]. To account for the effect of the aperture on the mode size, we also introduce a normalized aperture phase perturbation factor, $\phi_0 = 2\pi (n_0 - n_{de}) \xi_{de} \, t / \lambda$. In these expressions n_{de} is the index of the dielectric aperture, n_0 is the average index in the cavity, ξ_{de} is the standing-wave enhancement factor for the aperture (given by Eq. (2.5) with L_a replaced by t), and t is the aperture thickness.

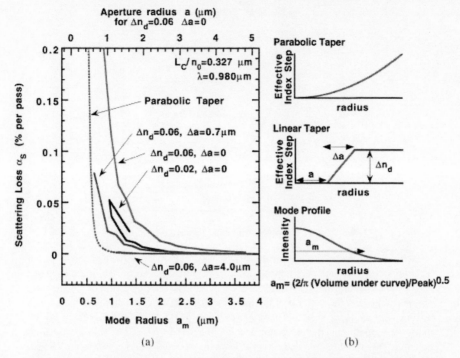

Figure 2.12. (a) Scattering loss from apertures with tapering over different distances Δa, and with different effective index steps Δn_d. The value $\Delta n_d = 0.06$ corresponds to an 80 nm thick oxide aperture in the first p-mirror layer of an 980 nm VCSEL with AlGaAs/GaAs mirrors. The value $\Delta n_d = 0.02$ corresponds to a 30 nm thick oxide aperture in a similar structure. The loss is plotted versus the mode radius in order to compare the apertures of different shapes. The upper axis indicates the aperture radius for an abrupt taper. The solid curve for the longer linear taper ($\Delta a = 4.0$ μm) initially overlaps the dashed curve for the parabolic taper; however, the aperture closes, and the long linear taper cannot produce the smallest mode sizes [4]. To generate the curves the inner aperture is changed in steps of 0.2 μm down to a radius of 0.2 μm. (b) Schematics of various tapered apertures and the mode profile.

The curves are an analytical empirical fit to the numerical calculation, which is very accurate for small ϕ_0. It is given by [4]

$$A_{\text{is}}(a_{\text{m}}, \Delta n, t, \lambda, L) = 2.28 \frac{\phi_0^{1.19}}{F^{1.3}} \exp\left[-\frac{0.206}{\phi_0 F} \right]. \tag{2.23}$$

Experimental data for devices with varying aperture diameters and thicknesses, as well as ones with different cavity lengths, are also illustrated on Figure 2.13 [27–29]. In all cases a good correspondence between theory and experiment is observed.

For the case of two apertures in the cavity, the amount of excess loss will depend upon the spacing of the apertures (see Figure 2.14). If the apertures are spaced by a cavity length or they are right next to each other, then they will effectively act like one thicker aperture, and consequently, scattering losses will be increased. However, when the apertures are evenly spaced (in the unfolded cavity), then the mode diffracts over only half a cavity length before

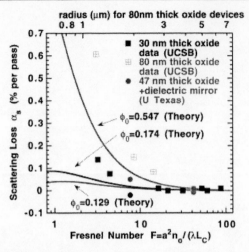

Figure 2.13. Universal curves for the single-pass scattering loss with laterally abrupt apertures. The estimate only depends on two parameters: the Fresnel number of the cavity (plotted along the x axis) and the effective single-pass phase shift of the aperture, ϕ_0. For reference, the radius a of the abrupt aperture is specified along the upper axis for a cavity with $n_0 = 3.52$. The upper curve $\phi_0 = 0.547$ corresponds to an 80 nm thick oxide aperture in the first p-mirror layer of a 980 nm VCSEL with AlGaAs/GaAs mirrors. The middle curve with $\phi_0 = 0.174$ corresponds to another 980 nm VCSEL with a 30 nm thick oxide aperture and AlGaAs/GaAs mirrors [4]. The lowest curve with $\phi_0 = 0.129$ corresponds to an oxide apertured VCSEL with a shorter (0.9 μm) cavity, which uses a dielectric mirror [28]. Equivalently, the curve corresponds to positioning the 30 nm thick oxide aperture closer to a standing-wave null.

being partially refocused by the aperture. One may treat two apertures in a cavity as two separate cavities of length L_s and $L_c - L_s$ with each cavity having an effective phase shift of $\phi_{1\text{eff}} = \phi_T L_s / L_C$ and $\phi_{1\text{eff}} = \phi_T (L_C - L_s)/L_C$ (where ϕ_T is the total effective phase shift of both apertures). The analysis ignores the variation of the aperture thickness but accounts for the variation in the spacing. To first order, we can add the losses from the two cavities, and then the total loss is given by

$$A_{\text{is}}(a_{\text{m}}, \Delta n, t, \lambda, L) = 2.28 \frac{\phi_T^{1.19}}{F^{1.3}} \exp\left[-\frac{0.206}{\phi_T F}\right]\left(\left[\frac{L_S}{L_C}\right]^{2.49} + \left[1 - \frac{L_S}{L_C}\right]^{2.49}\right).$$

(2.24)

Figure 2.14(c) plots the relative scattering losses for various aperture designs. By distributing the total effective phase shift over two apertures (instead of one) scattering losses can be more than halved. However, simply adding a second aperture to the cavity (and thereby doubling the total effective phase shift) generally gives higher losses unless the separation between apertures is close to half a cavity length.

2.3.3 Lateral Modes

The nature of the lateral optical confinement determines the possible lateral optical modes that might lase. Although VCSELs tend to lase in only a single axial mode, because the DBR

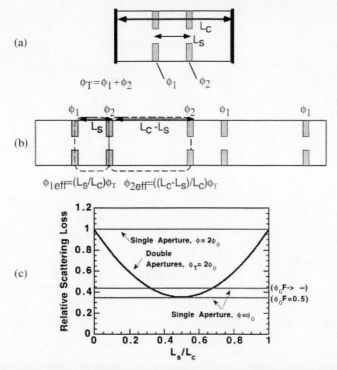

Figure 2.14. (a) Cavity with hard mirrors and two apertures. (b) An unfolded double-apertured cavity. The phase shifts can be distributed uniformly across two different cavities and then the scattering losses can be estimated from the single-aperture model. (c) A plot of the relative scattering loss for various single- and double-aperture designs.

mirrors fill much of the cavity, lateral modes can exist for device diameters down to less than a micron. The losses explicitly discussed above were those of the fundamental lateral mode, although the general discussion may be applied to losses for higher-order modes as well. In fact, the higher-order modes tend to experience higher size-dependent optical losses than the fundamental mode, and this is why it is possible to observe single-mode operation in devices that have diameters of several microns.

The lateral modes of gain-guided structures tend to vary with pumping level and thermal environment, and they are rather difficult to analyze analytically. However, the lateral modes of more strongly index-guided VCSELs with etched mesas or dielectric apertures can be well approximated as those of uniform step-index or graded index dielectric waveguides, provided the Fresnel number is sufficiently large ($F \sim 10$) so that significant diffraction does not occur in unguided sections. (See Figure 2.11.) Then, the modes can be determined from normalized dispersion charts, generally used for optical fibers. In both the etched-mesa and dielectrically apertured cases, the effective lateral index profile can be determined by dividing the net change in optical path length over the cavity length by the cavity length. Thus, the assumed uniform waveguide has a lateral index roll-off given by [4, 25]

$$\delta n_1(r) = -\phi_0(r)\lambda/(2\pi L) = -(n_0 - n_{de})\xi_{de}(r)t(r)/L. \tag{2.25}$$

For a step index resulting from an abrupt dielectric aperture or an etched mesa, the aperture thickness $t(r < a) = 0$ and $t(r > a) = t$. For the etched-mesa VCSEL, t is taken to be

L_{eff2} if the etch extends just through the top DBR mirror. For a tapered aperture, $t(r)$ varies with radius.

The change in the optical path length between the apertured and unapertured regions of the device gives rise to a change in the longitudinal resonant wavelength with radius [25]. Equation (2.17) gives the shift of the mode due to a change in a physical length, but it can be applied more generally:

$$
\begin{aligned}
\Delta\lambda(r) &= \lambda \frac{-(n_0 - n_{\text{de}})\xi_{\text{de}}(r)t(r)}{\tilde{n}_{\text{gl}}L_{\text{eff1}} + \tilde{n}_{\text{ga}}L_{\text{a}} + \tilde{n}_{\text{gp}}L_{\text{p}} + \tilde{n}_{\text{g2}}L_{\text{eff2}}} \\
&\approx \lambda \frac{-(n_0 - n_{\text{de}})\xi_{\text{de}}(r)t(r)}{n_0 L} = \lambda \frac{\delta n_1(r)}{n_0}.
\end{aligned}
\tag{2.26}
$$

Using Eq. (2.26) to determine the effective lateral index step is also possible, though it will not work if the shift becomes larger than the longitudinal mode spacing or the mirror stopband.

2.4 VCSEL Design

2.4.1 Modeling the P–I Characteristics

Using the models developed above, we can calculate the threshold currents, differential efficiencies, and thus P–I characteristics for some arbitrary VCSEL as a function of radius. All that we need are some experimentally or theoretically determined gain and loss numbers from which the analytical parameters can be de-embedded. Thus, we insert Eqs. (2.20) to (2.22) for the size-dependent carrier and optical losses into Eq. (2.8) to obtain the size-dependent output power,

$$
P_{01}(a, a_{\text{m}}) = F_1 \frac{h\nu}{q} \eta_{\text{i}} \frac{T_{\text{m}}}{A_{\text{i0}} + A_{\text{is}}(a_{\text{m}}) + T_{\text{m}}} [1 - I_{\text{th}}(a, a_{\text{m}})],
\tag{2.27}
$$

where,

$$
I_{\text{th}} = I'_{\text{th}} + I_{\text{ld}} + I_{\text{ls}} = \frac{J'_{\text{th}}}{\eta_{\text{i}}} \left[\pi a^2 + \left(K_{\text{ld}} + \eta_{\text{i}} \sqrt{\frac{\pi I_0}{\eta_{\text{i}} J'_{\text{th}}}} \right) a + \frac{\eta_{\text{i}} I_0}{2 J'_{\text{th}}} \right]
$$

and

$$
J'_{\text{th}} = N_{\text{w}} J_0 \exp\left[\frac{A_{\text{i0}} + A_{\text{is}}(a_{\text{m}}) + T_{\text{m}}}{g_0 \Gamma L} \right].
$$

Figure 2.15 gives example plots of some VCSEL characteristics using a typical set of parameters from practical devices. Figure 2.15(a) gives the differential efficiency and threshold gain (loss), and Figure 2.15(b), gives the threshold current. Using these same calculations it is also possible to construct design curves that plot the current required for threshold and that for a certain amount of power out as a function of the mean mirror reflection, R. Figure 2.16 gives plots for several materials systems of interest for a given device radius. Although the results are strongly dependent upon the choice of parameters, the general variations shown in Figures 2.15 and 2.16 will persist for nearly any reasonable set.

Given the projections of reduced loss in Figure 2.12 and recent experiments that approach these levels (Figure 2.13), we replot Figure 2.15(b) assuming a more optimistic set of parameters in Figure 2.17 [17] to obtain plots of the calculated threshold currents as a

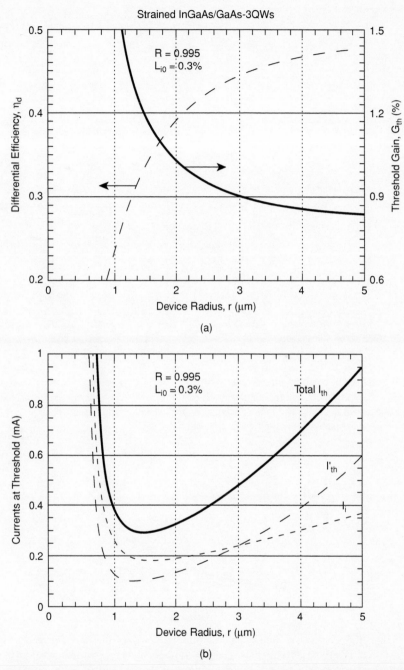

Figure 2.15. Theoretical predictions from the simple analytical model. (a) Differential efficiency and threshold gain versus VCSEL radius. (b) Current to reach threshold (including lateral leakage current) versus radius. Mean mirror reflectivity and broad area loss parameter are indicated. Gain parameters given in Table 2.1.

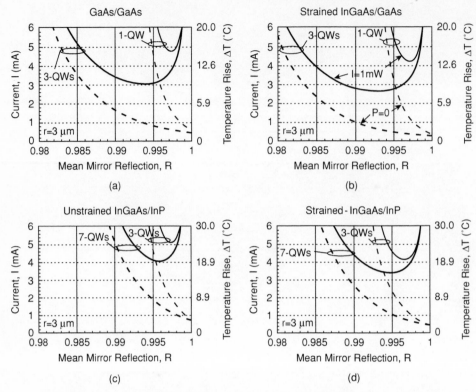

Figure 2.16. Theoretical current required to reach threshold (dash) and 1 mW out (solid) versus mirror reflectivity for quantum-well active regions of various compositions and number of wells from the simple analytical model. Device radius assumed to be 3 μm for all. Other parameters as in Figure 2.14 and Table 2.1.

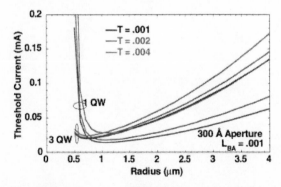

Figure 2.17. Calculated limits for threshold in low-loss dielectrically apertured VCSELs. Threshold is plotted versus device size accounting for contributions from the broad area gain level (due to mirror transmission T and round-trip loss $L_{BA} = 0.1\%$), diffusion in the quantum wells, ($D = \frac{5\text{cm}^2}{\text{sec}}$) and the size-dependent scattering losses from a 30 nm thick aperture as shown in Figure 2.13. Threshold in triple quantum well devices is less sensitive to the increased loss at small sizes. Carrier leakage contributes to about one-third the threshold current at the larger radii, but scattering losses dominate below 1 μm radius.

function of radius for three cases. Two are obtained by extrapolating experimental results, and one is for projected results from a tapered aperture VCSEL assuming a broad area loss equal to the lowest measured.

2.4.2 Power Conversion Efficiency

For many applications the overall power conversion or "wall-plug" efficiency is of paramount importance. VCSELs are attractive because high wall-plug efficiencies can be obtained with relatively low output powers. For example, in data links it is possible to obtain 1 mW of optical output power with only 2 or 3 mW of electrical power in. For high efficiency at low powers the VCSEL must have a low threshold current, a high differential efficiency, and a low excess series voltage drop. As discussed above, the VCSEL's small active region volume together with controlled losses can provide a low threshold current; however, its small size as well as the fact that the light travels perpendicular to the active layer also can lead to excess series resistance and voltage. As pointed out in Section 2.2 the current must either travel across the DBR mirror or along a transparent intracavity contact layer. In both cases a series resistance is added. For conduction across the multilayer mirror, we also have a series voltage due to the multiple heterobarriers. Thus, we model the power into the VCSEL as

$$P_{in} = I^2 R_s + I V_d + I V_s, \qquad (2.28)$$

where R_s is the series resistance, V_s is a current-independent series voltage, and V_d is the ideal diode voltage, which is equal to the quasi-Fermi level separation. V_d is clamped above threshold, and the combination of V_s and some of R_s models possible series diodes. For etched pillars, the device resistance scales inversely with the square of the radius, but for devices in which the injection is confined by an aperture and the contacts are much larger in area (either intracavity or top contacts) then the series resistance of devices scales roughly inversely with the aperture radius. Assuming the output-power versus current characteristics of the laser scaled perfectly (i.e., constant current density and slope efficiency), then one would always improve the wall-plug efficiency by operating smaller apertured devices at lower output power. In reality, parasitic leakage currents along with the diverging resistance cause the wall-plug efficiency to peak at a finite diameter even in the absence of scattering losses. To improve the efficiency of VCSELs at lower powers one must reduce lateral current leakage and optical scattering losses. The roll-over of the output power due to heating also diminishes the peak wall-plug efficiency, but typically the wall-plug efficiency is reduced by series resistance well before this happens.

2.4.3 Heat Flow

Although heat flow in a VCSEL is a complex three-dimensional problem, the average device temperature is relatively easy to model and measure. The power dissipated in the laser is

$$P_D = P_{in} - P_0 = P_{in}[1 - \eta], \qquad (2.29)$$

where η is the wall-plug efficiency, and the temperature rise is

$$\Delta T = P_D Z_T, \qquad (2.30)$$

where Z_T is the thermal impedance. For small VCSELs on a relatively thick substrate, a simple analytic expression for Z_T is useful [30]:

$$Z_T = \frac{1}{4\sigma_T a_{\text{eff}}} \tag{2.31}$$

where σ_T is the thermal conductivity of the substrate beneath the heat generating disk, and a_{eff} is the effective device radius. In the uncovered etched-mesa case, a_{eff} is approximately equal to a; in other cases it tends to be somewhat larger (but usually less than twice the current injection radius a) due to heat spreading in either surrounding epitaxial material or deposited heat spreaders. The reciprocal dependence on radius rather than area is due to a three-dimensional heat flow. In contrast, if the VCSEL is flip-chip bonded to a heat sink, a quasi-one-dimensional heat flow results, and then $Z_T = h/\sigma_T A$, where A is the effective area of the heat flow, h is the distance to the heat sink, and σ_T is the thermal conductivity of the material between the source and sink.

Although the thermal impedance typically increases as the device size shrinks, the temperature rise for a given current density actually can get smaller as device size is reduced. To examine the scaling of the temperature rise, let us assume a low scattering loss device in which the differential efficiency scales with device size. If we ignore the size dependence of V_s, but assume R_s scales as $1/a$ (as it does for a device with a large top contact and small aperture), then the temperature rise has the form:

$$\Delta T = C_2 J^2 a^2 + (C_1 J + C_{1\text{TH}} J_{\text{TH}})a, \tag{2.32}$$

where C_1, $C_{1\text{TH}}$, and C_2 are independent of device radius and current density. If the threshold current density can be held constant as device radius is reduced, then the temperature rise will be lower in smaller devices at the same current density. Consequently, smaller VCSELs can be driven to higher current densities before roll-over. For the case that threshold current density rises due to current spreading or excess losses, there will be an optimum device size for the minimum temperature rise.

VCSEL performance generally degrades with increasing temperature, though a deliberate misalignment of the gain from the mode at room temperature has been used to partially compensate this degradation [31]. It is possible to actually make the threshold decrease as the temperature is increased. This occurs because the gain moves to longer wavelengths (~ 3 Å/°C) about four times faster than the mode (~ 0.7 Å/°C in GaAs). This technique is now widely used to provide temperature-insensitive devices with threshold currents that remain nearly constant over a 70 to 80°C range.

Thermal crosstalk is another issue in array configurations. In this case minimizing the thermal impedance is not necessarily the best thing to do. For example, making use of lateral material to provide enhanced heat spreading for lower thermal impedance will actually increase thermal crosstalk between adjacent devices. Thus, as shown by more detailed numerical analysis [32], this has been a particular problem in planar proton-implanted structures. The best configuration for low thermal crosstalk is probably an etched-mesa design flip-chip bonded to some good heat sink, so that lateral heat spreading is short-circuited.

2.5 Dynamic Effects

2.5.1 Modulation Properties

The modulation properties of VCSELs can be predicted by the relaxation resonance. For bias currents sufficiently far above threshold to ensure effective carrier clamping, the frequency of this resonance is given by

$$f_R = \frac{1}{2\pi}\left[\eta_i \frac{\Gamma v_g}{qV}\frac{\partial g}{\partial N}(I - I_{th})\right]^{1/2} = \frac{1}{2\pi}\left[\eta_i \frac{\Gamma_{xy}\zeta v_g}{qL}\frac{\partial g}{\partial N}(J - J_{th})\right]^{1/2}, \qquad (2.33)$$

where the differential gain, $\partial g/\partial N$, is obtained from the gain versus carrier density at the bias point, g_{th}. Also, I_{th} in this derivation assumes a small β_{sp}, so if β_{sp} is large, one must substitute an effective threshold current equal to that expected with a small β_{sp} [33]. Provided the damping and parasitics are not too large, the 3 dB modulation bandwidth is about 1.55 f_R.

If β_{sp} is relatively large (i.e., >0.1) and nonradiative recombination small, the threshold current can be reduced significantly [34]. However, as cautioned above, this low threshold might not improve the frequency response at a given bias current. This is because the difference between the losses and the gain at threshold is then significant, and as more current is applied and more optical power is emitted, both the gain and the carrier density continue to increase. This lack of carrier clamping near threshold can result in a much reduced modulation frequency response. For very large β_{sp}, it becomes just that of a comparable LED (i.e., a cutoff given by the reciprocal of the carrier lifetime). Nevertheless, for pumping currents several times the low threshold value that accompanies the large β_{sp} (above where the threshold would have been with a small β_{sp}), the gain and carrier density are well clamped and the frequency response is again predicted by Eq. (2.33). As mentioned above, one must use the effective threshold corresponding to a small β_{sp} in this equation, and this value can be estimated by extrapolating the high-current slope of the $P-I$ curve (see Chapter 3) to zero power.

Aside from possible difficulties near threshold, the small active volume of VCSELs suggests that their potential modulation bandwidth may be quite high. In fact, if the other factors in Eq. (2.33) were similar, the VCSEL modulation bandwidth would be much higher than that of an edge-emitter. However, it is not possible to apply bias currents nearly as large as one can in edge-emitters because of the VCSEL's high thermal impedance and series resistance. Nevertheless, for the same small bias current above threshold, the VCSEL generally will have a much higher modulation bandwidth. This fact is illustrated by the small-signal modulation responses in Figure 2.18 [35]. As can be seen a -3 dB bandwidth over 15 GHz is observed for only 2.1 mA of total applied current (1.6 mA above threshold). This level of performance was enabled by low optical losses resulting from the use of a thin oxide aperture.

The bandwidth from the device in Figure 2.18 is limited by an avoidable parasitic capacitance across the thin oxide aperture. If this were removed by using a thicker oxide under the contacts, the bandwidth would have been about 19 GHz. A tapered dielectric aperture could provide the increased thickness under the contacts while preserving the desired low optical losses. Proton implantation also can be used to overcome this parasitic capacitance as shown from the results in Figure 2.19 [36]. With reduced capacitance, the -3 dB frequency is pushed out to 21.5 GHz.

After the parasitic capacitance, heating places the next most important limit on the maximum bandwidth. For the device in Figure 2.18, a maximum -3 dB bandwidth of 45 GHz is expected without heating (due only to gain compression) [36]. Heating causes the output power to roll over, but even before this happens the differential gain drops significantly. Figure 2.20 [37] illustrates the limits on the bandwidth imposed by parasitic capacitance and by heating of the devices from Figure 2.18. The curves are calculated

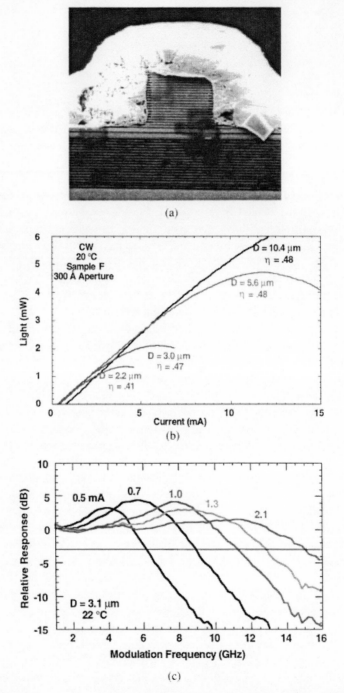

(a)

(b)

(c)

Figure 2.18. (a) A cross-sectional SEM of a dielectrically apertured VCSEL with p-intracavity contact and a thin oxide aperture. (b) Measured light versus current curves corresponding to devices from the same material. Note the good scaling of the differential efficiency. (c) Modulation response for intracavity-contacted oxide-apertured VCSEL at different biases. Because of its small size, the VCSEL is capable of providing high speed at low currents. In this case, a 15.2 GHz −3 dB bandwidth is achieved with only 2.1 mA of drive current [35].

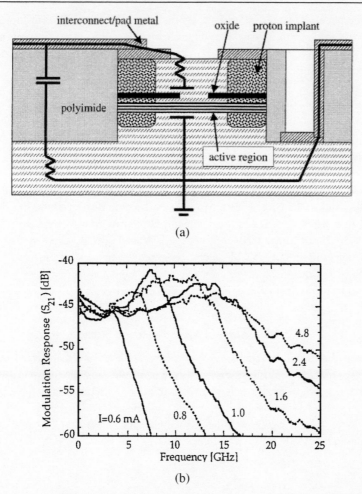

(a)

(b)

Figure 2.19. (a) Schematic of an oxide-apertured VCSEL proton implanted for reduced parasitic capacitance. The single-mode device has aperture opening of $\sim 4 \times 4$ μm with $I_{th} = 0.5$ mA. (b) Small signal frequency response, S_{21}, of the proton-implanted, oxide-apertured VCSEL at various bias currents. The initial slope is 14.2 GHz/(mA)$^{0.5}$ [36].

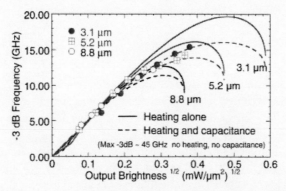

Figure 2.20. Plot of the -3 dB frequency predicted with heating and parasitic capacitance along with measured data from the devices shown in Figure 2.18. Device heating is the strongest limitation on bandwidth after parasitic capacitance [37].

from the temperature rise (as determined by measuring the shift of the lasing wavelength versus drive power) and the dependence of differential gain on temperature (computed using the approach outlined in Section 2.2). For $In_{0.2}Ga_{0.8}As$ quantum wells on GaAs, a fitting function was determined for the differential gain near room temperature [37]:

$$\frac{dg}{dN} = 2110 - 3.59T_j - (0.495 - 0.0082T_j)(g + 10\Delta T_j) \ [10^{-18} \text{ cm}^{-2}], \qquad (2.34)$$

where T_j is the junction temperature in kelvin, g is the material gain in cm^{-1}, and the $10\Delta T_j$ term was added to account for the shift of the gain peak relative to the mode.

As noted above smaller devices will generally have higher bandwidth for a given drive current; however, the right-hand side of Eq. (2.33) shows us that we will not get higher bandwidth from smaller devices driven at the same current density above threshold. If the maximum bandwidth occurred at the same current density in different size devices, then shrinking the VCSEL diameter would not help. What Eq. (2.33) does not reveal is that the maximum bandwidth of VCSELs can be improved by scaling of broad-area device characteristics to smaller sizes because the device temperature rise (at a given current density) can be lower in smaller devices as noted in Section 2.4.2.

2.5.2 Linewidth and Feedback Effects

The theoretical linewidth and sensitivity to feedback for VCSELs are given by the same expressions as for edge-emitting lasers. However, again one must be careful to avoid making some common assumptions. For either VCSELs or edge-emitters, the laser linewidth is given by the modified Schawlow–Townes linewidth multiplied by a carrier noise enhancement factor $(1 + \alpha^2)$, where α is the so-called linewidth enhancement factor. Thus, the full-width half-maximum linewidth can be written as [33]

$$\Delta\nu = \frac{\Gamma\beta_{sp}R_{sp}}{4\pi N_p}(1 + \alpha^2). \qquad (2.35)$$

This form suggests that the linewidth is directly proportional to β_{sp}. However, using the proportionality between spontaneous emission and gain given to the mode, $\beta_{sp}R_{sp}V = \Gamma g v_g n_{sp}$, and the fact that the modal gain clamps near the level of modal losses in typical lasers $(\beta_{sp} \ll 1)$, Eq. (2.35) can also be expanded in terms of physical device parameters using Eqs. (2.2) and (2.7) to give an expression that is seemingly independent of β_{sp}:

$$\Delta\nu \approx \frac{n_{sp}v_g\alpha_m h\nu}{4\pi\tau_p P_0}(1 + \alpha^2), \qquad \beta_{sp} \ll 1. \qquad (2.36)$$

Of course, this expression suffers from the same problems as does Eq. (2.33), if β_{sp} is small. That is, Eq. (2.35) is the asymptotic value that is approached once the gain is effectively clamped and the $P-I$ characteristic becomes linear, but for smaller drive currents the linewidth can be much larger as given by Eq. (2.35).

Measured linewidths are in rough agreement with the above predictions. Linewidths from VCSELs have ranged from 20 to 80 MHz at output powers ~ 1 mW [38–41]. Figure 2.21 illustrates one result. Relatively high internal losses existed in these devices. With higher single-mode powers expected from properly engineered oxide- or air-apertured devices, we expect linewidths to be considerably lower than this.

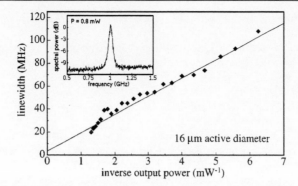

Figure 2.21. Plot of the measured linewidth versus inverse input power for a 16 μm diameter proton implanted VCSEL [41]. (Inset) Spectral power plotted against frequency.

The unwanted feedback effects that are problematic for single-frequency edge-emitters are expected to be similarly troubling for single-mode VCSELs. Again, the physics is the same. Nevertheless, the effects of feedback on the linewidth and noise of single-mode VCSELs may still be tolerable for many applications. For example, in free-space links it is generally desirable to use relatively high signal levels to reduce the complexity and power requirements of the receiver. Multimode VCSELs tend to be relatively insensitive to moderate feedback [42] since it has different effects on different modes, and on average the total output does not change much.

The importance of feedback is generally characterized by a feedback rate, κ_f, which is equal to the fractional increase in the field in the cavity due to the external feedback divided by the cavity round-trip time [33]. In VCSELs the level of feedback that makes it back into the cavity ($\propto T_m$) is lower than in an edge-emitter by about two orders of magnitude, but the cavity round-trip time ($\propto L$) tends to be smaller by about the same factor. Thus, again, similar effects are expected.

References

[1] L. A. Coldren and S. W. Corzine, Chapter 2 in *Diode Lasers and Photonic Integrated Circuits*, Wiley, New York, 1995.

[2] L. A. Coldren and S. W. Corzine, Appendix 5 in *Diode Lasers and Photonic Integrated Circuits*, Wiley, New York, 1995.

[3] R. H. Yan, Z. M. Chuang, S. W. Corzine, and L. A. Coldren, "Simultaneous Gain and Phase-Shift Enhancements in Periodic Gain Structures," *J. Appl. Phys.*, **67**, 4387–4389, 1990.

[4] E. R. Hegblom, D. I. Babic, B. J. Thibeault, and L. A. Coldren, "Scattering Losses from Dielectric Apertures in Vertical Cavity Lasers," *IEEE J. Selected Topics in Quantum Electron.*, **3**, 379–389, 1997.

[5] R. J. Ram, E. Goobar, M. G. Peters, L. A. Coldren, and J. E. Bowers, "Spontaneous Emission Factor in Post Microcavity Lasers," *IEEE Photon. Technol. Lett.*, **8**, 599–601, 1996.

[6] L. A. Coldren and S. W. Corzine, Chapter 4 in *Diode Lasers and Photonic Integrated Circuits*, Wiley, New York, 1995.

[7] L. A. Coldren and S. W. Corzine, Chapter 3 in *Diode Lasers and Photonic Integrated Circuits*, Wiley, New York, 1995.

[8] L. A. Coldren, "Lasers and Modulators for OEICs," in *Integrated Optoelectronics*, M. Dagenais, R. F. Leheny, and J. Crow, eds., Academic Press, New York, 1994.

[9] D. I. Babic and S. W. Corzine, "Analytic Expressions for the Reflection Delay, Penetration

Depth, and Absorptance of Quarter-Wave Dielectric Mirrors," *IEEE J. Quantum Electron.*, **28**, 514–524, 1992.

[10] S. W. Corzine, R. H. Yan, and L. A. Coldren, "A Tanh Substitution Technique for the Analysis of Abrupt and Graded Interface Multilayer Dielectric Stacks," *IEEE J. Quantum Electron.*, **27**, 2086–2090, 1991.

[11] J. W. Scott, *Design, Fabrication and Characterization of High-Speed Intra-Cavity Contacted Vertical Cavity Lasers*, Ph.D. Dissertation, ECE Technical Report #95-06, University of California, Santa Barbara, 1995.

[12] M. G. Peters, B. J. Thibeault, D. B. Young, J. W. Scott, F. H. Peters, A. C. Gossard, and L. A. Coldren, "Band-Gap Engineered Digital Alloy Interfaces for Lower Resistance Vertical-Cavity Surface-Emitting Lasers," *Appl. Phys. Lett.*, **63**, 3411–13, 1993.

[13] D. I. Babic, *Double-Fused Long-Wavelength Vertical Cavity Lasers*, Ph.D. Dissertation, University of California, Santa Barbara, 1995.

[14] R. P. Schneider, Jr. and J. A. Lott, "Cavity Design for Improved Electrical Injection in InAlGaP/AlGaAs Visible (639–661 nm) Vertical-Cavity Surface-Emitting Laser Diodes," *Appl. Phys. Lett.*, **63**, 917–19, 1993.

[15] B. Weigl, M. Grabherr, R. Jager, G. Reiner, and K. J. Ebling, "57% Wallplug Efficiency Wide Temperature Range 840 nm Wavelength Oxide Confined GaAs VCSELs," *Proc. 15th IEEE International Semiconductor Laser Conference*, Paper no. PDP2, 1996.

[16] B. J. Thibeault, T. A. Strand, T. Wipiejewski, M. G. Peters, D. B. Young, S. W. Corzine, L. A. Coldren, and J. W. Scott, "Evaluating the Effects of Optical and Carrier Losses in Etched Post Vertical Cavity Lasers," *J. Appl. Phys.*, **78**, 5871–5876, 1995.

[17] B. J. Thibeault, E. R. Hegblom, Y. A. Akulova, P. D. Floyd, J. Ko, R. Naone, and L. A. Coldren, "Electrical and Optical Losses in Dielectrically-Apertured Vertical-Cavity Lasers," *Proc. SPIE Photonics West '97*, **3003**, Paper 12, pp. 86–99, 1997.

[18] K. D. Choquette, M. Hong, R. S. Freund, J. P. Mannaerts, R. C. Wetzel, and R. E. Leibenguth, "Vertical-Cavity Surface-Emitting Laser Diodes Fabricated by In Situ Dry Etching and Molecular Beam Epitaxial Regrowth," *IEEE Photon. Technol. Lett.*, **5**, 284–7, 1993.

[19] P. D. Floyd, B. J. Thibeault, D. B. Young, L. A. Coldren, and J. L. Merz, "Vertical Cavity Lasers with Zn Impurity-Induced Disordering (IID) Defined Active Regions," *Proc. LEOS '96*, **1**, 207–208, 1996.

[20] J. W. Scott, R. S. Geels, S. W. Corzine, and L. A. Coldren, "Modeling Temperature Effects and Spatial Hole Burning to Optimize Vertical-Cavity Surface-Emitting Laser Performance," *IEEE J. Quantum Electron.*, **29**, 1295–1308, 1993.

[21] S. Y. Hu, S. W. Corzine, K.-K. Law, D. B. Young, A. C. Gossard, L. A. Coldren, and J. L. Merz, "Lateral Carrier Diffusion and Surface Recombination in InGaAs/AlGaAs Quantum-Well-Ridge-Waveguide Lasers," *J. Appl. Phys.*, **76**, 4479–4487, 1994.

[22] J. W. Scott, S. W. Corzine, D. B. Young, and L. A. Coldren, "Modeling the Current to Light Characteristics of Index-Guided Vertical-Cavity Surface-Emitting Lasers," *Appl. Phys. Lett.*, **62**, 1050–2, 1993.

[23] E. R. Hegblom, N. M. Margalit, B. J. Thibeault, L. A. Coldren, and J. E. Bowers, "Current Spreading in Apertured Vertical Cavity Lasers," *Proc. SPIE Photonics West*, **3003**, Paper 23, 176–179, 1997.

[24] D. I. Babic, Y. Chung, N. Dagli, and J. E. Bowers, "Modal Reflection of Quarter-Wave Mirrors in Vertical-Cavity Lasers," *IEEE J. Quantum Electron.*, **29**, 1950–1955, 1993.

[25] G. R. Hadley, "Effective Index Model for Vertical-Cavity Surface-Emitting Lasers," *Opt. Lett.*, **20**, 1483–1485, 1995.

[26] L. A. Coldren, B. J. Thibeault, E. R. Hegblom, G. B. Thompson, and J. W. Scott, "Dielectric Apertures as Intracavity Lenses in Vertical Cavity Lasers," *Appl. Phys. Lett.*, **68**, 313–315, 1996.

[27] P. D. Floyd, B. J. Thibeault, E. R. Hegblom, J. Ko, and L. A. Coldren, "Comparison of Optical

Losses in Dielectric Apertured Vertical-Cavity Lasers," *IEEE Photon. Technol. Lett.*, **8**, 590–592, 1996.

[28] D. L. Huffaker, L. A. Graham, H. Deng, and D. G. Deppe, "Sub-40 μA Continuous-Wave Lasing in an Oxidized Vertical-Cavity Surface-Emitting Laser with Dielectric Mirrors," *IEEE Photon. Technol. Lett.*, **8**, 974–976, 1996.

[29] B. J. Thibeault, E. R. Hegblom, P. D. Floyd, R. Naone, Y. Akulova, and L. A. Coldren, "Reduced Optical Scattering Loss in Vertical-Cavity Lasers Using a Thin (300 Å) Oxide Aperture," *IEEE Photon. Tech. Lett.*, **8**, 593–596, 1996.

[30] S. S. Kutateladze and V. M. Borishanski, *A Concise Encyclopedia of Heat Transfer*, Permagon, Oxford, 1966.

[31] D. B. Young, J. W. Scott, F. H. Peters, M. G. Peters, M. L. Majewski, B. J. Thibeault, S. W. Corzine, and L. A. Coldren, "Enhanced Performance of Offset-Gain High-Barrier Vertical-Cavity Surface-Emitting Lasers," *IEEE J. Quantum Electron.*, **29**, 2013–2021, 1993.

[32] M. Osinski and W. Nakwaski, "Thermal Analysis of Closely-Packed Two-Dimensional Etched-Well Surface-Emitting Laser Arrays," *IEEE J. Selected Topics in Quantum Electron.*, **1**, 681–96, 1995.

[33] L. A. Coldren and S. W. Corzine, Chapter 5 in *Diode Lasers and Photonic Integrated Circuits*, Wiley, New York, 1995.

[34] G. Bjork and Y. Yamamoto, "Analysis of Semiconductor Microcavity Lasers Using Rate Equations," *IEEE J. Quantum Electron.*, **27**, 2386–2396, 1991.

[35] B. J. Thibeault, K. Bertilsson, E. R. Hegblom, E. M. Strzelecka, Y. Akulova, and L. A. Coldren, "High Speed Characteristics of Low Optical Loss Oxide Apertured Vertical Cavity Lasers," *IEEE Photon. Technol. Lett.*, **9**, 11–13, 1997.

[36] K. L. Lear, M. Ochiai, V. M. Hietala, H. Q. Hou, B. E. Hammons, J. J. Banas, and J. A. Nevers, "High Speed Vertical Cavity Surface Emitting Lasers," *Proc. IEEE/LEOS Summer Topical Meeting*, Paper WA1, pp. 53–54, 1997.

[37] B. J. Thibeault, *High Efficiency Vertical Cavity Lasers Using Low-Optical Loss Intra-Cavity Dielectric Apertures,* Ph.D. Dissertation, Electrical and Computer Engineering, University of California, Santa Barbara, 1997.

[38] R. S. Geels, S. W. Corzine, and L. A. Coldren, "InGaAs Vertical-Cavity Surface-Emitting Lasers," *IEEE J. Quantum Electron.*, **27**, 1359–67, 1991.

[39] N. M. Margalit, J. Piprek, S. Zhang, D. I. Babic, K. Streubel, R. P. Mirin, J. R. Wesselmann, J. E. Bowers, and E. L. Hu, "64°C Continuous-Wave Operation of 1.5 μm Vertical Cavity Laser," *IEEE J. Selected Topics in Quantum Electron.*, **3**, 359–365, 1997.

[40] H. Tanobe, F. Koyama, and K. Iga, "Spectral Linewidth of AlGaAs/GaAs Surface Emitting Laser," *Electron. Lett.*, **25**, 1444–1446, 1989.

[41] W. Schmid, C. Jung, B. Weigl, G. Reiner, R. Michalzik, and K. J. Ebeling, "Delayed Self-Heterodyne Linewidth Measurement of VCSEL's," *Photon. Technol. Lett.*, **8**, 1288–1290, 1996.

[42] D. M. Kuchta and C. J. Mahon, "Mode Selective Loss Penalties in VCSEL Optical Fiber Transmission Links," *IEEE Photon. Technol. Lett.*, **6**, 288–90, 1994.

[43] E. R. Hegblom, B. J. Thibeault, R. L. Naone, and L. A. Coldren, "Vertical Cavity Lasers with Tapered Oxide Apertures for Low Scattering Loss," *Electron. Lett.*, **33**, 869–879, 1997.

3

Enhancement of Spontaneous Emission in Microcavities

E. F. Schubert and N. E. J. Hunt

3.1 Introduction and Historical Overview

One of the most fundamental processes in optoelectronic devices are radiative transitions, that is, the transitions of electrons from an initial quantum state to final state and the simultaneous emission of a light quantum. There are two distinct ways by which the emission of a photon can occur, namely by *spontaneous* and *stimulated* emission. The two distinct emission processes were first postulated by Einstein [1]. In the spontaneous emission process, the radiative transition occurs spontaneously, that is, without being induced by a stimulus. In the stimulated emission process, radiative transitions are induced by photons. The inducing photons and the stimulated photons have the same propagation direction and phase.

Einstein [1] modeled the spontaneous and stimulated recombination process in a system with two discrete atomic levels. He used the two proportionality constants A and B to calculate the spontaneous and stimulated recombination rates, respectively. The two constants A and B are known as the *Einstein A* and the *Einstein B* coefficients.

In semiconductors, radiative recombination occurs between two *groups* of levels, namely between the conduction band and the valence band. A simple model describing the spontaneous recombination in such a multilevel system is the Van Roosbroeck–Shockley model [2]. The model allows one to calculate the *bimolecular recombination coefficient*, which gives the equilibrium and nonequilibrium electron–hole radiative recombination rate in semiconductors. The bimolecular recombination coefficient of a multilevel system corresponds to the *Einstein A* coefficient in a two-level system.

Stimulated emission is employed in semiconductor lasers and superluminescent LEDs. It was realized in the 1960s that the stimulated emission mode can be used in semiconductors to drastically change the radiative emission characteristics. The efforts to harness stimulated emission resulted in the first room-temperature operation of semiconductor lasers [3] and the first demonstration of a superluminescent LED [4].

Spontaneous emission implies the notion that the recombination process occurs *spontaneously*, that is, without a means to influence this process. In fact, spontaneous emission has long been believed to be uncontrollable. However, research in microscopic optical resonators, where spatial dimensions are on the order of the wavelength of light, showed the possibility of controlling the spontaneous emission properties of a light-emitting medium. The changes of the emission properties include the spontaneous emission rate, spectral purity, and emission pattern. These changes can be employed to make more efficient, faster, and brighter optoelectronic semiconductor devices.

Microcavity structures have been realized with different active media and different microcavity structures. The first microcavity structure was proposed by Purcell [5] for emission frequencies in the radio frequency (rf) regime. Small metallic spheres were proposed as the resonator medium. However, no experimental reports followed Purcell's theoretical publication. In the 1980s and 1990s, several microcavity structures have been realized with different types of optically active media. The emission media included organic dyes [6, 7], semiconductors [8, 9], rare-earth-doped silica [10], and organic polymers [11, 12]. In these publications clear changes in spontaneous emission were demonstrated including spectral, spatial, and temporal emission characteristics.

At the beginning of the 1990s, current-injection microcavity LEDs were first demonstrated in the GaAs material system [13] and subsequently in organic light-emitting materials [11]. Both publications reported an emission line narrowing due to the resonant cavities. Resonant-cavity light-emitting diodes (RCLEDs) have many advantageous properties when compared to conventional LEDs, including higher brightness, increased spectral purity, and higher efficiency. For example, the spectral power density in RCLEDs was shown to be enhanced by more than one order of magnitude [14]. The properties of RCLEDs will be discussed later in this chapter.

The changes in optical gain in VCSELs due to the enhancement in spontaneous emission was analyzed by Deppe and Lei [15]. The comparison of a *macrocavity*, in which the cavity is much longer than the emission wavelength ($\lambda \ll L_c$), to a *microcavity* ($\lambda \approx L_c$) revealed that the gain can be enhanced by factors of 2–4 for typical GaAs emission linewidths at room temperature (50 nm). Thus the laser threshold currents can be lower in microcavity structures due to the higher gain.

It is important to distinguish between emission inside the cavity and emission out of the cavity. The enhancement of the spontaneous emission inside the cavity and emission through one of the mirrors out of the cavity can be very different. At moderate values of the finesse, the spontaneous emission inside and out of the cavity is enhanced. However, for very high finesse cavities (see, for example, Ref. [16], the overall emission out of the cavity can decrease [17]. In the limit of very high reflectivity reflectors ($R_1 = R_2 = 100\%$), the emission out of the cavity becomes zero. This effect will be discussed in the following section.

A device in which all the spontaneous emission occurs into a single optical mode has been proposed by Kobayashi et al. [18, 19]. This device has been termed *zero threshold laser* [20] and single-mode LED [21]. In a conventional laser, only a small portion of the spontaneous emission couples into a single state of the electromagnetic field controlled by the laser cavity. The rest is lost to free-space modes, which radiate out the side of the laser. The idea of a thresholdless laser is simple. It assumes a wavelength-size cavity in which only one optical mode exists. Thus spontaneous as well as stimulated emission couples to this optical mode. The thresholdless laser should lack the clear distinction between the spontaneous and the stimulated regime that is observed in conventional lasers. Clearly, the prospects of such a device is intriguing. Even though several attempts to demonstrate a thresholdless laser have been reported [20, 22, 23], a thresholdless laser has not yet been demonstrated.

3.2 Microcavity Fundamentals

3.2.1 Fabry–Perot Resonators

The simplest form of an optical cavity consists of two coplanar mirrors separated by a distance L_c. About one century ago, Fabry and Perot [24] were the first to build and analyze

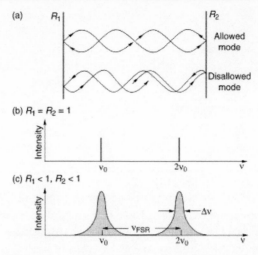

Figure 3.1. (a) Schematic illustration of allowed and disallowed optical modes in a Fabry–Perot cavity consisting of two coplanar reflectors with reflectivities R_1 and R_2. Mode density for a cavity with (b) zero-loss and (c) mirror losses.

optical cavities with coplanar reflectors. These cavities had a large separation between the two reflectors, that is, $L_c \gg \lambda$. However, if the distance between the two reflectors is on the order of the wavelength of light, new physical phenomena occur including the enhancement of the optical emission from an active material inside the cavity. Very small cavities, with typical dimensions of $L_c \sim \lambda$, will be denoted as *microcavities*.

Coplanar microcavities are the simplest form of optical microcavities and their properties will be briefly summarized below. For a detailed discussion of the optical properties of Fabry–Perot cavities, the reader is referred to the literature [25, 26]. A sketch of a Fabry–Perot cavity with two metallic reflectors with reflectivity R_1 and R_2 is shown in Figure 3.1. Plane waves propagating inside the cavity can interfere constructively and destructively resulting in stable (allowed) optical modes and attenuated (disallowed) optical modes, respectively. For lossless reflectors, the transmittance through the two reflectors is given by $T_1 = 1 - R_1$ and $T_2 = 1 - R_2$. Taking into account multiple reflections inside the cavity, the transmittance through a Fabry–Perot cavity can be expressed in terms of a geometric series. The transmitted light intensity (transmittance) is then given by

$$T = \frac{T_1 T_2}{1 + R_1 R_2 - 2\sqrt{R_1 R_2} \cos 2\phi}, \tag{3.1}$$

where ϕ is the phase change of the optical wave for a single pass between the two reflectors. The maxima of the transmittance occur if the condition of constructive interference is fulfilled, that is, if $2\phi = 0, 2\pi, \ldots,$. Insertion of these values into Eq. (3.1) yields the transmittance maxima as

$$T_{max} = \frac{T_1 T_2}{(1 - \sqrt{R_1 R_2})^2}. \tag{3.2}$$

For asymmetric cavities ($R_1 \neq R_2$), it is $T_{max} < 1$. For symmetric cavities ($R_1 = R_2$), the transmittance maxima are unity, $T_{max} = 1$.

Near $\phi = 0, 2\pi, \ldots$, the cosine term in Eq. (3.1) can be expanded into a power series $(\cos 2\phi \approx 1 - 2\phi^2)$. One obtains

$$T = \frac{T_1 T_2}{(1 - \sqrt{R_1 R_2})^2 + \sqrt{R_1 R_2} 4\phi^2}. \tag{3.3}$$

Equation (3.3) indicates that near the maxima, the transmittance can be approximated by a Lorentzian function. The transmittance T in Eq. (3.3) has a maximum at $\phi = 0$. The transmittance decreases to half of the maximum value at $\phi_{1/2} = (1 - \sqrt{R_1 R_2})/(4\sqrt{R_1 R_2})^{1/2}$. For high values of R_1 and R_2 (i.e., $R_1 \approx 1$ and $R_2 \approx 1$), it is $\phi_{1/2} = (1/2)(1 - \sqrt{R_1 R_2})$.

The cavity finesse, F, is defined as the ratio of the transmittance peak separation to the transmittance full-width at half-maximum:

$$F = \frac{\text{Peak separation}}{\text{Peak width}} = \frac{\pi}{2\phi_{1/2}} = \frac{\pi}{1 - \sqrt{R_1 R_2}}. \tag{3.4}$$

Inspection of Eq. (3.4) shows that the finesse becomes very large for high values of R_1 and R_2.

The wavelength and frequency of light are practically more accessible than the phase. Equations (3.1)–(3.4) can be converted to wavelength and frequency by using

$$\phi = 2\pi \frac{nL_c}{\lambda} = 2\pi \frac{nL_c \nu}{c}, \tag{3.5}$$

when L_c is the length of the cavity, λ is the wavelength of light in vacuum, ν is the frequency of light, and c is the velocity of light in vacuum. Figures 3.1(b) and (c) show the transmittance through a cavity with $R = 1$ and $R < 1$, respectively. In the frequency domain, the transmittance peak separation is called the *free spectral range* ν_{FSR}, as shown in Figure 3.1(c). The finesse of the cavity in the frequency domain is then given by $F = \nu_{FSR}/\Delta\nu$.

Frequently the *cavity quality factor* Q rather than the finesse is used. The cavity Q is defined as the ratio of the transmittance peak frequency to the peak width. Using this definition and Eq. (3.4), one obtains

$$Q = \frac{\text{Peak frequency}}{\text{Peak width}} = \frac{2nL_c}{\lambda} \frac{\pi}{1 - \sqrt{R_1 R_2}}, \tag{3.6}$$

where the peak width is measured in units of frequency.

Figure 3.2 shows an example of a reflectance spectrum of a microcavity consisting a 4-pair Si/SiO$_2$ distributed Bragg reflector (DBR) deposited on a Si substrate, a SiO$_2$ center region, and a 2.5-pair Si/SiO$_2$ DBR top reflector. The resonance wavelength of the cavity is approximately $1.0\,\mu m$. The reflectance of the cavity does not approach zero at the resonance wavelength because of the unequalness of the reflectivities of the two reflectors.

3.2.2 Reflectors

Different types of reflectors are shown in Figure 3.3, including metallic reflectors, distributed Bragg reflectors, hybrid metal-DBR reflectors, and reflectors based on total internal reflection (TIR). Metallic reflectors and hybrid reflectors are absorbing and cannot be used as light-exit reflectors. That is, the transmission is near zero, unless the thickness of the metal is very thin [27]. Total internal reflectors require that the angle of incidence be shallow in

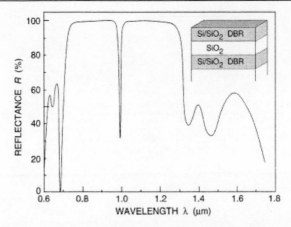

Figure 3.2. Reflectance of a Fabry–Perot cavity consisting of two Si/SiO$_2$ reflectors and a SiO$_2$ center region. At the resonance wavelength ($\lambda \approx 1,000$ nm), the reflectivity has a minimum.

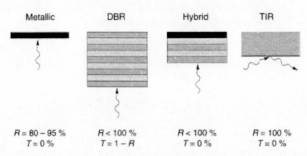

Figure 3.3. Schematic illustration of different types of reflectors including metallic reflector, distributed Bragg reflector, hybrid reflector, and total internal reflector.

order to achieve high reflectivity. TIR reflectors have been used in edge-emitting lasers [28] and in whispering gallery lasers [29].

The reflectance spectra of a metal mirror and a DBR are compared in Figure 3.4. Whereas metal reflectors exhibit a broad band with high reflectivity, DBRs display only a narrow band of high reflectivity denoted as *stop band*. The reflectance spectrum of the DBR displays interference fringes adjacent to the stop band. The properties of DBRs have been analyzed in detail by Coldren and Corzine [26], Yariv [30], and Björk et al. [31]. Here, only a brief summary will be given.

Consider a distributed Bragg reflector consisting of m pairs of two dielectric, lossless materials with refractive index n_1 and n_2. The thickness of the two layers is assumed to be a quarter wave, that is, $L_1 = \lambda_{\text{Bragg}}/4n_1$ and $L_2 = \lambda_{\text{Bragg}}/4n_2$. The period of the DBR is $L_1 + L_2$. The reflectivity of a single interface is given by Fresnel's equation for normal incidence

$$r = \frac{n_1 - n_2}{n_1 + n_2}. \tag{3.7}$$

Multiple reflections at the interfaces of the DBR and constructive interference of the multiple reflected waves increases the reflectivity with increasing number of pairs. The reflectivity

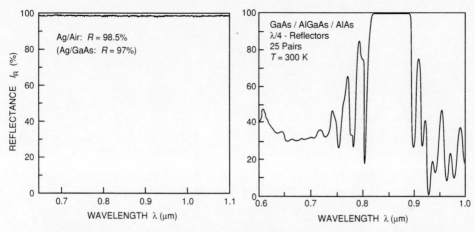

Figure 3.4. Reflectance of a metallic Ag/air reflector and an AlAs/GaAs distributed Bragg reflector.

has a maximum at the Bragg wavelength λ_{Bragg}. The reflectivity of a DBR with m quarter-wave pairs at the Bragg wavelength is given by [26]

$$R_{\text{DBR}} = |r_{\text{DBR}}|^2 = \left(\frac{1 - (n_1/n_2)^{2m}}{1 + (n_1/n_2)^{2m}} \right)^2 . \tag{3.8}$$

The *high-reflectivity* or *stop band* of a DBR depends on the difference in refractive index of the two constituent materials, Δn. The spectral width of the stop band is given by [30]

$$\Delta\lambda_{\text{stop band}} = \frac{2\lambda_{\text{Bragg}}\Delta n}{\pi n_{\text{eff}}}, \tag{3.9}$$

where n_{eff} is the effective refractive index of the mirror. It can be calculated by requiring the same optical path length normal to the layers for the DBR and the effective medium. The effective refractive index is then given by

$$n_{\text{eff}} = 2\left(\frac{1}{n_1} + \frac{1}{n_2} \right)^{-1} . \tag{3.10}$$

The length of a cavity consisting of two metal mirrors is the physical distance between the two mirrors. For DBRs, the optical wave penetrates into the reflector by one or several quarter-wave pairs. The penetration of the optical field is schematically shown in Figure 3.5(a). Only a finite number out of the total number of quarter-wave pairs are effective in reflecting the optical wave. The effective number of pairs seen by the wave electric field is given by [26]

$$m_{\text{eff}} \approx \frac{1}{2} \frac{n_1 + n_2}{n_1 - n_2} \tanh\left(2m \frac{n_1 - n_2}{n_1 + n_2} \right) . \tag{3.11}$$

For very thick DBRs ($m \to \infty$) the tanh function approaches unity and one obtains

$$m_{\text{eff}} \approx \frac{1}{2} \frac{n_1 + n_2}{n_1 - n_2} . \tag{3.12}$$

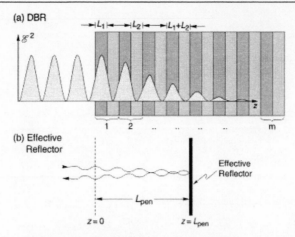

Figure 3.5. Illustration of the DBR penetration depth. (a) DBR consisting of two materials of thickness L_1 and L_2. (b) Ideal (metallic) reflector displaced by the penetration depth.

At the Bragg wavelength ($\lambda = \lambda_{Bragg}$), the phase change of the reflected wave is zero. In the vicinity of the Bragg wavelength ($\lambda \approx \lambda_{Bragg}$), the phase of the reflected wave changes *linearly* with wavelength. It is therefore possible to approximate a DBR with a metallike mirror located a distance L_{pen} behind the first dielectric interface, as shown in Figure 3.5. The reflection of the DBR can thus be expressed as

$$r_{DBR} \approx |r_{DBR}| e^{-2i(\beta - \beta_{Bragg})L_{pen}}, \tag{3.13}$$

where β is the average phase constant of the wave. The phase change at $z = 0$ (see Figure 3.5) of the wave reflected by the metal mirror is given by

$$r_{metal}|_{z=0} = |r_{metal}| e^{2i(2\pi/\lambda)L_{pen}}. \tag{3.14}$$

Equating the phase changes given by Eqs. (3.13) and (3.14) and using the phase changes of a DBR [26], we get for the penetration depth

$$L_{pen} = \frac{L_1 + L_2}{4r} \tanh(2mr). \tag{3.15}$$

For a large number of pairs ($m \to \infty$), the penetration depth is given by

$$L_{pen} \approx \frac{L_1 + L_2}{4r} = \frac{(L_1 + L_2)}{4} \frac{n_1 + n_2}{n_1 - n_2}. \tag{3.16}$$

Comparison of Eqs. (3.16) and (3.12) yields that

$$L_{pen} = \frac{1}{2} m_{eff} (L_1 + L_2). \tag{3.17}$$

The factor of $(1/2)$ in Eq. (3.17) is due to the fact that m_{eff} applies to the effective number of periods seen by the *electric field* whereas L_{pen} applies to the optical power. The optical power is equal to the square of the electric field and hence it penetrates half as far into the mirror. The effective length of a cavity consisting of two DBRs is thus given by the sum of the thickness of the center region plus the two penetration depths into the DBRs.

3.3 Calculation of Spontaneous Emission Based on Optical Mode Density

3.3.1 Optical Mode Density in a One-Dimensional Resonator

In this section, the enhancement of spontaneous emission will be calculated based on the changes of the *optical mode density* in a one-dimensional resonator, that is, a coplanar Fabry–Perot microcavity. We first discuss the basic physics causing the changes of the spontaneous emission from an optically active medium located inside a microcavity, and we will derive analytical formulas for the spectral and integrated emission enhancement. The spontaneous radiative transition rate in an optically active, homogeneous medium is given by (see, for example, Ref. [32])

$$W_{\text{spont}} = \tau_{\text{spont}}^{-1} = \int_0^\infty W_{\text{spont}}^{(\ell)} \rho(\nu_\ell) \, d\nu_\ell, \tag{3.18}$$

where $W_{\text{spont}}^{(\ell)}$ is the spontaneous transition rate into the optical mode ℓ, and $\rho(\nu_\ell)$ is the optical mode density. Assuming that the optical medium is homogeneous, the spontaneous emission lifetime, τ_{spont}, is the inverse of the spontaneous emission rate. However, if the optical mode density in the device depends on the spatial direction, as in the case of a cavity structure, then the emission rate given in Eq. (3.18) depends on the direction. Equation (3.18) can be applied to some small range of solid angle along a certain direction, for example the direction perpendicular to the reflectors of a Fabry–Perot cavity. Thus, Eq. (3.18) can be used to calculate the emission rate along a specific direction, in particular the optical axis of a cavity.

The spontaneous emission rate into the optical mode ℓ, $W_{\text{spont}}^{(\ell)}$, contains the dipole matrix element of the two electronic states involved in the transition [32]. Thus $W_{\text{spont}}^{(\ell)}$ will *not* be changed by placing the optically active medium inside an optical cavity. However, the optical mode density, $\rho(\nu_\ell)$, is strongly modified by the cavity. In this section, the changes in optical mode density will be used to calculate the changes in spontaneous emission rate.

We first compare the optical mode density in free space to the optical mode density in a microcavity. For simplicity, we restrict our considerations to the one-dimensional case, that is, to the case of a coplanar Fabry–Perot microcavity. Furthermore, we restrict our considerations to the emission along the optical axis of the cavity.

In a one-dimensional (1-D) homogeneous medium, the density of optical modes per unit length per unit frequency is given by

$$\rho^{1D}(\nu) = \frac{2n}{c}, \tag{3.19}$$

where n is the refractive index of the medium. Equation (3.19) can be derived using a formalism similar to that used for the derivation of the mode density in free space. The constant optical mode density given by Eq. (3.19) is shown in Figure 3.6(a).

In planar microcavities, the optical modes are discrete and the frequencies of these modes are integer multiples of the fundamental mode frequency. The optical mode density of a planar microcavity is schematically shown in Figure 3.6(a). The fundamental and first excited mode occur at frequency ν_0 and $2\nu_0$, respectively. For a cavity with two metallic reflectors (no distributed Bragg reflectors) and a zero or π phase shift of the optical wave upon reflection, the fundamental frequency is given by $\nu_0 = c/2nL_c$, where c is the velocity of light in vacuum and L_c is the length of the cavity. In a *resonant microcavity*, the emission

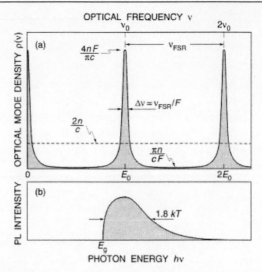

Figure 3.6. (a) Optical mode density of a one-dimensional planar microcavity (solid line) and of a homogeneous one-dimensional space. (b) Theoretical shape of the luminescence spectrum of bulk semiconductors.

frequency of an optically active medium located inside the cavity equals the frequency of one of the cavity modes.

The optical mode density along the cavity axis will be derived by using the relation between the mode density in the cavity and the optical transmittance through the cavity, $T(v)$:

$$\rho(v) = KT(v), \tag{3.20}$$

where K is a constant. The value of this constant will be determined by a normalization condition, (i.e., by considering a single optical mode). In the preceding section, the transmission through a Fabry–Perot cavity was given by

$$T(v) = \frac{T_1 T_2}{1 + R_1 R_2 - 2\sqrt{R_1 R_2}\cos(4\pi n L_c v/c)}. \tag{3.21}$$

The transmittance has maxima at $v = 0$, v_0, $2v_0$, $3v_0$, ... and minima at $v = v_0/2$, $3v_0/2$, $5v_0/2$, ... The Lorentzian approximation of a transmittance maximum at $v = 0$ is obtained by expanding the cosine term in Eq. (3.21) into a power series using $\cos x \approx 1 - x^2/2$. This gives

$$T(v) = \frac{T_1 T_2 (\sqrt{R_1 R_2})^{-1}(4\pi n L_c/c)^{-2}}{\frac{(1-\sqrt{R_1 R_2})^2}{(\sqrt{R_1 R_2})^{-1}(4\pi n L_c/c)^2} + v^2}. \tag{3.22}$$

Integrating $\rho(v)$ over all frequencies and the cavity length yields a single optical mode, that is,

$$K \int_0^{L_c} \int_{-\infty}^{\infty} \rho(v)\,dv\,dL = 1. \tag{3.23}$$

The lower and upper limit of the frequency integration can be chosen to be $\pm\infty$ since the Lorentzian approximation of Eq. (3.22) has only one maximum at $\nu = 0$. Using Eq. (3.20), Eq. (3.22), and the integration formula $\int_{-\infty}^{\infty}(a^2 + x^2)^{-1}\,dx = \pi/a$ yields

$$K = \frac{(R_1 R_2)^{3/4}}{T_1 T_2}\frac{4n}{c}(1 - \sqrt{R_1 R_2}). \tag{3.24}$$

Using Eq. (3.20), the optical mode density of a one-dimensional cavity for emission along the cavity axis is then given by

$$\rho(\nu) = \frac{(R_1 R_2)^{3/4}}{T_1 T_2}\frac{4n}{c}(1 - \sqrt{R_1 R_2})T(\nu). \tag{3.25}$$

Equation (3.25) allows one to calculate the density of optical modes at the maxima and minima. At the maxima, the mode density is given by

$$\rho_{\max} = \frac{(R_1 R_2)^{3/4}}{(1 - \sqrt{R_1 R_2})}\frac{4n}{c}. \tag{3.26}$$

Using $(R_1 R_2)^{3/4} \approx 1$ and the expression derived for the finesse F in the preceding section (see Eq. 3.4), one obtains an approximate expression for the mode density at the maxima:

$$\rho_{\max} \approx \frac{4}{\pi}\frac{nF}{c}. \tag{3.27}$$

That is, the mode density at the maxima is proportional to the finesse of the cavity. At the minima, the mode density is given by

$$\rho_{\min} = \frac{(R_1 R_2)^{3/4}(1 - \sqrt{R_1 R_2})}{(1 + \sqrt{R_1 R_2})^2}\frac{4n}{c}. \tag{3.28}$$

Using $(R_1 R_2)^{3/4} \approx 1$ and the expression derived for the finesse F in the preceding section (see Eq. 3.4), one obtains an approximate expression for the mode density at the minima:

$$\rho_{\min} \approx \pi\frac{n}{cF}. \tag{3.29}$$

That is, the mode density at the minima is inversely proportional to the finesse of the cavity. The comparison of the optical mode densities of a one-dimensional (1-D) free-space and a 1-D planar cavity is shown in Figure 3.6.

3.3.2 Spectral Emission Enhancement

Because the emission rate at a given wavelength is directly proportional to the optical mode density (see Eq. 3.8), the emission rate *enhancement spectrum* is given by the ratio of the 1-D cavity mode density to the 1-D free-space mode density. As calculated earlier, the cavity enhancement spectrum has a Lorentzian lineshape. The enhancement factor at the resonance wavelength is thus given by the ratio of the optical mode densities with and without a cavity, that is,

$$G_e = \frac{\rho_{\max}}{\rho^{1D}} \approx \frac{2}{\pi}F \approx \left(\frac{2}{\pi}\right)\frac{\pi(R_1 R_2)^{1/4}}{1 - \sqrt{R_1 R_2}}. \tag{3.30}$$

The equation shows that a strong enhancement of the spontaneous emission rate along the cavity axis can be achieved with microcavities.

Equation (3.30) represents the average emission rate enhancement out of both reflectors of the cavity. To find the enhancement out a single direction, we multiply the enhancement given by Eq. (3.30) by the fraction of the light exiting the mirror with reflectivity R_1 (i.e., $1 - R_1$) divided by the average loss of the two mirrors for one round trip in the cavity (i.e., $(1/2)[(1 - R_1) + (1 - R_2)]$). For large R_1 and R_2, this gives for the enhancement of the emission exiting R_1

$$G_e \approx \frac{2(1 - R_1)}{2 - R_1 - R_2} \frac{2F}{\pi} \approx \frac{1 - R_1}{1 - \sqrt{R_1 R_2}} \frac{2F}{\pi} \approx \left(\frac{2}{\pi}\right) \frac{\pi (R_1 R_2)^{1/4}(1 - R_1)}{(1 - \sqrt{R_1 R_2})^2}, \quad (3.31)$$

where we used the approximation $1 - \sqrt{R_1 R_2} \approx (1/2)(1 - R_1 R_2) \approx (1/2)(2 - R_1 - R_2)$. Equation (3.31) represents the emission rate enhancement out of a single reflector with reflectivity R_1.

Next we take into account the standing-wave effect, that is, the distribution of the optically active material relative to the nodes and antinodes of the optical wave. The antinode enhancement factor ξ has a value of 2, if the active region is located exactly at an antinode of the standing wave inside the cavity. The value of ξ is unity, if the active region is smeared out over many periods of the standing optical wave. Finally $\xi = 0$, if the active material is located at a node.

The emission rate enhancement is then given by

$$G_e = \left(\frac{\xi}{2}\right)\left(\frac{2}{\pi}\right) \frac{\pi (R_1 R_2)^{1/4}(1 - R_1)}{(1 - \sqrt{R_1 R_2})^2} \frac{\tau_{cav}}{\tau}, \quad (3.32)$$

where R_1 is the reflectivity of the light-exit mirror and therefore $R_1 < R_2$. Equation (3.32) also takes into account changes in the spontaneous emission lifetime in terms of τ, the lifetime without cavity, and τ_{cav}, the lifetime with cavity. The factor τ_{cav}/τ ensures that the enhancement decreases if the cavity lifetime is reduced as a result of the cavity. For planar microcavities, the ratio of the spontaneous lifetime with cavity, τ_{cav}, and the lifetime without cavity, τ, is $\tau_{cav}/\tau \geq 0.9$ [33]. Thus the emission lifetime is changed by only a minor amount in a planar microcavity.

3.3.3 Integrated Emission Enhancement

The total enhancement integrated over wavelength, rather than the enhancement at the resonance wavelength, is relevant for many practical devices. *On resonance*, the emission is enhanced along the axis of the cavity. However, sufficiently far *off resonance*, the emission is suppressed. Because the natural emission spectrum of the active medium (without cavity) can be much broader than the cavity resonance, it is, a priori, not clear if the integrated emission is enhanced at all. To calculate the wavelength-integrated enhancement, the spectral width of the cavity resonance and the spectral width of the natural emission spectrum must be determined. The resonance spectral width can be calculated from the finesse of the cavity or the cavity quality factor given in the preceding section.

The theoretical width of the emission spectrum of bulk semiconductors is $1.8\,kT$ (see, for example, Schubert, [34]), where k is Boltzmann's constant and T is the absolute temperature. At room temperature, $1.8\,kT$ corresponds to an emission linewidth of $\Delta\lambda_n = 31$ nm assuming an emission wavelength of 900 nm. For a cavity resonance width of 5–10 nm, one part of the spectrum is strongly enhanced, whereas the rest of the spectrum is suppressed. The integrated enhancement ratio (or suppression ratio) can be calculated analytically by assuming a Gaussian natural emission spectrum. For semiconductors at 300 K, the linewidth of the natural emission is, in the case of high-finesse cavities, much larger than the width of the cavity resonance. The Gaussian emission spectrum has a width of $\Delta\lambda_n = 2\sigma(2\ln 2)^{1/2}$ and a peak value of $(\sigma(2\pi)^{1/2})^{-1}$, where σ is the standard deviation of the Gaussian function. The integrated enhancement ratio (or suppression ratio) is then given by [35]

$$G_{\text{int}} = \frac{\pi}{2}G_e\Delta\lambda\frac{1}{\sigma\sqrt{2\pi}} = G_e\sqrt{\pi\ln 2}\frac{\Delta\lambda}{\Delta\lambda_n}, \tag{3.33}$$

where the factor of $\pi/2$ is due to the Lorentzian lineshape of the enhancement spectrum. Hence, the integrated emission enhancement depends on the natural emission linewidth of the active material. The value of G_{int} can be quite different for different types of optically active materials. Narrow atomic emission spectra can be enhanced by several orders of magnitude [10]. In contrast, materials having broad emission spectra, such as dyes or polymers [6, 7], may not exhibit any integrated enhancement at all. Equation (3.33) also shows that the width of the resonance has a profound influence on the integrated enhancement. Narrow resonance spectral widths, that is, high finesse values or long cavities [14], reduce the integrated enhancement.

As an example, we calculate the wavelength-integrated enhancement of a semiconductor microcavity structure by using Eqs. (3.32) and (3.33). With the reflectivities $R_1 = 90\%$ and $R_2 = 97\%$, an antinode enhancement factor of $\xi = 1.5$, and $\tau_{\text{cav}}/\tau \approx 1$, one obtains a finesse of $F = 46$ and a peak enhancement factor of $G_e = 68$ by using Eq. (3.32). Inserting this value into Eq. (3.33), using a cavity resonance bandwidth of $\Delta\lambda = 6.5$ nm [36] and the theoretical 300 K natural emission linewidth of $\Delta\lambda = 31$ nm, one obtains a theoretical integrated enhancement factor of $G_{\text{int}} = 13$. Experimental enhancement factors of five have been demonstrated [36] for the reflectivity values assumed above. The lower experimental enhancement is in part due to a broader natural emission linewidth, which exceeds the theoretical value of $1.8 kT$.

The spontaneous emission spectrum of a bulk semiconductor is schematically shown in Figure 3.6(b). For optimum enhancement, the cavity must be in resonance with the natural emission spectrum. Note that additional broadening mechanisms, such as alloy broadening, will broaden the natural emission spectrum over its theoretical value of $1.8 kT$. Quantum-well structures have an inherently narrower spectrum $(0.7 kT)$, owing to the step-function-like density of states. Low temperatures and excitonic effects can further narrow the natural emission linewidth. Thus, higher enhancements are expected for low temperatures and quantum-well active regions.

3.3.4 Device Design Rules

We next summarize several design rules to maximize the spontaneous emission enhancement in microcavity structures. These rules will also provide further insight in the fundamental operating principles of RCLEDs and the differences of these devices from VCSELs. The *first design criterion* for RCLEDs is that the reflectivity of the light-exit reflector, R_1, should be much lower than the reflectivity of the back reflector, that is,

$$R_1 \ll R_2. \tag{3.34}$$

This condition implies that light exits the device mainly through the reflector with reflectivity R_1. Equation (3.34) applies to the design of most RCLEDs including communication RCLED applications (where light should be emitted into the small core of a multimode fiber) and display RCLEDs (where light should be emitted toward the observer).

The *second design criterion* calls for the shortest possible cavity length L_c. To derive this criterion, the integrated enhancement is rewritten by using the expressions for the cavity F (see Eq. 3.4) and cavity Q (see Eq. 3.6), and by insertion of Eqs. (3.32) into Eq. (3.33), one obtains

$$G_{\text{int}} = \left(\frac{\xi}{2}\right)\left(\frac{2}{\pi}\right)\frac{1 - R_1}{1 - \sqrt{R_1 R_2}}\sqrt{\pi \ln 2}\frac{\lambda}{\Delta\lambda_n}\frac{\lambda_{\text{cav}}}{L_c}\frac{\tau_{\text{cav}}}{\tau}, \tag{3.35}$$

where λ and λ_{cav} is the active region emission wavelength in vacuum and inside the cavity, respectively. Since the operating wavelength λ and the natural linewidth of the active medium, $\Delta\lambda_n$, are given quantities, Eq. (3.35) shows that minimization of the cavity length L_c maximizes the integrated intensity.

The importance of a short cavity length is elucidated by the illustrations shown in Figure 3.7. The optical mode density of two different cavities, namely a short and a long cavity, are shown in Figure 3.7(a) and (b), respectively. Both cavities have the same mirror reflectivities and finesse. The natural emission spectrum of the active region is shown in Figure 3.7(c). The overlap between the resonant optical mode and the active region emission spectrum is best for the shortest cavity.

The largest enhancements are achieved with the shortest cavities, which in turn are obtained if the *fundamental* cavity mode is in resonance with the emission from the active medium. The cavity length is also reduced by using a DBR with a short penetration depth, that is, a DBR consisting of two materials with a large difference in the refractive index.

The *third design criterion* discussed here is the minimization of self-absorption in the active region. This criterion can be stated as follows: The reabsorption probability of photons emitted from the active region into the cavity mode should be much smaller than the escape

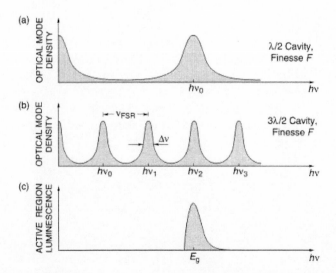

Figure 3.7. Optical mode density for (a) a short and (b) a long cavity with the same finesse F. (c) Spontaneous free-space emission spectrum of the active region.

probability of photons through one of the reflectors. Assuming $R_2 \approx 1$, this criterion can be written as

$$2\xi\alpha\ell < (1 - R_1), \tag{3.36}$$

where α and ℓ are the absorption coefficient and the length of the active area, respectively. If the criterion of Eq. (3.36) were not fulfilled, photons in the fundamental cavity mode would be reabsorbed by the active region. Subsequently, reemission will, with a certain probability, occur to the side (waveguiding modes), that is, not into the cavity mode. Another possibility is that the electron–hole pairs generated by reabsorption recombine nonradiatively. In either case, reabsorption processes occurring in high-finesse cavities reduce the cavity mode emission out of the cavity. Thus, if the condition of Eq. (3.36) is not fulfilled, the emission intensity of microcavities is lowered rather than enhanced.

Whereas the condition of Eq. (3.36) is fulfilled in RCLEDs, it is clearly not fulfilled in VCSELs. The spontaneous emission intensities of RCLEDs and VCSELs were compared by Schubert et al. [17]. In this comparison, the VCSEL and the RCLED were driven by an injection current of 2 mA, which is below the threshold current of the VCSEL of $I_{th} = 7$ mA. The spontaneous emission spectra of an RCLED and a VCSEL are shown in Figure 3.8. The VCSEL has an AlGaAs/GaAs quantum-well active region emitting at 850 nm. Both reflectors of the VCSEL are AlGaAs/AlAs DBRs. Figure 3.8 reveals that the emission intensity of the VCSEL in the spontaneous regime is more than a factor of ten lower than the emission intensity from the RCLED.

Because the magnitude of the maximum gain in semiconductors is always lower than the magnitude of the absorption coefficient in an unpumped semiconductor ($|g| < |\alpha|$), VCSELs could not lase if the condition of Eq. (3.36) were met. Thus, the spontaneous emission intensity in VCSELs is low and must be low to enable the device to lase. Figure 3.8 also reveals that the emission spectral linewidth of VCSELs is narrower than that of RCLEDs. The higher spectral purity is due to the higher values of R_1 and R_2 required in VCSELs.

The fulfillment of Eq. (3.36) by RCLEDs also implies that these devices cannot lase. As stated above, it is always $|g| < |\alpha|$. As a consequence, the mirror loss $(1 - R_1)$ is always larger than the maximum achievable round-trip gain $(2\xi g\ell)$. The fundamental inability of

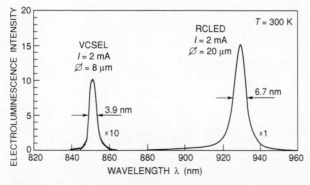

Figure 3.8. Spontaneous electroluminescence spectrum of a vertical-cavity surface-emitting laser (VCSEL) emitting at 850 nm and of a resonant-cavity light-emitting diode (RCLED) emitting at 930 nm. The drive currents for both devices are 2 mA. The VCSEL spectrum is multiplied by a factor of ten. The threshold current of the VCSEL is 7 mA.

RCLEDs to lase has been experimentally verified by high pulsed injecting currents without evidence of lasing.

The arguments used above imply that the spontaneous emission into the fundamental cavity mode in VCSEL structures is very low owing to reabsorption of photons by the active region. A reduction of the threshold current by increasing the reflectivity will be accompanied by a further decrease of the spontaneous emission below threshold. We therefore conclude that the so-called zero-threshold laser [18, 20] cannot be realized by a planar microcavity structure.

3.4 Calculation of Spontaneous Emission Based on Fermi's Golden Rule

3.4.1 Fermi's Golden Rule

The calculation of emission rate of an emitting dipole in a weak field can be given by Fermi's golden rule. Simply stated, the fractional emission rate for a single dipole into a particular photon mode is proportional to the square of the dot product of the dipole oscillation axis and the photon electric field at that position. The proportionality constants, which are not shown, include the local field factor [37] and oscillator strength for the electronic transition:

$$A_{\text{Rad,dipole}} \propto \sum_m |\hat{\mu} \cdot \vec{E}_m|^2. \tag{3.37}$$

To solve such a problem, one only needs to calculate the optical modes for the system, find the electric field at the dipole position, and sum the formula over all modes. The atomic lifetime is just the inverse of this value, or

$$\tau_{\text{dipole}} \propto \frac{1}{A_{\text{Rad, dipole}}}. \tag{3.38}$$

If the "system" under consideration is a large macroscopic region of space, then the optical mode density remains unchanged by an optical microcavity. The interaction of each of these modes with the atomic dipole, however, depends on Fermi's golden rule. This is in contrast to the argument of Section 3.3, where we modeled the change in emission by an optical microcavity as a change in the optical mode density of the microscopic system of the cavity itself. These two views are complementary.

We will consider the case of the planar microcavity by describing two popular calculation methods. The first is the cooperative dipole method, which treats the emission as a wave train that can interfere with itself. The second is the general calculation of both external and waveguiding modes.

3.4.2 The Cooperative Dipole Method

Other authors have described calculation methods for planar microcavities. These have included spectrum calculations [37, 38], emission lifetime methods involving perfect metallic reflectors [39], or equal reflectivity mirrors [20, 40, 41] or general nonabsorbing structures [37, 38]. Not everyone may need to calculate the emission into waveguide modes; they may just desire the emission rate and emission profile for emission around the normal axis. In this case, the cooperative dipole method [42] is attractive.

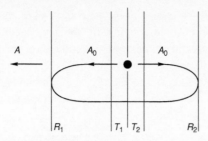

Figure 3.9. A dipole emitting waves in both the positive and negative directions toward two mirrors of same phase and reflectivity R_1 and R_2. We include the possibility of intracavity absorption regions with single-pass transmission T_1 and T_2. The emitted wave reflects back and forth, with the emitted wave being the geometric series of all waves emitted after each round-trip reflection.

One can assume that the macroscopic system has a uniform mode density, so that without interference effects, the atom would otherwise emit into all angles equally within the active media. The only effect to calculate, then, is the interference of the wave train with itself as it bounces back and forth in the cavity. The relative emission rate in any one direction can then be calculated by summing the waves emitting out of the planar mirrors. This interference can be thought of as a cooperative emission of an infinite series of image dipoles beyond the mirrors.

Figure 3.9 shows the situation. Although we will find a solution for all radiating angles, consider waves normal to the plane of the mirrors for simplicity. An emitting dipole would like to emit a wave $A_0 \exp[i(kx - \omega t)]$ in the positive x direction and a wave $A_0 \exp[-i(kx - \omega t)]$ in the negative x direction. However, owing to the reflections with the mirrors R_1 and R_2, the self-interference of the wave train modifies the intensity of the emitted wave. To find the emission through R_1 we have to sum a geometric series of the initial and all reflected waves.

Assume there is a phase change of δ_1 going through length x_1, and a phase change of δ_2 going through length x_2, and a phase change of δ_3 reflecting off mirror 1, and a phase change of δ_4 reflecting off mirror 2. Let the round-trip phase δ be

$$\delta = 2\delta_1 + 2\delta_2 + \delta_3 + \delta_4. \qquad (3.39)$$

Using the same notation as in Section 3.3, we define the normal-axis intensity enhancement factor out of mirror 1 to be G_e as compared to that in a bulk material with no reflections at the interface with air. If the intracavity transmission from the dipole to mirror 1 is T_1 and from the dipole to mirror 2 is T_2 (normally these are equal to 1), the G_e becomes

$$G_e = \frac{\tau_{\text{cav}}}{\tau_{\text{bulk}}} \left| \frac{A}{A_0} \right|^2 = \frac{\tau_{\text{cav}}}{\tau_{\text{bulk}}} \frac{T_1(1 - R_1)\left(1 + T_2^2 R_2 + 2T_2\sqrt{R_2}\cos(2\delta_2 + \delta_4)\right)}{1 + R_1 R_2 T_1^2 T_2^2 - 2T_1 T_2 \sqrt{R_1 R_2}\cos(\delta)}. \qquad (3.40)$$

If one is calculating the distributed reflection using a matrix method, one would actually be more likely to calculate the products $(T_1)^2 R_1$ and $(T_2)^2 R_2$ rather than separate parameters. As mentioned in Section 3.3, the ratio of the emission lifetimes, $\tau_{\text{cav}}/\tau_{\text{bulk}}$, is a result of energy conservation. For a given number of emitting dipoles, the total emitted power is constant. If the all-angle emission rate is modified by cavity effects, this can't affect the total power. Therefore the $\tau_{\text{cav}}/\tau_{\text{bulk}}$ factor is necessary when changing from emission rate calculations to intensity calculations for the same number of emitters. We have calculated [33, 43] that this fraction is about equal to 1.0 for our structure provided the resonance occurs at the peak of the natural emission spectrum. The calculation of the reflectivities and phases for arbitrary distributed mirrors can be done by any standard matrix method [44].

For an antinode position ($2\delta_2 + \delta_4 = 2m\pi, m \in I$), and no intracavity absorption ($T_1 = T_2 = 1$), and an on-resonance condition ($\delta = 2m\pi, m \in I$), this equation reduces to

$$G_e = \frac{\tau_{cav}}{\tau_{bulk}} \frac{(1 - R_1)(1 + \sqrt{R_2})^2}{(1 - \sqrt{R_1 R_2})^2}.\tag{3.41}$$

For off-normal emission at an angle θ_1 out of mirror 1, the normalized intensity for a dipole emitting at a wavelength λ_0 in air becomes (for a randomly polarized dipole)

$$I(\theta_1, \lambda_0)\, d\Omega_1 = \left(\frac{n_1}{n_a}\right)^2 \frac{\cos(\theta_1)}{\cos(\theta_a)} \frac{d\Omega_1}{4\pi} \sum_{j=1}^{2} \left(\frac{3}{2}G_{e,j}(\theta_1, \lambda_0)\right)\left(\frac{1}{3}\right).\tag{3.42}$$

This equation is normalized so that the integrated intensity for all angles out both mirrors is equal to 1. The subscript "a" denotes the values within the material holding the atom. The summation over j is over the S and P modes (TE and TM). The enhancement factor G_e is still from Eq. (3.40) calculated for the S and P modes. The ratios of the indices of refraction and angle cosines are due to the relative sizes of a solid angle of light in the atom material and in the outside world (material 1). By Snell's law, $\sin(\theta_a) = n_1 \sin(\theta_1)/n_a$. The final coefficient of (1/3) is for the probability of polarization in any direction for a random dipole. For a nonrandom dipole, this can be changed. For instance, an electron to heavy-hole transition in a III-V semiconductor quantum well would have a value of $(\cos(\theta_a))/2$, whereas an electron to light-hole transition would have a value of $(4\sin(\theta_a) + \cos(\theta_a))/6$.

3.4.3 The Photonic Modes of a Planar Microcavity

It is still useful to present a general emission rate calculation method using a more rigorous approach. The cooperative dipole method is not up to the task of true waveguiding modes, although it can work with absorbing materials quite well. The method presented here is general for any nonabsorbing structure and includes the possibility of waveguiding modes. Our calculation method would still work if one of the two mirrors were a highly reflecting metal. If both mirrors were metal, one could still approximately calculate the emission rate by assuming that all modes will be leaky waveguiding modes.

The Er-doped structure is shown in earlier sections. Consider the system in Figure 3.10, consisting of a microcavity, like our Er-doped ones, and a large-area substrate and an external world. There are ($s - 2$) thin layers between dielectric layers 1 and s. Consider a single-mode solution labeled j for the electomagnetic wave in the structure. The electric field in each layer t consists of a forward- and backward-going wave with complex amplitudes $C_{j,t}$ and $D_{j,t}$ respectively at angle θ_t from the normal. For any particular angle θ_t, the variation of the electric-field amplitude along the z direction can be modeled for a planar dielectric multilayer structure by using a matrix method to solve for the boundary conditions at the layer interface [45]. The index of refraction in any arbitrary layer f is given by n_f. The electric field vector for a particular optical mode j at a position z in layer f is given by

$$\boldsymbol{E}_{j,f}(x, y, z) = \left[C_{j,f} e^{ik_{j,f}z} \hat{u} + D_{j,f} e^{ik_{j,f}z} \hat{v} \right] e^{-i(k_{\beta j}x - \omega t)}.\tag{3.43}$$

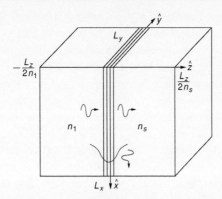

Figure 3.10. The system used for modeling spontaneous emission from planar layer structures. There are a total of s dielectric layers, with layer 1 and s being very large. The electric field for a particular propagation direction and polarization is calculated by using a matrix method to solve for the boundary conditions at the planar interfaces. The emission rate can then be calculated using Fermi's golden rule.

If k_0 is the magnitude of the free-space wavevector, then $k_{\beta,j}$ is the magnitude of the wavevector in the direction parallel to the layer planes for optical mode j. The value $k_{j,f}$ is the amplitude of the wavevector perpendicular to the layers in material f and is given by $k_{j,f} = [n_f^2 k_0^2 - k_{\beta,j}^2]^{1/2}$, where the argument (phase angle) of the complex bracketed term is defined to be in the range $-\pi/2$ to $3\pi/2$. The complex value $k_{j,f}$ will therefore have an argument in the range $-\pi/4$ to $3\pi/4$. The direction vectors u and v are along the z axis for TE-polarized modes and in the x–z plane for TM-polarized modes (also known as S and P modes). The u and v direction vectors are, of course, orthogonal to the direction of propagation of the wave, given by $k_{j,f}$. The bracketed expression in Eq. (3.43) defines for each layer the z dependence of the electric field for mode j and is denoted by $U_j(z)$.

One can now combine the solution for any particular optical mode with Fermi's golden rule for the spontaneous emission rate. Let the total spontaneous emission rate for a bulk material containing an emitting atom be given by A_a. If this emitting material is then placed in the planar microcavity, we can calculate the new radiative emission rate for an atom at position z_a within this material. The total radiative emission rate (not including waveguiding modes) out of a planar structure consisting of s layers is given by

$$A_{\mathrm{Rad}} = A_a \sum_{j=1}^{4} \left(\frac{1}{2}\right) \int_0^{\theta_{\mathrm{crit}}} \left[\frac{3}{2n_a} \varepsilon_0 M_j^2 |\hat{\mu} \cdot \vec{U}_j(z_a)|^2 \right]$$

$$\times \left(\frac{H(\theta_1) + H(\theta_s)}{H(\theta_1)\cos\theta_1 + H(\theta_s)\cos\theta_s} \right) n_m^2 \cos\theta_m \sin\theta_n \, d\theta_m, \tag{3.44}$$

$$H(\theta) = \begin{cases} 1, & \theta \in \{\Re e\}, \\ 0, & \theta \notin \{\Re e\}, \end{cases} \tag{3.45}$$

$$M_j^2 = \frac{L_z}{\int \varepsilon(z)|\vec{U}_j(z)|^2 \, dz} = \frac{2}{\varepsilon_0 \left[n_1 H(\theta_1)\left(C_{j,1}^2 + D_{j,1}^2\right) + n_s H(\theta_s)\left(C_{j,s}^2 + D_{j,s}^2\right) \right]}. \tag{3.46}$$

The vector μ is the dipole polarization. For a random dipole polarization, $|\mu \cdot U_j(z_a)|^2$ is equal to $|U_j(z_a)|^2/3$. The summation j is over the two polarization modes S and P, and over the two orthogonal solutions for any one angle and polarization. We numerically performed the calculation with layer m as the layer containing the atom, but it could be any layer providing the correct integration limits are used. In fact, choosing m as the higher index cladding layer is perhaps an even better choice [37]. The integration angle limit $\theta_m = \theta_{\text{crit}}$ is the angle at which total internal reflection occurs between the layer m and the larger index cladding layer, or $\pi/2$, whichever occurs first. The value M_i is a normalization constant, calculated assuming that the thicknesses of layers 1 and s are defined as $L_z/(2n_1)$ and $L_z/(2n_s)$. If θ_1 or θ_s are nonreal, then there is only one mode solution for each polarization, with boundary conditions $C_{j,1} = 0$ or $D_{j,s} = 0$ respectively. If both θ_1 and θ_s are real, then two solutions exist for each polarization mode. These can be found by setting $C_{j,s} = 0$ in one calculation and $D_{p,s} = 0$ in the second. One of the modes U_j is chosen to be the first solution, and the second orthogonal mode U_q is generated by adding a linear combination of the two solutions such that the following orthogonality condition is satisfied:

$$\int \varepsilon(z)\vec{U}_j^*(z) \cdot \vec{U}_q(z)\, dz = 0 \quad j \neq q. \tag{3.47}$$

If U_j is the one-mode solution, and V_p is the second solution for that polarization, then the second orthogonal mode is $U_q = bU_j + V_p$, where b is a complex number with the following value:

$$b = -\frac{(C_{j,1}^* C_{p,1} + D_{j,1}^* D_{p,1})}{\left(|C_{j,1}|^2 + |D_{j,1}|^2 + \left(\dfrac{n_s}{n_1}\right)|C_{j,s}|^2\right)}. \tag{3.48}$$

We performed all of our integrations over the angular range θ_m from 0 to $\pi/2$ assuming that layer m is the active region containing the emitting atoms. Some authors [37] prefer defining layer m to be either layer 1 or layer s, whichever has the higher index. That choice may be better, as it solves all modes other than true waveguiding modes. Using our choice of layer m as the emitting layer, if one of the cladding layers has a higher index than the emitting layer, modes in the cladding that couple evanescently with the active region will be missed. However, if these modes can be described as leaky waveguiding modes, their contribution to the total spontaneous emission rate can be included in the radiative emission rate A_{Wav} due to waveguiding modes:

$$A_{\text{Wav}} = A_a \sum_j \pi \frac{\Re e(k_{\beta,j})}{k_0^2} \Re e\left(\frac{dk_{\beta,j}}{dk_0}\right) \left[\frac{3}{2n_a}\varepsilon_0 M_j^2 |\hat{\mu} \cdot \vec{U}_j(z_a)|^2\right], \tag{3.49}$$

$$M_j^2 = \frac{1}{\int \varepsilon(z)|\vec{U}_j(z)|^2\, dz}. \tag{3.50}$$

The TE-polarized and TM-polarized waveguiding modes are both calculated [45] using the required boundary conditions $C_1 = 0$ and $D_s = 0$. This condition simply ensures that the waves decay as they extend out of the structure. The derivative $dk_{\beta,j}/dk_0$ is calculated numerically by solving for $k_{\beta,j}$ with two slightly different k_0 values. For a leaky mode, the mode wavevector $k_{\beta,j}$ is complex. When calculating the normalization constant, M_j^2, the integration is over all layers. If the structure supports leaky waveguide modes, the waves can freely propagate in one of the thick cladding layers, making the mode non-normalizable. Therefore when calculating M_j in Eq. (3.50), the integration is arbitrarily cut off 10 nm into the layer where the light freely propagates (the high-index layer). The total spontaneous emission rate for a particular wavelength and atom position is the sum of A_{Rad} and A_{Wav}.

3.4.4 The Main Lobe of the Planar Microcavity

For purposes of coupling light into a multimode fiber for optical communications, one is only interested in the main emission lobe of the planar microcavity. It is important to get a large intensity of light along the normal direction, where it best couples to a fiber. It is also important for this main lobe to have a large fraction of the total energy, but it is the light on axis that preferentially couples to the narrow acceptance angle of the fiber. A multimode fiber may have a numerical aperture of only 0.29. Coupling to a fiber could be done directly from the planar output surface or through an integrated or external lens. Either way, one will design a planar microcavity emitter such that the primary emission lobe is a maximum on-axis, or in other words so that the on-axis resonance wavelength approximately corresponds to the center wavelength of the spontaneous emission from a nonmicrocavity device. Experimental comparisons of different devices can be made by measuring the on-axis power per steradian for a given input current.

If the finesse of the cavity is large and has a value F, and the antinode enhancement factor is ξ, the average intensity enhancement $G_{\text{e,avg}}$ into both directions along the normal at the resonance peak wavelength is given by

$$G_{\text{e,avg}}(\lambda_{\text{peak}}, \theta = 0) = \frac{2F}{\pi} \frac{\xi \tau_{\text{cav}}}{\tau_0}. \tag{3.51}$$

For cavities resonant at the center wavelength of a natural spectrum, τ_{cav}/τ_0 roughly equals 1. We multiply $G_{\text{e,avg}}$ by $(2 - 2R_{\text{out}})/(2 - R_{\text{out}} - R_{\text{back}})$ to find the enhancement in the direction of the mirror R_{out}. This is fine for the emission enhancement at the peak resonance wavelength, but this number by itself is of little value since G_{e} only corresponds to the enhancement of spectral power density along the emission axis. One can calculate the wavelength-integrated intensity enhancement in the normal direction by integrating the enhancement over the range of the natural emission spectrum. If the natural emission spectrum is $N(\lambda_0)$, which has a width much smaller that the peak wavelength in vacuum λ_{peak}, then the average power enhancement into both substrate and air directions along the normal is given by

$$I_{\text{avg}}(\theta = 0) = \frac{\int N(\lambda_0) G_{\text{e,avg}}(\lambda_0, \theta = 0) \, d\lambda_0}{\int N(\lambda_0) \, d\lambda_0}. \tag{3.52}$$

Here λ_0 is the wavelength of light in vacuum. If $N(\lambda_0)$ were too large, one would have to include a term for the fractional energy per unit wavelength of the spectrum, but we ignore it in our calculations. This integral $I_{\text{avg}}(\theta = 0)$ is a maximum for high finesse values, where the enhancement spectrum $G_{\text{e,avg}}$ is much thinner than the natural spectrum. In this case, the $N(\lambda_0)$ term in the numerator can be brought out of the integral and becomes $N(\lambda_{\text{peak}})$. One also knows that, for narrow spectral widths, the $G_{\text{e,avg}}$ spectrum is of pseudo-Lorentzian shape. If the natural spectrum is then either Lorentzian or Gaussian with full width at half maximum $\Delta\lambda_{N,\text{Lorentz}}$ or $\Delta\lambda_{N,\text{Gauss}}$, then $I_{\text{avg,max}}(\theta=0)$ becomes, for the two cases,

$$I_{\text{avg,max}}(\theta=0) = \frac{\xi\tau_{\text{cav}}}{\tau_0}\frac{\lambda}{\pi L_{\text{cav}}}\frac{\lambda_0}{\Delta\lambda_{N,\text{Lorentz}}} \quad \text{or} \quad = \frac{\xi\tau_{\text{cav}}}{\tau_0}\frac{\sqrt{\ln 2}\,\lambda}{\sqrt{\pi}L_{\text{cav}}}\frac{\lambda_0}{\Delta\lambda_{N,\text{Gauss}}}. \quad (3.53)$$

As stated previously, we can multiply these factors by $(2-2R_{\text{out}})/(2-R_{\text{out}}-R_{\text{back}})$ to get the intensity enhancement factor $I_{\text{max}}(\theta=0)$ from the output mirror. For our 4-quantum-well semiconductor structures, the antinode enhancement factor ξ is about 1.75, whereas it is about 1.4 for the Er profile in our SiO_2/Si microcavities. The value τ_{cav}/τ_0 is about 0.98 for the semiconductor structure resonant at the peak of the natural spectrum, and it is about 0.88 in the Er-doped microcavity at $\lambda_0 = 1,535$ nm. For high drive currents, the InGaAs quantum wells have a Gaussian natural spectrum $\Delta\lambda_{N,\text{Gauss}}$ of 50 nm. We then have an integrated intensity enhancement out the output mirror of $I_{\text{max}}(\theta=0)=7.5$ for the semiconductor RCLED structure.

Notice that the maximum intensity enhancement depends inversely on cavity length and natural spectral width, not on finesse. Because the natural spectrum broadens at high currents, our 4-quantum-well RCLEDs are best operated at moderate current densities of less than 1 kA/cm²/well, or less than about 13 mA current for our 20 μm diameter devices. For an active region with index n ($n = 3.5$ for InGaAs), the integrated intensity enhancement value can be multiplied by the value $1/(4\pi n^2)$ to get the fractional power per steradian in air in the normal direction. For our structure, this is 0.86 per steradian along the normal axis. A sample far-field pattern is shown in Figure 3.11b. The exact shape depends on pump current and the match of the resonance wavelength with the peak of the natural spectrum.

What about the integrated intensity in the main output lobe of the emission spectrum? One can design the cavity to be on resonance at the peak of the natural resonance to achieve maximum intensity at zero angle. One could alternatively design the cavity just to a wavelength longer than the natural spectrum, similar to the case of Figure 3.11(a) for the Er microcavity. In this case of Figure 3.11(a), the power into the air is actually increased over a cavity resonant at 1,535 nm. The power can be increased by just less than a factor of two (it is less than two because of the decrease in τ_{cav}/τ_0 for this case). For the Er cavity, the integrated power from 0 to 20 degrees is 30% for a 1,550 nm resonance wavelength.

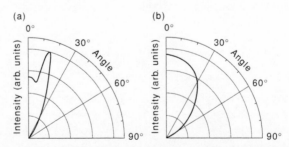

Figure 3.11. (a) The emission intensity as a function of angle for an Er-doped microcavity structure resonant on the long-wavelength peak of 1550 nm. The shorter wavelength 1535 nm peak emits at an angle away from the normal. (b) The emission intensity versus angle for an InGaAs/GaAs/AlGaAs microcavity LED resonant at the natural emission wavelength of 940 nm.

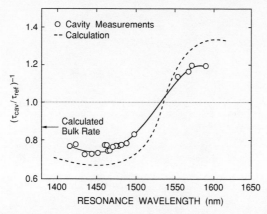

Figure 3.12. The theoretical and experimental emission rates as compared to reference structure with no top mirror for our Er-doped SiO$_2$/Si microcavities. The emission rate (or inverse lifetime) is plotted versus the central SiO$_2$ active region thickness as a multiple of light wavelength within the SiO$_2$. The theoretical implantation profile is taken into consideration when generating the theoretical curve. A thin cavity (smaller resonance wavelength) will not be resonant at any angle for any of the emitting wavlengths (but can emit into high-angle modes), thus reducing its emission rate. A cavity thick enough to support a resonant mode close to the normal for all emitting wavelengths has an enhancement of emission rate.

3.4.5 Lifetime Modification

Lifetime modification in 2-D and 3-D photonic confinement structures can be significant, but the effect of a planar microcavity on spontaneous emission lifetime is relatively small [20]. We have applied the model mentioned earlier to the Er-doped Si/SiO$_2$ cavity. The theoretical implantation profile of the Er in the SiO$_2$ active region was used to find an average emission lifetime versus active region thickness. The results are shown in Figure 3.12. At the resonant condition of thickness $= \lambda/2$, there is actually no change in spontaneous lifetime. This is because half the wavelengths are enhanced and half are suppressed. This is one reason why a device with maximum total emission must not be resonant at the emission peak, but rather to the long wavelength side. However, when the on-axis intensity is of importance, then a resonance exactly at an emission peak still makes sense. Various structures were deposited with different active region thickness, and the lifetimes were measured when optically pumped. The comparison with theory is qualitatively correct but the effect is less striking. This could be explained by a wider implantation profile than expected or by some nonradiative recombination.

Similar calculations for our RCLED structures show theoretical changes of plus or minus 5% with the resonance shifting from one side of the natural spectrum to the other. Therefore, in semiconductor structures, confinement better than that by a 1-D planar microcavity is needed to see large spontaneous emission lifetime changes.

3.5 Resonant-Cavity Devices

3.5.1 Resonant-Cavity Light-Emitting Diodes

The enhanced spontaneous emission occurring in microcavity structures can be beneficially employed in semiconductor and polymer light-emitting diodes (LEDs). Such devices, called

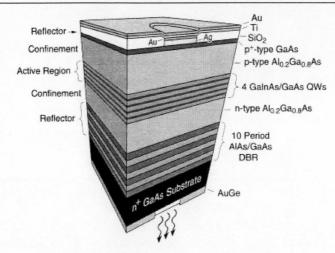

Figure 3.13. Structure of a InGaAs/GaAs resonant-cavity light-emitting diode with one AlAs/GaAs distributed Bragg reflector and one Ag reflector.

resonant-cavity light-emitting diodes (RCLEDs), were first realized in 1992 [13] in the GaAs material system. About a year later, RCLEDs were demonstrated in organic materials [11]. The main advantages of resonant-cavity devices over conventional LEDs are (i) higher emission intensities, (ii) higher spectral purity, and (iii) a more directed emission pattern.

Microcavity structures with enhanced spontaneous emission also include Er-doped micro-cavities [10]. Due to the inherently narrow luminescence line of the intra-atomic Er radiative transitions, there is a very good overlap between the cavity optical mode and the Er luminescence line. At the present time, no Er-doped current-injection devices exist. However, the great potential of Er-doped resonant cavities make the realization of Er-doped RCLEDs likely in the future. In this section, the design and key results of semiconductor RCLEDs and Er-doped Si/SiO$_2$ cavities will be reviewed.

The structure of an RCLED with an InGaAs active region is shown in Figure 3.13. The cavity is defined by one DBR and one metallic reflector. Also included are two confinement regions and a four-quantum-well active region. The heavily doped n-type substrate is coated with an antireflection ZrO$_2$ layer.

The motivation for the metal reflector is twofold. First, the metallic Ag reflector serves as a nonalloyed ohmic contact to the heavily doped p-type ($N_A \approx 5 \times 10^{19}$ cm^{-3}) GaAs top layer, thus effectively confining the pumped region to the area below the contact. Second, it was shown in the two preceding sections that the cavity length must be kept as short as possible to maximize the emission enhancement. Due to the lack of a penetration depth, metal reflectors allow for short cavity lengths. Cavities with two metallic reflectors have been reported [46]. However, optical absorption losses in the light exit mirror can be large in a double metal mirror structure, unless very thin metallic reflectors are used [27]. The lack of a p-type DBR also avoids the well-known problem of high resistance in p-type DBRs [47, 48]. It has been shown that parabolic grading yields the lowest ohmic resistance in DBRs. Such parabolic grading is suited to eliminating heterojunction band discontinuities [47].

The magnitude of the reflectivity of the DBR needs to be consistent with Eqs. (3.34) and (3.36). The Ag back mirror has a reflectivity of approximately 96%. According to Eq. (3.34), the DBR reflectivity must be <96%. The second criterion of Eq. (3.36) requires that $2\xi\alpha\ell < 1 - R_1$. Assuming $\xi = 1.3$, $\alpha = 10^4$ cm^{-1}, and $\ell = 400$ Å, one obtains the

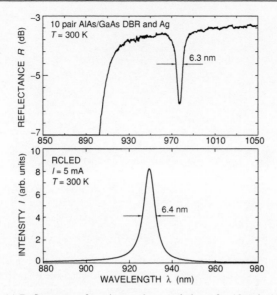

Figure 3.14. (a) Reflectance of a microcavity consisting of a 10-pair AlAs/GaAs distributed Bragg reflector and an Ag reflector. (b) Emission spectrum of a resonant-cavity light-emitting diode consisting of a 10-pair AlAs/GaAs distributed Bragg reflector and an Ag reflector.

condition $R_1 < 90\%$. Thus the mirror reflectivity of RCLEDs must be much lower than that of VCSELs. A high reflectivity would increase self-absorption and decrease the light output of the device as discussed earlier. De Neve et al. [49] used an extensive theoretical model to calculate the mirror reflectivity. The maximum efficiency was calculated at a reflectivity of $R_1 = 50$ to 60%.

The reflection and emission properties of the RCLED are shown in Figure 3.14(a) and (b). The reflection spectrum of the RCLED (Figure 3.14(a)) exhibits a highly reflective band for wavelengths >900 nm and a dip of the reflectivity at the cavity resonance. The spectral width of the cavity resonance is 6.3 nm. The emission spectrum of an electrically pumped device, shown in Figure 3.14(b), has nearly the same shape and width as the cavity resonance.

In conventional LEDs, the spectral characteristics of the devices reflect the thermal distribution of electrons and holes in the conduction and valence bands. The spectral characteristics of light emission from microcavities are as intriguing as they are complex. However, restricting our considerations to the optical axis of the cavity simplifies the cavity physics considerably. If we assume that the cavity resonance is much narrower than the natural emission spectrum of the semiconductor, then the on-resonance luminescence is enhanced whereas the off-resonance luminescence is suppressed. The on-axis emission spectrum should therefore reflect the enhancement, that is, the resonance spectrum of the cavity. The experimental results shown in Figure 3.14 confirm this conjecture.

The figure of merit of LEDs used in optical fiber communication systems is the photon flux density emitted from the diode at a given current, which, for a given wavelength, will be discussed in terms of microwatts per steradian. Note that the optical power coupled into a fiber is directly proportional to the photon flux density.

The intensity of an RCLED as a function of the injection current is shown in Figure 3.15. For comparison, we also show the calculated intensity of the *ideal isotropic emitter*, which is a hypothetical device. The ideal isotropic emitter is assumed to have an internal quantum

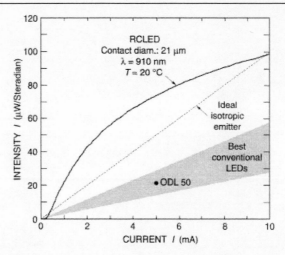

Figure 3.15. Light-versus-current curves of a resonant-cavity light-emitting diode and of the *ideal isotropic emitter*. The shaded region shows the intensities of the best commercial LEDs. The ODL 50 is a commercial communication LED sold by AT&T.

efficiency of 100% and the device is assumed to be clad by an antireflection coating providing zero reflectivity ($R = 0$) for all wavelengths emitted from the active region. If the photon emission inside the semiconductor is isotropic, then the optical power per unit current per unit solid angle normal to the planar semiconductor surface is given by

$$\frac{P_{\text{optical}}}{\Omega} = \frac{1}{4\pi n^2} \frac{2\pi \hbar c}{e\lambda},$$ (3.54)

where Ω represents the unit solid angle, n is the refractive index of the semiconductor, c is the velocity of light, e is the electronic charge, and λ is the emission wavelength in vacuum. Equation (3.54) is represented by the dashed line in Figure 3.15. Neither the 100% internal quantum efficiency nor such a hypothetical antireflection coating can be reduced to practice for fundamental reasons. Therefore, the ideal isotropic emitter represents an upper limit for the intensity attainable with any conventional LED. In fact, the best conventional LEDs have intensities lower than that of the ideal isotropic emitter. Also included in Figure 3.15 is the state-of-the-art ODL 50 GaAs LED used for optical fiber communication. All devices shown in Figure 3.15 have planar light-emitting surfaces, and no lensing is used.

Figure 3.15 reveals that the RCLED provides unprecedented intensities in terms of both absolute values and slope efficiencies. The slope efficiency is 7.3 times the efficiency of the best conventional LEDs and 3.1 times the calculated efficiency of the ideal isotropic emitter. At a current of 5 mA, the intensity of the RCLED is 3.3 times that of the best conventional LEDs including the ODL 50. The unprecedented efficiencies make the RCLED promising for optical interconnect and communication systems.

The higher spectral purity of RCLEDs reduces chromatic dispersion in optical fiber communications. The chromatic dispersion is directly proportional to the linewidth of the source. Since RCLEDs have linewidths 5–10 times narrower than conventional LEDs, the chromatic dispersion effects, which dominate at wavelengths of 800–900 nm, are reduced as well. Hunt et al. [35] showed that the bandwidth of an RCLED is a factor of 5–10

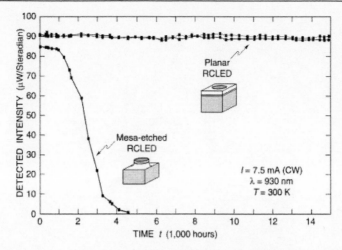

Figure 3.16. Room temperature reliability traces of two planar RCLEDs and one mesa-etched RCLED.

higher than that of conventional LEDs. Subsequently, Schubert et al. [17] demonstrated high modulation capability of RCLEDs. Eye diagram measurements with a random bit pattern generator revealed wide open eyes at 622 Mbit/sec.

Reliability measurements were also conducted at 300 K with planar and mesa-etched RCLEDs. The light output versus time curves of three devices driven by 7.5 mA (CW) are shown in Figure 3.16. Planar RCLEDs exhibit excellent reliability over a period of more than twenty months. The differences in reliability between planar and mesa-etched RCLEDs are due to surface recombination processes in mesa-etched devices [17].

It is important to note that the semiconductor RCLEDs are just one of potentially many applications for microcavity devices. These applications include microcavity detector structures [50] and Er-doped structures. The luminescence lines in Er-doped structures are particularly narrow, thus matching very well the resonance linewidth of high-finesse microcavities. Er-doped microcavities will be discussed below.

3.5.2 Er-Doped Microcavities

The rare earth element erbium (Er) is, as are many other rare earths, an optically active element. The main optical transition of Er occurs in the infrared spectral range at a wavelength of 1.55 μm, which coincides with the minimum attenuation wavelength of silica fibers. Because the radiative efficiency of Er in a silica host can be close to 100% [51], Er-doped silica fibers can be used as high-gain, low-noise optical amplifiers in the low-loss optical communication window at 1.55 μm [52].

The planar microcavities discussed here consist of a first Si/SiO$_2$ distributed Bragg reflector, a SiO$_2$ active region, and a second Si/SiO$_2$ DBR. The layers are grown by radio-frequency (rf) magnetron sputtering. The deposition rate was 5 Å/s and 30 Å/s respectively. The layer sequence of a cavity is shown in Figure 3.17 along with the Er-implantation profile. The bottom and top reflectors consist of 4 and $2\frac{1}{2}$ pairs of Si and SiO$_2$ quarter-wave layers. For a resonance wavelength of $\lambda = 1.55$ μm, the thicknesses of the Si and SiO$_2$ DBR layers are 1,150 Å and 2,700 Å, respectively.

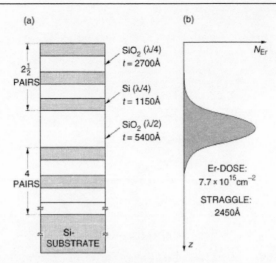

Figure 3.17. (a) Schematic illustration of a Er-doped resonant microcavity structure consisting of a four-pair bottom Si/SiO₂ DBR, an Er-doped SiO₂ active region, and a 2.5-pair top Si/SiO₂ DBR. (b) Calculated Er implantation profile.

After the growth of a full Si/SiO$_2$ microcavity, a 1 cm^2 area is implanted with Er using a dose of 7.7×10^{15} cm^{-2} and an implantation energy of 3.55 MeV. The projected range of the implant is 1.55 μm, so that the maximum of the Er concentration occurs in the center of the SiO$_2$ active region. The projected straggle of the implant is 2,450 Å. Postimplantation anneals at 700–900°C for 30 min are employed to optimize the radiative efficiency of Er.

The material system Si/SiO$_2$ is characterized by a large optical refractive index difference between the Si ($n \approx 3.4$) and the SiO$_2$ ($n \approx 1.5$) at a wavelength of 1.55 μm. Thus, only a few pairs of a Si/SiO$_2$ DBR are required for a high reflectivity mirror. The calculated peak reflectivities of the bottom and top reflectors are 99.8% and 98.5%, respectively.

The inherently narrow spectral linewidth of the Er intra-atomic transition makes Er an attractive candidate for resonant-cavity devices. The comparison of photoluminescence spectra measured with and without a resonant cavity is shown in Figure 3.18. The cavity structure (spectrum labeled "with cavity") is resonant at 1.54 μm. The no-cavity structure (spectrum labeled "without cavity") has just a bottom mirror but no top mirror. The top reflector was removed by selectively etching off the top distributed Bragg reflector. Comparison of the two photoluminescence spectra reveals that the peak intensity of the resonant cavity structure is a factor of 57 higher than the peak intensity of the no-cavity structure. The enhancement of the luminescence intensity has been studied on different samples and the intensity was found to be typically one to two orders of magnitude higher as compared to a no-cavity structure [10]. Note that the no-cavity structure has a bottom reflector that also enhances the photoluminescence intensity of that structure by a factor of two. Thus the enhancement of the cavity structure would be a factor of $2 \times 57 = 114$ when compared to a structure without any reflector.

The change in emission intensity is not due to different excitation conditions of the structures with and without a cavity. First, some of the exciting light ($\lambda = 980$ nm) is absorbed in the Si layers of the top mirror. The top mirror therefore causes a *weaker* excitation of the active region in the cavity as compared to the structure without a top reflector. Second, the

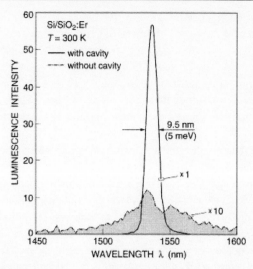

Figure 3.18. Room temperature photoluminescence spectrum of Er-doped SiO_2 with and without a resonant cavity.

excitation wavelength ($\lambda = 980$ nm) is far off resonance. Therefore the change in emission characteristic is entirely attributed to the effect of the resonant cavity.

In addition to the change in luminescence intensity, Figure 3.18 also reveals a significant change in the spectral purity of the Er emission. The photoluminescence has a clean, symmetric line shape, with a FWHM of 5 meV (9.5 nm). The Er emission spectrum of the structure without a cavity is broader, typically 20 meV wide, and has a double-peak structure. The change in spectral characteristics indicates that near normal emission is determined by the characteristics of the cavity and not by the inhomogeneously broadened emission of Er-doped SiO_2.

3.5.3 Emission Lifetimes

The spontaneous emission characteristics of an optically active medium located in a microcavity differ significantly from those without a cavity. The fundamental physical basis of these differences are the changes in the spontaneous emission probability. The probability of a spontaneous emission event in one atom per unit time is called the *emission rate* of that medium. The inverse of the emission rate is the *spontaneous emission lifetime* of the medium. For Er-doped SiO_2, typical emission lifetimes are 10 ms. We will next discuss and present experimental results on the changes of the spontaneous emission lifetime in Er-doped Si/SiO$_2$ microcavities, which have been first published by Vredenberg et al. [33].

After cavity effects at radio frequencies were first proposed by Purcell [5]. Goy et al. [53] demonstrated that confocal resonators can indeed drastically modify the spontaneous emission lifetime of an atom within a resonator of millimeter wavelengths. This concept has been extended to short planar cavities at optical wavelengths, where experiments have demonstrated lifetime changes by using flowing dye [6], dye-containing films [7], and semiconductors [9] as an active medium. None of these experimental results in planarcavities, however, was quantitatively compared to theoretical models. Nonradiative energy loss processes, pump-dependent spectra, and the effects of self-absorption of emitted light by active media can affect the measured lifetimes. In the experiments discussed here, the spontaneous lifetime changes of the $^4I_{13/2} \rightarrow {}^4I_{15/2}$ transition of Er^{3+} ions (emission wavelength $\sim 1,535$ nm) are discussed. The Er ions are implanted at low concentration into a thin (half-wavelength) SiO_2 film, surrounded by high-reflectivity Si/SiO_2 planar distributed Bragg reflectors. This material system is ideal for measuring cavity-induced lifetime changes because of the small self-absorption of the Er^{3+} ions and the narrow atomic emission spectrum.

The cavity is expected to change the total emission rate of the excited Er^{3+} ions, which can be probed with luminescence decay measurements. Thus for a cavity with λ_{res} larger than the average emission wavelength, the emission rate should increase, characterized by a shorter excited state lifetime, and vice versa for λ_{res} at shorter wavelengths. We first demonstrate this for the latter case. Time-resolved luminescence measurements after pulsed excitation were performed by mechanically chopping the exciting laser beam and monitoring the signal on a digital LeCroy 9410 oscilloscope. In Figure 3.19, the luminescence from a structure before top mirror deposition (curve a) shows single exponential decay, characterized by a lifetime of 9.5 ms. Deposition of the top DBR results in a 1.44 μm cavity, and the luminescence decay of this structure (curve b) has a lifetime of 13.7 ms, indeed longer than in the no-cavity structure. Finally, the top mirror was removed through selective wet chemical etching. The lifetime of this structure is 9.5 ms (curve c), comparable to the (similar) situation before the deposition of the top mirror. Since photon reabsorption effects can be excluded, the observed lifetime enhancement is indeed a real cavity effect and reflects a modified transition probability from the excited to the ground state due to a different coupling of the excited ion with light-wave modes in the cavity.

Vredenberg et al. [33] also compared the experimental lifetimes of Er-doped Si/SiO_2 microcavities with calculated lifetimes. The method employed by the authors is based on

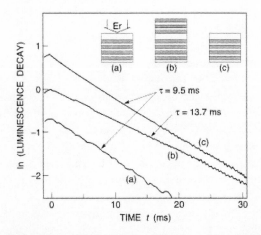

Figure 3.19. Radiative decay of Er (a) without, (b) with, and (c) again without a cavity. The measurements are performed (a) before deposition of the top reflector, (b) after deposition of the top reflector, and (c) after removal of the top reflector by etching. The cavity structure is resonant at 1.44 μm and hence inhibits a spontaneous emission from the Er-doped SiO_2 active region.

Fermi's Golden rule and is applicable to any low-absorption structure. The method also includes waveguiding modes. The calculated lifetimes were in excellent agreement with the experimental lifetimes.

We finally note that microcavities not only alter the spontaneous emission properties but also the absorption properties of an active material inside the cavity. This concept has been demonstrated with Er-doped cavities that are resonant with an absorption band of Er, namely at 980 nm [54].

3.6 Other Microcavity Structures

3.6.1 Multidimensional Photon Confinement

One can extend the concept of one-dimensional optical confinement to confinement in three-dimensions. The purpose of such structures might be to create new laser structures or to try to control the spontaneous emission of a device in three dimensions instead of just one.

If a planar microcavity is etched into a thin circular pillar, the high index change of the side walls will produce well-defined 3-D modes within the cavity. A thin pillar could theoretically emit a large fraction of its light into a single longitudinal mode, which would be advantageous for fiber coupling and optical communications. We present the theory behind such a device and discuss its usefulness.

Other approaches to enhanced spontaneous emission involve two- or three-dimensional photonic bandgap structures. We will discuss the possibility of using a 2-D array of pillars or holes to either suppress or enhance lateral emission. Another promising device to include is the photon-recycling LED, which, instead of suppressing lateral emission from an LED, absorbs the light and has a chance to reemit it in the desired direction. We will also mention microdisk lasers and RCLEDs made from a variety of material structures.

3.6.2 Emission Modes from a Dielectric Cylinder

The effect of a high-finesse, short-length, planar microcavity is to fix the wavevector of the light in the direction normal to the layer planes. The normal emission could then be described as consisting of a single longitudinal cavity mode. All the emitting wavelengths within some range of angles about the normal will then have approximately the same wavevector $k_\perp$ (equal to $n_c k_0$ where n_c is some averaged index of refraction of the cavity). The wavevectors parallel to the layer planes, however, are under no constraint. The total k vector of the resonance condition will increase with angle, resulting in a shift of the emission spectrum to shorter wavelength with angle from the normal.

If one wishes to couple the light from an LED into a single-mode fiber, or if one wishes to suppress the emission of light parallel to the layer planes, a logical step would be to consider a structure such as a pillar to provide lateral confinement of the light. The simplest of these structures from a theoretical point of view is a planar microcavity, with a cylindrical pillar etched from it. A cylinder by itself has well-defined lateral spacial modes, with specific lateral k-vectors. If the cylinder is etched from a planar microcavity, such a pillar would constrain the total k-vector into well-defined mode solutions for emitting angles close to the normal. We will first consider just the spontaneous emission along a cylindrical waveguiding pillar.

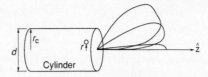

Figure 3.20. A depiction of emission lobes from a dielectric cylinder. The main lobe will have cylindrical symmetry, while the higher-order modes will have variation with rotation angle around the z axis. Consider the emitting atom being at some radius r from the center of the cylinder with diameter d. Emission will also occur out the sides of the cylinder and, at higher-order modes, out the ends. A thinner cylinder will have fewer modes.

A diagram of a cylinder and the emission from a number of modes is given in Figure 3.20. We can use Fermi's golden rule to determine the emission rate into all spatial modes j. The electric field profile of each mode of the cylinder throughout all positions r in a cross-section is denoted by $U_j(r)$. The total emission rate for the cylinder from a dipole at position r_a is,

$$A_{\text{cyl}}(\vec{r}_a) = A_a \sum_{j,\text{pol}} \frac{2\pi}{k_0^2} \frac{dk_{\beta,j}}{dk_0} \left[\frac{3}{2n_c} \varepsilon_0 M_j^2 |\vec{\mu} \cdot \vec{U}_j(\vec{r}_a)|^2 \right], \tag{3.55}$$

where the normalization constant for the modes is

$$M_j^2 = \frac{1}{\iint \varepsilon(r) |\vec{U}_j(r, \phi)|^2 \, dr \, d\phi}. \tag{3.56}$$

This formula includes emission in both the positive and negative z directions. The summation is over the spatial modes, and over both orthogonal polarizations. We redefine the value $k_{\beta,j}$ to mean the k-vector along the cylinder (z direction). The mode solutions can be found in references dealing with waveguiding in fibers [56]. The fundamental linearly polarized solution has two possible orthogonal polarizations (along x or y). The solution for polarization along the y axis is

$$\vec{U}_{0,\hat{y}} = J_0\left(\frac{ur}{d}\right)\hat{y}, \qquad u = r_c\sqrt{n_c^2 k_0^2 - k_{\beta 0}^2}. \tag{3.57}$$

Here J_0 means the Bessel function of the same name. For the fundamental mode, one can simplify the expression for A_{cyl} when the cylinder radius r_c is large enough to support at least a few modes. In this case, the value $dk_{\beta 0}/dk_0 \approx n_c^2 k_0 / k_{\beta 0} \approx n_c$. The value u approaches 2.405 for large cylinders (u approaches 3.83 for the second spatial mode). The value M_0^2 becomes $7.4218 g_1/(2\pi r_c 2n_c^2 \varepsilon_0)$, where g_1 describes the extent of the effective area extent of a wavefunction fully confined within the cylinder compared to the real one ($g_1 = 1$ for large cylinders, $1 \geq g_1 > 0$). If we assume a random dipole orientation, we get

$$\beta_{\text{cyl,pol}} = \frac{A_0(r_a)}{A_{\text{cyl}}(r_a)} = \frac{7.4218 g_1}{2r_c^2 n_c^2 k_0^2} J_0^2\left(\frac{ur_a}{a}\right) \frac{\tau_{\text{cyl}}}{\tau_0}. \tag{3.58}$$

Here $\beta_{\text{cyl,pol}}$ is the fractional spontaneous emission factor into the fundamental linearly polarized mode, into one of the two polarizations. The ratio τ_{cyl}/τ_0 is the spontaneous emission lifetime compared to a bulk material. For atoms spread out uniformly across the cylinder one can simplify the β value even further. This becomes

$$\beta_{\text{cyl,pol}} \approx \frac{g_1}{g_2} \frac{1}{r_c^2 n_c^2 k_0^2} \frac{\tau_{\text{cyl}}}{\tau_0}. \tag{3.59}$$

The value g_2 where $1 > g_2 > g_1 > 0$ is a new factor, but the ratio g_1/g_2 can now be thought of as the lateral confinement factor for the mode, which is equal to 1 for large cylinders.

Even though the fundamental lateral spatial mode is defined, the longitudinal k-vector is not constrained, meaning the emission spectrum contains approximately the same spectrum as the bulk emitter. These modes will exhibit some dispersion, with longer wavelengths emitting at larger angles.

3.6.3 Three-Dimensional Confinement: The Dielectric Pillar

We now combine the planar microcavity and the dielectric pillar to achieve three-dimensional confinement of the light, at least at angles close to the axis of the cylinder. Imagine the cylindrical device in Figure 3.20, but with the addition of planar dielectric mirrors at each end, forming a microcavity pillar. Such a device could be etched from a planar microcavity structure.

The total k-vector is well defined for the fundamental spatial mode. There will, of course, be a spread of wavelengths for a given mode owing to the finite Q of the microcavity. For a narrow enough pillar, with high enough Q, the higher order spacial modes will be spectrally separate from the main mode, that is, they will be shorter wavelength features. If one makes a pillar small enough that the first-order spatial mode is at a wavelength shorter than that emitted by the natural spectrum, then we will have succeeded in achieving a significant amount of spontaneous emission into a single spatial mode. This would be important where single-mode emission is required. The fraction of spontaneous emission into this fundamental mode can be referred to as the β factor.

The spontaneous emission factor β is approximately given by multiplying the microcavity power enhancement factor $I_{\text{avg}}(\theta_1 \approx 0)$ by β_{cyl} for the cylinder. This means that

$$\beta = 2\beta_{\text{pol}} \approx 2\beta_{\text{cyl,pol}} I_{\text{avg}}(\theta_1 \approx 0). \tag{3.60}$$

For the specific case of a randomly oriented dipole, where the atoms are evenly distributed across the cross section of the pillar, and where the natural spectrum is a Lorentzian shape, the spontaneous emission factor for a single polarization becomes

$$\beta_{\text{pol}} \approx \zeta \frac{g_1}{g_2} \frac{\tau_{\text{pillar}}}{\tau_0} \frac{\lambda_0^2}{4\pi^2 n_c^2 r_c^2} \frac{\lambda}{\pi L_{\text{cav}}} \frac{\lambda_0}{\Delta \lambda_{N,\text{Lorentz}}}. \tag{3.61}$$

The total β factor for both polarizations is just double this value. For a Gaussian natural spectrum, replace $1/(\pi(\Delta\lambda_{N,\text{Lorentz}}))$ by $(\ln(2)/\pi)^{0.5}/(\Delta\lambda_{N,\text{Gauss}})$. When ζ and g_1/g_2 and $\tau_{\text{pillar}}/\tau_0$ are all equal to one, this Eq. 3.61 becomes the well-known one [57] for the average emission fraction per mode of an arbitrary emitting volume.

Consider one of our InGaAs/GaAs/AlGaAs RCLED devices, designed so that the fundamental spatial mode occurred at the peak of the natural spectrum. If we design the pillar narrow enough such that the second spatial mode occurs at a wavelength 30 nm shorter than the fundamental, where the natural spectrum is weak, then almost no emission will occur into the second spatial mode. One does have to worry about other longitudinal modes overlapping our natural spectrum (since our cavity has an effective length of 2.7λ), and at high emission angles, the DBR mirrors of our cavity don't have good reflectivity any more. This highlights the fact that we will never be able to suppress high-angle modes. The best we can do is to optimize the emission from the fundamental mode.

To achieve a 30 nm shift in the second-order mode compared to the first, one would need a pillar radius of $r_c = 0.5\,\mu$m. Even assuming a confinement factor of about 1 and ξ of 1.3, the spontaneous emission factor β is only 0.1. Once the pillar diameter drops below $\lambda/2$, the confinement factor and effective mirror reflectivities drop dramatically, making $r_c \approx 0.125\,\mu$m an approximate lower limit for size. We cannot estimate the coefficients well enough to make a meaningful estimate of β at this size.

Fabricating such an electrically pumped semiconductor LED at this small size would be virtually impossible with current technology. The optical scattering at the edges of the etched pillar, the problems of making electrical contact, electrical passivation of the semiconductor active region side-walls, and the difficulty in removing heat are only some of the obstacles. One might consider using metalized pillars to preserve the lateral confinement factor and provide suppression of lateral modes, thus enhancing the β of the fundamental mode. Unfortunately, metal has a skin depth, and the resulting absorption would drastically reduce the efficiency of the cavity. The dielectric microcavity pillar by itself does little to suppress lateral emission; it just helps enhance the desired single-mode emission and reduces the total number of emitting modes because of the small device volume.

3.6.4 Thresholdless Lasers

There has been a number of papers describing a thresholdless laser as a possible communications device [6]. This would essentially be an LED that emitted most or all of its light into the fundamental spatial cavity mode. In this way, one can achieve the lasing condition of there being one photon in the cavity almost without gain. If the β of the LED was almost 1, then the light versus current curve of the device would be linear and indistinguishable from a laser with no threshold. A comparison of the light versus current for a conventional laser, a high-β laser, and a truly thresholdless laser are given in Figure 3.21. If one were to envision a thresholdless device as fabricated from a microcavity pillar, there are a number of problems besides the ones of fabrication. Simply having a good subthreshold intensity, giving only a small kink in the $L-I$ curve between subthreshold and lasing, is not good enough for a thresholdless laser. The output modulation of such a device would be extremely slow below threshold compared to the lasing regime. For there to be lasing, there must be some gain, because the β will never be 1. For there to be gain, we must pump the active region past transparency first, since it is absorbing under no pump. This gain condition requires a certain carrier density, which will, in fact, determine the threshold.

One can imagine a device such that all other optical modes but the fundamental are highly suppressed, and the internal quantum efficiencies are nearly perfect. In this case, with even a small current, the carrier densities within the device would slowly build until the active region became transparent, and some light could get out through the fundamental optical mode. Such a device would have a very small threshold, but one would have to be careful not to ever drive the device below its threshold even momentarily, because of the long time required to build the necessary space charge inside the device for transparency. Also, if the fundamental emission was suppressed at low carrier densities, it is unlikely that the other modes would be suppressed by an even higher degree, which is what would be required to

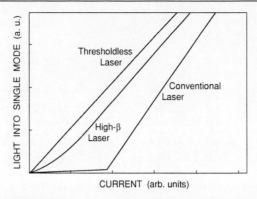

Figure 3.21. Qualitative light-output versus current curves for the single spatial-mode emission from a conventional laser, a high β-factor laser, and a thresholdless laser. The conventional laser has a distinct current threshold. The high β-factor laser has a threshold, but it is much less distinct in the I–V curve. It would be noticeable in the spectrum and device modulation speed, however. A theoretical thresholdless laser would have a β close to 1 and would somehow suppress all other lossy emission until the carrier density required for gain (or at least transparency) was achieved.

avoid losing the injected carriers. In fact, it is the difficulty in suppressing any side emission that makes semiconductor thresholdless lasers impractical.

One could, of course, make a laser with a very small threshold by making it very small. A very small device would have a very small threshold. The problem is that such a device would have a very small power output also. For a hypothetical 4-quantum-well single-mode LED, which does not rely on gain but achieves a β close to 1, the current densities cannot be more than about 50 $\mu A/\mu m^2$. Using the formulas for β for a dielectric pillar, and this current density, one can find that less than 2 μW could be emitted into a single mode for an RCLED structure such as ours. This is clearly too small for high-speed communications in the 100–650 Mbit/sec regime, where 10 μW would be considered a minimum for an emitter. An RCLED with a very short carrier lifetime owing to extreme enhancement of the desired emission could have higher pump currents, but no semiconductor design can do this. If the device is a laser, higher powers could be drawn because of the reduced carrier lifetime, but the device sizes are incredibly small for real fabrication or for any sort of heat dissipation.

The conclusion to this is that semiconductor RCLEDs should remain devices larger than about 5 or 10 microns in diameter, and they should remain as multimode emitters. Semiconductor vertical-cavity lasers are best fabricated at sizes that make sense technologically, without regard to the β factor of the spontaneous emission. A high-β laser may not be desirable anyway, since the spontaneous emission will introduce excess noise into the emission.

3.6.5 Large-Area Photon Recycling LEDs

Instead of trying to confine the lateral emission, one could instead try to recycle it. By reabsorbing much of the lateral emission, which doesn't exit the face of a planar device, one can recapture the energy in the active region, giving it another chance to emit in the desired direction. We will present examples of such a device by two different groups.

The first example of such a device is an optically pumped semiconductor structure by Schnitzer et al. [58]. Consider the structure as given in part (a) of Figure 3.22. The back side

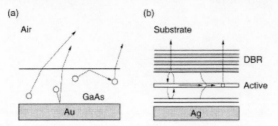

Figure 3.22. Two approaches to photon recycling LEDs. (a) A bulk semiconductor epilayer is placed on top of gold. Most spontaneous emission that does not escape into air is reabsorbed and has a chance to emit again. (b) A semiconducor quantum-well microcavity is designed with a waveguiding active region. The waveguiding light is reabsorbed after some tens of microns and will have a chance to re-emit out of the device.

of a thin semiconductor layer is coated with gold. If one optically pumps the sample at low intensities, the active region remains absorbing. Most of the emission (95%) that hits the semiconductor/air interface will totally internally reflect and stay within the semiconductor. The gold is also a good reflector at infrared wavelengths. The semiconductor is still absorbing at the emission wavelength, and so after some average absorption length L_{abs} the trapped light has a chance to reemit. This device exhibited a 72% external efficiency, which required a 99.7% internal quantum efficiency. The fabrication of such a device required the active epilayer to be etched and floated off of its substrate. An electrical device fabricated in such a manner would have reliability problems, but a more practical photon recycling device has been made by Blondelle et al. [59].

The concept of photon recycling was used by DeNeve et al. [49] in an InGaAs/GaAs/AlGaAs RCLED of similar layer structure to our devices. A simplified diagram of their structure is shown in part (b) of Figure 3.22. If the graded-composition carrier confinement region is thick enough around the quantum wells, it also acts as a graded-index waveguide. DeNeve et al. calculated that 30% of the light emitted by the active region goes into this waveguide mode. This will get reabsorbed after some tens of microns, allowing the photons another chance to re-emit out the top of the structure. In this way, about a 30% increase in external quantum efficiency can be attained. The active region of our RCLEDs were not thick enough to support a strongly guided mode. One might think that a strong waveguiding mode would steal power away from the normal emission, but this does not appear to be the case. The energy tends to be stolen from other high-angle modes instead, meaning that just modifying the waveguiding mode does little to change the external efficiency. Making use of this high-angle light by achieving photon recycling of the waveguiding emission is an attractive option.

One drawback to photon recycling is the that it requires a device with large enough diameter that multiple reabsorption events can occur within the emitting area of the device. This makes them less attractive for fiber-optic communications, where small diameters couple better to fibers, especially when coupling lenses are used. Another drawback is that self-absorption necessarily increases the lifetime of the spontaneous emission, slowing down the maximum modulation rate of the devices.

In fact, the DeNeve device is shown to work at highest efficiencies at large diameters anyway, and they are not targeting it for fiber communications. At high current densities, the carrier confinement in the quantum wells is reduced, and efficiency decreases, and so a larger device tends to be more efficient. Also, their device is resonant to the long wavelength side of the natural emission peak, rather than at the peak. This shifts the maximum intensity of the main emission lobe to some angle, rather than being on axis, but it maximizes the total amount of emission into the main lobe. The Blondelle and DeNeve device is therefore designed for maximization of total emission from the top of the device, rather than for the specific needs of fiber coupling. Display devices or free-space communications devices could benefit from this approach. They achieved a maximum of 16% external quantum efficiency, as compared to the theoretical 2% for an ideal bulk emitter of the same index.

3.6.6 Other Novel Confined Photonic Emitters

A complete discussion of other emitting structures that confine the photons is outside the scope of this chapter. It is useful, however, to discuss the properties of other confined-photon emitters.

One such structure is the microdisk laser [27], which is fabricated as a thin dielectric disk that couples light out the edges of the disk. Lasing modes can be described by a mode number M, where $\exp(iM\phi)$ is the form of the electric field around the cylindrical disk. Because waves can propagate both ways, M can be positive or negative. The disk can be fabricated with a thickness such that the emission perpendicular to the disk is suppressed. Small disks will only support a small number of modes, and therefore they can have a high spontaneous emission factor β. The Q of these modes are also high enough to achieve lasing. One attractive aspect to such disks is that the lasing emission occurs in the plane of the sample, from a very small device. This could be useful for integration of photonic devices on a single wafer. However, the output is difficult to efficiently couple into waveguides and fibers, as it only couples evanescently. Advances have been made to improve the longevity, operating temperature range, and active-region passivation of such devices [60]. Room-temperature CW electrical pumping remains a problem, however.

Photonic bandgap structures may someday yield practical photonic devices. These structures involve 2-D or 3-D photon confinement by periodic patterning of the emitting material. An example of a 2-D confined structure is given by Baba et al. [61]. Such a structure could consist of a series of rods or holes in a regular pattern, such as a hexagonal close-packed array. An example of this is given in Figure 3.23. The periodicity of the array creates a bandgap for lateral emission at certain emission energies and one or both polarizations. By suppressing the lateral emission, the rod structure will have a large bandgap for TM emission and a smaller gap for TE emission but not at the same emission energies. However, if the emitting region had a dipole oriented mainly along the rods (such as quantum-well electron/light–hole recombination), the lateral emission could be efficiently suppressed. The structure with the patterned holes will have a smaller bandgap than that of the rod structure, but it has the advantage that it can have a true bandgap for both polarizations of light. By themselves, or combined with a planar microcavity, longitudinal emission could theoretically be enhanced.

3.6.7 Other Novel RCLED Materials and Devices

It is likely that the future of microcavity LEDs and confined-photon emitters will involve new materials and applications from those currently used in communications and display systems. Confined photonic spontaneous emitting devices will always have competition from lasers, conventional LEDs, and other forms of display devices. It is only a matter of time, however, before the right materials combination at the right wavelengths, for the right

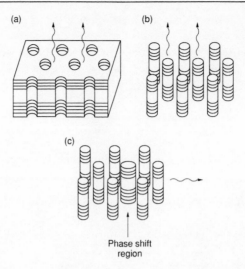

Figure 3.23. An example of two-dimensional photonic bandgap structures. Semiconductors with a regular hexagonal array of holes or pillars can possess bandgaps, suppressing lateral emission at certain wavelengths. This effect could be utilized to improve the performance of vertically emitting RCLEDs. As an alternative, by introducing a phase-shift region or defect into the lattice, the lateral emission can be enhanced, giving a two-dimensional microcavity.

application, will make comercial devices a reality. It is therefore worthwhile to mention recent confined-photonic LED materials systems.

Wilkinson et al. [46] have fabricated an AlGaAs/GaAs thin film emitter with metal mirrors on silicon substrate, emitting at 880 nm. Pavessi et al. [62] have fabricated porous silicon RCLEDs at 750 nm wavelength. Fisher et al. [63] have investigated a conjugate polymer RCLED designed for 650 nm. Hadji et al. [64] have realized a CdHgTe/HgTe RCLED operating at 3.2 μm wavelength.

A structure of particular note [65] is a GaAs/Al$_x$O$_y$ RCLED operating at 950 nm, where the aluminum oxide in the output mirror has been converted from AlAs by oxidation. This composition of output mirror allows the effective cavity length to remain as small as possible, maximizing the possible output enhancement in a semiconductor RCLED. The back mirror is Ag, as with our design. Another particularly interesting device is the broadly tunable RCLED by Larson et al. [66]. The top mirror is a deformable membrane that can be moved by electrostatic forces. Tunable emission is shown from a range of 938 nm to 970 nm.

References

[1] Einstein A. Z., *Physics*, **18**, 121, 1917.

[2] Van Roosbroeck W. and Shockley W. "Photon-radiative recombination of electrons and holes in germanium," *Phys. Rev.*, **94**, 1558, 1954.

[3] Hayashi I., Panish M. B., Foy P. W., and Sumski S., *Appl. Phys. Lett.*, **17**, 109, 1970.

[4] Hall R. N., Fenner G. E., Kingsley J. D., Soltys T. J., and Carlson R. O., *Phys. Rev. Lett.*, **9**, 366, 1962.

[5] E. M. Purcell, *Phys. Rev.*, **69**, 681, 1946.

[6] De Martini F., Innocenti G., Jacobovitz G. R., and Mataloni P. *Phys. Rev. Lett.*, **59**, 2955, 1987.

[7] Suzuki M., Yokoyama H., Brarson S. D., and Ippen E. P., *Appl. Phys. Lett.*, **58**, 998, 1991.

[8] Yablonovitch E., Gmitter T. J., and Bhat R. *Phys. Rev. Lett.*, **61**, 2546, 1988.

[9] Yokoyama H., Nishi K., Anan T., Yamada H., Boorson S. D., and Ippen E. P., *Appl. Phys. Lett.*, **57**, 2814, 1990.

[10] Schubert E. F., Vredenberg A. M., Hunt N. E. J., Wong Y. H., Poate J. M., Jacobson D. C., Feldman L. C., and Zydzik G. J., "Giant enhancement in luminescence intensity in Er-doped Si/SiO$_2$ resonant cavities," *Appl. Phys. Lett.*, **61**, 1381, 1992.

[11] Nakayama T., Itoh Y., and Kakuta A., "Organic photo- and electroluminescent devices with double mirrors," *Appl. Phys. Lett.*, **63**, 594, 1993.

[12] Dodabalapur A., Rothberg L. J., and Miller T. M., "Color variation with electroluminescent organic semiconductors in multimode resonant cavities," *Appl. Phys. Lett.*, **65**, 2308, 1994.

[13] Schubert E. F., Wang Y. H., Cho A. Y., Tu L. W., and Zydzik G. J., "Resonant cavity light emitting diode," *Appl. Phys. Lett.*, **60**, 921, 1992.

[14] Hunt N. E. J., Schubert E. F., Logan R. A., and Zydzik G. J. "Enhanced spectral power density and reduced linewidth at 1.3 μm in an InGaAsP quantum well resonant cavity light emitting diode," *Appl. Phys. Lett.*, **61**, 2287, 1992.

[15] Deppe D. G. and Lei C., "Spontaneous emission and optical gain in a Fabry–Pérot microcavity," *Appl. Phys. Lett.*, **60**, 527, 1992.

[16] J. L. Jewell, Y. H. Lee, S. L. Mc Call, J. P. Harbison, and L. T. Florez, "High-Finesse.(Al, Ga) As Interference Filters Grown by Molecular Beam Epitaxy," *Appl. Phys. Lett.*, **53**, 640–642, 1988.

[17] Schubert E. F., Hunt N. E. J., Malik R. J., Micovic M., and Miller D. L., "Temperature and modulation characteristics of resonant-cavity light-emitting diodes," *IEEE J. Lightwave Commun.*, **14**, 1721, 1996.

[18] Kobayashi T., Segawa T., Morimoto A., and Sueta T., *Meeting of the Jpn. Soc. of Appl. Phys.*, Tokyo, 1982; see also Ref. 20.

[19] Kobayashi T., Morimoto A., Sueta T. *Meeting of the Jpn. Soc. of Appl. Phys.*, Tokyo, 1985; see also Ref. 20.

[20] Yokoyama H., "Physics and device applications of optical microcavities," *Science*, **256**, 66, 1992.

[21] Yablonovitch E., personal communication, 1994.

[22] Yokoyama H. and Ujihara K. *Spontaneous Emission and Laser Oscillation in Microcavities*, CRC Press, Boca Raton, 1995.

[23] Numai T., Kosaka H., Ogura I., Kurihara K., Sugimoto M., and Kasahara K., "Indistinct threshold laser operation in a pnpn vertical to surface transmission electrophotonic device with a vertical cavity," *IEEE J. of Quantum Electron.*, **29**, 403, 1993.

[24] Fabry G. and Pérot A., *Ann. Chim. Phys.*, **16**, 115, 1899.

[25] Saleh B. E. A. and Teich M. C., *Fundamentals of Photonics*, Wiley, New York, 1991.

[26] Coldren L. A. and Corzime S. W., *Diode Lasers and Photonic Integrated Circuits*, Wiley, New York, 1995.

[27] Tu L. W., Schubert E. F., Kopf R. F., Zydzik G. J., Hong M., Chu S. N. G., and Mannaerts J. P., "Vertical cavity surface emitting lasers with semitransparent metallic mirrors and high quantum efficiencies," *Appl. Phys. Lett.*, **57**, 2045, 1990.

[28] Smith G. M., Forbes D. V., Coleman J. J., Verdeyen J. T., "Optical properties of reactive ion etched corner reflector strained-layer InGaAs-GaAs-AlGaAs quantum-well lasers," *IEEE Photonics Technology Lett.*, **5**, 873, 1993.

[29] McCall S. L., Levi A. F. J., Slusher R. E., Pearton S. J., and Logan R. A., "Whispering-gallery mode microdisk lasers," *Appl. Phys. Lett.*, **60**, 289, 1992.

[30] Yariv A., *Quantum Electronics*, 3rd ed., Wiley, New York, 1989.

[31] Björk G., Yamamoto Y., and Heitmann H., in *Confined Electrons and Photons*, Burstein E. and Weisbuch C., eds., Plenum Press, New York, 1995.

[32] Yariv A., *Theory and Applications of Quantum Mechanics*, Wiley, New York, 1982, p. 143.

[33] Vredenberg A. M., Hunt N. E. J., Schubert E. F., Jacobson D. C., Poate J. M., and Zydzik G. J., *Phys. Rev. Lett.*, **71**, 517, 1993.

[34] Schubert E. F., *Doping in III–V Semiconductors*, Cambridge University Press, Cambridge U.K., 1993, p. 512.

[35] Hunt N. E. J., Schubert E. F., Kopf R. F., Sivco D. L., Cho A. Y., and Zydzik G. J., "Increased fiber communications bandwidth from a resonant cavity light-emitting diode emitting at $\lambda = 940$ nm," *Appl. Phys. Lett.*, **63**, 2600, 1993.

[36] Schubert E. F., Hunt N. E. J., Micovic M., Malik R. J., Sivco D. L., Cho A. Y., and Zydzik G. J., *Science*, **265**, 943, 1994.

[37] Bjork G., Machida S., Yamamoto Y., and Igeta K., "Modification of spontaneous emission rate in planar dielectric microcavity structures," *Phys. Rev. A.*, **44**, 669–681, 1991.

[38] Deppe D. G. and Lei C., "Spontaneous emission from a dipole in a semiconductor microcavity," *J. Appl. Phys.*, **70**, 3443–3448, 1991.

[39] Brorson S. D., Yokoyama H., and Ippen E. P., "Spontaneous emission rate alteration in optical waveguide structures," *IEEE J. Quantum Electron.*, **27**, 1492–1499, 1990.

[40] Ujihara K., "Spontaneous emission and the concept of effective area in a very short optical cavity with plane-parallel dielectric mirrors," *Jpn. J. Appl. Phys.*, **30**, L901, 1991.

[41] Feng, X.-P., "Theory of a short optical cavity with dielectric multilayer film mirrors," *Opt. Commun.*, **83**, 162–176, 1991.

[42] Dowling J. P., Scully M. O., and De Martini F., "Radiation pattern of a classical dipole in a cavity," *Opt. Commun.*, **82**, 415–419, 1991.

[43] Hunt N. E. J., Vredenberg A. M., Schubert E. F., Becker P. C., Jacobson D. C., Poate J. M., and Zydzik G. J., "Spontaneous emission control in planar structures: Er^{3+} in Si/SiO_2 microcavities," in E. Burstein and C. Weisbuch, eds., *Confined Electrons and Photons*, Plenum Press, New York, pp. 715–728, 1995.

[44] Born M. and Wolf E., *Principles of Optics*, Pergamon Press, Oxford, 1980.

[45] Walpita L. M., "Solutions for planar optical waveguide equations by selecting zero elements in a characteristic matrix," *J. Opt. Soc. Am. A.*, **2**, 595–602, 1985.

[46] Wilkinson S. T., Jokerst N. M., and Leavitt R. P., "Resonant cavity enhanced thin film AlGaAs/GaAs/AlGaAs LEDs with metal mirrors," *Appl. Optics*, **34**, 8298, 1995.

[47] Schubert E. F., Tu L. W., Zydzik G. J., Kopf R. F., Benvenuti A., and Pinto M. R., "Elimination of heterojunction band discontinuities by modulation doping," *Appl. Phys. Lett.*, **60**, 466, 1992.

[48] K. L. Lear and R. P. Schneider Jr, "Unparabolic mirror grading for vertical cavity surface emitting lasers," *Appl. Phys. Lett.*, **68**, 605–607, 1996.

[49] DeNeve H., Blondelle J., Baets R., Demeester P., Van Daele P., and Borghs G., "High efficiency planar microcavity LEDs: Comparison of design and experiment," *IEEE Photonics Technol. Lett.*, **7**, 287, 1995.

[50] Ünlü M. S. and Strite S., "Resonant cavity enhanced photonic devices," *J. Appl. Phys.*, **78**, 607, 1995.

[51] Miniscalo W. J., "Erbium Doped Glasses for Fiber Amplifiers at 1500 nm," *J. Lightwave Technology*, **9**, 234–250, 1991.

[52] Desurvire E., *Erbium Doped Fiber Amplifiers*, Wiley, New York, 1994.

[53] Goy P., Raimond J. M., Gross M., and Haroche S., *Phys. Rev. Lett.*, **50**, 1903, 1983.

[54] Schubert E. F., Hunt N. E. J., Vredenberg A. M., Harris T. D., Poate J. M., Jacobson D. C., Wong Y. H., and Zydzik G. J. *Appl. Phys. Lett.*, **63**, 2603, 1993.

[55] Schubert E. F., and Hunt N. E. J., "15,000 hrs stable operation of resonant cavity light emitting diodes," *Appl. Phys. A.*, **66**, 319 1993.

[56] H.G. Unger, *Planar Optical Waveguides and Fibres*, Clarendon, Oxford, 1977.

[57] Baba T., Hamano T., Koyama F., and Iga K., "Spontaneous emission factor of a microcavity DBR surface-emitting Laser," *IEEE J. Quantum Electron.*, **27**, 1347–1358, 1991.

[58] Schnitzer I., Yablonovitch E., Caneau C., and Gmitter T. J., "Ultra-high spontaneous emission quantum efficiency, 99.7% internally and 72% externally, from AlGaAs/GaAs/AlGaAs double heterostructures," *Appl. Phys. Lett.*, **62**, 131–133, 1993.

[59] Blondelle J., De Neve H., Demeester P., Van Daele P., Borghs G., and Baets R., "16% External quantum efficiency from planar microcavity LED's at 940 nm by precise matching of the cavity wavelength," *Electron. Lett.*, **31**, 1286–1288, 1995.

[60] Mohideen U., Hobson W. S., Pearton J., Ren F., and Slusher R. E., "GaAs/AlGaAs microdisk lasers," *Appl. Phys. Lett.*, **64**, 1911–1913, 1993.

[61] Baba T., and Matsuzaki T., "GaInAsP/InP 2-dimensional photonic crystals," in Rarity J. and Weisbuch C., eds., *Microcavities and Photonic Bandgaps*, Kluwer Academic Publishers, Dordrecht, Netherlands, pp. 193–202, 1996.

[62] Pavesi L., Guardini R., and Mazzoleni C., "Porous silicon resonant cavity light emitting diodes," *Solid State Commun.*, **97**, 1051–1053, 1996.

[63] Fisher T. A., Lidzey D. G., Pate M. A., Weaver M. S., Whittaker D. M., Skolnick M. S., Bradley D. D. C., "Electroluminescence from a conjugated polymer microcavity structure," *Appl. Phys. Lett.*, **67**, 1355–1357, 1995.

[64] Hadji E., Bleuse J., Magnea N., and Pautrat J. L., "3.2 μm Infrared resonant cavity light emitting diode," *Appl. Phys. Lett.*, **67**, 2591–2593, 1995.

[65] Huffaker D. L., Lin C. C., Shin J., and Deppe D. G., "Resonant cavity light emitting diode with an Al_xO_y/GaAs reflector," *Appl. Phys. Lett.*, **66**, 3096–3098, 1995.

[66] Larson M. C. and Harris J. S. Jr., "Broadly tunable resonant-cavity light emission," *Appl. Phys. Lett.*, **67**, 590–592, 1995.

4

Epitaxy of Vertical-Cavity Lasers

R. P. Schneider Jr. and Y. H. Houng

Epitaxial growth of vertical-cavity surface-emitting lasers (VCSELs) represents possibly the greatest hurdle for the development of these devices for real-world applications. The combination of the challenging nature and the high degree of complexity of the epitaxial structure (often with one hundred or more discrete epitaxial layers, including complex composition and doping grading schemes), along with the precision and uniformity required (usually to better than 1% absolute in optical thickness) has posed strict limitations on the process that, at least in the early days of VCSEL research, strongly curtailed broad-based research efforts and thus slowed development of the devices. It is interesting to note that many of the early "breakthroughs" in device performance benchmarks were closely linked to advances in the underlying epitaxial technology or approaches. In this chapter the key issues surrounding growth of all-epitaxial VCSELs (that is, those in which both distributed Bragg reflectors are grown epitaxially) will be reviewed.

This chapter will be divided into several parts. In the first part, the relevant epitaxy technologies, molecular beam epitaxy (MBE), gas-source molecular beam epitaxy (GSMBE), and metalorganic vapor-phase epitaxy (MOVPE) will be described, and their respective roles in early VCSEL development will be briefly overviewed. Then the critical issues that uniquely challenge VCSEL epitaxy will be overviewed: 1) epitaxial design and engineering issues unique to VCSELs – that is, those unique VCSEL device issues that are most dependent on the capabilities and limitations of the epitaxial process and the properties of the materials – specifically including growth of VCSELs with DBR composition profiles and doping profiles designed to yield the highest efficiency and 2) VCSEL epitaxy manufacturing technology issues, including both those reactor design-dependent issues that influence manufacturing efficiency within a growth run (uniformity yield and reactor capacity) and those that affect run-to-run yield (process stability and reproducibility, including in situ monitoring and control). Throughout these first several parts of the chapter, the discussion will focus primarily on the development of "conventional" AlGaAs-based VCSELs (operating in the near-IR 780–980 nm range). In the final part, the issues surrounding incorporation of new materials and heterostructure designs for new wavelengths (e.g., red VCSELs) and other performance advantages will be discussed.

4.1 Role of Epitaxy in Early VCSEL Development

The first reported VCSELs, in the mid-1980s, were not all-epitaxial structures, but rather included one or more dielectric mirror stacks on either side of epitaxial active regions. Included among these is the first reported room-temperature CW GaAs VCSEL [1]. The

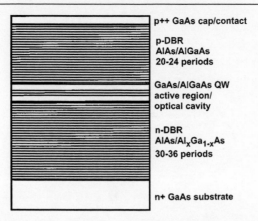

p++ GaAs cap/contact

p-DBR
AlAs/AlGaAs
20-24 periods

GaAs/AlGaAs QW
active region/
optical cavity

n-DBR
AlAs/Al$_x$Ga$_{1-x}$As
30-36 periods

n+ GaAs substrate

Figure 4.1. Schematic illustration of the epitaxial AlGaAs-based VCSEL structure. Included are the bottom (n) distributed Bragg reflector (DBR), the active region optical cavity, and the top (p) DBR.

desire to create VCSEL structures in which the mirror stacks were deposited epitaxially, as illustrated in Figure 4.1, originated in the obvious simplification of the postgrowth processing steps that would be required to build the devices. With epitaxially integrated mirrors that could also conduct current from metal contacts, VCSELs could be fabricated with planar batch-style processes that were already in use in the microelectronics industry. Such processing would dramatically reduce the cost of the devices and make true mass manufacturing a reality. However, refining the epitaxial technologies to enable growth of the entire structure was, and to some extent remains, the limiting factor in VCSEL development. In this chapter, only the growth of all-epitaxial VCSELs will be considered.

In practice, three different epitaxy techniques have been used for growth of VCSELs: molecular beam epitaxy (MBE), gas-source molecular beam epitaxy (GSMBE), and meta-lorganic vapor-phase epitaxy (MOVPE). Other conventional compound semiconductor epitaxy techniques such as liquid-phase epitaxy (LPE) and hydride vapor-phase epitaxy (HVPE) do not exhibit the requisite precision or control. In fact, precision and stability issues in epitaxy may have been the principal limitation to the initial widespread development of VCSELs. Since the wavelength of operation of a VCSEL diode is directly related to the thickness of the layers, overall system stability on the order of a percent or so is required to properly match the gain wavelength with the Fabry–Perot mode (depending of course on the uniformity across the wafer). In the following section, we briefly review the fundamental principles and technology of MBE, GSMBE, and MOVPE, particularly as relevant to the growth of VCSELs.

4.1.1 Description of the Epitaxy Techniques

4.1.1.1 Overview of MBE Growth

Several lengthy and detailed reviews of the technology and fundamental principles of MBE have been written [2–4]. The basic MBE process achieves epitaxial growth in an ultra-high-vacuum (UHV) (total pressure $<10^{-10}$ torr) environment through the reaction of multiple molecular beams of differing flux density and chemistry with a heated single crystal substrate. The process is illustrated schematically in Figure 4.2, which shows the essential

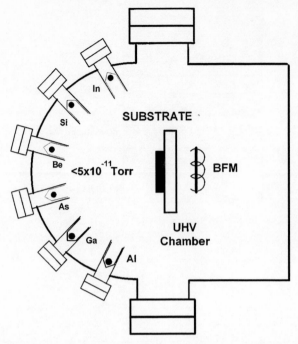

Figure 4.2. Schematic illustration of the MBE process for doped AlGaAs. Each furnace contains a crucible that contains one of the constituent elements of the desired film, including the host matrix and the dopant atoms.

elements for MBE of doped AlGaAs. Each furnace contains a crucible that contains one of the constituent elements of the desired film, be it part of the matrix of the host crystal or one of the dopants. The temperature of each source is chosen so that films of the desired composition may be obtained. The sources are arranged around the heated crystalline substrate in such a way as to ensure optimum film uniformity both of composition and thickness, and the rotation of substrate has greatly facilitated this criterion. Additional control over the growth process is achieved by inserting mechanical shutters between each individual source. As the name molecular beam epitaxy implies, the flow of components from the source to the substrate is in the molecular and not hydrodynamic flow region. Thus the beams can be considered, for all practical purposes, as unidirectional with negligible interaction within them. The interposition of a mechanical shutter will then effectively stop the beam from reaching the substrate. Operation of these shutters permits rapid changing of the beam species to abruptly alter the composition and/or doping the growing layer.

Crystal growth during molecular beam epitaxy is governed predominantly by kinetics as opposed to the quasi-equilibrium conditions existing during LPE or vapor-phase epitaxy (VPE) growth. The fundamental studies of MBE growth kinetics [5, 6] demonstrated that achieving a stoichiometric III–V epitaxial growth lies in the surface chemical dependence of the sticking coefficient of the group V elements. At temperatures at which epitaxial growth occurs, only that amount of the group V element is adsorbed that satisfies the available group III bonding orbitals at the surface. The growth rate is therefore determined by the arrival rate of the group III elements while the condition of stoichiometry is satisfied simply by growing in an excess flux of the group V elements. For example, the growth rate of

MBE GaAs films is entirely controlled by the flux density of the Ga beam impinging on the substrate surface, and the sticking coefficient of Ga is close to unity at typical MBE growth temperatures (450–620°C) under As-stabilized growth conditions. This implies that nearly all the Ga atoms incident on the surface are incorporated into the growing epitaxial layer.

The growth model established from kinetic measurements [6] is not unique to GaAs but is also valid for AlAs, InP, for a number of other III–V compounds, and with minor modifications also for ternary III–III–V alloys. In practical terms, $Al_xGa_{1-x}As$ films having good crystallographic perfection can be grown by simply directing beams of Ga, Al, and excess arsenic simultaneously to the growth interface. Since the sticking coefficient of Al on GaAs is unity, good compositional control of the growing film can be achieved by adjusting the flux densities of the impinging group III beams. Difficulties only arise when the substrate temperature is too high. A thorough kinetic study of the involved surface processes [6] has shown that the principal limitation to the growth of III–III–V alloys by MBE is the thermal stability of the less stable of the two III–V compounds of which the alloy is to be composed. At higher temperatures preferential desorption of the more volatile group III element occurs. Thus, the surface composition of the alloy reflects the relative flux ratio of the group III elements only if growth is carried out at temperature below which GaAs (in the case of $Al_xGa_{1-x}As$), InAs (in the case of $In_xGa_{1-x}As$), or InP (in the case of $In_xGa_{1-x}P$) are thermally stable.

In the case of compounds involving more than one group V element, such as GaAsP, InAsP, and GaInAsP, precise ratios of the group V beam fluxes are required to grow the compounds with desired mole fractions. This is necessary because the sticking coefficients of group V elements are quite different at the growth temperatures of >450°C. The sticking coefficient decreases in the order of Sb, As, and P, and tetramers have a smaller sticking coefficient than dimers. In the case of InGaAsP, precise control of the flux ratios of the elements is required for lattice matching and energy gap control. Although MBE has been successful for the growth of arsenic and antimony compounds, the growth of phosphorus compounds by conventional solid source MBE (SSMBE) has been hampered by the high vapor pressure and the allotropic property of solid phosphorus. The gas-source molecular beam epitaxy (GSMBE) technique, in which the elemental As and P sources are replaced by gaseous AsH_3 and PH_3, respectively, represents an alternative approach, as will be discussed in the next section.

4.1.1.2 Overview of GSMBE Growth

In GSMBE [7], hydride gases (AsH_3 and PH_3) are delivered into the growth chamber and are thermally cracked through a high-temperature (>900°C) cracker with high cracking efficiency (>99.9%) to produce dimeric group V species. Group III sources in GSMBE are elemental Ga, In, and Al contained in high-temperature effusion cells similar to that of SSMBE. GSMBE uses the growth system design similar to SSMBE with modifications in group V source delivery and UHV pumping methods. The growth chamber pressure is maintained at $<10^{-4}$ torr to ensure the growth processes are carried out under the molecular flow conditions. Therefore, the growth mechanisms in GSMBE are similar to that of the conventional SSMBE as described in the previous section. However, the use of group V dimers increases the sticking coefficient and so allows for epitaxial growth at higher temperatures and growth rates. Other reported benefits of growth with group V dimers include improved photoluminescence and minority carrier lifetimes [8, 9], reduced electron trap concentrations [10], improved AlGaAs and InGaAs surface morphology [11, 12], lower Ga-related "oval" defects [9], and reduced unintentional carbon incorporation [13].

In GSMBE, the group V hydrides are delivered through mass-flow controllers or pressure-controlled leak-valves. Precise amounts of hydrides are injected into the hydride crackers and are nearly 100% cracked before reaching substrate surface. Unlike MOCVD, in which the decomposition of hydrides takes place at the substrate surface and is strongly temperature dependent, the desired fluxes of both As_2 and P_2 can be obtained in GSMBE at the substrate surface and are independent of substrate temperature. The growth of high-quality lattice-matched $Ga_x In_{1-x} As_y P_{1-y}$ on GaAs over the entire composition range $(0 < y < 1)$ and multiple quantum well (MQW) lasers using GSMBE have been demonstrated [14, 15].

4.1.1.3 Overview of MOVPE Growth

Fundamental principles of MOVPE growth are overviewed in several review articles and books [16–18]. While most MBE growth systems share many common features, MOVPE growth platforms in general exhibit less similarity and consistency in design. The MOVPE process itself is much more complex than that of MBE, and the lack of clear consensus on reactor system design principles mirrors this complexity. However, with the increased complexity brings also a greater degree of flexibility, and it was the ability of MOVPE to readily grow phosphorous-containing materials (historically difficult using MBE) that led to its widespread use in the mid 1980s. A simplified schematic illustration of a modern MOVPE system is shown in Figure 4.3. Included are several subsystems: the gas handling system, in which the reactants are metered and mixed; the reaction cell, which may be vertical or horizontal in design; and the pressure control system, including the vacuum pump(s), throttle valve, other exhaust components, and finally the exhaust treatment facility (scrubber).

Source materials are chemical compounds and are transported to the heated substrate in the reactor cell via a carrier gas stream, usually high-purity Pd-diffused H_2. Comparatively high process pressures are used, usually on the order of 1/10 atmosphere (76 torr), though pressures anywhere between 20 torr to 760 torr are not uncommon. Group III sources are metalorganic compounds, including trimethylgallium (TMGa), trimethylaluminum (TMAl), and trimethylindium (TMIn), whereas group V species are typically hydride gases, including arsine (AsH_3) and phophine (PH_3). The metalorganic (MO) compounds, in solid or liquid form, are contained in stainless-steel containers called "bubblers" and are transported via a carrier gas through the bubbler. The hydride gases are usually in the form of compressed gases and are metered directly into the reactor gas stream.

These reactants decompose in the heated zone above the susceptor, where they react to form (e.g.) GaAs, InP, etc. The MOVPE growth process is mass transport limited, so that, to first order, growth rate and group III composition control is dependent mainly on the partial pressure of the MO reactants present in the reactor [17]. Furthermore, since the group V species have very high vapor pressures and their incorporation is self-limiting, group V partial pressure has very little influence on the growth rate; and growth usually takes place with V/III ratios $\gg 1$ (e.g., 50–100). Unlike MBE growth, growth rate and composition control (on the group III sublattice) is relatively insensitive to other process variables including growth temperature. (Note that recent studies employing extremely accurate in situ growth rate monitors have indicated that group V partial pressure may influence growth rate more strongly than initially believed [19]; further study will likely elucidate the relative role of reactor design, etc.) Thus, growth rates and ternary and quaternary compositions are all

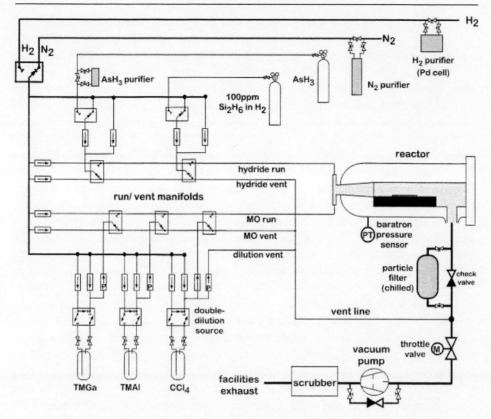

Figure 4.3. A schematic illustration of an MOVPE system for doped AlGaAs. Included are several subsystems: the gas blending system, in which the reactants are metered and mixed; the reaction cell, which may be vertical or horizontal in design (horizontal channel reactor is shown); and the pressure control system, including the vacuum pump(s), throttle valve, other exhaust components, and the exhaust treatment facility (scrubber). Additional descriptions in the text.

controlled by precisely controlling the partial pressure of the reactants in the reactor, and the precision with which these partial pressures can be controlled dictates in large part the stability and precision of the epitaxy process. This topic will be discussed in greater detail in Section 4.4.

Design principles for MOVPE gas handling systems, the subsystem in which the reactants are metered and mixed, are reasonably well understood and exhibit some uniformity. Some common design features are noted in Figure 4.3. In the following, the process flow is described with reference to Figure 4.3. H_2 carrier gas is purified with a heated Pd cell, and N_2 with either a resin-type purifier or getter-type purifier. The main carrier valve switches flow to the sources and reactor between H_2 (for growth) and N_2 (for purging). Three-way valves are used to open and bypass the MO bubblers. Source MFC controls the flow rate of carrier gas through the bubbler (Q_S). Dilution MFC allows constant flow (Q_D) in the source line, independent of source flow. The pressure controller maintains constant pressure over the source, requisite for reproducible growth rate control. On double-dilution source, an additional inject MFC controls flow (Q_I) to the reactor, downstream of source and dilution MFCs, while the excess flow is routed to a separate vent line. Double-dilution sources enable several orders of magnitude in useful dynamic range by scaling the source flow to the reactor by the relation $[Q_S/(Q_S + Q_D)] * Q_I$. Double-dilution is often included on dopant source lines to enable doping control over several orders of magnitude (e.g., $\sim 10^{16}\,\text{cm}^{-3}$ to $\sim 10^{19}\,\text{cm}^{-3}$), and even on group III sources to enable the partial pressure to be controlled downstream of the bubbler, thus avoiding possible transients in the bubbler itself. Hydride lines are usually supplied as compressed gases; these include pure arsine and a dilute mixture of silane in hydrogen (e.g., 100 ppm). Resin-type purifiers are commonly employed on the arsine lines. Gases are switched in and out of the reactor using continuously purged "5/2-way" valve configurations in the run–vent manifold. Care is taken to eliminate the pressure difference between the run and vent manifolds, to avoid switching transients that blur interfaces and perturb flow characteristics in the reactor, enabling growth of very thin quantum-well structures. Growth takes place at the heated susceptor in the reactor cell.

The design issues associated with the reactor cell itself remain somewhat unresolved, with several fundamentally different approaches in use, each of which exhibit unique advantages and disadvantages. The most commonly used MOVPE reactors fall into two general design classifications: horizontal reactors, including channel reactors and radial-flow planetary reactors, and vertical reactors, including rotating-disk reactors (RDRs) and vertical stagnation flow reactors. These reactor designs will be overviewed in more detail, and compared in terms of wafer uniformity and relevance for VCSEL manufacturing, in Section 4.3.

The pressure control system of MOVPE reactors is very similar in design. Gases are exhausted through a particulate filter, and a downstream throttle valve controls the process pressure. A mechanical pump provides pressure control from 1 atm to $\sim 1/10$ atm (76 torr), the typical operating pressure in modern MOVPE reactors. A bypass with an appropriate check valve provides pressure relief. Finally the exhausted gases are routed through some kind of scrubber system to remove the toxic gases before general exhaust in the facility exhaust system.

4.1.2 Early Development of VCSEL Epitaxy Processes

Early all-epitaxial VCSEL development focused on the use of molecular-beam epitaxy for growth, despite the fact that MOVPE had, by the end of the 1980s, begun to be regarded as the epitaxial technique of choice for production of light-emitting diodes (LEDs) and laser diodes (LDs), due to the high optical efficiency of the materials as well as the high throughput of the process. The critical issue favoring MBE for VCSELs was the demonstrated superior precision and control. In fact, MBE is not necessarily a significantly more stable process than modern-day MOVPE. However, not only did early MOVPE reactors exhibit much more unstable behavior, but almost since its inception MBE has benefited from an in situ probe that is routinely used to calibrate the growth rate prior to growth, namely, reflective high-energy electron diffraction (RHEED). Such a growth rate probe ensures that most of the growth runs result in usable wafers, as run-to-run fluctuations in the growth rate can be compensated and effectively eliminated. Not until recently has there been a useful in situ growth-rate monitor for MOVPE (reflectance), and several engineering issues (discussed later) continue to limit its widespread implementation.

The first all-epitaxial VCSEL diodes were reported by 1989 [20–23]. In these first devices MBE-grown material was used. However, demonstrating low series resistance in these early VCSELs proved challenging. Early VCSEL diodes suffered from high series resistance originating in the large heterobarriers to hole transport in the distributed Bragg reflectors, primarily in the top (p) DBR. A number of schemes have been developed over the years to address this problem, primarily related to tailoring the composition [22, 24–28] and doping profiles [28–31] at the interfaces between the DBR layers to reduce the height of the barriers. In this regard conventional MBE is limited. Since the growth rate and the composition of MBE-grown AlGaAs alloys are primarily dependent on the mass flux from heated effusion cells, and the flux is in turn exponentially dependent on the temperature of the cells, arbitrary composition profiles, including continuous grading, are difficult to realize. In addition, MBE suffers from limited dopant and materials flexibility. Conventional p- and n-dopants are Be and Si, respectively, and both of these species exhibit composition-dependent activation, which is of particular concern in distributed Bragg reflectors. Additionally, Be is prone to redistribution under some growth conditions [32]. Carbon, provided from a solid source (a heated filament) is not in widespread use, though it is known to yield lower resistance in p-DBRs [32].

GSMBE employs the flux from elemental group III metals, but it differs from MBE in that the group V species are supplied as hydride gases. In addition, some dopants (including carbon) can be supplied as gases. Since the growth rate and composition of AlGaAs is still determined by controlling heated effusion cells, GSMBE suffers from the same group III composition-control limitations as MBE. However, GSMBE has significant advantages in materials and dopant flexibility, due to the greater ease with which species such as phosphorous and carbon can be supplied as gases. Early reports of 980 nm VCSELs grown using GSMBE [33, 34] took advantage of this dopant flexibility to enable high-efficiency devices with C-doping from CBr_4. More recently, record-high operating efficiencies have been achieved in GSMBE-grown 850 nm VCSELs with C-doped p-DBRs [35].

The primary motivation for the initial use of MOVPE for all-epitaxial VCSELs early in their development was the greater ease with which the alloy compositions could be continuously graded at the DBR interfaces, to reduce the series resistance. Since MOVPE growth is fundamentally a mass transport–limited process, the growth rate and composition of the material is readily controlled by mass flow controllers and pressure transducers, which can in turn be easily controlled according to any arbitrary analog signal with a high degree of accuracy. Additional motivation for the use of MOVPE for VCSELs was the ready availability of C as the p-type dopant, using CCl_4 [36, 37]. Zhou et al. [38] first reported an all-epitaxial VCSEL diode grown using MOVPE with continuous grading in the distributed Bragg reflectors and extensive use of C for p-doping. These devices exhibited state of the art operating voltage, resistance, and power efficiency for top-emitters. With additional refinement of the process, several groups have since demonstrated further improvements in device performance using MOVPE [39, 40], and today many performance benchmarks have been established using MOVPE-grown epitaxy.

VCSEL performance benchmarks have evolved rapidly in the past few years, as epitaxy processes have matured and selective oxidation processes have been developed. MBE, GSMBE, and MOVPE have all been shown to be relevant to VCSEL research and continue to enjoy wide popularity. In the following section, the key epitaxial design and engineering issues that limited early VCSEL development will be reviewed in the context of all three growth techniques, and those issues still outstanding will be described.

4.2 Epitaxial Engineering for High-Efficiency VCSELs

A number of unique epitaxial design and engineering issues (those issues most strongly related to and dependent upon the growth techniques and capabilities and the materials properties) contribute to the ultimate performance of VCSELs. Included are the growth of DBRs designed to provide the lowest resistance current transport (discussed in Section 4.2.1), realization of doping profiles designed to provide the lowest resistance and the lowest losses as well as optimal current spreading in top-emitters (Section 4.2.2), and the growth of DBRs enabling oxide confinement (Section 4.2.3). In these sections the discussion will be limited to the case of n-substrate top-emitting VCSELs. In Section 4.2.4, the special considerations for p-substrate top-emitters will be discussed.

4.2.1 Composition Engineering for High-Efficiency n-Substrate VCSELs

Distributed Bragg reflectors (DBRs) represent the most unique and challenging aspect of VCSELs from an epitaxial design perspective. They serve a dual role, defining the laser cavity as highly reflective mirrors, as well as providing the pathway for current injected from the contact layers. The principle design issues associated with the DBRs are thus a trade-off between providing sufficiently high reflectivity for light generated in the cavity along with the lowest possible resistance for both electrons and holes transported to the active region. In this section, a brief description of a typical DBR structure will be followed by a discussion of the epitaxial design and growth issues associated with realizing low-resistance DBRs required for high-efficiency devices.

Typical DBRs consist of 20 to 40 pairs of alternating low- and high-index $Al_xGa_{1-x}As$ "quarter-wave" ($\lambda/4$) layers, whose thickness is determined by $d = \lambda/4n$, where λ is the design wavelength, d is the quarter-wave layer thickness, and n is the index of refraction of the alloy at the design wavelength. The total number of periods in the DBR determines the reflectivity of the structure. The bottom DBR (high reflector) is usually designed for a total reflectivity of $>99.99\%$, while the top DBR (output coupler) is designed with a reflectivity of 99.6–99.9%, depending on the desired device characteristics and other details of the device design (e.g., the number of QWs). For a typical infrared (IR) VCSEL, the design wavelength is 850 nm, and the indices of refraction of the $Al_xGa_{1-x}As$ alloys are in the range 3.0 to 3.5, so that $\lambda/4$ layer thicknesses are in the range 600–750 Å. Since the center wavelength of the stopband is directly related to the thickness of the quarter-wave layers, precise control of the thickness of each of the layers (ideally, to better than 1%, or two to three monolayers) is required to achieve optimal performance at the desired wavelength. The low-index layer is usually AlAs, or an $Al_xGa_{1-x}As$ alloy with $x > 0.9$, while the composition of the high-index layer is usually defined by transparency of the material to the emitted light (e.g., $x \sim 0$ for 980 nm VCSELs, $x \sim 0.15$ for 850 nm VCSELs, $x \sim 0.25$ for 780 nm VCSELs, and $x \sim 0.5$ for "red" VCSELs emitting at 650–680 nm). Some kind of composition grade is almost always included between the endpoint compositions of the layers, adding to the complexity of the structure, and often the doping profile is modulated as well. For VCSELs emitting at 850 nm, the total thickness of both n- and p-DBR structures exceeds 6 μm, making this a particularly thick epitaxial structure, which, along with the relatively high Al composition required in much of the DBR, can also make growing very smooth surfaces more difficult. In addition, the range of compositions required for these structure is unique among optoelectronic structures and adds significantly to the complexity to the growth, as growth and doping characteristics are also a strong function of alloy composition, especially in MOVPE growth.

Although achieving sufficiently high reflectivity ($>99.9\%$) in epitaxially grown DBR structures was shown to be straightforward in very early work [41, 42], demonstrating low-resistance current transport in DBRs proved much more difficult. For a typical "square" DBR composition profile, in which the low- and high-index layers are AlAs and GaAs, respectively, heterobarriers in excess of 500 meV can be created, leading to dramatically high series resistance in a single device. However, it was recognized early in VCSEL development that composition grading and dopant modulation at the DBR interfaces could be used to reduce the resistance of the stack [22, 24, 29, 38, 43]. For the purposes of this chapter, only a brief discussion of the salient points will be presented for illustration purposes.

4.2.1.1 Design of Low-Resistance DBRs

Design of the optimal DBR composition profile depends on a number of desired device parameters and usually is subject to trade-offs between abrupt profiles (for the highest reflectivity per interface) and graded profiles (for the lowest series resistance in the stack). Results of example calculations illustrating the influence of composition grading on the valence band profile are given in Figure 4.4, for the case of an $AlAs/Al_{0.5}Ga_{0.5}As$ p-DBR

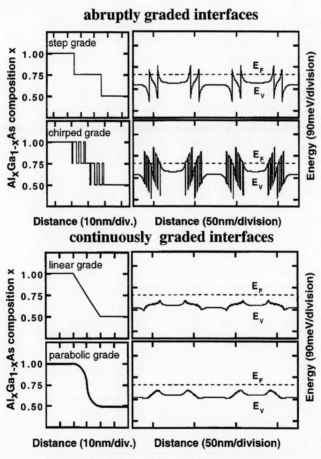

Figure 4.4. Calculated valence band profiles for different composition grading profiles in $AlAs/Al_{0.5}Ga_{0.5}As$ p-DBRs designed for use in a red light emitting VCSEL. Shown are abruptly graded (including step-graded and chirped-graded) and continuously graded (including linearly graded and parabolically graded) DBRs. Continuously graded interfaces provide the flattest valence band profiles and thus the lowest barriers to hole transport.

designed for use in a red-emitting VCSEL. Of course, this choice of DBR composition, for a very specific wavelength of operation, will give lower barriers to hole transport relative to DBRs designed for longer-wavelength operation, in which a larger composition range is required. Nevertheless this comparison serves to illustrate the trends. Valence band profiles are shown for four different composition profiles. Abruptly graded profiles include a single step grade, with a 20 nm wide $Al_{0.75}Ga_{0.25}As$ "step," and a digital "chirped" grade, intended to approximate a linear profile but with minimum layer thicknesses on the order of 1 nm. The latter profile is often employed in MBE-grown device structures (e.g., ref. [44]). Continuously graded interfaces shown here include a linear profile (over 20 nm) and a parabolic profile (over 20 nm). Clearly, the smoothest valence band profiles, and thus the lowest DBR resistance, are obtained using a continuous grade at the DBR interfaces. There is little difference in the total barrier height between the two kinds of abrupt grades, nor between the two kinds of continuous grades. However, the continuously graded designs provide a total barrier height reduction of ~150 meV relative to the abruptly graded interfaces.

As the interface design will also influence the reflectivity at that interface, for comparison the total calculated reflectivity achieved with these designs is given as a function of the number of DBR pairs in the stack in Figure 4.5. The highest reflectivity per interface is obtained for a square profile. The parabolic profile yields the closest approximation to the "square limit" in reflectivity, as it most closely approximates the square composition profile. There is an additional drop-off in the reflectivity for linearly graded profiles and then a substantial reduction in the reflectivity per interface for the step-graded profile. Besides the higher reflectivity, parabolically graded DBR designs should provide for lower resistance at cryogenic temperatures, owing to the smoother valence band profile relative to the linearly graded design [45].

The principal goal of much of the DBR modeling such as that outlined above is the design of DBR structures that exhibit the lowest possible resistance while taking advantage of the greatest possible composition range in the DBR (limited only by absorption of the

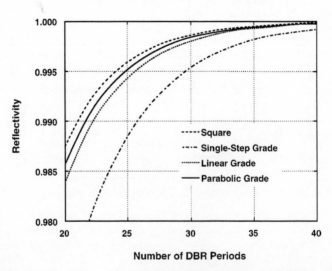

Figure 4.5. Calculated reflectivity achieved with DBRs grown with different interface composition profiles, given as a function of the number of DBR pair in the stack.

emitted light). In this way, the reflectivity per interface is maximized, and the number of required DBR pairs is minimized. However, alternate approaches have instead employed intermediate-composition endpoint layers to reduce the heterobarrier height, at the cost of lower reflectivity per interface [46–48]. Devices employing this epitaxial design approach exhibited some of the best performance metrics of early VCSELs, particularly with respect to threshold voltage, operating voltage, resistance, and wall-plug efficiency. However, the design compromises left the overall device performance still short of today's metrics. Such an approach is used less often in current VCSEL designs, in which effective grading schemes have been developed to take better advantage of the greatest possible composition range in the DBRs while still allowing low series resistance.

4.2.1.2 Growth of VCSELs with Composition-Engineered DBRs

While band structure modeling clearly indicated the utility of some form of interface composition grading for reductions in device resistance early in the development of VCSELs, the epitaxy technology required to realize such structures lagged behind. One of the principal reasons for this delay is the relative difficulty of realizing continuous composition grading using MBE. Early attempts at grading the DBR composition using MBE instead employed step grading and digital grading (including the chirped grade in Figure 4.4) [20, 24, 26, 27, 28, 34]. The effectiveness of these approaches, independent of accompanying modulation doping over the graded portion, is limited when compared to the continuous grading approaches. However, at such an early stage of VCSEL development the improvements in series resistance demonstrated by these approaches were well received.

There have been relatively few reports of continuous grading using MBE. One of the most effective approaches was outlined by Chalmers, Lear, and Killeen [25], who carefully ramped the effusion cell temperatures in the MBE chamber to provide continuous piecewise linear grading of the Al composition in the DBRs. Such continuous ramping is complicated by the exponential dependence of growth rate on cell temperature and the fact that cooling of the cell cannot be controlled actively. These devices yielded the lowest resistance and operating voltages yet achieved in 980 nm VCSELs when reported in 1993. However, due to the inherent technical difficulty of continuously ramping the cell temperatures in a reproducible fashion, this approach never met with widespread acceptance.

The primary motivation for the initial use of MOVPE was to more easily employ continuous grading schemes at the DBR interfaces. In MOVPE the growth rate and composition is controlled solely by mass transport of gaseous reactants to the substrate [17]. The mass transport is controlled by high-precision mass flow controllers and pressure transducers, which can in turn be easily controlled according to any arbitrary analog signal with a high degree of accuracy. Thus realization of even highly complex continuous grading schemes is straightforward, leading to obvious benefits in epitaxial design flexibility for VCSELs. Zhou et al. [38] were the first to demonstrate a VCSEL diode grown using MOVPE with continuous grading in the DBRs. A 12 nm thick linearly graded portion was inserted at each DBR interface, with no special provision for modulating the doping profile. Whereas C was used for doping the p-DBR structure, the contact/cap layer was doped using Zn. The series resistance, operating voltages, and power efficiency of these devices were the best that had been reported to that date for top-emitting VCSELs, grown by either MBE or MOVPE, indicating the effectiveness of continuous grading for reducing the p-DBR resistance. Based on the calculations of the previous section, some performance improvements should be

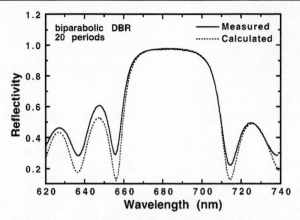

Figure 4.6. Measured and calculated reflectance spectra for an MOVPE-grown AlAs/Al$_{0.5}$Ga$_{0.5}$As DBR with parabolic interface grading. (After Schneider et al. [39].)

possible using parabolic rather than linear grading. The ability of MOVPE to grow DBR structures with parabolic grading is illustrated in Figure 4.6 [39]. Good agreement between the measured and calculated reflectivity is observed, confirming accurate growth despite the complexity inherent in such an epitaxial structure.

Since these initial demonstrations there have been a number of reports of state-of-the-art VCSELs grown with continuously graded interfaces with MOVPE. 980 nm VCSELs were reported [49] that took advantage of the continuous grading capabilities of MOVPE with a unique "uniparabolic" grading design [50]. This design combined continuous linear and parabolic grading at the DBR interfaces for relatively flat valence band profiles with additional advantages in minimizing the thermal resistance and maximizing the reflectivity per interface. Additional epitaxial design refinements relative to previous VCSEL structures included the use of C throughout the p-DBR and including the contact layers, in which very high doping (approaching $p \sim 10^{20}$ cm^{-3}) was employed for additional reduction of the device resistance. The low resistance of the resulting devices manifested in record low operating voltages and record high wall-plug efficiencies, exceeding 20% for the first time. More recently there have been numerous reports of low-resistance, high-efficiency VCSELs emitting from the near-IR into the red, grown with MOVPE and employing continuously graded AlGaAs DBRs (including both linear and parabolic designs) [39, 40, 45, 51–54]. It is noteworthy that, at present, many of the best reported device performance metrics for both "conventional" (proton-implanted) and oxide-confined devices emitting at the technologically relevant wavelengths of 650–680 nm, 780 nm, 850 nm, and 980 nm have been achieved using MOVPE growth and some form of continuous grading in the DBR interfaces. One notable exception is the recent report of record performance in 850 nm oxide-confined VCSELs grown using GSMBE [35, 55]; this exception will be discussed in greater detail in Section 4.2.2.2.4 of this chapter.

4.2.2 Doping Engineering for Efficient n-Substrate VCSELs

The resistance of vertical-cavity lasers has both vertical and lateral components. Besides the vertical injection though the DBRs, current must spread from the edges of the metal contact to uniformly fill the laser aperture. While composition engineering may have profound effect on the former irrespective of the doping profile, it has limited effect on the latter. Doping plays a critical role not only in reducing spreading resistance, but also in reducing vertical transport for cases in which composition engineering flexibility is limited. In addition, doping can strongly affect the efficiency of the devices through free-carrier absorption loss. Doping engineering refers to modulation of the doping profile, as well as to the appropriate choice of dopant species, for specific effects. In the following, the key engineering issues associated with engineering the doping for low-resistance p-DBRs in both MBE- and MOVPE-grown devices will be discussed.

4.2.2.1 Dopant Profile Design Issues

4.2.2.1.1 Band Structure Engineering for Low-Resistance p-DBRs

Precise modulation of the dopant profile can also be used to engineer flat valence band profiles for low-resistance p-DBRs, relaxing the demands on composition profiles [26, 29–31]. For example, modulating the doping such that a much higher charge concentration is placed precisely at the heterojunction (so-called delta doping) can have a significant influence on the relative barrier height at the DBR interface and lead to a substantial reduction in series resistance for the DBR stack (e.g., refs. [35, 44]). This added degree of freedom in designing DBR heterojunctions has enabled significant progress toward the realization of low-voltage diodes using MBE despite the difficulty in continuously grading the interfaces.

However, just as engineering issues limit the extent to which composition grading schemes can be adopted in epitaxy processes, dopant modulation schemes are likewise limited by both engineering constraints and fundamental dopant properties. Issues influencing the ability of the given epitaxy processes to realize the doping profiles generated through band structure engineering optimization modeling will be described in Section 4.2.2.2.

4.2.2.1.2 Current Spreading

The geometry of top-emitting VCSELs grown on absorbing substrates requires that the current be injected uniformly across an emitting aperture that has a diameter of at least 10 to 20 μm and is placed only 2–3 μm above the laser active region as defined by either proton implant or native oxide. Proton-implanted VCSELs are primarily gain-guided devices, so that the current injection profile in the laser aperture strongly influences the properties of the device, including the near-field, far-field, and modal characteristics of the emitted light. Any enhancements in the current spreading resistance is likely to significantly affect the device's performance and beam characteristics. In proton-implanted VCSELs the current-spreading problem is exacerbated by the damage trail of the protons, which inhibits current spreading in the DBR. Indeed, most implanted VCSEL geometries employ a 1–2 μm overlap of contact metal over the implant aperture to improve current injection. Bottom-emitting geometries are usually superior to top-emitting in this regard, as current is injected straight down from a metal contact that completely covers the laser aperture, and the bottom-side current spreading is enhanced by the much greater thickness of the substrate (>100 μm even after thinning). However, most technologically relevant VCSELs today are top-emitters, as 980 nm emitting VCSELs have few relevant applications, and GaAs substrates are absorbing in the 650–850 nm range. One notable exception is the long-wavelength VCSELs emitting at 1.3–1.55 μm; however, these are still at a relatively early stage of development.

Current spreading can be enhanced in several ways. Including thicker contact/current-spreading layers on the top of the VCSEL structure may be effective; however, substantial thickness increases may be necessary, placing greater burden on the epitaxy process, and furthermore such thick layers raise the risk of greater absorption of the emitted light, especially important for the shorter-wavelength devices (e.g., red VCSELs). Another approach is to increase the hole concentration in the contact/current spreading layers, to $p \gg 10^{19}$ cm^{-3}, to decrease the spreading resistance. This approach does not require increases in the layer thickness or process time. However, it does place more stringent demands on the doping control, requiring that very high doping concentrations be placed in a relatively thin layer. The relevant issues affecting the incorporation of such high concentrations of acceptors will be overviewed in the next section (4.2.2.2).

Another method for enhancing current spreading is to employ higher-doped regions at abrupt DBR interfaces to enhance the lateral mobility of the carriers, as in the uniparabolic DBRs proposed by Lear and Schneider [50]. Here again, greater demands are placed on the ability to modulate the composition and the doping over a relatively narrow region. Whereas the abrupt part of the DBR interface is readily achieved in MBE, the same is not true for the parabolic part of the interface grade. Using MOVPE both parts of the grade are straightforward. Doping characteristics that influence the ability to employ appropriate doping profiles also play a role in the ability of the epitaxy process to realize such structures; these will be overviewed in the next section.

4.2.2.2 p-Doping Issues for VCSELs

VCSELs present unique doping challenges and complications for several reasons: 1) the wide composition range required in the DBRs leads to similarly widely varying doping characteristics, including composition- and process-dependent dopant incorporation, activation and passivation, as well as composition- and process-dependent etchback effects; 2) modulation of the doping profile in the DBRs, as well as high doping concentrations in the current-spreading (cap) layers, both required for reduced device resistance, requires that the dopant species be controllable over a wide range of doping concentration; and 3) the long growth times associated with such thick structures can exacerbate dopant diffusion. Because of the unique challenges VCSELs present to both p- and n-type doping, a general discussion of doping issues relevant to VCSELs is warranted. In the following, key doping issues for both MOVPE and MBE growth of VCSELs will be overviewed, with particular attention to those areas in which VCSELs present unique challenges. First p-type doping in the context of n-substrate VCSELs will be overviewed, followed by the effects of unintentional dopants, and, in Section 4.2.4, n-type doping in the context of p-substrate devices will be discussed.

4.2.2.2.1 p-Dopant General Characteristics

For MBE growth, the choice of p-dopant has historically been limited, as Be is the only p-dopant readily suited for UHV growth, and it has performed suitably for most applications. For the cases of MOVPE and GSMBE, the ability to use gaseous sources enables the use of a wider range of acceptor dopant species, which exhibit widely varying doping characteristics. Indeed, incorporation of acceptors in $Al_x Ga_{1-x} As$ grown using MOVPE is a highly complex process, strongly dependent on a variety of factors including: growth temperature, growth rate, substrate orientation, dopant species, alloy composition, dopant

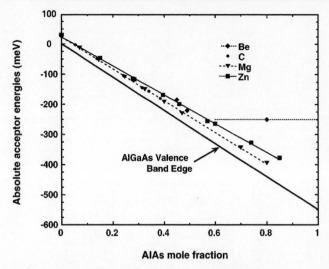

Figure 4.7. The dependence of acceptor activation energy on $Al_x Ga_{1-x} As$ composition x for some commonly used acceptor dopant species. (After Drummond et al. [56].)

source (either metalorganic or hydride), and V/III ratio. It is not the intent of the present review to describe these dependencies for every dopant species of interest, except as relevant for the case of VCSEL growth. Essentially all reported VCSEL work has made use of Be (in the case of MBE and GSMBE growth), C (for MBE, GSMBE, and MOVPE growth), and Mg and Zn (in the case of MOVPE growth); so this review will focus on these dopants. Most emphasis will be placed on C, since it seems to exhibit the best overall dopant properties and also because it is rapidly gaining wide acceptance.

The dependence of activation energy on $Al_x Ga_{1-x} As$ composition x for these most commonly used dopant species are compared in Figure 4.7 [56]. C and Mg exhibit the shallowest activation energy over the entire composition range, whereas Zn and Be tend to go deeper in higher Al composition alloys. In fact, the data of Figure 4.7 indicate that the activation energy of Be exceeds 100 meV in AlAs. If not properly compensated, the resultant low hole concentration could lead to higher DBR resistance, with the consequent increase in both forward voltage and device heating. Figure 4.8 gives the effect of dopant specie on the valence band profile, given identical doping profiles and acceptor concentrations. Even though the acceptor profiles are identical, the activation varies as a function of the acceptor energy, and so the valence band profiles vary as a function of dopant specie. Be exhibits the deepest activation energy of the dopants, and thus p-DBRs doped with Be exhibit the greatest barriers to hole transport. Indeed, this may be one of the reasons Be-doped devices grown by MBE have historically exhibited higher resistance than MOVPE devices. In contrast, the low activation energy observed for C and Mg over the entire AlGaAs composition range implies that high hole concentrations can be readily obtained even in the limit of high Al composition, making the growth of low-resistance DBR structures more straightforward.

4.2.2.2.2 p-Doping Issues in MBE/GSMBE

Beryllium is the most commonly used p-type dopant in MBE. Other solid p-type dopant sources, such as Zn and Mg, are not suitable for MBE/GSMBE growth owing to very low

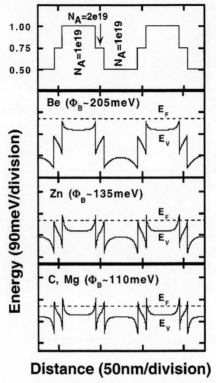

Distance (50nm/division)

Figure 4.8. The effect of dopant specie on the calculated valence band profile. Identical doping profiles, acceptor concentrations, and composition profiles were used in each case; only the composition-dependent acceptor activation characteristics were varied, as a function of the dopant species.

incorporation probability under normal growth conditions [57–60]. Be fluxes are generated from elemental Be in a high-temperature PBN effusion cell. Be forms a shallow acceptor in GaAs with 28.0 meV binding energy [61]. Be appears to be incorporated with a sticking coefficient of unity and near-unity electrical activation over the usual range of substrate temperature, arsenic pressure, and Be incident flux. Maximum doping concentrations of 5×10^{19} cm^{-3} are achieved. The doping levels in Al$_x$Ga$_{1-x}$As, at a given Be flux, were found to be practically independent of x in the composition range of $0 < x < 0.33$ [61]. However, anomalous diffusion and surface segregation of Be in GaAs and AlGaAs epitaxial layers at high doping levels ($>10^{19}$ cm^{-3}) is detrimental for device applications [32, 62, 63]. To suppress the diffusion of Be, low substrate temperature and high arsenic fluxes are used. However, under these growth conditions both the optical and electrical characteristics of the films are degraded. Since high growth temperatures are usually needed to obtain high-quality AlGaAs materials, suppression of Be diffusion imposes severe constraints on the doping level of Be in the AlAs/AlGaAs p-DBR of a VCSEL structure.

The complications with Be doping at high concentrations have motivated the development of various carbon doping sources for both SSMBE and GSMBE. Besides the very low diffusion coefficient and potentially very high doping level, other C doping characteristics in MBE Al$_x$Ga$_{1-x}$As were found to be very attractive. The incorporation and electrical activation of C are essentially the same over the entire AlAs mole fraction range. The free-carrier concentration increases slightly with x and peaks at around $x = 0.5$ [64]. Carbon doping using resistively heated graphite filaments was developed in SSMBE systems [65, 66], and doping levels in excess of $p \sim 5 \times 10^{19}$ cm^{-3} were demonstrated. However, for hole concentrations greater than $p \sim 5 \times 10^{18}$ cm^{-3} the material degrades, resulting in photoluminescence (PL) intensities that are more than ten times weaker than for a similarly Be-doped samples [67]. It has been proposed [64] that the solid source carbon doping technique may promote the formation of nonradiative recombination centers because the evaporated carbon has a large thermal energy and therefore is more likely to chemically react on the substrate surface. Another drawback using graphite filaments is that the operating temperature of the filament is typically >2,000°C and consumes more than 500 W. This raises concerns of the reliability of graphite filaments, the contamination due to outgassing of cell components, and thermal crosstalk between source cells.

The diffusion coefficient of Be was found to be strongly dependent on Be concentration, and the diffusion process can be described by the interstitial-substitution mechanism [63]. The diffusion coefficient at $\sim$700°C was found to be less than 3×10^{-16} cm^{-3}/s for doping level less than 1×10^{18} cm^{-3} and is 8×10^{-15} cm^{-3}/s for doping level greater than 1×10^{19} cm^{-3}. By using an order of magnitude larger As$_4$ flux during the MBE growth, the Be diffusion coefficient was lower by an order of magnitude.

Among the various gaseous carbon doping sources, such as TMGa, CCl$_4$, and CBr$_4$, CBr$_4$ has become the most attractive C doping source for use in both SSMBE and GSMBE because of its very low vapor pressure and high doping efficiency [68] GaAs films having doping level higher than 1×10^{20} cm^{-3} with a electrical activation of nearly 100% have been grown with good electrical and optical properties in SSMBE [33, 69] and GSMBE [33, 68] systems. The vapor from high-purity CBr$_4$ is typically sublimed from a stainless-steel bottle directly into growth chamber without the use of a carrier gas. For a CBr$_4$ flux necessary for a doping level of 1×10^{20} cm^{-3}, the beam equivalent pressure of CBr$_4$ is only $\sim 5 \times 10^{-8}$ torr. This high doping efficiency is compatible with the capabilities of conventional SSMBE systems. Carbon incorporation from CBr$_4$ is insensitive to growth temperature in the range of 500–650°C [68]. Above 650°C the incorporation of C starts to decrease with increasing temperature. A 50% reduction in doping efficiency has been found at 700°C and is due to the dopant desorption at the substrate surface. In addition to its favorable temperature dependence, CBr$_4$ is less sensitive to V/III ratio than other gaseous carbon sources. It has been found in GSMBE [68] that the carbon incorporation from CBr$_4$ is approximately proportional to the square root of the AsH$_3$ flow rate. Due to the very high incorporation efficiency of C, only very small CBr$_4$ fluxes are needed to obtain high doping level in MBE/GSMBE. Furthermore, there is no measurable reduction in growth rate, or etchback effects, in GaAs layers grown at temperatures as high as 700°C.

4.2.2.2.3 p-Doping Issues in MOVPE

General Characteristics In MOVPE most acceptor species are provided from pure liquid or solid sources, in conventional metalorganic bubblers: C from CCl$_4$ and CBr$_4$; Zn from DEZn or DMZn; and Mg from bis-cyclopentadienyl magnesium (Cp$_2$Mg). All of these sources provide for adequate delivery of the dopant molecules to the substrate surface, allowing dopant incorporation in excess of 10^{18} cm^{-3} (in some cases much higher).

C incorporation, in doped or nominally undoped material, is strongly dependent on a wide variety of growth conditions (e.g., see ref. 70). In general, it is enhanced when using

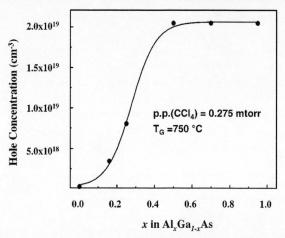

Figure 4.9. Dependence of hole concentration on $Al_x Ga_{1-x} As$ composition x, for constant CCl_4 partial pressure and growth temperature. (After Hou et al. [71].)

TMGa (rather than TEGa), at lower V/III ratios, at lower growth temperatures, and for AlGaAs alloys with higher Al mole fraction. CCl_4 was the first well-behaved carbon source that was met with widespread acceptance [36, 37]. Doping concentrations from 10^{16} cm^{-3} to $> 10^{19}$ cm^{-3} were readily achieved for $Al_x Ga_{1-x} As$ alloys across the composition range, though lower temperatures are necessary for efficient doping of alloys with $x < 0.2$. The strong dependence of hole concentration on $Al_x Ga_{1-x} As$ composition x is illustrated in Figure 4.9, for constant CCl_4 partial pressure and growth temperature [71]. CCl_4 was used for the p-DBR in early reports of MOVPE-grown VCSELs [38, 39]. Besides CCl_4 and CBr_4, several alternate carbon sources have been investigated, including TMGa and TMAs [70]; however, to date none have been as effective. CCl_4 is now banned for use because of environmental concerns, and so many groups have instead turned to CBr_4 for C doping, with similar success [72, 73] despite the comparatively lower vapor pressure of the solid source. In addition, it is well known that C can be readily incorporated in AlGaAs without intentionally doping with a source, as will be described in Section 4.2.2.3. With an appropriate choice of growth conditions, it may be possible to achieve sufficient hole concentrations ($p > 10^{18}$ cm^{-3}) in nominally undoped structures, given high enough Al mole fractions in the layers.

Both CCl_4 and CBr_4 have the somewhat unique characteristic that under typical MOVPE growth conditions their decomposition products lead to pronounced etchback of the growing material [71, 73]. This etchback is strongly dependent on growth temperature and dopant flow rate, and to a lesser degree on V/III ratio and alloy composition, and can be as significant as 10–20% of the growth rate or more. Because C incorporates much more readily in alloys with high Al content, and much less so in alloys with a composition x near 0, etchback effects tend to be more pronounced in n_H layers that require higher dopant flow rates. Even at constant CCl_4 flow rate a pronounced composition dependence of etchback rate has been observed. The composition dependence of CCl_4 etchback in $Al_x Ga_{1-x} As$ at constant growth temperature and dopant flow rate is given in Figure 4.10 [71]. Controlling etchback effects is most difficult in 980 nm VCSELs incorporating binary GaAs DBR layers. Etchback effects are almost negligible in red VCSELs, with an n_H layer composition of $x \sim 0.5$. This composition- and temperature-dependent etchback effect represents an additional complication to an already complex epitaxial process. Since etchback is thermally

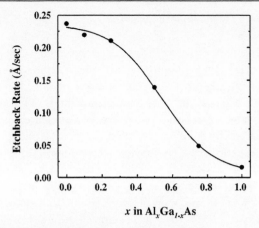

Figure 4.10. The dependence of CCl_4 etchback on $Al_xGa_{1-x}As$ composition x, for constant temperature and CCl_4 flow rate. (After Hou et al. [71].)

activated, achieving good temperature uniformity across an epitaxial wafer is critical for CCl_4- or CBr_4-doped VCSELs. Temperature nonuniformity can lead to etchback nonuniformity and consequently thickness uniformity in the structure.

Zn has been a widely used dopant in MOVPE for many years. Bass and Oliver [74] studied Zn incorporation from DMZn into GaAs and found a linear dependence of hole concentration on DMZn partial pressure. Glew [75] compared DMZn and DEZn over a wide temperature range and found decreasing incorporation efficiency at higher growth temperatures. Hole concentrations of $> 10^{19}$ cm^{-3} have been demonstrated using either source. While Zn-doped layers are not subject to etchback as for C-doped layers, Zn is more prone to diffusion, as will be discussed.

Mg has also been used extensively in MOVPE of AlGaAs and related alloys. In early studies [76, 77], it was established that controllable doping was possible over a wide concentration range from 10^{15} cm^{-3} to 10^{19} cm^{-3}, using (bis) cyclopentadienyl magnesium (Cp_2Mg). Mg incorporation depends on growth temperature, with increasing temperature resulting in lower hole concentrations. Mg may enable higher doping concentrations, particularly in materials such as AlGaInP necessary for red VCSELs. It too is not subject to etchback but may exhibit significant diffusion.

H_2 Passivation In MOVPE growth, acceptor activation depends not only on the activation energy but also on the presence of hydrogen in the growth ambient. It has been well documented that the presence of H^+ in the MOVPE ambient, present primarily through the decomposition of hydrides AsH_3 and PH_3, leads to partial passivation of acceptors [78–81]. The extent of passivation is somewhat sensitive to growth parameters such as the V/III ratio and the growth temperature. Passivation effects can be at least partially reversed through the use of a postgrowth high-temperature anneal in an atmosphere containing no H^+ (i.e., N_2), as H^+ is very mobile in the semiconductor lattice and will diffuse rapidly into the ambient [81–84]. Reversing passivation effects is most critical in systems in which achieving high hole concentrations is difficult (e.g., AlGaInP doped with either Zn or Mg).

Incorporation Limits High doping concentrations, in the absence of diffusion, are particularly useful for VCSELs. For example, appropriate delta-doping within the DBR structure can lead to reduced vertical and lateral (spreading) resistance, as described in the previous section. In addition, very high doping concentrations ($p \gg 10^{19}$ cm^{-3}) in the contact layers can allow straightforward fabrication of ohmic contacts even with nonalloyed contact metals such as TiPtAu, in addition to reducing lateral resistance for enhanced current spreading and concurrently providing more uniform current injection in the laser aperture. Acceptor incorporation, activation, and passivation combine to determine the maximum achievable hole concentration. In addition, there are fundamental limits to the total achievable incorporation, above which there can be adverse effects on morphology.

Of the MOVPE dopants for AlGaAs, C exhibits the greatest solid solubility and the fewest adverse morphological effects at high doping concentrations. Through appropriate doping and reactivation using C in MOVPE growth, hole concentrations in excess of 10^{20} cm^{-3} have been achieved in carbon-doped GaAs [85–89]. In contrast, Mg and Zn cannot be used to achieve such high hole concentrations. However, both can be used to achieve maximum hole concentrations of $p \sim 1$–2×10^{19} cm^{-3}, sufficient for most laser applications.

Diffusion   Dopant diffusion can affect the epitaxial design of the VCSEL in two ways. First, extensive dopant diffusion through the lattice during the relatively long growth times required for VCSELs could potentially disturb the doping profile, and in the worst case could displace the p-n junction, compromising device performance. Additionally, some dopant species tend to disorder the lattice as they diffuse. Any dopant diffusion, even on a smaller length scale than would affect the p-n junction position, can "blur" precise doping modulation profiles of interest for engineering the valence band structure in the DBR. Though dopant diffusion is generally undesirable, the use of dopant diffusion disordering has been investigated for reducing the series resistance of AlGaAs p-DBRs by introducing intermixing at the DBR interfaces [90].

Diffusion processes can be highly complex and variable, accounting for some inconsistencies in measured diffusion coefficients. Diffusion of Zn and Mg acceptors can occur via a "kick-out" mechanism, in which acceptor and group III interstitials compete for group III sites (e.g., ref. [91]). Alternately, they may diffuse via a substitutional-interstitial mechanism, in which group III vacancies are annihilated by acceptor interstitials. The diffusion characteristics are strongly dependent on the concentration and type of charged point defects (e.g., the concentration of group III vacancies depends on the V/III ratio during growth), and generation of Ga self-interstitials occurs via surface pinning of the Fermi level during growth. As a result, diffusion can be strongly influenced by dopant concentration: Zn and Mg acceptors in InP often exhibit a dramatic increase in diffusion at concentrations exceeding $\sim 1 \times 10^{18}$ cm^{-3} (e.g., ref. [92]). Reported diffusion coefficients for Zn, Mg, and C in GaAs are given in Table 4.1.

In some reports, Mg is thought to exhibit a smaller diffusion coefficient that Zn (e.g., see refs. [76, 77]). However, reports exhibit some inconsistency, and in some cases Mg is observed to diffuse more rapidly [93]. The inconsistencies are presumably due to differences in the doping concentration, doping profiles, dopant passivation, and point defect concentrations. In addition, incorporation of Mg is known to be more variable than that of Zn [92]. This may be due in part to the very high reactivity of Mg with residual O or H_2O in the growth system, which cannot only lead to highly variable doping effects but system- and process-dependent passivation effects as well. In cases in which doping accuracy is required and concentration-dependent diffusion cannot be tolerated, Zn may be a safer choice between the two, if other device requirements do not favor use of Mg (e.g., doping in AlInP).

Table 4.1. *Diffusion characteristics for common acceptors in GaAs.*

Acceptor	Temp. (°C)	Diffusion coefficient (cm²/s)	Reference
Mg	900	1.4×10^{-13}	[77]
Mg	850	5×10^{-14}	[77]
Mg	900	6.8×10^{-14}	[93]
Zn	900	2.2×10^{-14}	[93]
Zn	800	6×10^{-14}	[94]
Zn	900	3.2×10^{-8}	[77]
Zn	625[a]	7×10^{-17}	[95]
Zn	900	$0.5\text{–}1.0 \times 10^{-12}$	[95]
C	825	$1\text{–}2 \times 10^{-16}$	[96]
C	950	$\sim 10^{-16}$	[97]
Be	800	1×10^{-15}	[61]

[a] During epitaxial growth.

Unlike Mg and Zn, C incorporates on the group V sublattice. One of the principal advantages of C doping is that it exhibits very low diffusion over the entire range of doping concentration to $p \gg 10^{19}$ cm^{-3}, enabling pinpoint control of dopant placement in even the most complex DBR designs incorporating narrow delta-doped regions.

Substrate Orientation　Acceptor incorporation depends strongly on the wafer orientation or "offcut." Whereas MBE is usually performed on exactly oriented (100) substrates, it is common in MOVPE for some slight misorientation to be used (e.g., 2° toward the nearest ⟨110⟩). The origin of this practice lies in the reported improvement in surface morphology (e.g., see ref. [98]), though reports are somewhat scattered and inconsistent. There are however other benefits to using highly misoriented substrates, including the possibility of polarization locking in VCSELs [99, 100], reduction of oxygen incorporation [101, 102], suppression of ordering in AlGaInP alloys used for red lasers (e.g., ref. [103], and references therein), and an enhancement in acceptor incorporation. The latter has been particularly useful in AlGaInP-based red lasers, in which incorporation of Mg and Zn is difficult. Given the technological relevance for use of highly misoriented substrates for VCSELs, it is worth considering the specific doping trends in more detail.

The dependence of dopant incorporation on substrate misorientation has been the subject of several studies (e.g., refs. [104–108]). In general, Zn and Mg (both group II acceptors) incorporation increases significantly (by 5–10 times) as the misorientation is increased from (100) toward the (311)A face, in GaAs and AlGaInP, and decreases in a similar fashion as the misorientation is increased to the (311)B face. In contrast, incorporation of Se, S, and Te (group VI donors) is observed to increase as the misorientation is increased toward (311)B and decrease as the misorientation is increased toward (311)A. This opposing dependence for acceptors and donors has been used to fabricate lateral p-n junctions on nonplanar substrates [109, 110].

Si is a group IV donor, and C is a group IV acceptor, and so the dependence described above is not necessarily expected to apply as for group II and VI dopants. In fact, reports are somewhat conflicting. Kondo et al. [106, 107] observed very little dependence of Si incorporation on crystal orientation, using both SiH_4 and Si_2H_6 in AlGaInP, and using Si_2H_6 in GaAs. Some reduction in incorporation efficiency was observed for increasing angle toward (311)A using SiH_4 in GaAs. Bhat et al. [104] observed a similar reduction in incorporation efficiency for increasing angle toward (311)A, also using SiH_4 in GaAs, and furthermore reported increasing incorporation toward (311)B, the behavior expected for group VI donors (i.e., preferential incorporation on the B faces, and supression on the A faces). In the case of C, Caneau, Bhat, and Koza [105] report increasing incorporation on (311)A relative to (100) and decreasing incorporation on (311)B relative to (100), similar to the trends observed for group II acceptors. Kondo and Tanahashi [108] studied many more substrate orientations and found a much more complex dependence in both GaAs and AlGaAs. C incorporation was observed to decrease with increasing angle from (100) to (411)A, then increase abruptly relative to (100) on the (311)A face, and then remain relatively constant with increasing angle toward (111)A. In the B direction, C incorporation was observed to decrease almost monotonically as the angle was increased from (100) to (311)B. Further work is necessary to elucidate the mechanisms of dopant incorporation on crystal orientation, including the role of the Fermi level (e.g., see ref. 111, and references therein).

4.2.2.2.4 p-Doping Summary

VCSELs require dopant species that: 1) can be well controlled for precise doping profiles in the DBRs; 2) are effective over the entire AlGaAs composition range in the DBRs; and 3) have very high achievable concentrations ($\gg 10^{19}$ cm^{-3}) in GaAs for low current-spreading resistance. In addition, desirable characteristics include relative ease of doping calibration over the alloy compositions and growth conditions of interest and effectiveness in other In-containing alloys (e.g., AlGaInP, GaInAsP) that may be used in the active region heterostructure. The analysis for MBE/GSMBE and MOVPE dopants is somewhat different because of the difference in available dopant species as well as the structural considerations that arise from the different growth techniques.

MOVPE   Because of the straightforward implementation of continuous grading in MOVPE, VCSELs grown with this technique may be less sensitive to the choice of dopant species than are MBE-grown devices. Indeed, as indicated in Section 4.2.2.2.1, C, Mg, and to a lesser extent Zn, all exhibit excellent activation throughout the AlGaAs composition range. Zn and Mg are prone to diffusion and are not amenable to the precise concentration modulation control possible with C. In addition, Mg is subject to long turn-on and turn-off transients as a result of strong memory effects arising from the tendency of Mg to adsorb to the reactor walls [112]. However, the compositional grading flexibility afforded by MOVPE reduces the demand on dopant modulation techniques for band structure engineering. Severe diffusion may also be avoided if care is taken to minimize the doping concentration and growth temperature, consistent with low resistance and sufficiently high-quality material. While C exhibits nearly ideal characteristics for the DBRs,there may be instances for which Zn or Mg doping may be preferred. For example, C is ineffective in doping AlGaInP or GaInAsP alloys. Red VCSELs, which employ AlGaInP-based active regions, and "Al-free" near-IR and long-wavelength (1.3–1.55 μm) VCSELs containing GaInAsP active regions may require Mg or Zn doping in the cavity to optimize placement of the p-n junction at the QWs. In these cases, it is still possible to use C in the AlGaAs DBR; however, simplicity in

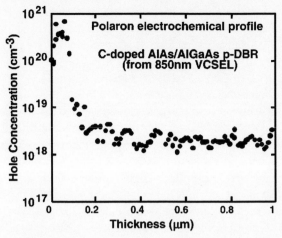

Figure 4.11. Electrochemical profile through an MOVPE-grown n-substrate VCSEL structure. Shown is the measured hole concentration in the p^{++} GaAs cap layer and in several p-DBR periods. (From Schneider and Figiel [113].)

growth or system source capacity may dictate that the same dopant be used for both active region and DBR. It is likely that in these cases, either Zn or Mg can be used appropriately to realize high-performance VCSELs.

Choice of MOVPE p-dopant species may be more critical for the current spreading/cap layer(s) and for growth of p-substrate (n-up) VCSELs. In the case of the former, C is clearly the superior choice due to the ability to dope to concentrations approaching 10^{20} cm^{-3} with no danger of diffusion. An electrochemical profile of the current-spreading and p-DBR layers in an n-substrate (p-up) VCSEL structure is given in Figure 4.11 [113]. Hole concentrations in excess of 10^{20} cm^{-3} are readily obtained in the cap layers, facilitating ohmic contacts with nonalloyed TiPtAu contacts and reducing spreading resistance. Even with such high doping concentrations, no evidence of diffusion into the DBR, doped at the 10^{18} cm^{-3} level, is observed. Use of Zn or Mg limits the maximum hole concentration to $\sim 10^{19}$ cm^{-3} or less, and, aside from the potential for excessive diffusion, the resultant increase in spreading resistance may limit ultimate device performance. Growth of p-substrate VCSELs may also favor use of C doping, because of the superior diffusion and "switch-off" characteristics. It has been reported that memory effects and long switch-off concentration tails may arise when Mg is used, due partly to its tendency to adsorb to the walls of the delivery tubing in the gas mixing system [112]. Such behavior may result in difficulty in precise placement of the p-n junction, as well as less precision in the dopant modulation profiles. Finally, the additional time-at-temperature for p-dopants in n-up epitaxial structures may lead to additional diffusion; in this case C is again the preferred dopant owing to its stability against diffusion.

MBE/GSMBE MBE/GSMBE p-DBR structures differ from those that are grown using MOVPE in the difficulty in realizing continuously graded DBR interfaces. This places somewhat greater demands on doping flexibility, as the dopant profile must be better controlled for engineering flat band profiles for low-resistance carrier transport. For example, typical MBE/GSMBE-grown VCSEL structures employ a digitally graded "chirped" interface structure in the DBRs. When such structures are grown with uniform doping profiles, they do not yield flat valence band profiles, as indicated in Figure 4.4. To achieve the flattest band profile and thus the lowest resistance with such a DBR structure, the doping must be modulated over a relatively broad range of concentration within a relatively narrow region.

Because the activation energy for Be in AlGaAs increases as the Al composition is increased, higher doping concentrations are required to achieve acceptable hole concentrations ($>10^{18}$ cm^{-3}) in high-Al composition alloys compared to low-Al composition alloys. Not only must the dopant concentration therefore be varied over a relatively wide range just to achieve uniform hole concentrations across the wide composition range of the DBR, but these higher dopant atom concentrations may lead to enhanced dopant diffusion and/or segregation in the structure. These effects combine to limit the control with which the hole concentration can be modulated, thus limiting the degree to which the DBR interface band structure can be engineered to yield flat band profiles and low resistance to hole transport. Additionally, it has been reported that the Be segregation may further complicate the growth of p-substrate (n-up) devices due to resultant residual doping through the active region and into the n-DBR [32, 114]. While there are no reports of n-up devices grown using Be as the p-dopant, Be has been used extensively with some success in much of the MBE VCSEL literature. However, it is noteworthy that currently Be-doped VCSELs do not exhibit state-of-the-art device characteristics.

Carbon has proven a very effective p-dopant in both MBE and GSMBE, using CBr$_4$ to supply the dopant atoms [34, 35, 44]. C doping can be controlled precisely over a very wide range, and its relative stability to diffusion allows an additional degree of freedom in band structure engineering in digitally graded DBRs. In addition, the very high ultimate doping concentration possible with C is useful for the contact/current-spreading layer to enable further reductions in device resistance.

It is useful to compare the DBR resistance achieved in MBE-grown devices using Be and C and similar step-graded DBR designs. Although there are no reports comparing these dopants directly, in the same growth chamber and using the same DBR design, broad comparisons are still possible. Houng et al. reported growth of 980 nm VCSELs with very low resistance using GSMBE at a relatively early stage in VCSEL development [34]. These devices employed digital interface grading in the DBRs and C doping throughout, with dopant modulation for increased hole concentration in the step-graded portion of the structure and very high doping ($p \sim 10^{20}$ cm^{-3}) in the contact/current-spreading layer. The operating voltages and wall-plug efficiencies reported exceeded those achieved in any other MBE or GSMBE-grown VCSELs, for which Be doping was used. More recently, low operating voltages and record-high wall-plug efficiencies were obtained in 850 nm oxide-confined VCSELs grown by MBE using C as the sole p-dopant [35, 55]. In this case, delta-doping was used in conjunction with step grading at the DBR interfaces. The device characteristics were similar to, or exceeded, the best results demonstrated in MOVPE-grown VCSELs using continuous DBR interface grading. This group also reported significant improvements in device performance after switching from Be to C doping. Indeed, while Be doping continues to enjoy broad use, devices grown using digital interface grading and Be doping do not match the performance levels achieved from VCSELs grown with carbon doping. Although such broad comparisons cannot conclusively point to the role of C in improving the performance of the devices, the superior dopant characteristics of C seem particularly well suited to MBE/GSMBE-grown devices that by necessity employ noncontinuous interface grading.

4.2.2.3 Unintentional Doping

While optimizing intentional doping profiles are clearly of primary importance for improving VCSEL performance, unintentional dopants may counter these efforts and adversely affect device performance. Unintentional dopants in MOVPE are incorporated principally from the metalorganic sources and are dominated by C and O. MBE and GSMBE usually exhibit lower unintentional dopant concentrations than MOVPE materials, partly because the sources are in elemental form. However, C and O may also be incorporated as background impurities with these techniques, and if their concentrations are excessive, similar deleterious effects may be observed.

In MOVPE, carbon is incorporated as a decomposition product of the metalorganic compounds. Surface chemistry and growth studies (e.g., see ref. 70) have shown that the methyl compounds (TMGa, TMAl, etc.) represent the richest source of carbon, which incorporates easily as the methyl compound decomposes. Lower background concentrations are possible using ethyl sources (e.g., TEGa), from which less C is incorporated owing to the β-hydride elimination reaction that reduces the amount of residual carbon at the surface [115]. As is typical of MOVPE-grown materials, the incorporation of C is strongly dependent upon a number of growth parameters, including alloy composition, growth temperature, and V/III ratio [17, 116]. Al exhibits the strongest affinity for C, Ga less so, and In least of all. For this reason, AlGaAs compounds exhibit relatively high p-type backgrounds due to C doping, with doping concentrations increasing as the Al composition is increased (typically from $<10^{15}$ cm^{-3} for GaAs, to $>10^{18}$ cm^3 for AlAs). This is illustrated in Figure 4.12, a SIMS profile of an n-DBR from a typical VCSEL epitaxial wafer from which high-performance devices were fabricated. The DBR was intentionally doped only with Si (from disilane) and

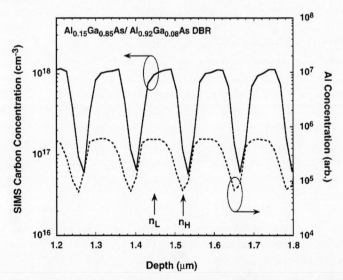

Figure 4.12. SIMS profile of an n-DBR from a typical p-substrate 850 nm VCSEL epitaxial wafer from which high-performance devices were fabricated, showing the variation of unintentional C incorporation with Al composition. The DBR was intentionally doped only with Si (from disilane) and grown holding all other growth parameters (temperature, growth rate, V/III ratio, etc.) constant throughout the layers. The n_H layer composition is Al$_{0.15}$Ga$_{0.85}$As, and the n_L layer composition is Al$_{0.92}$Ga$_{0.08}$As.

grown holding all other growth parameters (temperature, growth rate, V/III ratio, etc.) constant throughout the layers. The unintentional C concentration varies from $< 10^{17}$ cm^{-3} in the n_H layers (Al$_{0.15}$Ga$_{0.85}$As), to $> 10^{18}$ cm^{-3} in the n_L layers (nominally Al$_{0.92}$Ga$_{0.08}$As). On the other extreme, In-containing compounds on GaAs (AlGaInP, GaInAsP) never exhibit p-type backgrounds from C because of the difficulty of incorporating C into In-containing alloys.

C incorporation also depends strongly on V/III ratio, so that the p-type background concentration can be varied over a wide range by varying the AsH$_3$ flow rate [116–118]. This property of MOVPE-grown AlGaAs may be used to advantage in p-DBRs, especially for red VCSELs, which require relatively high Al composition (Al$_{0.5}$Ga$_{0.5}$As) even in the n_H layers. Using lower V/III ratios increases the hole concentration and reduces the reliance on carbon doping sources. However, the high background concentrations limit the range of doping possible in the n-DBRs. It is often desirable to reduce the doping concentration in the DBR layers immediately adjacent to the optical cavity to the mid- to high-10^{17} cm^3 range, to reduce free-carrier absorption losses. C acceptor concentrations can easily exceed 10^{18} cm^3 in AlAs, and the resulting compensation of donor species effectively limits the n-doping range to $n > 10^{18}$ cm^3. C incorporation also shows a strong dependence on growth temperature: Background concentrations in AlGaAs decrease as the temperature is raised.

C may also incorporate in GaAs/AlGaAs QW active regions, in both the QWs and barrier layers, and the results can be more damaging than its incorporation into the DBRs. Excessive p-type backgrounds may displace the position of the p-n junction by compensating donors on the n side of the active region. Excessive doping in the active region can also increase optical losses through free carrier absorption and lead to additional scattering losses, reducing the device efficiency.

Unintentional oxygen plays a major role in VCSEL performance and reliability. Oxygen is a deep nonradiative recombination center in AlGaAs. Thus, in excessive concentrations in the active region, it can reduce optical efficiency and lead to higher lasing thresholds [17, 116, 119–121]. O is responsible for compensating shallow donors such as Si [122] and Se [123] and shallow acceptors such as C and Zn [123]. The compensation of intentionally introduced Si is enhanced with the gas-phase reaction of silane and disilane with oxygen, which further depletes the Si supply. Thus, in high concentrations O can lead to increased DBR resistance. O is also thought to contribute to device degradation. Phonon emission at nonradiative recombination centers leads to dislocation movement and gives rise to so-called dark-line defects [124, 125], which ultimately lead to device failure. Indeed, VCSEL degradation modes have been observed to originate in the DBRs, where the Al mole fraction is highest [126]. Finally, excessive O can also manifest itself in poor morphology of the epitaxial wafer [116, 127].

In MOVPE, oxygen incorporation depends on a number of growth parameters, including composition, growth temperature, and source purity. In modern leak-tight MOVPE systems, O is incorporated primarily from the Al source TMAl, owing to the presence of O-containing alkoxide (Al$_x$O$_y$) impurities, which are difficult to remove in the source purification process [128]. Additional O may be present if leak integrity of the gas-handling system has been compromised, or if other system components (e.g., graphite susceptors) are contaminated. Al has a very strong affinity for O, and O concentrations increase with Al composition. Figure 4.13 shows a SIMS profile taken from within the upper DBR of two different VCSEL epitaxial wafers from which high-performance devices were fabricated. The O concentration in the n_H layer (Al$_{0.15}$Ga$_{0.85}$As) is $< 10^{17}$ cm^{-3}, while that in the n_L

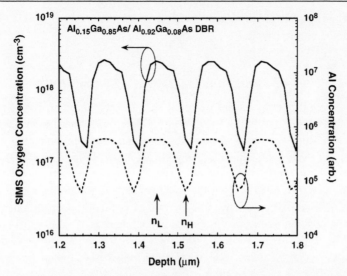

Figure 4.13. SIMS profile of an n-DBR from a typical p-substrate 850 nm VCSEL epitaxial wafer from which high-performance devices were fabricated, showing the variation of unintentional O incorporation with Al composition. The DBR was intentionally doped only with Si (from disilane) and grown holding all other growth parameters (temperature, growth rate, V/III ratio, etc.) constant throughout the layers. The n_H layer composition is $Al_{0.15}Ga_{0.85}As$, and the n_L layer composition is $Al_{0.92}Ga_{0.08}As$.

layer (nominally $Al_{0.92}Ga_{0.08}As$) is in the low- to mid-10^{18} cm^{-3} range. O concentrations exceeding 10^{19} cm^{-3} are quite common in MOVPE-grown AlAs.

Reducing the O concentration is clearly a key to improving VCSEL performance and reliability. In recent years significant progress has been made in source purification processes. "Low-alkoxide" grade TMAl sources have been available for several years and generally result in high-purity material [102]. Another path to lower O concentration in AlGaAs is the use of high growth temperatures [129], which reduce the O background owing to the increasing volatility of the Al suboxide [17]. Temperatures exceeding 725°C are commonly used to produce AlGaAs with the highest PL efficiency and smoothest surface morphology. The drawback of this approach is the potential for increased dopant diffusion.

Another approach to reducing background concentrations of both O and C in AlGaAs is the use of alternate Al sources. Growth using alternate sources such as trimethylamine alane (TMMAl) and ethyldimethylamine alane (EDMAAl) has been shown to result in lower background concentrations of both C and O. This is due to the lack of a direct Al–C bond in the source molecule and the relative absence of alkoxide impurities. Several groups demonstrated lower C and O impurity levels in AlGaAs alloys grown using TMAAl compared to those grown using TMAl [130–133]. Later work established the effectiveness of EDMAAl for growth of higher-purity AlGaAs alloys [134]. Unfortunately, TMAAl and EDMAAl are relatively unstable compared to TMAl, and they decompose at relatively low temperatures. It is therefore very difficult to achieve acceptable wafer uniformity in horizontal reactors, even when very high flow velocities are used to reduce depletion effects. Its use may be better suited for vertical reactors, especially those incorporating cooled injection plenums. There is only one report of the growth of VCSEL structures using either of these sources. Schneider et al. [133] grew an optically pumped 850 nm VCSEL using

TMAAl in the GaAs/AlGaAs QW active region for improved active region efficiency from lower C and O impurity levels.

4.2.3 Considerations for Oxide Confinement

As is detailed in other chapters in this book, VCSEL fabrication processes based on wet oxidation have revolutionized the field, resulting in dramatic performance improvements for VCSELs emitting over a broad wavelength range from the red to the infrared. The requirements for oxide confinement have significant implications for the DBR design. Most popular designs for the DBRs in oxide-confined VCSELs follow the original work of Choquette et al. [135], as illustrated in Figure 4.14. Precise placement of a single oxide layer in the DBR stack is enabled by taking advantage of the strong dependence of oxidation rate on Al composition. For example, the oxidation rate of an $Al_{0.98}Ga_{0.02}As$ layer is some three times faster than that for an $Al_{0.96}Ga_{0.04}As$ layer, and ~ten times faster than for a $Al_{0.92}Ga_{0.08}As$ layer. By placing an n_L layer with a higher Al endpoint composition nearest the cavity, an effective oxide aperture is formed while still allowing for adequate current spreading in the remainder of the DBR stack.

The design of the oxide layer has evolved somewhat since the earliest reports. For example, enhanced single-mode output can be achieved by engineering the placement, thickness, and even the tip profile of the oxide layer [35, 136]. Such designs place greater demands on the thickness and composition control of the epitaxy process, as well as on the continuous grading capabilities. It is likely that these parameters also affect the reliability of oxide-confined VCSELs. In the future, even more complex layer designs may be expected.

The principle challenges with incorporating such composition profiles into the DBR include adequate control of the composition grading at the oxide layer/semiconductor interface, accurate composition calibrations in the limit of high Al composition x, thickness control, and reproducibility, particularly of the oxide layer composition. MBE is somewhat limited in its ability to create the required composition profiles, because of the previously discussed difficulty in continuously grading the compositions, especially in the limits nearest the binary endpoints. Most reported MBE-grown oxide-confined VCSELs employ digital composition grading around the oxide layer. MOVPE may offer some advantages

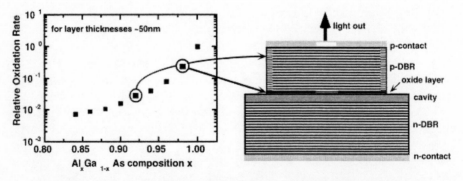

Figure 4.14. Schematic illustration of oxide-confined VCSEL DBR design for fabrication of oxide-confined VCSELs. Precise vertical placement of the oxide layer is accomplished by taking advantage of the highly sensitive dependence of oxidation rate on Al composition. (After Choquette et al. [135].)

in continuous grading capability. However, the extreme sensitivity of the oxidation rate on Al composition is a particularly stringent test for any epitaxy technique, and at this time it is unlikely that these techniques can provide the composition reproducibility required for oxide VCSEL manufacturing. This topic will be discussed in greater detail in Section 4.3.1.

4.2.4 Considerations for p-Substrate Geometries

Though *p-substrate (n-up)* VCSEL structures have received relatively little attention in the literature they nevertheless pose significant and unique technological challenges. Anode-common arrays are receiving increasing attention for applications requiring high modulation speeds from npn transistors. In this section, the design consequences of the requirement for a p-substrate geometry will be overviewed. Design considerations for n-DBRs in p-substrate geometries are different than those for n-DBRs in n-substrate geometries. In n-substrate designs, current spreading resistance in the n-DBR is enhanced by the relatively large area, while in p-substrate geometries, current is necessarily funneled into a smaller area, and vertical transport across n-DBR interfaces contributes more to the series resistance of the device.

The transport properties of n-type $Al_xGa_{1-x}As$ are somewhat unique and special considerations must be made for n-DBR design. $Al_xGa_{1-x}As$ alloys exhibit a direct–indirect (Γ–X) conduction band crossover point at a composition of $x \sim 0.45$. The dependence of the energy of these two bands, and the near-lying L band, is plotted as a function of composition in Figure 4.15. Continuously grading the DBR interfaces from high to low Al composition does not continuously grade the electron energy from high to low energy, as is the case for p-DBRs and hole energy. Compounding this difficulty is the complex dependence of donor activation energy on composition. The donor binding energy, measured with temperature-dependent Hall effect, varies strongly with $Al_xGa_{1-x}As$ composition with minima at the binary endpoints and a maximum of ~ 160 meV near the Γ–X crossover point ($x \sim 0.45$) [137, 138]. The energy of this so-called DX center is constant with respect to the L band.

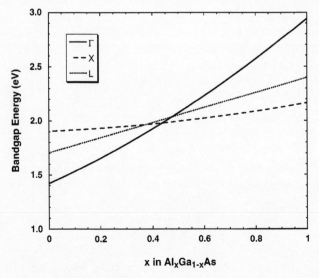

Figure 4.15. The dependence of the Γ, X, and L conduction band energies on $Al_xGa_{1-x}As$ composition x.

In Si-doped material, the room-temperature carrier concentration is roughly equivalent to the donor concentration near the binary endpoints, while the electron concentration is some two orders of magnitude lower than the donor concentration at Al compositions near $x \sim 0.45$. The dependence of donor activation energy, measured using deep-level transient spectroscopy (DLTS), is dependent on the donor species: Si exhibits a significantly deeper activation energy (430 meV) than either Se (280 meV) or Te (280 meV) [139].

Other donor properties are also relevant to p-substrate VCSEL design. Besides exhibiting the highest activation energy, Si is amphoteric and begins self-compensating at concentrations in excess of $\sim 3 \times 10^{18}$ cm^3, making higher doping concentrations impossible. However, Si is a well-behaved dopant in other respects. In MBE and GSMBE, it is supplied from elemental form and is easily controlled over a broad concentration range from mid-10^{14} cm^{-3} to the 10^{18} cm^{-3} range. In MOVPE it is supplied from a dilute mixture of silane or disilane in hydrogen. Incorporation of Si from disilane exhibits less temperature dependence than that from silane [140] and is thus the preferred source. It incorporates linearly as a function of partial pressure and exhibits little dependence on composition (other than the C-compensation effects observed in high Al containing alloys). It can also be controlled precisely: It has a low diffusion coefficient and no memory effect and so can be placed precisely over a narrow thickness range. Using double-dilution, the concentration can be controlled over a wide range, from the 10^{14} cm^{-3} range to the low-10^{18} cm^{-3} range.

Whereas Si is used extensively in MOVPE and MBE epitaxy, Se and Te are unique to MOVPE and are different from Si in many other respects. The typical sources of Se and Te are from hydrogen selenide (H$_2$Se) and dimethyltelluride (DMTe), respectively, usually supplied in a dilute compressed-gas mixture with H$_2$. Neither is amphoteric – they are classic group VI donors, which reside on group V sites. This enables higher achievable electron concentrations than is possible using Si: mid-10^{18} cm^3 to low-10^{19} cm^3 concentrations have been reported in many GaAs using either dopant [141, 142]. The smaller activation energy associated with the Se and Te DX center in AlGaAs alloys results in lower resistivity and brighter PL compared to Si-doped alloys. However, unlike the case for Si, Se incorporation is superlinear with increasing H$_2$Se partial pressure [143]. In addition, both Se and Te exhibit memory effects, which manifests as long "turn-on" and "turn-off" tails in the as-grown heterostructure [142, 144]. Such behavior effectively limits the precision with which the dopant may be placed in the heterostructure, and consequently the design flexibility possible with n-DBRs. Even with these shortcomings, Se has successfully been implemented in delta-doped structures [141] and is often preferred for MOVPE-grown optoelectronic devices (see, for example, ref. 145).

There have been reports of high-performance p-substrate VCSELs using both Si and Se doping for the n-DBR. Low-voltage performance has been achieved [146] from p-substrate VCSELs grown with Se doping and conventional linear composition grading schemes in the n-DBR. Lear et al. [114] reported lower operating voltages in p-substrate VCSELs, compared to similarly designed n-substrate VCSELs, grown using Si as the n-dopant and conventional linear DBR grading. The improved performance was attributed to the reduction in the lateral (spreading) resistance arising from the relatively higher mobility of n-type material. Schneider, Hou, and Corzine [147] measured record-low operating voltage ($V_F < 1.65$ V at 2 mW output power) p-substrate oxide-confined VCSEL arrays employing Si doping and a novel composition grading scheme intended to address the unique problems presented by n-DBRs. The composition and doping grading schemes used are illustrated in Figure 4.16. The composition is graded linearly from the n_L (high x)

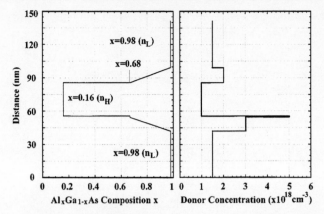

Figure 4.16. Schematic of Si-doping and composition grading scheme adopted for the n-DBR in p-substrate 850 nm VCSELs exhibiting record-low operating voltages. (After Schneider, Hou and Corzine [147].)

endpoint composition to an x value significantly higher than that of the DX center; then an abrupt composition step is employed to the n_H ($Al_{0.15}Ga_{0.85}As$) layer. This avoids the high resistance associated with the intermediate Al-composition alloys and furthermore provides higher carrier mobility at the abrupt interface for improved lateral resistance. The key to successful implementation of this scheme is precise control of the dopant modulation profile, particularly the δ-doped layer. Such profiles may not be possible using Se or Te doping.

4.3 VCSEL Manufacturing Issues

In the last section we reviewed the ways in which epitaxy processes have been refined to an extent that has enabled the growth of high-performance VCSELs emitting over a broad portion of the spectrum, from red to near-IR. It is noteworthy that much of the work that led to these breakthroughs was a product of research laboratories, for which manufacturing and production issues were of distinctly lower priority than demonstrating significant leaps in device performance. Indeed, given the goal of "pushing the envelope" in device performance, and imperfect epitaxial tools, it can be argued that there was little motivation to improve the processes. Wafer nonuniformity could be used to compensate for poor reproducibility, ensuring that some area on each wafer exhibited the proper cavity mode wavelength. However, now that VCSELs are entering data communications markets, with the potential for making inroads into very high volume consumer-driven markets in the near future, much greater emphasis is being placed on innovations that improve VCSEL epitaxy manufacturing efficiency.

Manufacturing issues for VCSELs are somewhat unique among optoelectronic devices because the combination of device structure complexity and the stringent tolerances on layer thickness and composition affect process yield to a degree not encountered in any other high-volume edge-emitting laser or LED epitaxy process. In the following section, epitaxy technology issues relevant to the high-efficiency manufacturing of VCSEL epitaxy will be overviewed. After a description of the ways in which the properties of VCSELs uniquely challenge the efficiency of manufacturing the devices, the section will address two distinct technology areas affecting overall manufacturing efficiency: reactor design, as it affects wafer uniformity (*uniformity yield*) and wafer throughput, and process stability and control, affecting *run-to-run yield*.

4.3.1 Yield Issues for VCSELs

While it can be said that achieving highly reproducible and uniform growth is important for virtually every optoelectronic device, some special properties associated with VCSELs make them particularly sensitive to deviations from exact layer specifications and wafer nonuniformity. The thickness of the DBR layers and the optical cavity determine the wavelength of the Fabry–Perot (cavity) mode, which determines the wavelength of emission for the laser. Any thickness nonuniformity will lead to a proportional wavelength nonuniformity. Not only does this affect the yield of the process via the emission wavelength specification, it also influences other properties of the laser in more profound and complex ways, compounding the yield problem. This is so because many properties of the laser depend very sensitively on the alignment between the wavelength of the cavity mode and that of the gain medium.

Compared to the cavity mode wavelength, the wavelength of emission of typical quantum-well active regions, with quantum well thickness between 6 and 10 nm, is relatively insensitive to thickness variations on the order of several percent around the design specification. Whereas the active region gain spectrum usually exhibits a relatively broad maximum, even a 2% variation in cavity thickness may result in a "detuning" of the Fabry–Perot mode by >15 nm (i.e., 2% at 850 nm) relative to the gain. The threshold and efficiency of VCSELs are extremely sensitive to the offset between the gain and the cavity mode wavelengths. Any misalignment between the two forces lasing at a wavelength different from that at the peak of the gain curve, with corresponding compromises in the device performance.

This dependence is illustrated in Figure 4.17 using wafers exhibiting some appreciable nonuniformity. Plotted are the dependence of threshold current, peak power, and threshold voltage on the wavelength of emission for devices tested along the radius of a wafer

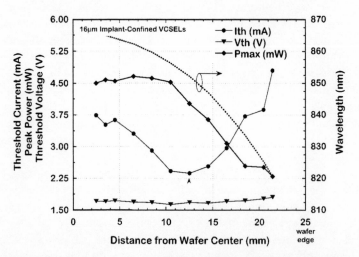

Figure 4.17. The dependence of threshold current, peak power, threshold voltage, and wavelength of emission on wafer position for devices tested along the radius of a wafer exhibiting nonuniformity of ~6% (total cavity mode wavelength variation ~50 nm).

exhibiting nonuniformity of ~6% (total cavity mode wavelength variation ~50 nm). For this wafer, the variation in the cavity mode wavelength is much greater than that of the active region gain (<4 nm across the wafer). A broad minimum in threshold is observed at that wavelength at which the gain and cavity mode are best aligned near room temperature. A broad maximum in the peak power corresponds to that wavelength at which the gain and cavity mode are best aligned at the elevated temperatures attained at higher drive currents. The relative insensitivity of the threshold voltage on the wafer position offers reassurance that the observed dependence probably does not result from a variation in the electrical properties of the epitaxial layers.

Not only does the relative alignment between the cavity mode and gain wavelengths depend strongly on their relative uniformity across the wafer, but they also depend on temperature, as their temperature coefficients differ considerably. This is of paramount importance given that most standards specify performance from 15°C (or lower) to 70°C (or higher). In addition, VCSELs are subject to significant temperature increases (as much as 30–40°C or more) that arise from relatively high operating current densities [148] and that vary with injected current. The temperature coefficient of the gain medium (primarily a bandgap shrinkage effect) is ~0.15 nm/°C, while that for the Fabry–Perot mode (primarily a second-order index effect) is ~0.05 nm/°C. The gain and the cavity mode wavelengths therefore align perfectly only at a unique temperature, at which the laser threshold is lowest. This temperature is a function of both the ambient temperature and the drive current. In cases in which the cavity mode wavelength is blue-shifted with respect to the gain wavelength, the threshold increases and the efficiency decreases with increasing temperature, as the cavity mode and the gain peak move further out of alignment. For cases in which the cavity mode is red-shifted with respect to the gain, the threshold first decreases as the temperature increases and the gain and cavity mode move into alignment. As the temperature is further increased, the gain and cavity mode move further out of alignment and the threshold increases again. By tuning the offset appropriately, it is possible to achieve a relatively flat threshold and efficiency response over a typical operating temperature range (e.g., 15 to 85°C) [149]. However, to achieve such a flat response only a very small variation in the specified gain/cavity mode offset can be tolerated – preferably on the order of only a few nanometers or less. This places further demands on the requirement for highly uniform and highly precise epitaxy.

In contrast, for edge-emitting laser and LED wafers, the variation in the wavelength can be very small (negligible) even with a comparable thickness nonuniformity of 5–6% across a wafer, since the gain wavelength is relatively insensitive to thickness nonuniformities of this magnitude. Assuming no other variation in doping, etc., very little variation in device properties would be observed over the wafer area, and yield could in fact be high. Clearly, manufacturing efficiency and yield for VCSEL epitaxy pose a much more difficult problem than for conventional optoelectronic device technologies: Even very moderate deviations from precise thickness specifications cannot be tolerated.

The situation is further complicated when manufacturing of oxide-confined VCSELs is considered. As described in Section 4.2, the oxidation rate of $Al_xGa_{1-x}As$, in the limit of x near 1, is extremely sensitive to the composition. Previous reports [135] have indicated that a difference in the composition of only $x = 0.98$ to $x = 0.96$ can yield a threefold difference in the oxidation rate. In a conventional oxide VCSEL fabrication scheme, the oxide aperture size is controlled by a timed oxidation, which necessarily depends on a precise knowledge of the oxidation rate. To achieve an oxide aperture reproducibility of ±5%, an acceptable composition reproducibility and uniformity would be only $\pm 5 \times 10^{-4}$. Such a composition variation is difficult to measure. And, given the natural process instabilities expected from either MOVPE or MBE techniques, it may be unrealistic to expect that such high reproducibility is feasible in the absence of active process control. Use of the binary AlAs alloy instead of the ternary $Al_{0.98}Ga_{0.02}As$ may be a desirable approach from a process control perspective; however, it is often avoided because of concerns about device reliability [150].

4.3.2 Reactor Design Issues: Uniformity Yield and Throughput

In this section the efficiency of the VCSEL epitaxial process within a growth run will be overviewed. This is dependent both on the uniformity of the epitaxial wafers and the wafer capacity of the reactor, both of which are clearly related to the reactor design. MBE and MOVPE reactor design fundamentals relevant to uniformity yield and wafer capacity for VCSEL epitaxy will be overviewed.

For the purposes of this discussion, a "production-scale" reactor will be defined as one that has a capacity of five 3-inch wafers. Results achieved in reactors with smaller and larger capacity will be reviewed as appropriate. Of course, the reactor capacity required depends on the market pull and the process yield. Data communications markets alone may not justify such large reactors for VCSEL manufacturing, even in the absence of improved process control. However, one of the great promises for VCSELs lies in the potential for LED-like mass manufacturing, which would open up much larger consumer-driven laser markets, including laser printing, DVD, bar-code scanning, plastic optical fiber (POF) communications, etc. Although acceptance into these markets is by no means certain at this time, this review is intended to evaluate the readiness of VCSEL epitaxial technology to meet the demands such large markets might make.

4.3.2.1 MBE and GSMBE

State-of-the-art MBE systems share many fundamental features, but with some different engineering approaches in implementation of certain system configurations and components to meet specific application needs. While the specific design of MBE system is variable, depending upon the materials and device structures to be grown and the in situ analysis to be carried out, there are certain features that most systems should have in common. The basic elements of a modern MBE growth system are UHV chambers, molecular beam effusion cells, substrate holder and heater assembly, and analytical instrumentation. Some important points are described below.

4.3.2.1.1 Reactor Chamber

Cylindrical stainless-steel vacuum systems offer the best versatility and ruggedness and are used in all modern MBE reactor systems. MBE systems with independently pumped multiple UHV chambers separated by gate valves are commonly used to satisfy the requirements of an ultraclean growth environment and high wafer throughput. The typical configuration comprises a wafer loading/unloading chamber, an intermediate UHV buffer chamber (for pregrowth preparation), and a deposition UHV chamber. The base pressure of the deposition chamber is typically less than 5×10^{-11} torr. The primary pumping arrangement is commonly a combination of ion and sorption pumps, and/or a combination of an alternative trapped turbomolecular pump, trapped diffusion pump, or He cryopump. When reactive or high-vapor-pressure materials such as P, S, Te, and CBr_4 are used, or in GSMBE where large continuous gas loads are used (e.g., AsH_3 and PH_3), then a throughput pumping system (e.g., diffusion and turbomolecular pumps) is preferred. Secondary pumping in the near-vicinity of the growth environment using a large liquid-nitrogen cryopanel is essential for the production of high-purity semiconductor material. More critical than the ultimate vacuum levels are the vacuum conditions maintained during growth operation. These levels are not just a function of pumping speeds but also of the choice of materials used in the high-temperature components, the effectiveness of thermal shielding, heat dissipation, and the purity of the source materials. High-temperature source assemblies should be housed inside liquid-nitrogen-cooled cryoshrouds, with their respective orifices as close as practically possible, and individually isolated to prevent intercontamination and thermal crosstalk. The effects of this combination of pumping systems and cryopaneling is to significantly reduce the residual gas level, especially the residual active gas species (e.g., O_2, CO, H_2O, CO_2) to less than 1×10^{-11} torr. Cryopaneling has been shown to be particularly effective for the growing Al-containing semiconductors, since it reduces the partial pressure of oxygen- and carbon-containing species, which are known to react strongly with Al.

4.3.2.1.2 MBE Sources

The effusion cell design is a key element in MBE systems. For a true Knudsen effusion cell, molecular effusion obeys a cosine law and so a similar distribution must be expected across the substrate surface. In reality the sources used in MBE are far from ideal Knudsen cells but provide many of the control aspects. The MBE cells employed large orifices to obtain enhanced growth rates at the lowest operating temperature as well as to improve uniformity at the substrate. Crucial requirements for the design of MBE effusion cells are for a low outgassing rate from the cell material and a negligible reaction between the source charge and the cell material, in order to minimize the concentration of impurities in the molecular beams. Chemically stable pyrolitic boron nitride (PBN) has become the material of choice for the construction of effusion cells. The crucible sits in a heater winding, which is well shielded by multiple layers of thin tantalum foil. A thermocouple sits in close proximity to the crucible and, with a controller, regulates the temperature. The dual-filament heating configuration, or so-called hot-lip effusion cell, has become very popular for Ga and In source cells in MBE. It uses a straight-walled crucible instead of the conventional conical crucible to increase the useful capacity of the source charge, but with a necked-down small orifice or a conical insert to eliminate the "shadowing" of the melt by the straight-wall crucible that reduces growth uniformity. It has been demonstrated that this configuration reduces flux burst transients and provides long-term temperature stability and fast response times. Oval defect generation, by the gallium source, is also reduced, specifically, defect densities of <25 defects/cm^2 for $\sim 1\,\mu$m thick GaAs layers and <300 defects/cm^2 for $\sim 19\,\mu$m thick GaAs layers have been demonstrated [151]. However, the "hot-lip" effusion cell is not suitable for the Al source because the wetting and overflow of Al melt at high temperatures can cause damage of the crucible. Therefore, for aluminum cells, heating should be intentionally nonuniform to keep a negative gradient of temperature between the bottom and the lip of the crucible ("cold-lip" crucible configuration). To reduce flux transients for Al cells without heat-shielding inserts, the position of the cell relative to the shutter has be optimized so it will be less sensitive to changes in the radiative shielding provided by the shutter [44].

The group V dimer sources offer advantages over teramers and are likely to be accepted as the standard form of the SSMBE group V sources. Cracker cells are used to produce dimers from solid group V charges (arsenic and phosphorous). The cracker cell is a modified version of the Knudsen effusion cell in which the effusion beams are directed from a conventional Knudsen crucible enclosure (sublimator) heated to $<500°C$ via a higher-temperature ($800–1,000°C$) cracker region onto the substrate. The recent introduction of valved cracking effusion cells to supply As_2 and P_2 fluxes from solid sources allowed for the precise control of group V fluxes during MBE. The stepper-motor controlled automated valve positioners can be used to abruptly initiate or terminate the dimer fluxes or provide instantaneous temporal response with a nearly linear relationship between the valve position and dimer flux. With the introduction of valved crackers, MBE has since been used to grow arsenide/phosphide heterostructures with abrupt interfaces and $Ga_xIn_{1-x}As_yP_{1-y}$ films lattice matched to InP. These advances have led to the development of SSMBE-grown quaternary containing $1.35\,\mu$m and $1.55\,\mu$m laser diodes that had threshold current densities comparable to devices grown by other techniques [152, 153].

The cells for dopants must match specific requirements that are different than those of the host matrix materials, i.e., very low quantities of materials are used, relatively high temperatures are required, and very rapid reaction to a set point variation must be obtained. It is therefore better to use a cell especially designed for that purpose. The construction of a cell with low thermal inertia, and specially configured heating elements, and thermal shielding provides a rapid response to a temperature change. Using low load, but having a large diameter/large angle crucible installed in those cells, will provide uniform flux for desired doping uniformity.

4.3.2.1.3 GSMBE Sources

In GSMBE AsH_3 and PH_3 are used as alternative group V sources. The gas sources are undiluted high-purity gases stored in stainless-steel cylinders outside the UHV chamber and are regulated by electronic mass-flow controllers or pressure-controlled UHV leak valves before being injected into the UHV chamber through a high-temperature ($900–1,000°C$) hydride cracker. Two types of hydride cracker have been used in GSMBE [154]: a high pressure gas source (HPGS) that cracks the hydrides at pressure in the hundreds of torr range and a low pressure gas source (LPGS) that cracks the gases in the range of tenths of torr or less. Both sources are capable of cracking AsH_3 and PH_3 to produce useful fluxes of the group V elements as primarily dimers. The use of tantalum in the low pressure cracker provides catalytic cracking of the gases and produces not only dimers of As_2 and P_2 but also significant amounts of monomers. Since the flux is controlled in all instances by metering gas flow instead of the normal temperature ramping of a solid source effusion cell, rapid changes in the dimer or monomer flux can be facilitated and thus act as an ideal source for the growth of quaternary $Ga_xIn_{1-x}As_yP_{1-y}$ by GSMBE.

CBr_4 is the most attractive source for p-type dopant used in either SSMBE or GSMBE for the growth of VCSEL structures. The vapor from high-purity CBr_4 is sublimed from a stainless-steel container directly into the growth chamber without the use of a carrier gas.

A typical CBr_4 delivery system uses a closed-loop pressure controller placed upstream to a fixed-orifice UHV leak valve to adjust the CBr_4 flux injected into the system through a low-temperature ($\sim 150°C$) gas injector cell. Maintaining the gas injector at a moderate temperature ($< 300°C$) prevents condensation of CBr_4 before reaching the substrate surface but does not crack the gas, while the dissociation of CBr_4 takes place at the heated substrate surface. A vent–run configuration is also employed to improve the abruptness of the doping profile.

4.3.2.1.4 Substrate Holder and Uniformity Issues

The major requirement of a substrate holder suitable for the production of uniform and reproducible layers is that the temperature across the substrate is uniform and reproducible to within $\pm 5°C$. This is especially critical for the growth of III–V–V and In-containing compounds. The assembly, which is constructed with refractory materials (Mo, Ta, and PBN), is heated either resistively or by radiation. Temperature measurement is achieved using either a thermocouple in contact with the substrate holder and/or an infrared pyrometer for direct measurement of the substrate surface temperature by viewing the substrate through a viewport. The substrate holder should be able to rotate in a plane orthogonal to the direction of the incident beams to ensure that deviations in thickness uniformity and composition arising from the overlapping consinusoidal distributions from each Knudsen source are averaged out. For the growth of VCSEL structures, it is necessary to rotate the substrate with a rotation period compatible with the time for monolayer deposition and so remove possible alloy composition fluctuations in the growth direction.

The cell–substrate evaporation geometry of the growth chamber should be optimized for thickness, composition, and doping uniformity, as well as for reducing surface defects. All cell ports should face the substrate holder at the same angle and distance and are symmetrically distributed around the source flange. There are two different configurations of arranging the source assembly relative to the substrate surface. In the vertical evaporation system the source assembly is placed at the bottom of the deposition chamber and the molecular beams direct vertically upward toward the substrate. In this configuration the position of the beam sources can be properly adjusted for a good overlap of the molecular beams at the substrate surface, and the available cell volume may be exploited with maximum charges, thus reducing the frequency of refilling cycles. In the horizontal evaporation system the source assembly is positioned approximately in a horizontal geometry with respect to the growth surface; a controlled distribution of solid and liquid source materials into the individual cells, inclined either upwards or downwards with respect to the horizontal plane, is required. A distinct advantage of horizontal evaporation systems is that it minimizes contamination of cell charges by falling flakes, which have been deposited on the shutters and inner walls of the vacuum system in previous growth runs.

4.3.2.1.5 Analytical Instrumentation

Although many different types of analytical instruments are available and used in MBE, the essential probes for a production MBE system are beam flux monitors, RHEEDs, pyrometers, and quadrupole mass spectrometers (QMSs). The beam flux monitor and RHEED are used for pregrowth calibrations and setup, and the pyrometer is used for accurate substrate temperature measurements, while QMS is employed for monitoring system vacuum integrity and background gases and is usually not used during the normal growth operation.

4.3.2.1.6 Production-Scale Reactors

To meet volume production needs the essential design criteria for high throughput multi-wafer MBE reactor systems are reliability, reproducibility, flexibility, and full automation. Most MBE manufacturers can specify the necessary system characteristics needed to produce very uniform, high-quality films. The main issue, however, is reliability. High reliability requires ingenious simplicity in the design of critical moving parts and makes use of rugged components. The number of moving parts under UHV should be minimized. For example, the need to grow very thin epitaxial layers for superlattice structures, such as with a VCSEL where thousands of shutter actuations are performed per growth run, places a very stringent requirement for reliability and performance of shutters. Most commercial systems now have soft action, magnetically coupled shutters in which the velocity of the shutter approaches zero at the endstops. The swishing mechanism increases reliability so that the shutter's lifetime is extended from only a couple of thousand actuations to over a million. It also eliminates mechanical shock and reduces vibration-related particles.

The requirement of high uniformity places stringent conditions on the design of production MBE systems by imposing larger components such as evaporation chambers and effusion cells and increased separation from source to substrate to achieve the required uniformity. These system constraints in turn require higher cell temperatures to preserve an acceptable growth rate and this last requirement naturally leads to the use of even larger crucibles to contain larger amounts of source material to avoid rapid depletion. Large capacity effusion cells for major source materials, as large as 700 cm^3, have been used in the high-throughput MBE system to improve the up time, increase the wafer throughput, and improve the stable flux over extended periods.

With the implementation based on key design parameters for scale-up, MBE systems' manufacturers are offering high performance multiwafer production systems to meet the growing need for high-throughput machines. To date there are more than thirty high-throughput multiwafer MBE systems installed for the productions of electronic and photonic devices. Most of these systems are capable of automatically handling as many as thirteen 2-inch wafers, five 3-inch wafers, or three 4-inch wafers simultaneously on a single 9- to 10-inch platen carrier. More than ten of these platens can be loaded at once through a cassette into the loading chamber, and these can transferred into the unloading chamber after batch processing. A fully automated system of both platen transfers and growth permits continuous processing of all these platens, without opening, loading, or unloading chambers. By automating virtually all pumpdown, growth, and transfer procedures, and by using large capacity source cells and rugged system components, an up-time of approximately 90% for a multiwafer MBE system has been demonstrated. With the optimized deposition geometry, uniformities of better than $\pm 1.5\%$ in thickness, doping, and alloy composition across all wafers over the entire wafer platen are routinely obtained [155]. For the growth of 980 nm VCSEL structures on a five 3-inch wafers platen, as shown in Figure 4.18, the variation of Fabry–Perot wavelength is $\leq 0.25\%$ across all wafers of an entire 5×3-inch platen [156].

4.3.2.2 MOVPE

Unlike MBE and GSMBE, MOVPE encompasses a wide variety of fundamentally different reactor designs. These include horizontal reactors (including both channel reactors and

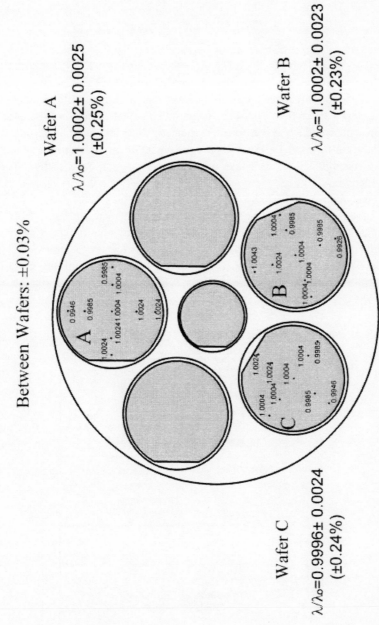

Whole Platen: ±0.25%

Between Wafers: ±0.03%

Wafer A

$\lambda/\lambda_o = 1.0002 \pm 0.0025$ (±0.25%)

Wafer B

$\lambda/\lambda_o = 1.0002 \pm 0.0023$ (±0.23%)

Wafer C

$\lambda/\lambda_o = 0.9996 \pm 0.0024$ (±0.24%)

Figure 4.18. Wavelength uniformity for 980 nm VCSELs grown in a production-scale multiwafer GSMBE reactor. (From Houng [157].)

radial-flow planetary reactors), vertical reactors (including rotating disk reactors (RDRs)), and vertical stagnation flow reactors. In the following section, the design, operating principles, and performance characteristics of these reactors will be described separately, in the context of high-yield VSCEL growth.

4.3.2.2.1 Horizontal Reactors

Channel Reactors   Horizontal reactors are in practice some of the simplest of the MOVPE reactor technologies and have been used to grow essentially every III–V compound. There have been several detailed design reviews of horizontal MOVPE reactors (e.g., refs. [17, 157, 158]). The simplest reactors consist of no more than a simple quartz channel through which the reactants are passed. The wafer is placed on a heated graphite susceptor in the channel, which defines the reaction zone. A schematic illustration of a horizontal channel reactor is shown in Figure 4.19. For low-pressure operation, an outer tube with a circular cross section encompasses an inner liner tube with a rectangular cross section. The outer tube is usually purged so that its pressure is slightly higher than the pressure in the inner liner tube. The circular tube is strong enough to tolerate the pressure drop for low-pressure operation, while the rectangular tube provides better flow characteristics.

Demonstrating good wafer uniformity in horizontal reactors has proven to be challenging. Many of the challenges lie in the design of the reactor, while others are intrinsic to the process itself. The primary barrier to achieving very high uniformity in horizontal channel reactors is fundamental to the process: Reactants are depleted as the gas stream passes over the wafer from the front to the back of the susceptor. The front of the wafer

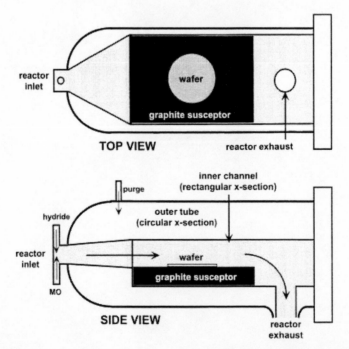

Figure 4.19. Schematic illustration of a typical low-pressure horizontal channel reactor for MOVPE.

is exposed to a higher concentration of reactants, and therefore it will experience a higher growth rate than the back part of the wafer. Depletion effects can be effectively compensated through the use of substrate rotation (to average the growth rate over the front and back portions of the wafer) and to a limited extent through reactor design. However, other limitations of the process adversely affect the gas flow characteristics and reactant distribution in horizontal channel reactors and thus limit the maximum achievable wafer uniformity.

Establishing smooth, laminar flow characteristics strongly depends on many reactor design parameters and process conditions (e.g., see refs. [17, 158]). Nonoptimal flow characteristics result in degraded wafer uniformity and unintentional composition gradients. Undesirable recirculation, return flow, and transverse and longitudinal "rolls" result from improper reactor geometry (e.g., incorporating right angles or "dead spaces") [158], from buoyancy effects arising from large thermal gradients, including those between the heated susceptor and the reactor sidewalls [159], and from inappropriate flow and pressure parameters. In addition, friction along the reactor sidewalls results in gradients in the flow velocity across the width of the reactor channel and a corresponding variation in the reactant depletion effects, symmetric about the longitudinal centerline of the channel. Reactant injection schemes can likewise lead to lateral variations in flow velocity and reactant concentration. Reactant and carrier gases are often injected from a relatively small orifice into a relatively wide channel, and flow visualization techniques have shown that jetting can result. Improved reactor and process design has been employed to address these shortcomings, with limited success. Streamlining the reactor geometry, to avoid abrupt and large discontinuities in the channel profile, and reducing the growth pressure, have led to marked improvements in the uniformity by improving the flow characteristics and eliminating eddy currents. Establishing turbulent flow upstream of the susceptor has proved beneficial for reducing jetting [160]. Using sophisticated inserts to divert the gas along the entire width of the channel has further improved the uniformity [161]. With these improvements, horizontal channel reactors have been shown to be capable of thickness uniformity of 1–2% across a 2-inch wafer, and somewhat worse for larger wafer sizes.

Horizontal channel reactors do not scale well to production capacities. For more than a single wafer, depletion and channel underfilling are too severe to overcome effectively. However, much of the early MOVPE VCSEL work was demonstrated using single-wafer horizontal channel reactors, and this important work has set the stage for the demonstration of larger-scale (production) epitaxy in the future, albeit using different reactor geometries. Zhou et al. [38] reported the first MOVPE growth of state-of-the-art VCSELs in 1991, using a horizontal flow channel reactor. In this case, wafer nonuniformity of "several percent" was reported. In later work Lear et al. [49] used MOVPE-grown material to establish long-lasting performance standards for 980 nm VCSELs. This material was grown in a horizontal channel reactor with a total capacity of one 2-inch wafer, for which nonuniformity of 6% over a 2-inch wafer was reported. Additional early MOVPE work [39, 40, 45, 51, 54] on high-performance near-IR VCSELs also made use of horizontal channel reactors. The Honeywell team used a reactor with a capacity of one 4-inch wafer, for which a wafer nonuniformity of ~4% over most of a 3-inch wafer was reported [40]. It is significant that, to date, many of the performance records for MOVPE VCSELs emitting over the 650–980 nm wavelength range have been achieved using single-wafer horizontal channel reactors. However, the reported nonuniformity and capacity for such reactors is clearly insufficient for a viable high-volume production process.

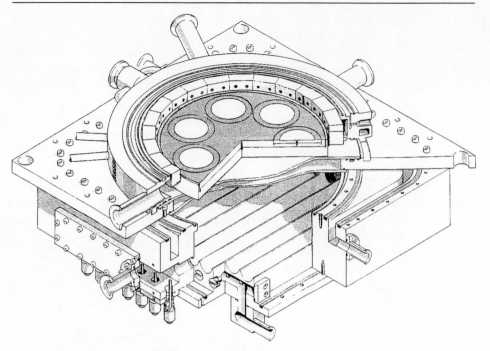

Figure 4.20. A cutaway view of a typical commercially available radial-flow planetary reactor. (Courtesy Aixtron Inc.)

Radial-Flow Planetary Reactors   A recent innovation in horizontal reactor geometries enables high-capacity, high-uniformity growth. Multiwafer radial-flow planetary reactors initially developed at Phillips [162] are centrosymmetric rather than linear. The gas is injected downwards to the center of a large circular rotating susceptor (the *planet*) and directed radially outwards toward the edge of the susceptor. The wafers (*satellites*) are positioned around the injector and are themselves rotated about their axes (this double-rotation action is the origin of the designation "planetary"). Figure 4.20 shows a cutaway view of a typical commercially available planetary reactor. Though this reactor bears little resemblance to channel reactors, it is a true horizontal reactor because the wafers are exposed to the gas as it travels radially from the center of the susceptor to the outer edge. In a vertical reactor, the gas impinges directly on the wafer surface from above.

This design offers some key advantages over horizontal channel reactors. Most importantly, its geometry removes the effects of reactor sidewalls, including both the friction and temperature effects that promote poor flow characteristics (as described previously). In addition, the geometry effectively eliminates injection jetting by combining radial injection with the rotation of the planet. These features reduce the challenge of achieving very high wafer uniformity as well as depletion effects. To accomplish this, the individual wafers (satellites) are rotated about their axes, to average the depletion from the point nearest the center of the planet to the point nearest the outer edge of the planet.

Deposition profiles in planetary reactors have been modeled as a function of total flow rate, pressure, and rotation condition [163]. The radial depletion profile, without satellite rotation, was found to be relatively linear for all but the very lowest total flow rates. Such a profile, due in part to the decreasing flow velocity along the planet radius [164], is

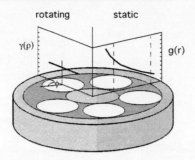

Figure 4.21. A schematic illustration of the radial-flow planetary reactor, showing the calculated dependence of layer thickness on radial position on the planet with and without satellite rotation. (Courtesy Aixtron Inc.)

advantageous as it is easier to compensate effectively through wafer rotation. The calculated thickness uniformity, with satellite rotation, was found to be insensitive to growth pressure over the range 100 mbar to atmospheric pressure. Achieving such high uniformity in layers grown even at atmospheric pressure is possible because of the absence of eddy or recirculation currents in the reactor [163]. A schematic of the planetary reactor, showing the calculated dependence of layer thickness on position within the reactor with and without satellite rotation, is given in Figure 4.21. Recent modeling of the temperature distribution within planetary reactors, including control of the ceiling temperature, has led to additional improvements in the wafer uniformity [165].

A key advantage of the planetary reactor is its relatively high growth efficiency (i.e., the ratio of material deposited on the wafer to reactant present in the gas phase), both in terms of growth rate as a function of MO (group III) concentration and material quality as a function of hydride (group V) partial pressure. This is due in part to the strong reactant depletion across the wafer. Compared to a horizontal channel reactor with a capacity of one 3-inch wafer, planetary reactors with a capacity of five 3-inch wafers (a 500% increase in capacity) may require only ∼50%–100% greater total flow to achieve similar growth rates. Hydride consumption is very efficient, partly because the hydride gas encounters the heated susceptor, aiding in its decomposition, prior to reaching the substrate, thus increasing the effective V/III ratio at the wafer surface. This is particularly advantageous for phosphide materials, because of the relatively high thermal stability of phosphine and the high vapor pressure of P in the growing layer.

Susceptor temperature uniformity is also important for achieving highly uniform epitaxy, especially as it affects dopant incorporation. Planetary reactors may offer some advantages in this area. Conventional planetary reactors make use of linear lamp heater strips, over which the planet is rotated. The combination of the double circular rotation of both the planet and the individual satellites over heaters with a linear geometry enables a high degree of temperature averaging and likely much improved temperature uniformity across the wafers. Temperature uniformity of 0.8°C over seven 2-inch wafers at a temperature of 700°C has been measured in such reactors [166].

There have been few reports of the use of planetary reactors for growth of VCSELs; however, the little information available is encouraging. Figure 4.22 shows a plot of the variation of the cavity mode wavelength in the directions normal to and parallel to the major flat, for a 3-inch diameter 850 nm VCSEL wafer grown in a planetary reactor with a capacity

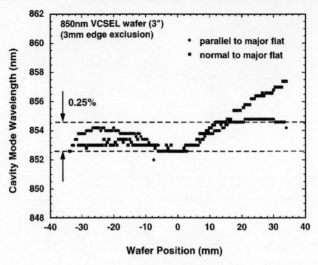

Figure 4.22. Variation of the cavity-mode wavelength over a 3-inch VCSEL wafer grown in a planetary reactor with a capacity of eight 3-inch wafers.

of eight 3-inch wafers (five 4-inch wafers). The total wavelength (thickness) variation is less than 0.25% for most of the wafer, and within 0.5% for the entire wafer (with 3 mm edge exclusion), comparable to the best reported results for any reactor geometry. Broadening the analysis to include all of the wafers on the susceptor, the total thickness nonuniformity is still less than 1%. The combination of high wafer capacity and excellent uniformity, along with high growth efficiency, may make planetary reactors a viable alternative for VCSEL production.

4.3.2.2.2 Vertical Reactors

Rotating Disk Reactors   Vertical rotating disk reactors (RDRs) represent a distinct class of MOVPE reactors that are subject to much different operating principles than their horizontal reactor counterparts. There have been several detailed theoretical and experimental treatments of these reactors [167–169]. An idealized schematic of an RDR, showing some of the key design features, is given in Figure 4.23. RDRs make use of well-characterized and stable natural flow patterns resulting from the centrifugal pumping action of a susceptor ("disk") rotating at very high speeds (~1,000 rpm) to define a compressed and uniform boundary layer at the wafer surface. The centrifugal pumping action of this reactor is a product of viscous drag near the surface of the rapidly spinning disk, which induces a natural laminar flow pattern, pulling the gas straight down toward the disk and then redirecting the flow radially to the edges of the disk. To establish ideal flow patterns, and hence provide the most uniform deposition, the gas velocity at the inlet must match the pumping capacity of the disk, itself a function of rotation rate, disk temperature, process pressure, and other system variables. This *velocity matching* condition can be readily predicted using simplified mathematical models with a high degree of accuracy so long as a number of boundary conditions describing the geometry and thermal profile of the reactor are maintained [168]. Indeed, the ease of modeling flow characteristics in RDRs is a key advantage of their design. Some of the reactor design variables that must be accounted for in the model include the

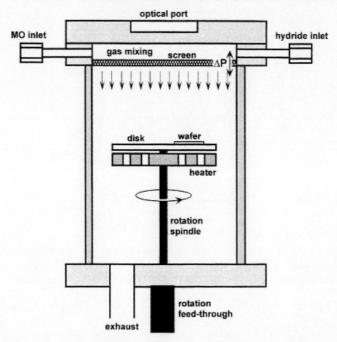

Figure 4.23. Schematic illustration of a rotating disk reactor (RDR).

disk diameter, the disk-to-inlet spacing, the temperature of the inlet and reactor sidewalls (ideally they are cooled), and the design of the inlet to the extent that uniform plug flow across the diameter of the reactor chamber is achieved. Some flexibility in these variables is permitted so long as the natural flow patterns induced by the rotating disk are not disturbed.

Under flow conditions that are not consistent with the velocity-matching requirement, recirculation currents may result that re-entrain reactants and reactant decomposition byproducts after they have already made contact with the disk (and wafer), leading to more nonuniform deposition and imprecise heterostructure interface definition. This behavior has been confirmed experimentally, using novel flow visualization techniques on RDR reactors with transparent cell walls [168, 170, 171]. RDRs differ fundamentally from vertical reactors often described in early MOVPE literature (e.g., see ref. 16), in which slow rotation rates were commonly used and inlet conditions were nonoptimal, leading to unstable flow patterns, including recirculation cells, and typically resulting in nonuniform material.

Using proper design and operating parameters, RDRs can produce extremely uniform VCSEL material. To date, the most uniform material has been reported for smaller reactors. Wang et al. [172] reported very uniform (<1% nonuniformity) AlGaAs deposition over a single 2-inch GaAs wafer, using an early RDR operated under conditions consistent with optimal flow characteristics. Very high uniformity AlGaAs DBR growth has been demonstrated using a custom-built RDR that strictly adhered to the design criteria described above [173]. The reactor, with a capacity of one 2-inch wafer, included several key design features: water cooling in the reactor sidewalls; uniform plug flow from an inlet constructed to provide a suitable pressure differential for uniform distribution across the entire diameter of the inlet; and a reactor geometry conducive to laminar flow generated from the rotating disk. Using a total inlet flow velocity calculated to match the pumping capacity of the

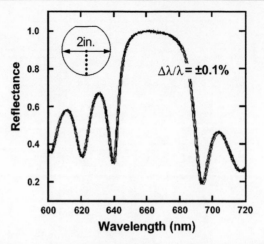

Figure 4.24. Reflectance spectra obtained at different points along the radius of a 2-inch wafer grown in a single-wafer RDR, indicating thickness uniformity on the order of ±0.1%. (From Killeen, Schneider and Figiel [173].)

rotating disk (a function of the pressure, rotation rate, and geometrical and temperature considerations), DBR wafers exhibiting wavelength (thickness) uniformity of ±0.1% were grown. Reflectance spectra obtained at different points along the radius of the wafer are presented in Figure 4.24. This is believed to be the highest uniformity yet achieved in a heteroepitaxial AlGaAs structure; indeed, it is somewhat difficult to measure nonuniformity in this limit. Such a high degree of uniformity would meet any VCSEL manufacturing specifications, given adequate process control to enable precise wavelength targeting.

Of course, such small reactors are of limited utility in VCSEL manufacturing; however, in theory, RDRs scale very well to arbitrarily large sizes of interest for production, so long as the boundary conditions are met – this is a key advantage of this reactor design. A schematic of a production-scale RDR, available commercially, is given in Figure 4.25. It is important to note that, unlike production-scale radial-flow planetary horizontal reactors, the total carrier and reactant flow rates must scale with the disk size to maintain optimal flow characteristics in the reactor. Alternatively, the growth pressure must be reduced. For a small reactor with enough capacity for one 4-inch wafer (a disk diameter of 7 inches) a typical total flow rate might be on the order of 20–30 slm at a pressure of 30–50 torr. To scale this same reactor for a total capacity of five 3-inch wafers would require a disk diameter of approximately 12 inches. Scaling the total flow rate to the reactor (inlet) area yields a required total flow of ~60 slm, including a factor of ~3 increase in reactant flows.

There have been a number of reports of the growth of highly uniform DBRs and VCSELs using commercial RDRs with multiple wafer capacity. Hou et al. [174] reported high-yield epitaxial growth of VCSELs on a commercial RDR with a 1×4-inch or 3×2-inch wafer capacity, including wavelength uniformity of 0.5% across a single 3-inch wafer. Tompa et al. [175] reported growth of uniform AlGaAs DBRs (±0.9% wavelength variation over 4-inch GaAs wafers) on a production-scale RDR with a disk diameter of 300 mm and a capacity of 4×4-inch wafers or 17×2-inch wafers. In accordance with the established scaling rules for such reactors, the results were obtained using higher total flow rates (~50 slm) and lower process pressures (15–30 torr) than necessary for smaller-scale reactors.

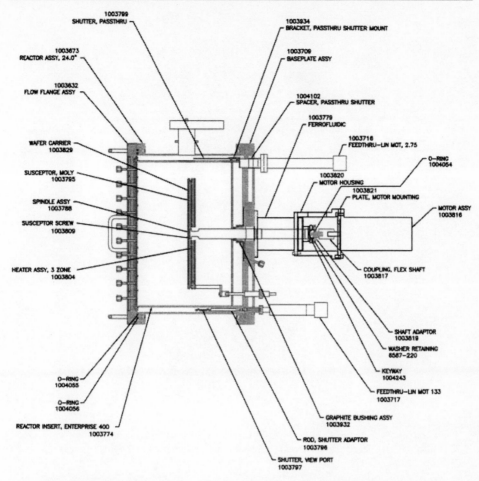

Figure 4.25. Schematic drawing of a commercially available production-scale RDR. (Courtesy Emcore Corp.)

A difficulty in scaling RDRs lies in the design of the heater. In conventional RDRs, the heater consists of a single resistive graphite coil. As the disk diameter is increased, achieving highly uniform temperature distribution becomes more challenging. Gurary et al. [176] demonstrated improved temperature uniformity on an RDR with a 12-inch disk diameter by employing a three-zone concentric heater under the rotating disk. By properly balancing the different zones, temperature uniformity on a 2-inch Si wafer was improved from $\pm 5.6°C$ (using a single-zone heater at a nominal temperature of 650–675°C) to $\pm 2.5°C$ (using the three-zone heater at a nominal temperature of 650–675°C).

There are as yet no published reports of highly uniform VCSELs grown on large production-scale RDRs. However, given the availability of large-scale commercial RDR systems, the excellent results demonstrated on the smaller systems, and the inherent scalability of RDRs, it is likely that high-yield VCSEL epitaxy processes on such systems is feasible.

Stagnation Flow Reactors One relatively recent (and ongoing) development in epitaxial growth tools is the newest generation of the vertical stagnation flow reactor. While stagnation

Figure 4.26. Schematic illustration of a commercially available stagnation flow reactor with a close-spaced "shower head" inlet configuration. (Courtesy Thomas Swan Ltd.)

flow reactors have in general been known for years [177], the first commercial reactors, available only recently, exhibit a number of unique features that merit special attention. Figure 4.26 gives a schematic illustration of a commercially available stagnation flow reactor. The injection "shower head" consists of a dense network of hundreds of capillary injection tubes (diameter ∼5 mils) brazed together, through which the reactants enter the reaction chamber. Two injection plenums, usually one for hydride gases and one for MO sources, are separated and use different independent capillary networks. The injection head is water cooled, allowing it to be placed very close to the growth surface and reducing the possibility of reactant decomposition prior to arriving at the boundary layer of the heated wafer. In usual practice the injection head is placed only 10–20 mm from the heated susceptor, so that the cooled injection head defines the vertical extent of the boundary layer. The susceptor is rotated; however, slow rotation rates (<200–300 rpm) are commonly used, as these reactors do not rely on viscosity pumping through high rotation rates to define the boundary layer and provide uniform deposition.

The stagnation flow design has several advantages. First, the gas flow characteristics, and thus the wafer uniformity, are relatively insensitive to the growth parameters such as rotation rate, total flow rate, growth pressure, etc., that more strongly influence growth in a conventional rotating disk reactor. Instead, the uniformity of the wafer more closely

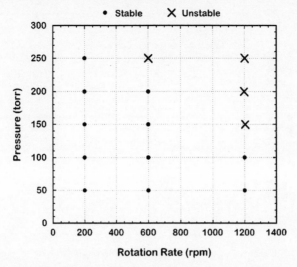

Figure 4.27. "Stability map" indicating the process parameter regimes for which stable flow characteristics are achieved in a vertical stagnation flow reactor with a "shower head" inlet. (After Hummel, Joh, and Bhat [178].)

mirrors the gas distribution profile arising from the inlet configuration. Figure 4.27 illustrates the relative insensitivity of the growth process to process variables. This "stability map" indicates the process parameter regimes for which stable flow characteristics are achieved (that is, with no recirculation currents) [178]. Stable characteristics are predicted for a very wide range of rotation rates and process pressures – even to pressures as high as 250 torr, given low enough rotation rates. In contrast, RDRs usually require high rotation rates and low pressures for stable flow characteristics, and even then, as described above, these variables are closely coupled and must be uniquely defined. The relative insensitivity of the stagnation flow reactors to these process parameters is of particular interest for VCSELs, as it implies a more robust, reproducible (and thus more manufacturable) process.

Another advantage of the reactor design lies in the design of the inlet: It allows one to keep group III and group V species completely separate until they are mixed in the boundary layer itself. This may be particularly useful for materials such as the nitrides, for which very strong prereactions have been noted. However, given the enhanced sensitivity of wafer uniformity to inlet design in stagnation flow reactors, such a complex design may make more difficult the realization of highly uniform wafers. Achieving uniform plug flow across the diameter of the inlet is of utmost importance, possibly even more so than for RDRs, for which the natural pumping action of the disk may compensate for imperfect inlet conditions. In addition, since the boundary layer is defined by the position of the cooled injection head, extreme care must be taken in manufacturing to ensure that the boundary layer is uniform. This too is unlike the case for RDRs, in which the thickness of the boundary layer is defined naturally by the process parameters.

There have been few demonstrations of growth of devices using this newest generation of stagnation flow reactors. Excellent uniformity of emission from 780 nm AlGaAs QW active regions grown in a stagnation flow reactor with a capacity of one 2-inch wafer is shown in Figure 4.28 [179]. In a larger reactor, with a capacity for three 2-inch wafers, wavelength (thickness) uniformity of $\pm 2\%$ has been demonstrated on AlGaAs DBRs [180]. To date there

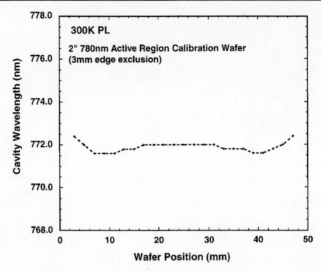

Figure 4.28. Photoluminescence wavelength as a function of wafer position for 780 nm AlGaAs QW active region grown in a stagnation flow reactor with a close-spaced "shower head" inlet. (From Hummel [179].)

are no published reports of the growth of high-performance VCSEL diodes in the higher-capacity reactors. Presumably, as the reactor design and the process are further refined, even better uniformity will be possible in the multiwafer reactors. In theory these reactors should scale very well. Since the wafer uniformity depends primarily on the uniformity of the gas flow profile as it is injected through the shower-head plenum, the uniformity possible with the larger multiwafer systems may be limited mainly by the uniformity with which such complex injectors, with many hundreds or even thousands of independent brazed capillaries, can be manufactured.

4.3.3 VCSEL Epitaxy Process Stability and Control (Run-To-Run Yield)

In this section the effect of "run-to-run" yield on manufacturing efficiency is discussed in light of recent advances in the development of in situ monitoring tools designed to allow more precise and reproducible thickness and composition control. Epitaxy process stability has been an issue more critical for development of VCSELs than for any other optoelectronic device. Demands on thickness precision and uniformity are so stringent as to effectively limit the widespread development of VCSELs until just recently, and even now critical issues remain. As suggested earlier, the choice of MBE as the epitaxy technique of choice for VCSELs was initially driven by the perception of superior process control. This was due in part to the ready availability of RHEED, which allowed pregrowth growth rate calibrations and thus improved process yield. Only in the past few years has MOVPE demonstrated sufficient stability for growth of VCSELs. However, even after these advances, it is clear that currently no epitaxy process is sufficiently stable, in the absence of appropriate process control methods, to enable very high yield manufacturing of VCSEL epitaxy. In this part of the chapter, epitaxy technology relevant to VCSEL development will be overviewed, with particular attention to thickness control, including development of in situ monitoring techniques.

4.3.3.1 Process Stability in MBE/GSMBE

Since the MBE/GSMBE is entirely a kinetically controlled growth process as opposed to that of the quasi-thermal equilibrium growth process of MOVPE, the growth of III–V material in MBE is less complex than in MOVPE. For the growth of $III_x^{(a)}$–$III_{1-x}^{(b)}$–V type alloys, the composition is effectively fixed by the group III element fluxes, since the sticking coefficients of the group III elements are all unity under normal growth conditions. In practice this means that composition can be controlled to <0.5% in the value of x. For the case of III–$V_y^{(c)}$–$V_{1-y}^{(d)}$ alloy films the situation is much more complex. Not only are the sticking coefficients of the incident group V elements less than unity, but complex interactions occur between them, and also between them and the particular group V element of which the substrate is composed. Therefore, the overall process control for both thickness and composition in MBE/GSMBE depends on the control of source molecular beam fluxes and substrate temperature.

A basic advantage of MBE/GSMBE over most other forms of epitaxy is the ability to include facilities for monitoring, control, and surface analysis in the growth chamber. The UHV environment of the MBE system allows the whole range of surface analytical probes to be employed so that both the chemical and structural properties of the growing epitaxial layers can be monitored and/or controlled before, during, and after growth. An ion gauge mounted on the substrate-holder carousel can be moved into the substrate position for a rapid, simple, and reliable method for direct measurement of beam fluxes and is invaluable for growth rate and alloy composition control. Beam intensity control of better than $\pm 1\%$ can readily be achieved using the temperature feedback control system. Reflection high-energy electron diffraction (RHEED) is essential because it gives information about substrate cleanliness and growth conditions. Its intensity oscillation can be used to measure the growth rate down to atomic layer accuracy. Furthermore, most of the state-of-the-art MBE systems are also equipped with an optical viewport at the line of sight of the growing substrate surface; thus external optical monitoring equipment, such as pyrometric interferometry or reflectance spectroscopy, can be readily implemented. Infrared pyrometer is commonly used to measure and control the substrate temperature with excellent accuracy, typically $<\pm 2°C$. These and other facilities enable either real-time or immediate postgrowth assessment of structural, chemical, and electrical properties related to growth parameters. Information thus gained can then be rapidly fed back for growth parameter control. An example of typical long-term "baseline" source stability is given in Figure 4.29, which shows the source temperature required to achieve a calibrated growth rate (determined from pyrometry) [44]. The data shown include both the GaAs growth rate and the AlGaAs alloy composition obtained daily using the pyrometric interferometry monitoring technique in a span of six months between source recharges. Using an ion gauge flux monitor and pyrometry to calibrate the source flux activation energies and rate of changes, excellent reproducibility was achieved. The daily target value for GaAs is 1 μm/h, and the average value for six months is 1.0025 μm/h with a variation of ± 0.012 μm/h. The target value for AlGaAs composition is 0.30 and the average value obtained is 0.294 ± 0.012. However, a wide range of source temperatures were required, between 3–10% of the mean value, which would lead to a large growth rate variation if left uncalibrated.

For the growth of VCSEL structures, because of its complexity and stringent requirements in thickness and composition controls, not only the long-term beam flux drifts but also short-term flux transients have to be controlled to less than 1% variations. The standard

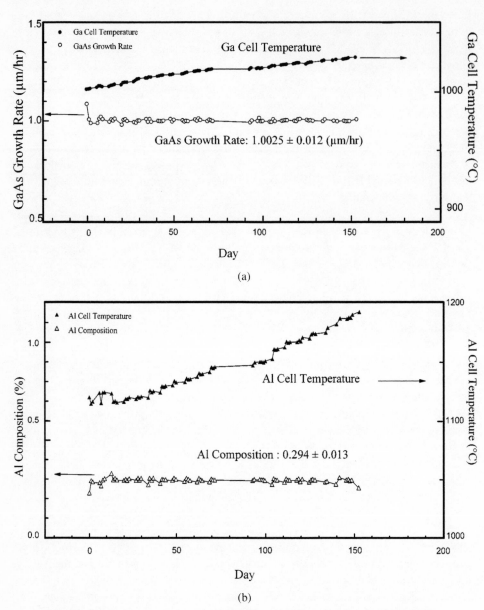

Figure 4.29. Plots of (a) the daily preset temperature of the Ga cell and the growth rate of GaAs and (b) the daily preset temperature of the Al cell and the alloy composition of AlGaAs grown in a GSMBE system. The preset teperatures were calibrated daily using interferometric pyrometry; the variation of the preset temperature with time, to achieve a constant growth rate, is an indication of the "baseline" (no control) process stability for this reactor.

growth-rate calibrations are not entirely sufficient; further improvement can be achieved by in situ monitoring and control technique as will be discussed latter in this section.

4.3.3.2 Process Stability in MOVPE

Metalorganic vapor-phase epitaxy differs from MBE in many significant respects and presents many different kinds of process control challenges. Since it is a "high-pressure" process (usually $\sim$.1 atmosphere), the electron diffraction techniques that have proven so useful in MBE are not possible. Even maintaining optical lines of sight has proven to be difficult, owing to interference from plentiful reaction by-products that deposit on most of the internal surfaces of the reactor. Furthermore, the requirement for easy accessibility into the reactor for frequent component cleaning, etc. make precision alignment of components difficult, especially in manufacturing environments. MOVPE is also a fundamentally more complex process than MBE, with more than 100 distinct process control points such as flow controllers, pressure transducers, and valves and strict requirements on gas handling design and reaction chamber geometry. This inherent complexity is one of the reasons for the relatively slow development of MOVPE relative to MBE for VCSEL epitaxy.

Given the importance of growth rate control and stability in VCSEL growth, it is useful to consider the factors influencing system stability in greater detail. Since MOVPE is mass-transport limited, growth rate and composition control are directly dependent on the control of the reactant partial pressures in the reactor cell. These in turn depend upon a number of factors, including: control of the MO reactant vapor pressures (dependent on bubbler temperature and pressure); control of the delivery rate of the carrier gas to the reactor (dependent on the mass flow controllers associated with a particular source); and control of total flow rate and pressure in the reactor cell (dependent on essentially every other transducer in the gas-handling system). Typical vapor pressures of the MO sources in their bubblers are in the range of 1–100 torr and depend exponentially on temperature. For example, the vapor pressure of TMGa is described using the relation $\log_{10} p(\text{torr}) = 8.495 - 1824/T\,(\text{K})$, which yields a vapor pressure of $\sim$36 torr at a typical bubbler temperature of $-10°\text{C}$. A variation of the bubbler temperature of only $\pm 0.5°\text{C}$ leads to a variation in the vapor pressure of $\Delta p \sim 2.2$ torr, or 6% of absolute. All other process parameters being constant, this leads to a corresponding 6% variation in the growth rate in the reactor. To achieve a minimum acceptable vapor pressure (growth rate) variation of 0.5%, the temperature stability must be better than $\pm 0.05°\text{C}$. To account for additional sources of variation in the reactant partial pressure in the reactor (e.g., output variations on *any* of the 40 or more transducers on a typical modern reactor), significantly better temperature control is desired. The situation is more complicated for ternary and quaternary alloys, for which multiple bubblers are used, and for doped layers, for which dopant species such as CBr_4 may strongly influence growth rate through etchback as described earlier. Furthermore, if AlGaAs composition control for oxide-confined VCSELs is considered, the acceptable composition reproducibility of 5×10^{-4} derived previously translates into a minimum acceptable temperature variation of $<\pm 0.003°\text{C}$ on each of two bubblers (TMGa and TMAl).

Despite the early lines of thinking that MOVPE was inherently too unstable for growth of such complicated devices as VCSELs, improvements in process design and the performance of key components such as temperature baths, mass flow controllers, and pressure transducers has led to significant improvements in the short-term "baseline" process stability (that is, the stability expected immediately after careful system calibration, with no

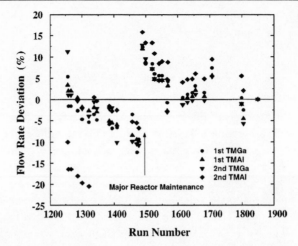

Figure 4.30. Growth rate normalized to source flow rate, as a function of run number. Data taken over the course of several months is indicative of "baseline" (no control) stability of an MOVPE reactor. Total deviations on the order of 15% are observed. (After Breiland et al. [181].)

external means of monitoring and control). The long-term baseline stability of an RDR MOVPE system used for VCSEL growth is indicated in Figure 4.30 [181]. Shown is the source flow rate required to deliver a calibrated growth rate as a function of run number, with the data taken over the period of several months. Long-term drift on the order of $\pm15\%$ is indicated. Short-term stability of this reactor is much better. Figure 4.31 shows the deviation of cavity mode wavelength for 850 nm and 780 nm VCSELs, grown over the course of 20 or 30 growth runs [174]. Total deviation of $\pm0.3\%$ is observed over this short time scale. Excellent short-term stability of 780 nm VCSEL growth in a stagnation flow reactor has also been observed [179]. Figure 4.32 shows in situ reflectance traces taken from six consecutive 780 nm VCSEL growth runs. The traces are so close as to overlap, making them indistiguishable. The inset of this figure shows a close-up of the reflectance trace at the very end of the run, where cumulative differences should be most apparent; even at this higher resolution, it is difficult to distinguish between the different runs. Longer-term process stability from this reactor has not yet been reported.

Despite the progress made in process stability and the implications for enabling small-scale VCSEL research and development, the presently achievable baseline process stability is not sufficient for high-yield and cost-efficient VCSEL epitaxy manufacturing, especially when oxide VCSEL manufacturing is considered. While short-term "baseline" process stability (always measured immediately after careful system calibration) can be excellent, longer-term stability (over weeks or months of operation) is highly variable. It is certain that, if VCSELs are to have the maximum impact on any number of optoelectronics markets, additional process control techniques must be implemented.

4.3.3.3 Development of In Situ Tools for Process Control

Implementation of on-line monitoring for process control has always been of interest to the compound semiconductor epitaxy community; however, the unique demands placed upon epitaxy processes by VCSELs have led to redoubled efforts in this area. A number of

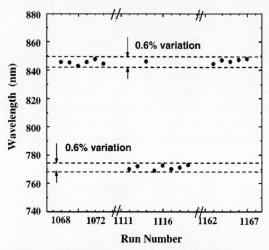

Figure 4.31. Wavelength reproducibility for 780 nm and 850 nm VCSELs grown in an RDR MOVPE reactor, obtained over the course of 20 or 30 growth runs. Total deviation is ~±0.3%. (After Hou et al. [174].)

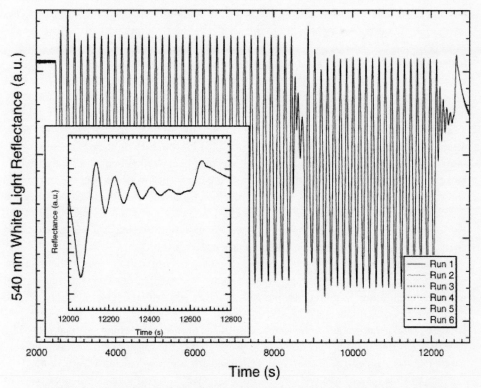

Figure 4.32. Single-wavelength reflectance traces measured in situ from six consecutive 780 nm VCSEL runs. Inset shows higher resolution of the traces over the last several layers in the structure. All six traces overlay and are indistinguishable, indicating a very high level of process stability. These structures were grown in a stagnation flow vertical reactor. (From Hummel [179].)

techniques have been used to monitor MBE and MOVPE in situ, including surface-sensitive techniques such as ellipsometry and reflectance difference spectroscopy (RDS) (e.g., see refs. 182, 183). However, only a few in situ techniques have been used expressly for the purpose of VCSEL growth or show particular promise as a manufacturing solution for on-line monitoring and control of VCSEL epitaxy. These include two techniques that rely on optical interference from the growing layer(s) to measure the growth rate: reflectance and pyrometric interferometry. Another, phase-locked epitaxy, takes advantage of RHEED in an MBE environment to measure the growth rate, monolayer by monolayer, at the wafer surface during the growth of the VCSEL structure. Finally, optical flux monitoring precisely measures the flux of the elemental species in the effused beam during MBE by monitoring the resonant optical absorption of the elements. Implementation of these techniques run the full range from periodic growth rate calibration to on-line closed-loop control during VCSEL growth.

4.3.3.3.1 Reflectance

Single-wavelength reflectance has received increasing attention in both MBE and MOVPE techniques in recent years [181, 184] and is now available commercially for MOVPE. In its simplest form, a laser is directed to the surface of the growing wafer, preferably at normal incidence, and the intensity of the reflected light is monitored. As the layer grows, the intensity of the reflected light will exhibit oscillations whose period is described by the expression $\Gamma(\text{sec}) = \lambda(2nr_g)^{-1}$, where λ is the wavelength of the reflected light in nm, n is the temperature- and composition-dependent index of refraction of the material, and r_g is the growth rate of the layer (in nm/sec). Thus, by accurately measuring the period of the intensity modulation, and with a precise knowledge of the index of refraction of the material, very accurate growth rates are obtained. Normal-incidence reflectance is relatively insensitive to incidence angle, so that some amount of "wobble" in wafer rotation can be tolerated. Reflectance is also insensitive to polarization, simplifying the optical setup with inexpensive components, and reducing sensitivity to optical window quality. These attributes make reflectance well suited for manufacturing environments, in which robust components and ease of use are prerequisite. In fact, reflectometry has found widespread application for the deposition of many materials aside from AlGaAs, including GaInAsP/InP [185], HgCdTe [186], tungsten [187], Si [188], and diamond [189], and it has even attracted attention for controlling precision III-V etching processes [190].

The simplicity and robustness of the experimental setup is one of the principal attractions of reflectance [181]. Figure 4.33 gives a schematic of the setup, in a conventional vertical reactor configuration. Required components include a white-light or laser light source, a Si photodetector, miscellaneous optical components (such as beamsplitters, filters, and fiber couplers), and a data acquisition system. One of the only system requirements imposed by reflectance is that there be clear optical lines of sight in the reactor cell. This is relatively straightforward in vertical reactors, for which the downflow of reactants tends to keep the top flange of the reactor relatively clean. Purged optical ports offer even more assurance of clear optical access to the wafer surface, and these are now available on typical commercial vertical reactors. However, in horizontal reactors the ceiling of the inner quartz channel is quickly coated with opaque reaction by-products that block the light. One effective solution to this problem is to include a purged optical port on the outer tube and a very small hole directly above the wafer on the inner tube (growth channel), as illustrated in Figure 4.34. If

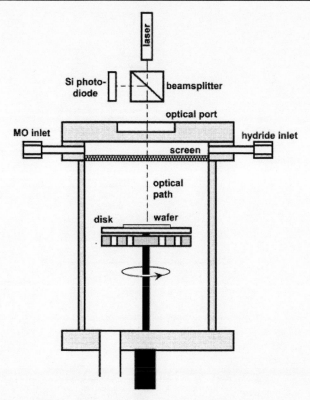

Figure 4.33. A schematic illustration of the setup for in situ reflectivity measurements in a conventional vertical reactor configuration. For spectral reflectance, the laser is replaced with a lamp, and a spectrometer is inserted in front of the detector. (After Killeen and Breiland [192].)

the hole in the inner tube is sufficiently small, and if the pressure in the outer tube is slightly higher than in the inner tube, then the perturbation to the growth process will be minimal. Thus, clear optical line of sight is preserved throughout a growth run. Such arrangements are available on commercial horizontal reactors.

Despite the relative simplicity of the experimental setup, data analysis can be quite complex. One of the principal difficulties is the requirement for precise knowledge of the index of refraction of growing material, a quantity that is dependent on both composition and temperature. In addition, complex structures such as DBRs are composed of multiple layers for which interfaces are often ambiguously defined. Defining the "start" of a layer may be difficult, as is obtaining a precise description of the underlying interface structure, which necessarily influences the analysis of the upper layers in the structure. Breiland and Killeen [191] simplified the analysis considerably by developing a method by which growth rates and optical constants could be extracted simultaneously from a growing layer, with only a knowledge of the starting (substrate) reflectance required. The technique, employing *virtual interface* methodology, was shown to provide very accurate and reproducible growth rate data regardless of the definition of the "starting" interface in the structure. Thus, growth rate and optical constant data for arbitrarily complex device structures, with arbitrary composition profiles, can be determined precisely. An example of the reflectivity data obtained

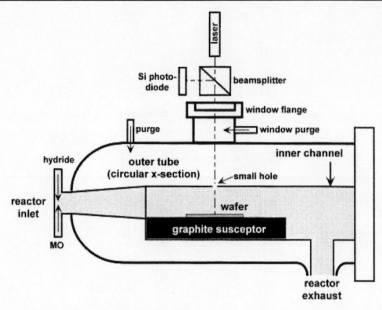

Figure 4.34. A schematic illustration of the setup for in situ reflectivity measurements in a conventional horizontal channel reactor. A small hole in the inner reactor liner tube provides clear optical access to the wafer surface.

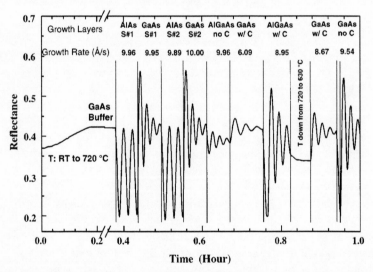

Figure 4.35. Example of in situ single-wavelength reflectivity trace obtained from a typical $Al_xGa_{1-x}As$ growth rate calibration structure, including several different $Al_xGa_{1-x}As$ compositions, growth temperatures, and C doping levels (to calibrate etch-back effects from CCl_4). (After Hou et al. [174].)

from a calibration structure, including several different alloy compositions and doping concentrations, is given in Figure 4.35. By applying the virtual interface method to this data, precise growth rates for several different layer compositions (including any CCl_4 or CBr_4 etchback effects) are readily obtained in a single growth run [181, 174].

Another innovation in reflectance technology has been the development of spectral reflectance, in which a broadband light source is used and the reflected light is spectrally resolved [192]. Significantly more information is available by using a wide range of wavelengths, or even by using several preselected wavelengths. Besides a more accurate measure of the growth rate, spectral reflectance can be used to extract compositional information from the growing layers. One of the drawbacks of spectral reflectance is simply the enormity of the data collected during a growth run. For a VCSEL growth run, a complete reflectance spectrum is measured at every acquisition cycle (usually every 1 or so). However, by accounting for the temperature dependence of the indices of the materials in the structure, such complete data can provide a very accurate picture of the evolution of the structure as it grows, allowing the development of more advanced on-the-fly control techniques.

Implementation of reflectance for control of MOVPE has thus far been primarily in the form of pregrowth growth rate calibration. Frateschi et al. [193] reported the use of in situ reflectance to calibrate the growth rates of GaAs and AlAs just prior to growing VCSELs in an atmospheric pressure reactor. Wide variations in the growth rates ($\pm 10\%$) were detected before the pregrowth calibration routine was enacted; afterwards the workers reported reproducibility of $<1\%$ and accuracy to within 2%. The authors attributed the small remaining error to temperature drift in the bubblers during the course of the growth run. In subsequent work Hummel et al. [194] automated the measurement and correction procedure at the initiation of the growth of the structure. More recently, Hou et al. [175] employed periodic (every 20–30 runs) in situ growth rate calibrations using reflectance to improve wavelength control in VCSEL epitaxy. Reproducibility of $\pm 0.3\%$ in VCSEL Fabry–Perot wavelength was achieved over a period of 100 growth runs, despite the fact that different DBR and active region designs were incorporated into the VCSELs. It is likely that even better short-term and long-term reproducibility could be realized by incorporating the calibration layers *within* every VCSEL growth run and automating the correction procedure. Ultimately even more sophisticated process control, including real-time (*on-the-fly*) closed-loop control as the VCSEL is growing, is desirable. Feasibility for such an approach has been demonstrated using pyrometric interferometry in GSMBE [44, 195, 196], as will be described in the following section.

Reflectance has also been implemented in MBE of VCSELs. Chalmers and Killeen [184] reported the use of reflectance to measure the optical cavity thickness partway through its growth. In this way appropriate corrections could be made prior to the completion of the growth. Although improved reproducibility for the Fabry–Perot wavelength was achieved, this procedure requires the undesirable step of interrupting growth midway through the structure. Jackson et al. [197] reported improved control over layer thickness in MBE using a reflectance spectroscopy technique in which spectra of distributed Bragg reflector (DBR) mirrors were obtained during growth to enable mid-run correction of growth rate drifts.

4.3.3.3.2 Pyrometric Interferometry

Pyrometric interferometry enables the measurement of the substrate temperature as well as the determination of growth rate and optical thickness of the growing layer from the well-known-oscillations in temperature radiation of the heated wafer during the growth of heterostructures [198]. This technique uses an optical pyrometer to measure the periodic modulation in emitted light intensity of the growing epitaxial layer surface. This periodic modulation is due to changes in interference between light reflected from the top and bottom

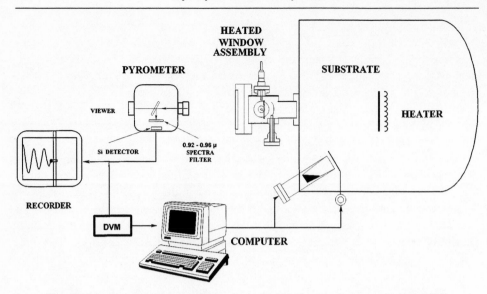

Figure 4.36. Typical experimental setup for in situ interferometric pyrometry in an MBE chamber. (After Houng et al. [195].)

of the epilayer as the epilayer thickness changes. By monitoring the endpoint intensity and phase of the signal one can determine the optical thickness of the growing layer. As only a viewport in front of the substrate is required, this technique can be easily adapted to most epitaxial growth systems. Moreover, it is relatively insensitive to vibration, rotation, and misalignment of the wafer. A typical experimental setup is shown in Figure 4.36.

For the growth of VCSEL and related structures, it is highly desirable to a develop real-time closed-loop control system based on in situ monitoring methods that yield information on the growing layer thickness as well as the cumulative thickness of the deposited materials. Through a real-time computation algorithm, one can control each growing layer thickness precisely, and any deviations from the target growth rate can then be compensated for the growth of the subsequent layers, such as $\lambda/4$ DBR layers and QW layers in the active region of a VCSEL structure. The growth of highly reproducible 980 nm VCSEL wafers using this technique was first reported in 1993 [34]. Since then, several papers have been published on the theoretical treatments and refinements of the technique for the growth of 850 and 980 nm VCSELs [34, 44, 195, 196, 200, 201]. The reproducibility of the VCSEL structures with a variation of the Fabry–Perot wavelength of $\pm0.4\%$ and the gain peak of the active region with a run-to-run variation of less than 0.3% has been demonstrated using this technique [34, 44]. Although pyrometric interferometry has been demonstrated more widely in a GSMBE environment, this technique is also compatible with MOVPE.

4.3.3.3.3 Phase-Locked Epitaxy

It is well known that one cycle of intensity oscillations seen in a reflection high-energy electron diffraction (RHEED) pattern of the growing MBE surface corresponds to the growth of one atomic layer of material deposited. This idea forms the basis of the phase-locked epitaxy technique (PLE) [202], in which RHEED oscillations are used to control the MBE group III source shutters during growth and thereby grow layers of thickness corresponding

to a precise number of atomic layers. This technique has produced a $(AlAs)_6(GaAs)_3/$
$(AlAs)_1(GaAs)_4$ 35.5-period Bragg reflector with a 1% maximum variation in layer to layer
thickness. Additionally, 980 nm VCSELs were grown entirely by PLE [46, 203]. However,
there are several limitations associated with this technique. The Al mole fraction of the layers is limited to $0 \leq x \leq 0.75$ because of the difficulty in maintaining sufficiently smooth
surfaces for RHEED oscillations at high Al mole fractions. In addition, because the substrate cannot be rotated during growth to accommodate PLE, variations in layer thickness
of 5%/cm are present.

4.3.3.3.4 Optical Flux Monitoring

While the techniques described above measure the growth rate of the layers at the growing
surface, optical flux monitoring (OFM) measures the rate of delivery of the elemental
source to the growing wafer [204]. The growth rate is then calculated from the known
semiempirical relation between source flux and growth rate. The key to this technique lies
in a highly sensitive resonant absorption measurement capable of detecting relative changes
in the transmitted light of 0.01%. The molecular beam flux is then calculated from a Beer's
law relationship between the atomic beam density and the transmission of the light. Hollow-
cathode lamps were employed as the light source and were operated using feedback from
a reference photodiode for enhanced stability.

OFM was used with automatic closed-loop feedback to control the center wavelength of
several AlGaAs DBR structures, grown over the course of several days and with different
effusion cell temperatures [204]. Wavelength reproducibility with accuracy of $<\pm0.3\%$ was
achieved, compared to a typical reproducibility of $\pm5\%$ with no control and no effort to
change the cell temperatures. An additional demonstration of the effectiveness of OFM was
reported in subsequent work [205], in which the PL emission of GaInAs multiple quantum-
well structures grown with and without OFM control was compared. Even though the Ga
effusion cell temperature was deliberately varied during the course of the MQW growth, a
single narrow PL line was obtained when OFM control was activated, implying thickness
variations of $<0.8\%$. In contrast, a similar structure grown with the same deliberate Ga
growth rate ramping scheme and with conventional timed shutter control (no active OFM
control) exhibited multiple PL peaks, with energy position consistent with a thickness
variation of 13% between the thinnest and thickest QWs.

4.4 New Materials and Wavelengths

In the previous sections of this chapter, epitaxial growth issues associated with the growth
of AlGaAs-based VCSELs were overviewed in detail. To date AlGaAs has been the most
technologically relevant materials system for VCSELs, enabling the growth of devices emitting in the range 780 nm to 980 nm (the latter incorporating InGaAs QWs). However, this
materials system does not lend itself to a great deal of design flexibility, either for new operating wavelengths outside this narrow range or for further enhanced device performance
within this range. Indeed, there are a wide range of III–V alloys that may be grown on GaAs
and InP substrates. Many of these alloys offer advantages in performance and wavelength
accessibility. Figure 4.37 shows the families of ternary and quaternary alloys that are technologically relevant for VCSELs on GaAs and InP substrates, indicating the dependence
of bandgap energy (wavelength) on lattice mismatch relative to these substrates.

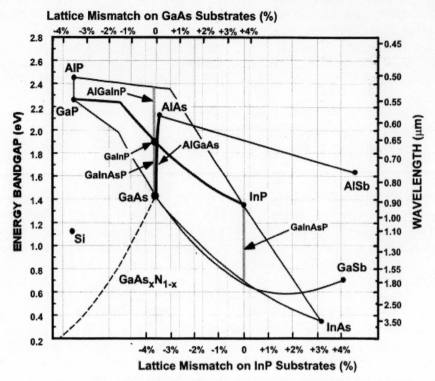

Figure 4.37. A plot of the dependence of bandgap energy (emission wavelength) on lattice mismatch relative to GaAs and InP substrates, for the most important ternary and quaternary alloy families of interest for VCSELs.

In this section some recent developments in the growth of VCSELs using different materials and epitaxy approaches will be reviewed. It is noteworthy that such innovations in materials and epitaxy technology form the foundation for future progress in VCSEL technology, impacting devices, systems, and applications. Topics to be discussed here include: AlGaInP/AlGaAs-based red (650–680 nm) VCSELs, GaInAsP/AlGaAs-based near-IR (780–850 nm) VCSELs, AlGaInAs-based 850 nm VCSELs, GaInAsN-based long-wavelength (1.3–1.55 μm) VCSELs, Sb-based VCSELs grown by MBE, application of selective-area epitaxy to VCSELs for multiple-wavelength arrays, and finally III–V nitride-based visible VCSELs. These topics are not intended to encompass the spectrum of innovations in materials and epitaxy for VCSELs; nor do they offer a complete review of progress within the topic areas; rather they give some sense for the degree to which materials and epitaxy drive further progress in the VCSEL field.

4.4.1 AlGaInP-Based Red VCSELs

VCSELs that emit in the red portion of the spectrum, between 630 and 680 nm, are of interest for a number of applications, including plastic optical fiber (POF) communications, laser printing, laser pointing and telemetry, bar code scanning, and digital video disks (DVDs). If developed in a timely manner, red VCSELs may enjoy the broadest application of any VCSELs. These devices are described in detail in Chapter 7 of this book, as well as in another review [103]. In the following, some of which the unique epitaxy issues associated with red VCSELs will be overviewed.

The principal epitaxial challenge facing the development of red VCSELs was the necessity of incorporating ternary/quaternary AlGaInP heterostructures into the active region/optical cavity of a structure otherwise containing only AlGaAs. AlGaInP, principally used for high-brightness LEDs and red edge-emitting lasers, is grown almost exclusively using MOVPE, which seems to produce the highest quality material. Incorporating these heterostructures presents a number of epitaxial growth and design challenges, related to the added complexity of the growth, the necessity of using MOVPE (at a time in which MBE was the only "proven" growth technique for VCSELs), and other more fundamental properties of AlGaInP materials and heterostructures.

AlGaInP laser heterostructures are known to be subject to some very difficult epitaxy design challenges in themselves (e.g., see the review by Bour [206]), even before the additional complexity associated with incorporating AlGaAs DBR structures is considered. For example, AlGaInP materials are known to spontaneously order on the group III sublattice, reducing the energy bandgap and optical efficiency. These undesirable properties can be avoided only through the use of high growth temperatures and highly misoriented substrates. Additional key challenges posed by the laser heterostructures include very shallow confinement potentials and relatively high thermal resistance, which combine to make them particularly sensitive to heating effects, including thermally activated carrier leakage. Given the high operating current densities in VCSELs, heating effects are likely to be even more critical for red VCSELs compared to edge-emitters, and thus they are a key consideration in epitaxial design. A key issue in the growth of AlGaInP laser heterostructures is p-doping. While C has been shown to be an almost ideal p-dopant for AlGaAs DBRs, it is ineffective in AlGaInP alloys. Both Mg and Zn can be used for p-doping of AlGaInP; however, they exhibit high activation energies in AlGaInP, resulting in relatively low hole concentrations ($<10^{18}$ cm^{-3}), which may exacerbate current leakage. In addition, as described in Section 4.2.2, these dopants are also prone to diffusion, particularly during the long growth times and high temperatures required for VCSELs.

The first demonstration that AlGaAs DBR and AlGaInP active region heterostructures could be successfully integrated into a VCSEL structure in a single growth sequence was reported in 1992. Schneider et al. [207] reported photopumped lasing from an undoped structure with GaInP/AlGaInP strained QWs in the active region and Al$_{0.5}$Ga$_{0.5}$As/AlAs DBRs for the mirrors. Fabricating an electrically injected red VCSEL proved more difficult because of the challenges outlined above. Several epitaxial design approaches were taken to address these issues.

Early work focused on the use of Mg-doped AlInP cladding layers in the optical cavity, to provide the greatest possible electron confinement from the combination of the enhanced p-doping capabilities of Mg and the inclusion of the largest bandgap material in this alloy system (AlInP). Mg diffusion, resulting in dopant pileup at the AlGaInP/AlGaAs (cavity/DBR) interfaces, complicated the growth, and in this early work necessitated the use of very thick (8λ) optical cavities. This basic structure resulted in the first reports of room-temperature pulsed [208, 209] and then continuous wave [210, 211] operation from a red VCSEL. However, the high thermal resistance and high free-carrier absorption losses associated with such a thick AlGaInP cavity are clearly undesirable and may have compromised the ultimate performance of the devices.

A novel epitaxial design solution to this problem incorporated AlAsP grading between the AlGaInP cavity and the AlGaAs DBRs to enable C doping very close to the QWs [212]. This approach not only eliminates Mg and Zn doping altogether, it also minimizes the quantity of AlGaInP materials in the cavity to 100–200 nm, which should yield a significant reduction in the thermal resistance of the device structure. The drawback is greater thermally activated carrier leakage caused by lower confinement potentials available in the AlGaAs DBR that is placed in closer proximity to the active region, as indicated by higher T_0 measured in edge-emitters with the same epitaxial design. This structure provided the best performance reported to date in a red VCSEL, with CW output powers as high as 8 mW (multimode) and 2 mW (single mode) at $\lambda = 680$–690 nm [213].

It is likely that red VCSEL epitaxial designs will continue to evolve, and future designs will incorporate features from these and other approaches. Clearly, materials and epitaxy design issues will continue to play a key role in establishing the performance metrics that ultimately will enable this technology to move into real-world applications. Perhaps just as critical to the development of these devices are manufacturing issues, which also exhibit unique characteristics relative to AlGaAs-based VCSELs.

Red VCSELs exhibit greater sensitivity to process instabilities than the longer-wavelength AlGaAs-based VCSELs. The optical cavity of a typical red VCSEL consists of an active region heterostructure with three different AlGaInP alloy compositions: $Ga_{1-x}In_xP$ strained QWs, $(Al_{0.5}Ga_{0.5})_{0.5}In_{0.5}P$ barriers, and $(Al_{0.7}Ga_{0.3})_{0.5}In_{0.5}P$ or $Al_{0.5}In_{0.5}P$ spacer layers, along with doping from either Zn or Mg. In addition, the DBRs of the red VCSEL are composed of n_L and n_H layers and continuously graded segments as described in Section 4.2. The added complexity of the optical cavity in this structure relative to the simple AlGaAs active regions in near-IR VCSELs makes the red VCSEL a significantly more challenging structure to manufacture. Short-term baseline stability for the growth of Red VCSELs in a simple horizontal channel reactor is illustrated in Figure 4.38. Despite the added complexity, reproducibility to within ~1% can be achieved over the course of several weeks with appropriate system calibration. However, another complication to red VCSEL epitaxy manufacturing is related to fundamental design constraints. $Al_{0.5}Ga_{0.5}As/AlAs$ DBRs exhibit a smaller index contrast than DBRs used in near-IR VCSELs, and so typical DBR stopband widths are only ~40 nm at a design wavelength of 670 nm, about half that of typical near-IR VCSELs. More importantly, the reflectivity of the stopband varies more rapidly with wavelength near the center of the stopband. The properties of red VCSELs will thus be more sensitive to slight mismatch between the cavity wavelength and the DBR stopband center, as well as the offset between the cavity and active region gain wavelengths. Along with additional progress in epitaxial growth and device design, process control and yield issues discussed in Section 4.3 are critical to establishing a manufacturable red VCSEL epitaxy process.

4.4.2 GaInAsP-Based Near-IR VCSELs

AlGaAs alloys and heterostructures have formed the foundation for near-IR VCSEL work for the duration of VCSEL development. This alloy system is the most straightforward to implement in VCSELs, owing to its relative ease of growth, lattice-matching throughout the composition range, and optical properties appropriate for growth of active regions and DBRs operating in the technologically relevant near-IR wavelength range. However, despite the relative ease with which AlGaAs is incorporated in VCSELs, these alloys and heterostructures also impose a host of design constraints on the devices. Principal among these include

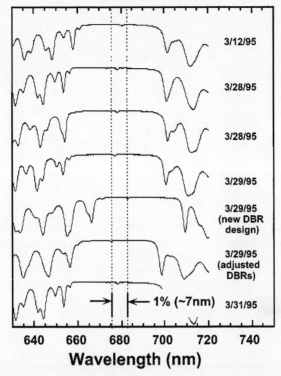

Figure 4.38. Reflectance spectra measured from a series of red VCSELs, grown over the course of almost one month in a simple horizontal channel reactor, illustrating baseline process stability on the order of 1%. (From Schneider and Figiel [113].)

the inability to use strain for performance enhancements through band structure engineering (it may be used only for moderate compressive strains, at the expense of longer wavelength) and the requirement that significant fractions of Al be present in the active region to achieve wavelengths $\lambda < 830$ nm (eg., $Al_xGa_{1-x}As$ with $x = 0.12$ is required for $\lambda = 780$ nm, a technologically relevant wavelength of interest for printing and DVD applications).

An alternate materials system, GaInAsP, may provide a number of advantages. This quaternary system is outlined in Figure 4.37. Note that, for quaternary alloys lattice matched to GaAs, the wavelength range 850 nm (GaAs) to 650 nm ($Ga_{0.5}In_{0.5}P$) can be accessed. Furthermore, complete flexibility in using tensile or compressive strain is possible, enabling a greater range of wavelength and/or performance benefits than possible with lattice-matched alloys. This raises the possibility of employing strained quantum wells for 850 nm and 780 nm, with the option of using strain-compensated barriers for improved structural stability and enhanced gain. Photoluminescence spectra obtained from AlGaAs- and GaInAsP-based active regions designed to emit at 780 nm and 850 nm are shown in Figure 4.39 [113]. The structures were grown using MOVPE, in the same growth system. Comparable room-temperature optical efficiency is achieved in both cases. Note that in the 850 nm structures, the QW active region was strained ($\varepsilon \sim 1\%$), which should provide significant performance enhancements relative to unstrained GaAs QWs. An added benefit is the ability to completely avoid the inclusion of Al in the active region for the entire wavelength range from 980 nm to 650 nm. This should enhance the reliability of the devices by removing the nonradiative

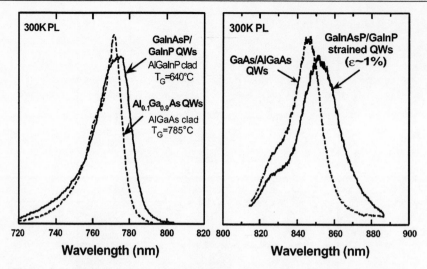

Figure 4.39. Room-temperature photoluminescence spectra for 780 nm (left) and 850 nm (right) active regions. Conventional AlGaAs-based active regions are compared with GaInAsP-based active regions. The GaInAsP-based 850 nm QWs are grown with 1% compressive strain. (From Schneider and Figiel [113].)

recombination centers associated with oxygen impurities that are often found in higher concentrations in Al-containing alloys. Reliability should be further enhanced due to the use of InGaAsP strained quantum wells in the active region, which is thought to pin dislocation movement and thus suppress the formation of dark-line defects [214]. Numerous workers have reported on the growth and fabrication of high-performance "Al-free" edge-emitting lasers employing GaInAsP active regions (e.g., see refs. [215, 216]) and have in fact observed improved resistance to dark-line defect formation even in lattice-matched structures [217].

Despite the apparent advantages of this alloy system, to date there has been only one report of GaInAsP-based near-IR VCSELs. Schneider and Crawford [218] reported a structure with an active region containing several nominally lattice-matched GaInAsP QWs designed for emission at 780 nm, with $Ga_{0.5}In_{0.5}P$ barrier layers and $(Al_{0.5}Ga_{0.5})_{0.5}In_{0.5}P$ cladding layers. The $(Al_{0.5}Ga_{0.5})_{0.5}In_{0.5}P$ alloy was used to provide the largest possible confining energy for injected carriers and thus the potential for improved higher-temperature performance. These initial structures exhibited reasonable performance; simple proton-implanted devices achieved CW lasing at room temperature with thresholds in the range 3–7 mA, with peak output powers of 2–5 mW. However, substantial performance improvements relative to AlGaAs-based devices were not demonstrated. In the future, by employing strain-compensated active regions, and more closely studying device characteristics such as the over-temperature performance and reliability, it is likely that such performance improvements will be realized.

4.4.3 AlGaInAs-Based 850 nm VCSELs

An alternate approach to incorporating strained QWs for 850 nm uses AlGaInAs QWs, in which the Al and In concentrations are balanced to achieve compressively strained active regions emitting near 850 nm [219, 220]. This approach has been used to fabricate 850 nm oxide-confined VCSELs, and threshold currents of 156 μA were achieved in devices with 2.8 μm apertures. The improved performance relative to conventional GaAs QW structures grown in the same reactor was attributed to the increased gain and reduction in transparency current associated with strained QWs. Postgrowth rapid thermal annealing was found to be critical for low-threshold operation, presumably to anneal nonradiative point defects. These threshold values are comparable to the best achieved with conventional lattice-matched GaAs/AlGaAs QW active regions grown by MOVPE (although efficiency is still not as good). Disadvantages of this approach relative to GaInAsP include the necessity of including Al in the QWs, which may compromise the reliability of the devices, and the inability to use strain compensation. However, this alloy combination is relatively straightforward using MBE.

4.4.4 GaInAsN-Based 1.3 μm VCSELs

Efforts at realizing 1.3 μm VCSELs have received intense attention for several years, owing to enormous potential markets in telecommunications and data communications. The fundamental difficulty in the epitaxial design is the lack of materials lattice-matched to InP that exhibit sufficient index contrast to create high-reflectivity DBRs or, conversely, the lack of materials lattice-matched to GaAs substrates that exhibit sufficient gain at long wavelengths to create viable laser active regions. One of the most promising approaches is that of fusion bonding active regions of GaInAsP/InP grown on InP substrates and GaAs/AlGaAs DBRs grown on GaAs substrates. This approach, is described in detail in Chapter 8 of this book, seeks to marry the best features of both alloy families by directly integrating them from their respective substrates. Other approaches instead seek to extend the properties of the alloys available on these substrates to enable growth of the entire structure in a single epitaxial run. Of the latter, GaInNAs QW active regions appear the most promising.

It has recently been shown that the huge band bowing present in the GaAsN system leads to a strong suppression of the GaAs bandgap with the addition of small amounts ($\sim$few percent) of N [221]. The expected dependence of bandgap on alloy composition is illustrated in Figure 4.38. The addition of N in GaInAs is expected to not only suppress the bandgap energy of GaInAs but also to compensate the compressive strain from In, enabling a greater concentration of In to be incorporated without relaxation, further increasing the wavelength. Besides allowing 1.3 μm AlGaAs-based DBRs to be integrated with 1.3 μm active regions in a single epitaxial growth run, it would also provide advantages in electron confinement in the active region for better high-temperature performance (i.e., higher T_0) [221], which would provide a distinct advantage over wafer-fusion processes.

Kondow et al. [221] proposed this system for application to 1.3 μm laser diodes and VCSELs. They later demonstrated growth of GaNAs alloys with up to 10% nitrogen incorporation [222], with photoluminescence at around 1.3 μm. Later this same group reported the first edge-emitting laser diodes, operating pulsed at room temperature [223], and then CW [224]. Besides relatively low thresholds of 0.8 kA/cm^2, the group measured record-high characteristic temperature of 127 K, consistent with earlier predictions. While the initial reports of GaInNAs lasers were made with GSMBE-grown material, more recently GaInNAs-based long-wavelength laser diodes have also been demonstrated with MOVPE-grown material [225].

Progress in realizing 1.3 μm VCSELs with this approach has been rapid. The Hitachi group reported CW photopumped lasing at 1.22 μm and 1.25 μm in a VCSEL structure containing a $Ga_{0.7}In_{0.3}N_{0.007}As_{0.993}$ QW active region and AlGaAs-based DBRs [226, 227]. More recently, this group reported pulsed electrically injected lasing from a similar doped structure [228]. As DBR designs are further refined, and active region optical efficiency is optimized, this approach could well become the new standard for long-wavelength VCSELs on GaAs substrates, at 1.3 μm and, perhaps, beyond.

4.4.5 Sb-Based VCSELs

Another approach to achieving long-wavelength (1.3 μm) VCSELs is to incorporate Sb-based alloys into the structure. On InP substrates, the alloys are incorporated into the DBRs to extend the available index contrast for high-reflectivity mirrors. On GaAs substrates, it may be possible to incorporate Sb into the active region to extend the accessible gain wavelengths. Historically, MBE and GSMBE have shown greater proficiency in growing antimonides; however, recently there has been substantial progress in the growth of Sb-based laser materials using MOVPE [229–231].

AlAsSb and AlPSb, both lattice-matched to InP, exhibit both large (indirect) bandgaps and relatively low indices of refraction, making them viable alternative for the low-index (n_L) layer in a DBR structure. AlAsSb and the quaternary system AlGaAsSb are indicated in Figure 4.38. An optical index difference as high as $\Delta x \sim 0.6$ is possible in this materials system, making possible the growth of high-reflectivity DBRs. GSMBE has been used to grow AlPSb/GaPSb DBRs exhibiting high reflectivity at 1.55 μm [232, 233]. No VCSELs have yet been reported with such DBR structures. AlGaAsSb-based DBRs exhibiting high reflectivity at 1.3 μm [234] and 1.55 μm [235, 236] have been reported by several groups. Blum et al. [237] have recently reported an optically pumped VCSEL emitting at 1.55 μm and employing AlGaAsSb/AlAsSb DBRs. These structures, grown using MBE on InP substrates, included conventional GaInAsP/InP active regions. To date there are no reports of electrically injected VCSELs operating between 1.3 and 1.55 μm using antimonides in the DBRs. However, Te doping experiments have indicated that low-resistance n-type DBRs are feasible [238, 239]. Furthermore, oxidation experiments on AlAsSb alloys show that, while not as straightforward as AlAs oxidation, the AlAsSb oxidation process has potential for application to oxide-confined long-wavelength VCSELs on InP substrates, with all the anticipated performance benefits [240, 242].

A recent innovation in 1.3 μm VCSELs instead incorporates Sb-containing heterostructures into the active region of structures grown on GaAs substrates. It was proposed that using strained GaAsSb-GaInAs/GaAs "bi-layer" QWs might enable wavelengths as long as 1.3 μm to be accessed with compressive strains of only ~2% (similar to those required for 980 nm active regions) [243]. Photoluminescence wavelengths as long as 1.33 mm were demonstrated at room temperature. More recently PL has been demonstrated at wavelengths as long as 1.32 μm with GaAsSb/GaAs strained QWs [244]. Furthermore electrically injected lasing from broad-area edge-emitting lasers was achieved, with threshold current densities of 3 kA/cm^2 at a wavelength of 1.22 μm. VCSELs employing these active regions have not yet been demonstrated.

4.4.6 Growth Techniques for Multiple-Wavelength Arrays

Wavelength-division multiplexing using VCSELs emitting at different and precisely controlled wavelengths has received intense interest recently due to the bandwidth enhancements possible in both data and telecommunications applications. Fabrication of arrays of VCSELs emitting at controllable wavelengths presents some interesting epitaxy challenges. Several approaches have been reported, using both MBE and MOVPE techniques, and substantial progress has been made toward demonstrating arrays emitting over a broad spectral range.

One of the most promising approaches takes advantage of selective growth characteristics in MOVPE [245–247]. Many groups have reported on the selectivity of growth rate on substrate surface preparation. This includes the enhancement of growth rate on unmasked regions of a surface patterned with SiO_2 or Si_3N_4, proportional to the ratio of masked to unmasked area. Because this effect is dependent on the mobility of reactant species on the masked areas, it can also be used to give lateral control of the alloy composition and QW emission wavelength [248–250]. Growth rate also depends strongly on the growth planes, so that appropriate etching to produce nonplanar surfaces provides an opportunity for three-dimensional control of the epitaxial structure. This technique has been used for the growth of quantum wire lasers [251] and low-threshold quantum-well lasers [252–254]. Interestingly, just as doping depends on substrate orientation, it also depends strongly on growth plane in nonplanar growth [104, 109]. This property has been used to fabricate lateral p-n junctions [104, 109, 110].

Growth rate can also be controlled laterally on the surface simply by creating topographical relief on the surface of the wafer [255]. It is this approach that has been used to fabricate arrays of VCSELs emitting at different wavelengths [256–258]. By varying the width of etched channels relative to that of adjoining mesas, the growth rate can be modified in a controllable manner. Ortiz et al. [258] found the growth rate to be enhanced on the surface of the mesas and reduced in the adjoining channels. The wavelength of emission for the VSCEL was found to be linearly dependent on the *spatial duty factor*, a measure of the ratio between the channel and mesa widths. VCSEL emission was obtained over a wavelength range exceeding 20 nm (30 nm by photopumping).

While nonplanar and selective growth techniques are not readily achieved in MBE growth, other approaches have proven effective for lateral control of VCSEL emission wavelength. By patterning the backside of a wafer to impart controlled temperature variation on the top surface of the wafer, it is possible to vary the growth rate laterally in a controlled fashion [259–262]. Total wavelength emission range of greater than 60 nm was reported, with a distribution of 117.14 nm/mm and a small standard deviation of 4.1%.

Another approach for wavelength control of VCSELs requires micromachining to control the wavelength of emission of the VCSEL [263–266]. A micromachined suspended deformable membrane functions as the top mirror for the VCSEL laser cavity. By applying a bias between the membrane and the cavity, the air gap thickness is modulated, adding a phase shift that effectively modulates the resonance wavelength of the VCSEL. Although not an epitaxy problem per say, the materials challenges presented with this approach make it appropriate for mention in this review. A tuning range of greater than 30 nm (multimode) has been demonstrated [265, 267].

4.4.7 III–V Nitrides

The recent demonstration of CW room-temperature lasing in GaN-based blue edge-emitting lasers [267] raises the possibility for nitride-based blue VCSELs. Such devices would have application in high-density memory devices such as those using the digital video disk (DVD) format. Further development of longer-wavelength nitride-based VCSELs, including green VCSELs, may lead to the development of laser-based RGB displays. While little effort in the development of nitride-based VCSELs has been reported, it is likely that this field will attract increasing attention as the epitaxy issues that have plagued nitride materials technology are resolved. Although it is not at all certain that all-epitaxial approaches will be preferred (i.e., whether or not dielectric mirrors will be preferred to epitaxial ones), there have been reports of the growth and characterization of AlGaN-based DBRs. Fritz and Drummond [269] reported the MBE growth of AlN/GaN DBRs and measured the index contrast in the system. More recently, Redwing et al. [270] reported the MOVPE growth of an all-epitaxial blue VCSEL with $Al_{0.40}Ga_{0.60}N$-$Al_{0.12}Ga_{0.88}N$ DBRs. The structure was made to lase at 363 nm with photopumping. This first demonstration of photopumped lasing in an all-epitaxial GaN-based blue VCSEL may set the stage for further development in the future.

4.5 Summary

Epitaxial design and growth issues critical to the development of high-performance vertical-cavity surface-emitting lasers were reviewed. The epitaxial design approaches that are necessary for achieving low resistance and high efficiency in AlGaAs-based near-IR VCSELs were overviewed for growth by both molecular beam epitaxy (MBE) and metalorganic vapor phase epitaxy (MOVPE). The influence of composition engineering and dopant engineering on the resistance of p-DBRs was described for both growth techniques. In MBE, choice of acceptor dopant species was shown to be critical for the growth of low-resistance p-DBRs owing to the limited flexibility in composition grading. In MOVPE, device resistance is less sensitive to choice of dopant species because of the continuous composition grading capabilities of the growth technique. However, doping issues, including those surrounding unintentional dopants, were shown to affect epitaxial design and device performance in varied and complex ways. This section was concluded with a summary of epitaxial design issues associated with oxide-confined VCSELs and p-substrate VCSELs.

VCSELs were shown to present a number of difficult challenges for robust epitaxial manufacturing, owing to the extreme sensitivity of their properties to even minute thickness and composition nonuniformities and deviations from design specifications. Manufacturing issues relevant to VCSEL production were addressed, including those associated with uniformity yield and throughput (dependent primarily on reactor design) and run-to-run yield (dependent on process stability and control). High wavelength uniformity (better than 1%) has been achieved on production-scale MOVPE reactors (both radial-flow planetary systems and rotating disk reactors) as well as multiwafer GSMBE systems. Although process control strategies are still in their infancy, a number of in situ monitoring techniques show a great deal of promise. Chief among them are optical interferometric techniques, based on reflectometry and pyrometry.

In the final part of this chapter, we reviewed the more recent developments in materials and epitaxy technology that underly development of devices with improved performance and wider wavelength accessibility and that form the foundation for future advances in the field. Included were AlGaInP-based red VCSELs, GaInAsP-based near-IR VSCELs, GaInNAs-based long-wavelength (1.3 μm) VCSELs, Sb-based VCSELs for long-wavelength applications, growth techniques for multiple-wavelength arrays, and III–V nitride VCSELs. As the materials and epitaxy foundation for these devices continues to mature over the coming years, VCSELs will no doubt have an even more profound impact on optoelectronics and photonics systems and applications.

Acknowledgments

RPS would like to acknowledge past and present colleagues from Sandia National Laboratories and Hewlett-Packard Labs, who have contributed much to the author's understanding as well as to the richness of this field: J. Tsao, H. Hou, K. Killeen, J. Figiel, T. Brennan, W. Brieland, J. Lott, H. Chui, S. Chalmers, K. Choquette, K. Lear, T. Drummond (all from SNL); S. Corzine, M. Tan, D. Babic, K. Killeen, A. Tandon, S. Steward, S. Hummel (from HPL); and R. Herrick (HP-CSSD). YHH acknowledges valuable discussions with D. Mars (HPL) and B. Liang (HP-CSSD).

References

[1] Koyama, F., Kinoshita, S., and Iga, K., "Room temperature CW operation of GaAs vertical cavity surface emitting laser," *Trans. Inst. Electron. Inf. Commun. Eng. E*, **E71**, 1089–90 (1988).

[2] Parker, E. H. C., ed., *The Technology and Physics of Molecular Beam Eptiaxy*, Plenum Press, New York (1985).

[3] Herman, M. A. and Sitter, H., *Molecular Beam Epitaxy, Fundamentals and Current Status*, Springer-Verlag, Berlin (1989).

[4] Tsao, J. Y., *Materials Fundamentals of Molecular Beam Epitaxy*, Academic Press, San Diego (1993).

[5] Arthur, J. R., "Interaction of Ga and As_2 molecular beams with GaAs surface," *J. Appl. Phys.*, **39**, 4032–4 (1968).

[6] Foxon, C. T., and Joyce, B. A., "Surface processes controlling the growth of $Ga_x In_{1-x}As$ and $Ga_x In_{1-x}P$ alloy films by MBE," *J. Cryst. Growth*, **44**, 75–83 (1978).

[7] Panish, M. B. and Temkin, H., *Gas Source Molecular Beam Epitaxy*, Springer-Verlag, Berlin (1993).

[8] Kunzel, H., Knecht, J., Jung, H., Wunstel, K., and Ploog, K., "The effect of arsenic vapor species on electrical and optical properties of GaAs grown by molecular beam epitaxy," *Appl. Phys.*, **A28**, 168–73 (1982).

[9] Sacks, R. N., Eichler, D. W., Pastorello, R. A., and Colombo, P., "Evaluation of a new high capacity, all-tantalum molecular-beam-epitaxy arsenic cracker furnace," *Vac. Sci. Technol.*, **B8**(2), 168–71 (1990).

[10] Neave, J. H., Blood, P., and Joyce, B. A., "A correlation between electron traps and growth processes in n-GaAs prepared by molecular beam epitaxy," *Appl. Phys. Lett.*, **36**(4), 311–12 (1980).

[11] Foxon, C. T., Blood, P., Fletcher, E. D., Hilton, D., Hulyer, P. J., and Vening, M., "Substrate temperature dependence of SQW alloy and superlattice lasers grown by MBE using As_2," *J. Cryst. Growth*, **111**, 1047–51 (1991).

[12] Pao, Y. C. and Harris, J. S., Jr., "Molecular beam epitaxial growth and structure design of $In_{0.52}Al_{0.48}As/In_{0.53}Ga_{0.47}As/InP$ HEMTs," *J. Crystal Growth*, **111**, 489–94 (1991).

[13] Chand, N., Harris, T. D., Chu, S. N. D., Fitzgerald, E. A., Lopata, J., Schnoes, M., and Dutta, N. K., "Performance of a valved arsenic cracker source for MBE growth," *J. Cryst. Growth*, **126**, 530–38 (1993).

[14] Zhang, G., Ovtchinnikov, A., Nappi, J., Hakkarainen, T., and Asonen, H., "GSMBE growth of GaInAsP on GaAs substrates and its application to 0.98 μm lasers," *J. Cryst. Growth* **127**, 1033–6 (1995).

[15] Pessa, M., Hakkarainen, T., Keskinen, J., Rakennus, K., Salokatve, A., Zhang, G., and Asonen, H., "Current state of gas-source molecular beam epitaxy for growth of optoelectronic materials," *Proc. SPIE*, **1361**, 529–42 (1991).

[16] Ludowise, M. J., "Metalorganic chemical vapor deposition of III-V semiconductors," *J. Appl. Phys.*, **58**, R31–55 (1984).

[17] Stringfellow, G. B., *Organometallic Vapor Phase Epitaxy: Theory and Practice,* Academic Press, New York (1989).

[18] Razhegi, M., *The MOCVD Challenge: Volume 2 A Survey of GaInAsP-GaAs for Photonic and Electronic Device Applications.* Inst. of Physics (1995).

[19] Chui, H. C., Biefeld, R. M., Hammons, B. E., Breiland, W. G., Brennan, T. M., Jones, E. D., Moffat, H. K., Kim, M. H., Grodzinski, P., Chang, K. H., and Lee, H. C., "Tertiarybutylarsine for metalorganic chemical vapor deposition growth of high purity, high uniformity films," *J. Electron. Mater.*, **26**, 37–42 (1997).

[20] Lee, Y. H., Jewell, J. L., Scherer, A., McCall, S. L., Harbison, J. P., and Florez, L. T., "Room-temperature continuous-wave vertical-cavity single-quantum-well microlaser diodes," *Electron. Lett. (UK)*, **25**, 1377–8 (1989).

[21] Lee, Y. H., Tell, B., Brown-Goebeler, K., Jewell, J. L., and Hove, J. V., "Top-surface-emitting GaAs four-quantum-well lasers emitting at 0.85 μm," *Electron. Lett. (UK)*, **26**, 710–11 (1990).

[22] Geels, R. S., Corzine, S. W., Scott, J. W., Young, D. B., and Coldren, L. A., "Low threshold planarized vertical-cavity surface-emitting lasers," *IEEE Photonics Technol. Lett.*, **2**, 234–6 (1990).

[23] Geels, R. S., Corzine, S. W., and Coldren, L. A., "InGaAs vertical-cavity surface-emitting lasers," *IEEE J. Quantum Electron.*, **27**, 1359–67 (1991).

[24] Tai, K., Yang, L., Wang, Y. H., Wynn, J. D., and Cho, A. Y., "Drastic reduction of series resistance in doped semiconductor distributed Bragg reflectors for surface-emitting lasers," *Appl. Phys. Lett.*, **56**, 2496–8 (1990).

[25] Chalmers, S. A., Lear, K. L., and Killeen, K. P., "Low resistance wavelength-reproducible p-type (Al, Ga)As distributed Bragg reflectors grown by molecular beam epitaxy," *Appl. Phys. Lett.*, **62**, 1585–7 (1993).

[26] Zeeb, E. and Ebeling, K. J., "Potential barriers and current-voltage characteristics of p-doped graded AlAs-GaAs heterojunctions," *J. Appl. Phys.*, **72**, 993–9 (1992).

[27] Kurihara, K., Numai, T., Ogura, I., Yasuda, A., Sugimoto, M., and Kasahara, K., "Reduction in the series resistance of the distributed Bragg reflector in vertical cavities by using quasi-graded superlattices at the heterointerfaces," *J. Appl. Phys.*, **73**, 21–7 (1993).

[28] Reiner, G., Zeeb, E., Moller, B., Ries, M., and Ebeling, K. J., "Optimization of planar Be-doped InGaAs VCSEL's with two-sided output," *IEEE Photonics Technol. Lett.*, **7**, 730–2 (1995).

[29] Schubert, E. F., Tu, L. W., Zydzik, G. J., Kopf, R. F., Benvenuti, A., and Pinto, M. R., "Elimination of heterojunction band discontinuities by modulation doping," *Appl. Phys. Lett.*, **60**, 466–8 (1992).

[30] Peters, M. G., Thibeault, B. J., Young, D. B., Scott, J. W., Peters, F. H., Gossard, A. C., and Coldren, L. A., "Band-gap engineered digital alloy interfaces for lower resistance vertical-cavity surface-emitting lasers," *Appl. Phys. Lett.*, **63**, 3411–13 (1993).

[31] Babic, D. I., Dohler, G. H., Bowers, J. E., and Hu, E. L., "Isotype heterojunctions with flat valence or conduction band," *IEEE J. Quantum Electron.*, **33**, 2195–8 (1997).

[32] Kopf, R. F., Schubert, E. F., Downey, S. W., and Emerson, A. B., "n- and p-Type dopant profiles in distributed Bragg reflector structures and their effect on resistance," *Appl. Phys. Lett.*, **61**, 1820–2 (1992).

[33] Houng, Y. M., Lester, S. D., Mars, D. E., and Miller, J. N., "Growth of high-quality p-type GaAs epitaxial layers using carbon tetrabromide by gas source molecular-beam epitaxy and molecular-beam epitaxy," *J. Vac. Sci. Technol.*, **B11**(3), 915–18 (1993).

[34] Houng, Y. M., Tan, M. R. T., Liang, B. W., Wang, S. Y., Yang, L., and Mars, D. E.,

"InGaAs(0.98 μm)/GaAs vertical cavity surface emitting laser grown by gas-source molecular beam epitaxy," *J. Cryst. Growth*, **136**, 216–20 (1994).

[35] Weigl, B., Grabherr, M., Jung, C., Jager, R., Reiner, G., Michalzik, R., Sowada, D., and Ebeling, K. J., "High-performance oxide-confined GaAs VCSELs," *IEEE J. Sel. Topics Quantum Electron*, **3**, 409–15 (1997).

[36] Cunningham, B. T., Haase, M. A., McCollum, M. J., Baker, J. E., and Stillman, G. E., "Heavy carbon doping of metalorganic chemical vapor deposition grown GaAs using carbon tetrachloride," *Appl. Phys. Lett.*, **54**, 1905–7 (1989).

[37] Cunningham, B. T., Baker, J. E., and Stillman, G. E., "Carbon tetrachloride doped $Al_xGa_{1-x}As$ grown by metalorganic chemical vapor deposition," *Appl. Phys. Lett.*, **56**, 836–8 (1990).

[38] Zhou, P., Cheng, J., Schaus, C. F., Sun, S. Z., Zheng, K., Armour, E., Hains, C., Hsin, W., Myers, D. R., and Vawter, G. A., "Low series resistance high-efficiency GaAs/AlGaAs vertical-cavity surface-emitting lasers with continuously graded mirrors grown by MOCVD," *IEEE Photonics Technol. Lett.*, **3**, 591–3 (1991).

[39] Schneider, R. P., Jr., Lott, J. A., Lear, K. L., Choquette, K. D., Crawford, M. H., Kilcoyne, S. P., and Figiel, J. J., "Metalorganic vapor phase epitaxial growth of red and infrared vertical-cavity surface-emitting laser diodes," *J. Cryst. Growth*, **145**, 838–45 (1994).

[40] Hibbs-Brenner, M. K., Morgan, R. A., Walterson, R. A., Lehman, J. A., Kalweit, E. L., Bounnak, S., Marta, T., and Gieske, R., "Performance, uniformity, and yield of 850-nm VCSELs deposited by MOVPE," *IEEE Photonics Technol. Lett.*, **8**, 7–9 (1996).

[41] Van der Ziel, J. P. and Ilegems, M., "Multilayer GaAs-AlGaAs dielectric quarter-wave stacks grown by molecular beam epitaxy," *Appl. Opt.*, **14**, 2627–30 (1975).

[42] Thornton, R. L., Burnham, R. D., and Streifer, W., "High reflectivity GaAs-AlGaAs mirrors fabricated by metalorganic chemical vapor deposition," *Appl. Phys. Lett.*, **45**, 1028–30 (1984).

[43] Yoffe, G. W., "Rectification in heavily doped p-type GaAs/AlAs heterojunctions," *J. Appl. Phys.*, **70**, 1081–3 (1991).

[44] Houng, Y. M. and Tan, M. R. T., "MBE growth of highly reproducible VCSELs," *J. Cryst. Growth*, **175/176**, 352–58 (1997).

[45] Lu, B., Luo, W.-L., Hains, C., Cheng, J., Schneider, R. P., Choquette, R. P., Lear, K. L., Kilcoyne, S. P., and Zolper, J. C., "High-efficiency and high-power vertical-cavity surface-emitting laser designed for cryogenic applications," *IEEE Photonics Technol. Lett.*, **7**, 447–8 (1995).

[46] Walker, J. D., Kuchta, D. M., and Smith, J. S., "Vertical-cavity surface-emitting laser diodes fabricated by phase-locked epitaxy," *Appl. Phys. Lett.*, **59**, 2079–81 (1991).

[47] Kawakami, T., Kadota, Y., Kohama, Y., and Tadokoro, T., "Low-threshold current low-voltage vertical-cavity surface-emitting lasers with low-Al-content p-type mirrors grown by MOCVD," *IEEE Photonics Technol. Lett.*, **4**, 1325–7 (1992).

[48] Peters, M. G., Young, D. B., Peters, F. H., Scott, J. W., Thibeault, B. J., and Coldren, L. A., "17.3% Peak wall plug efficiency vertical-cavity surface-emitting lasers using lower barrier mirrors," *IEEE Photonics Technol. Lett.*, **6**, 31–3 (1994).

[49] Lear, K. L., Schneider, R. P., Choquette, K. D., Kilcoyne, S. P., Figiel, J. J., and Zolper, J. C., "Vertical cavity surface emitting lasers with 21% efficiency by metalorganic vapor phase epitaxy," *IEEE Photonics Technol. Lett.*, **6**, 1053–5 (1994).

[50] Lear, K. L. and Schneider, R. P., Jr., "Uniparabolic mirror grading for vertical cavity surface emitting lasers," *Appl. Phys. Lett.*, **68**, 605–7 (1996).

[51] Lear, K. L., Choquette, K. D., Schneider, R. P., Jr., Kilcoyne, S. P., and Geib, K. M., "Selectively oxidised vertical cavity surface emitting lasers with 50% power conversion efficiency," *Electron. Lett.*, **31**, 208–9 (1995).

[52] Morgan, R. A., Hibbs-Brenner, M. K., Marta, T. M., Walterson, R. A., Bounnak, S., Kalweit, E. L., and Lehman, J. A., "200 Degrees C, 96-nm wavelength range, continuous-wave lasing from unbonded GaAs MOVPE-grown vertical cavity surface-emitting lasers," *IEEE Photonics Technol. Lett.*, **7**, 441–3 (1995).

[53] Morgan, R. A., Hibbs-Brenner, M. K., Walterson, R. A., Lehman, J. A., Marta, T. M., Bounnak, S., Kalweit, E. L., Akinwande, T., and Nohava, J. C., "Producible GaAs-based MOVPE-grown vertical-cavity top-surface emitting lasers with record performance," *Electron. Lett.*, **31**, 462–4 (1995).

[54] Schneider, R. P., Jr., Tan, M. R. T., Corzine, S. W., and Wang, S. Y., "Oxide-confined 850 nm vertical-cavity lasers for multimode-fibre data communications," *Electron. Lett.*, **32**, 1300–2 (1996).

[55] Jager, R., Grabherr, M., Jung, C., Michalzik, R., Reiner, G., Weigl, B., and Ebeling, K. J., "57% Wallplug efficiency oxide-confined 850 nm wavelength GaAs VCSELs," *Electron. Lett.*, **33**, 330–1 (1997).

[56] Drummond, T. J., Gee, J., Terry, F. L., and Weng, R., "Application of InAlAs/GaAs superlattice alloys to GaAs solar cells," *Conference Record of the Twenty First IEEE Photovoltaic Specialists Conference–1990*, **1**, 105–10 (1990).

[57] Naganuma, M., and Takahashi, K., "Ionized Zn doping of GaAs molecular beam epitaxial films," *Appl. Phys. Lett.*, **27**, 342–4 (1975).

[58] Bean, J. C. and Dingle, R., "Luminescent p-GaAs grown by zinc ion doped MBE," *Appl. Phys. Lett.*, **35**, 925–7 (1979).

[59] Wood, C. E. C., Desimone, D., Singer, K., and Wicks, G. W., "Magnesium- and calcium-doping behavior in molecular-beam epitaxial III–V compounds," *J. Appl. Phys.*, **53**, 4230–355 (1982).

[60] Cho, A. Y. and Panish, M. B., "Magnesium-doped GaAs and $Al_xGa_{1-x}As$ by molecular-beam epitaxy," *J. Appl. Phys.*, **43**, 511823 (1972).

[61] Ilegems, M., "Beryllium doping and diffusion in molecular-beam epitaxy of GaAs and $Al_xGa_{1-x}As$," *J. Appl. Phys.*, **48**, 1278–87 (1977).

[62] Pao, Y. C., Hierl, T., and Cooper, T., "Surface effect-induced fast Be diffusion in heavily doped GaAs grown by molecular-beam epitaxy," *J. Appl. Phys.*, **60**, 201–4 (1986).

[63] Miller, J. N., Collins, D. M., and Moll, N. J., "Control of Be diffusion in molecular beam epitaxy GaAs," *Appl. Phys. Lett.*, **46**, 960–62 (1985).

[64] Ito, H., Nakajima, O., and Ishibashi, T., "Carbon doping for AlGaAs/GaAs heterojunction bipolar transistors by molecular–beam epitaxy," *Appl. Phys. Lett.*, **62**, 2099–101 (1993).

[65] Malik, R. J., Nottenburg, R. N., Schubert, E. F., Walker, J. F., and Ryan, R. W., "Carbon doping in molecular beam epitaxy of GaAs from a heated graphite filament," *Appl. Phys. Lett.*, **53**, 2661–63 (1988).

[66] Hoke, W. E., Lemonias, P. J., Lyman, P. S., Hendriks, H. T., Weir, D., and Colombo, P., "Carbon doping of MBE and $Ga_{0.7}Al_{0.3}As$ films using a graphite filament," *J. Cryst. Growth*, **111**, 269–73 (1991).

[67] Hoke, W. E., Lemonias, P. J., Weir, D. G., Brierley, S. K., Hendriks, H. T., Adlerstein, M. G., and Zaitlin, M. P., "Molecular-beam epitaxial growth of heavily acceptor doped GaAs layers for GaAlAs/GaAs heterojunction bipolar transistors," *J. Vac. Sci. Technol.*, **B10**(2), 856–58 (1992).

[68] De Lyon, T. J., Buchan, N. I., Kirchner, P. D., Woodall, J. M., Scilla, G. J., and Cardone, F., "High carbon doping efficiency of bromomethanes in gas source molecular beam epitaxial growth of GaAs," *Appl. Phys. Lett.*, **58**, 51719 (1991).

[69] Lemonias, P. J., Hoke, W. E., Weir, D. G., and Hendriks, H. T., "Carbon p^+ doping of molecular-beam epitaxial GaAs films using carbon tetrabromide," *J. Vac. Sci. Technol.*, **B12**(2), 1190–92 (1994).

[70] Kuech, T. F. and Redwing, J. M., "Carbon doping in metalorganic vapor phase epitaxy," *J. Crystal Growth*, **145**, 382–389 (1994).

[71] Hou, H. Q., Hammons, B. E., and Chui, H. C., "Carbon doping and etching of $Al_xGa_{1-x}As$ $(0 < x < 1)$ with carbon tetrachloride in metalorganic vapor phase epitaxy," *Appl. Phys. Lett.*, **70**, 3600–2 (1996).

[72] Richter, E., Kurpas, P., Gutsche, D., and Weyers, M., "Carbon doped GaAs grown in low

pressure-metalorganic vapor phase epitaxy using carbon tetrabromide," *J. Electron. Mater.*, **24**, 1719–22 (1995).

[73] Tateno, K., Kohama, Y., and Amano, C., "Carbon doping and etching effects of CBr_4 during metalorganic chemical vapor deposition of GaAs and AlAs," *J. Cryst. Growth*, **172**, 5–12 (1997).

[74] Bass, S. J. and Oliver, P. E., *Inst. Phys. Conf. Ser.* **33b**, 1 (1977).

[75] Glew, R. W., "Zinc doping of MOCVD GaAs," *J. Cryst. Growth*, **68**, 44–7 (1984).

[76] Lewis, C. R., Dietze, W. T., and Ludowise, M. J., "The growth of magnesium-doped GaAs by the OM-VPE process," *J. Electron. Mater.*, **12**, 507–24 (1983).

[77] Kozen, A., Nojima, S., Tenmyo, J., and Asahi, J., "Metalorganic-vapor-phase-epitaxial growth of Mg-doped $Ga_{1-x}Al_xAs$ layers and their properties," *J. Appl. Phys.*, **59**, 1156–9 (1986).

[78] Johson, N. M., Burnham, R. D., Street, R. A., and Thornton, R. L., "Hydrogen passivation of shallow-acceptor impurities in p-type GaAs," *Phys. Rev. B*, **33**, 1102–5 (1986).

[79] Pan, N., Bose, S. S., Kim, M. H., Stillman, G. E., Chambers, F., Devane, G., Ito, C. R., and Feng, M., "Hydrogen passivation of C acceptors in high-purity GaAs," *Appl. Phys. Lett.*, **51**, 596–8 (1987).

[80] Kozuch, D. M., Stavola, M., Pearton, S. J., Abernathy, C. R., and Hobson, W. S., "Passivation of carbon-doped GaAs layers by hydrogen introduced by annealing and growth ambients," *J. Appl. Phys.*, **73**, 3716–24 (1993).

[81] Pearton, S. J., "Hydrogen in III–V compound semiconductors," *Mater. Sci. Forum*, **148–149**, 393–480 (1994).

[82] Stockman, S. A., Hanson, A. W., Lichtenthal, S. M., Fresina, M. T., Hofler, G. E., Hsieh, K. C., and Stillman, G. E., "Passivation of carbon acceptors during growth of carbon-doped GaAs, InGaAs, and HBTs by MOCVD," *J. Electron. Mater.*, **21**, 1111–18 (1992).

[83] Ishibashi, A., Mannoh, M., and Ohnaka, K., "Annealing effects on hydrogen passivation of Zn acceptors in AlGaInP with p-GaAs cap layer grown by metalorganic vapor phase epitaxy," *J. Cryst. Growth*, **145**, 414–19 (1994).

[84] Li, G. and Jagadish, C., "Effect of low temperature postannealing on the hole density of C delta-doped GaAs and $Al_{0.3}Ga_{0.7}As$," *Appl. Phys. Lett.*, **69**, 2551–3 (1996).

[85] Hobson, W. S., Ren, F., Abernathy, C. R., Pearton, S. J., Fullowan, T. R., Lothian, J., Jordan, A. S., and Lunardi, L. M., "Carbon-doped base GaAs-AlGaAs HBT's grown by MOMBE and MOCVD regrowth," *IEEE Electron Device Lett.*, **11**, 241–3 (1990).

[86] Enquist, P. M., "p-Type doping limit of carbon in organometallic vapor phase epitaxial growth of GaAs using carbon tetrachloride," *Appl. Phys. Lett.*, **57**, 2348–9 (1990).

[87] Chen, H. D., Chang, C. Y., Lin, K. C., Chan, S. H., Feng, M. S., Chen, P. A., Wu, C. C., and Juang, F. Y., "Carbon incorporation during growth of GaAs by $TEGa-AsH_3$ base low-pressure metalorganic chemical vapor deposition," *J. Appl. Phys.*, **73**, 7851–6 (1991).

[88] Hanna, M. C., Lu, Z. H., and Majerfeld, A., "Very high carbon incorporation in metalorganic vapor phase epitaxy of heavily doped p-type GaAs," *Appl. Phys. Lett.*, **58**, 164–6 (1991).

[89] Stockman, S. A., Hofler, G. E., Baillargeon, J. N., Hsieh, K. C., Cheng, K. Y., and Stillman, G. E., "Characterization of heavily carbon-doped GaAs grown by metalorganic chemical vapor deposition and metalorganic molecular beam epitaxy," *J. Appl. Phys.*, **72**, 981–7 (1992).

[90] Khan, A., Woodbridge, K., Ghisoni, M., Parry, G., Beyer, G., Roberts, J., Pate, M., and Hill, G., "Application of intermixing to p-type GaAs/AlAs distributed Bragg reflectors for series resistance reduction in vertical cavity devices," *J. Appl. Phys.*, **77**, 4921–6 (1995).

[91] Uematsu, M., Wada, K., and Gosele, U., "Non-equilibrium point defect phenomena influencing beryllium and zinc diffusion in GaAs and related compounds," *Appl. Phys. A, Solids Surf.*, **A55**, 301–12 (1992).

[92] Veuhoff, E., Baumeister, H., Rieger, J., Gorgel, M., and Treichler, R., "Comparison of Zn and Mg incorporation in MOVPE InP/GaInAsP laser structures," *Proceedings from the Third International Conference on Indium Phosphide and Related Materials*, 72–5 (1991).

[93] Nordell, N., Ojala, P., van Berlo, W. H., Landgren, G., and Linnarsson, M. K., "Diffusion of

Zn and Mg in AlGaAs/GaAs structures grown by metalorganic vapor-phase epitaxy," *J. Appl. Phys.*, **67**, 778–86 (1990).

[94] Enquist, P., Hutchby, J. A., and de Lyon, T. J., "Growth and diffusion of abrupt zinc profiles in gallium arsenide and heterojunction bipolar transistor structures grown by organometallic vapor phase epitaxy," *J. Appl. Phys.*, **63**, 4485–93 (1988).

[95] Hobson, W. S., Pearton, S. J., Schubert, E. F., and Cabaniss, G., "Zinc delta doping of GaAs by organometallic vapor phase epitaxy," *Appl. Phys. Lett.*, **55**, 1546–8 (1989).

[96] Cunningham, B. T., Guido, L. J., Baker, J. E., Major, J. S., Jr., Holonyak, N., Jr., and Stillman, G. E., "Carbon diffusion in undoped, n-type, and p-type GaAs," *Appl. Phys. Lett.*, **55**, 687–9 (1989).

[97] Abernathy, C. R., Pearton, S. J., Caruso, R., Ren, F., and Kovalchik, J., "Ultrahigh doping of GaAs by carbon during metalorganic molecular beam epitaxy," *Appl. Phys. Lett.*, **55**, 1750–2 (1989).

[98] Johnson, E. S. and Legg, G. E., "Critical misorientation morphology in AlGaAs and GaAs grown by atmospheric-pressure MOCVD on misoriented substrates," *J. Cryst. Growth*, **88**, 53–66 (1988).

[99] Takahashi, M., Vaccaro, P., Fujita, K., Watanabe, T., Mukaihara, T., Koyama, F., and Iga, K., "An InGaAs-GaAs vertical-cavity surface-emitting laser grown on GaAs(311)A substrate having low threshold and stable polarization," *IEEE Photonics Technol. Lett.* (1996).

[100] Ju, Y.-G., Lee, Y.-H., Shin, H.-K., Kim, I., "Strong polarization selectivity in 780-nm vertical-cavity surface-emitting lasers grown on misoriented substrates," *Appl. Phys. Lett.*, **71**, 741–743 (1997).

[101] Suzuki, M., Itaya, K., Nishikawa, Y., Sugawara, H., and Hatakoshi, G., "Effects of substrate misorientation on reduction of deep levels and residual oxygen incorporation in InGaAlP alloys," *Gallium Arsenide and Related Compounds 1992. Proceedings of the Nineteenth International Symposium*, 465–70 (1992).

[102] Islam, M. R., Chelakara, R. V., Neff, J. G., Fertitta, K. G., Grudowski, P. A., Holmes, A. L., Ciuba, F. J., Dupuis, R. D., and Fouquet, J. E., "The growth and characterization of AlGaAs double heterostructures for the evaluation of reactor and source quality," *J. Electron. Mater.*, **24**, 787–92 (1995).

[103] Schneider, R. P., Jr., Lott, J. A., Hagerott Crawford, M., and Choquette, K. D., "Epitaxial design and performance of AlGaInP red (650–690 nm) VCSELs," *Int. J. High Speed Electron. Syst.*, **5**, 625–66 (1994).

[104] Bhat, R., Caneau, C., Zah, C. E., Koza, M. A., Bonner, W. A., Hwang, D. M., Schwarz, S. A., Menocal, S. G., and Favire, F. G., "Orientation dependence of S, Zn, Si, Te, and Sn doping in OMCVD growth of InP and GaAs: application to DH lasers and lateral p-n junction arrays grown on non-planar substrates," *J. Cryst. Growth*, **107**, 772–8 (1991).

[105] Caneau, C., Bhat, R., and Koza, M. A., "Dependence of doping on substrate orientation for GaAs:C grown by OMVPE," *J. Cryst. Growth*, **118**, 467–9 (1992).

[106] Kondo, M., Anayama, C., Okada, N., Sekiguchi, H., Domen, K., and Tanahashi, I., "Crystallographic orientation dependence of impurity incorporation into III-V compound semiconductors grown by metalorganic vapor phase epitaxy," *J. Appl. Phys.*, **76**, 914–27 (1994).

[107] Kondo, M., Anayama, C., Tanahashi, T., and Yamazaki, S., "Crystal orientation dependence of impurity dopant incorporation in MOVPE-grown III-V materials," *J. Cryst. Growth*, **124**, 449–56 (1994).

[108] Kondo, M. and Tanahashi, T., "Dependence of carbon incorporation on crystallographic orientation during metalorganic vapor phase epitaxy of GaAs and AlGaAs," *J. Cryst. Growth*, **145**, 390–6 (1995).

[109] Bhat, R., Zah, C. E., Caneau, C., Koza, M. A., Menocal, S. G., Schwarz, S. A., and Favire, R. J., "Orientation-dependent doping in organometallic chemical vapor deposition on nonplanar InP substrates application to double-heterostructure lasers and lateral p-n junction arrays," *Appl. Phys. Lett.*, **56**, 1691–3 (1990).

[110] Anayama, C., Sekiguchi, H., Kondo, M., Sudo, H., Fukushima, T., Furuya, A., and Tanahashi, T., "One-step-metalorganic-vapor-phase-epitaxy-grown AlGaInP visible laser using simultaneous impurity doping," *Appl. Phys. Lett.*, **63**, 1736–8 (1993).

[111] Tandon, A., *Incorporation and diffusion of dopants in GaAs, InP and InGaAs: implications for devices*. Ph.D. Thesis, University of Utah (1998).

[112] Kondo, M., Anayama, C., Sekiguchi, H., and Tanahashi, T., "Mg-doping transients during metalorganic vapor phase epitaxy of GaAs and AlGaInP," *J. Cryst. Growth*, **141**, 1–10 (1994).

[113] Schneider, R. P. and Figiel, J. J., Unpublished data (1995).

[114] Lear, K. L., Hou, H. Q., Banas, J. J., Hammons, B. E., Furioli, J., and Osinski, M., "Vertical cavity lasers on p-doped substrates," *Electron. Lett.*, **33**, 783–4 (1997).

[115] Kobayashi, N., Makimoto, T., *Jpn. J. Appl. Phys.*, **24**, L824 (1985).

[116] Kuech, T. F., Wolford, D. J., Veohoff, E., Deline, V., Mooney, P. M., Potemski, R., and Bradley, J., "Properties of high-purity $Al_xGa_{1-x}As$ grown by the metal-organic vapor-phase-epitaxy technique using methyl precursors," *J. Appl. Phys.*, **62**, 632–43 (1987).

[117] Nakanisi, T., "The growth and characterization of high quality MOVPE GaAs and GaAlAs," *J. Cryst. Growth*, **68**, 282–94 (1984).

[118] Hata, M., Fukuhara, N., Zempo, Y., Isemura, M., Yako, T., and Maeda, T., "Residual impurities in epitaxial layers grown by MOVPE," *J. Cryst. Growth*, **93**, 543–9 (1988).

[119] Kisker, D. W., Miller, J. N., and Stringfellow, G. B., "Oxygen gettering by graphite baffles during organometallic vapor phase epitaxial AlGaAs growth," *Appl. Phys. Lett.*, **40**, 614–16 (1982).

[120] Mihashi, Y., Miyashita, M., Kaneno, N., Tsugami, M., Fujii, N., Takamiya, S., and Mitsui, S., "Influence of oxygen on the threshold current of AlGaAs multiple quantum well lasers grown by metalorganic chemical vapor deposition," *J. Cryst. Growth*, **141**, 22–8 (1994).

[121] Schwartz, B. D., Setzko, R. S., Mott, J. S., Macomber, S. H., and Powers, J. J., "Oxygen incorporation, photoluminescence, and laser performance of AlGaAs grown by organometallic vapor phase epitaxy," *J. Electron. Mater.*, **24**, 1687–90 (1995).

[122] Goorsky, M. S., Kuech, T. F., Cardone, F., Mooney, P. M., Scilla, G. J., and Potemski, R. M., "Characterization of epitaxial GaAs and $Al_xGa_{1-x}As$ layers doped with oxygen," *Appl. Phys. Lett.*, **58**, 1979–81 (1991).

[123] Huang, J. W., Bray, K. L., and Kuech, T. F., "Compensation of shallow impurities in oxygen-doped metalorganic vapor phase epitaxy grown GaAs," *J. Appl. Phys.*, **80**, 6819–26 (1996).

[124] Petroff, P. M., "Physics and materials issues behind the lifetime problem in semiconductor lasers and light emitting diodes," *Proc. SPIE – Int. Soc. Opt. Eng.*, **2683**, 52–8 (1996).

[125] Ishii, M., Kan, H., Susaki, W., Nishiura, H., and Ogata, Y., "Reduction of crystal defects in active layers of GaAs-AlGaAs doubel-heterostructure lasers for long-life operation," *IEEE J. Quant. Electron.*, **QE-13**, 600–4 (1977).

[126] Herrick, R. W., Cheng, Y. M., Beck, J. M., Petroff, P. M., Scott, J. W., Peters, M. G., Robinson, G. D., Coldren, L. A., Morgan, R. A., and Hibbs-Brenner, M. K., "Analysis of VCSEL degradation modes," *Proc. SPIE – Int. Soc. Opt. Eng.*, **2683**, 123–33 (1996).

[127] Redwing, J. M., Nayak, S., Savage, D. E., Lagally, M. G., Dawson-Elli, D. F., and Kuech, T. F., "The effect of controlled impurity incorporation on interfacial roughness in $GaAs/Al_xGa_{1-x}As$ superlattice structures grown by metalorganic vapor phase epitaxy," *J. Cryst. Growth*, **145**, 792–8 (1994).

[128] Kuech, T. F., Potemski, R., Cardone, F., and Scilla, G., "Quantitative oxygen measurements in OMVPE $Al_xGa_{1-x}As$ grown by methyl precursors," *J. Electron. Mater.*, **21**, 341–6 (1992).

[129] Tsai, M. J., Tashima, M. M., and Moon, R. L., "The effects of the growth temperature on $Al_xGa_{1-x}As$ ($0 \leq x \leq 0.37$) LED materials grown by OM-VPE," *J. Electron. Mater.*, **13**, 437–46 (1984).

[130] Hobson, W. S., Harris, T. D., Abernathy, C. R., and Pearton, S. J., "High quality $Al_xGa_{1-x}As$ grown by organometallic vapor phase epitaxy using trimethylamine alane as the aluminum precursor," *Appl. Phys. Lett.*, **58**, 77–9 (1991).

[131] Hobson, W. S., van der Ziel, J. P., Levi, A. F. J., O'Gorman, J., Abernathy, C. R., Geva, M., Luther, L. C., and Swaminathan, V., "Low-threshold GaAs/AlGaAs quantum-well lasers grown by organometallic vapor-phase epitaxy using trimethylamine alane," *J. Appl. Phys.*, **70**, 432–5 (1991).

[132] Roberts, J. S., Button, C. C., David, J. P. R., Jones, A. C., and Rushworth, S. A., "MOVPE growth of AlGaAs using trimethylamine alane," *J. Cryst. Growth*, **104**, 857–60 (1991).

[133] Schneider, R. P., Jr., Bryan, R. P., Jones, E. D., Biefeld, R. M., and Olbright, G. R., "Trimethylamine alane for MOVPE of AlGaAs and vertical-cavity surface-emitting laser structures," *J. Cryst. Growth*, **123**, 487–94 (1992).

[134] Hou, H. Q., Breiland, W. G., Hammons, B. E., Biefeld, R. M., Baucom, K. C., and Stall, R. A., "Growth study of AlGaAs using dimethylethylamine alane as the aluminum precursor," *J. Electron. Mater.*, **26**(10), 1178–83 (1997).

[135] Choquette, K. D., Schneider, R. P., Jr., Lear, K.L., and Geib, K. M., "Low threshold voltage vertical-cavity lasers fabricated by selective oxidation," *Electron. Lett.*, **30**, 2043–4 (1994).

[136] Hegblom, E. R., Babic, D. I., Thibeault, B. J., and Coldren, L. A., "Estimation of scattering losses in dielectrically apertured vertical cavity lasers," *Appl. Phys. Lett.*, **68**, 1757–9 (1996).

[137] Chand, N., Henderson, T., Klem, J., Masselink, W. T., Fischer, R., Chang, Y.-C., and Morkoc, H., "Comprehensive analysis of Si-doped $Al_xGa_{1-x}As$ ($x = 0$ to 1): theory and experiments," *Phys. Rev. B*, **30**, 4481–92 (1988).

[138] Mooney, P.M., "Deep donor levels (DX centers) in III-V semiconductors," *J. Appl. Phys.*, **67**, R1–26 (1990).

[139] Kumagai, O., Kawai, H., Mori, Y., and Kaneko, K., "Chemical trends in the activation energies of DX centers," *Appl. Phys. Lett.*, **45**, 1322–3 (1984).

[140] Kuech, T. F., Meyerson, B. S., and Veuhoff, E., "Disilane: a new silicon doping source in metalorganic chemical vapor deposition of GaAs," *Appl. Phys. Lett.*, **44**, 986–8 (1984).

[141] Fujiwara, Y., Fukumoto, Y., Kobayashi, T., and Hamakawa, Y., "Se Atomic planar doping to GaAs by MOCVD and its application to two-dimensional electronic devices," *Extended Abstracts of the 19th Conference on Solid State Devices and Materials (25–27 Aug. 1987)*, 159–62 (1988).

[142] Houng, Y.-M. and Low, T. S., "Te doping of GaAs and $Al_xGa_{1-x}As$ using diethyltellurium in low pressure OMVPE," *J. Cryst. Growth*, **77**, 272–80 (1986).

[143] Glew, R. W., "H_2Se doping of MOCVD grown GaAs and GaAlAs," *J. Phys. Colloq.*, **43**, 281–6 (1982).

[144] Lewis, C. R. Ludowise, M. J., and Dietze, W. T., "H_2Se 'memory effects' upon doping profiles in GaAs grown by metalorganics chemical vapor deposition in (MO-CVD)," *J. Electron. Mater.*, **13**, 447–61 (1984).

[145] Ogasawara, N., Karakida, S., Miyashita, M., Hayafuji, N., Tsugami, M., Mihashi, Y., and Murotani, T., "Diffusion of p- and n-type dopants in GaAs/AlGaAs DH structure grown by MOCVD," *Advanced III-V Compound Semiconductor Growth, Processing and Devices Symposium, Mater. Res. Soc.*, 733–8 (1992).

[146] Lei, C., Hodge, L. A., Dudley, J. J., Keever, M. R., Liang, B., Bhagat, J. K., and Liao, A., "High performance and highly reliable 850 nm VCSELs on both n and p-type substrates for fiber optic links," *OSA Trends in Optics and Photonics (TOPS)*, **15** (Advances in Vertical Cavity Surface Emitting Lasers), 150–155 (1997).

[147] Schneider, R., Hov, H., and Corzine, S., Unpublished work (1997).

[148] Hasnain, G., Tai, K., Yang, L., Wang, Y. H., Fischer, R. J., Wynn, J. D., Weir, B., Dutta, N. K., and Cho, A. Y., "Performance of gain-guided surface emitting lasers with semiconductor distributed Bragg reflectors," *IEEE J. Quantum Electron.*, **27**, 1377–85 (1991).

[149] Young, D. B., Scott, J. W., Peters, F. H., Thibeault, B. J., Corzine, S. W., Peters, M. G., Lee, S.-L., and Coldren, L. A., "High-power temperature-insensitive gain-offset InGaAs/GaAs vertical-cavity surface-emitting lasers," *IEEE Photonics Technol. Lett.*, **5**, 129–32 (1993).

[150] Choquette, K. D., Geib, K. M., Chui, H. C., Hammons, B. E., Hou, H. Q., Drummond, T. J.,

and Hull, R., "Selective oxidation of buried AlGaAs versus AlAs layers," *Appl. Phys. Lett.*, **69**, 1385–7 (1995).

[151] Sacks, R. N., Colombo, P., Patterson, G. A., and Stair, K. A., "Improved MBE-grown GaAs using a novel, high-capacity Ga effusion cell," *J. Cryst. Growth*, **175**, 66–71 (1996).

[152] Toivonen, M., Salokatve, A., Jalonen, M, Nappi, J., Asonen, H., Pessa, M., and Morison, R., "All solid source molecular beam epitaxy growth of 1.35 μm wavelength strained-layer GaInAsP quantum well laser," *Electron. Lett.*, **31**, 797–99 (1995).

[153] Johnson, F. G., King, O., Seifrth, F., Stone, D. R., Whaley, R. D., Dagenais, M., and Chen, Y. J., "Solid source molecular beam epitaxy of low threshold 1.55 μm wavelength GaInAs/GaInAsP/InP semiconductor lasers," *J. Vac. Sci. Technol.*, **B14**, 2753–55 (1996).

[154] Panish, M. B., "Gas source molecular beam epitaxy of GaInAs(P): gas source, single quantum wells, supperlattice p-i-n's and bipolar transistors," *J. Cryst. Growth*, **81**, 249–60 (1987).

[155] Kerr, T., "Picogiga multiwafer MBE System," *III-Vs Review*, **4**(3), 20–21 (1991).

[156] Houng, Y. M., Unpublished data. (Epitaxial Wafers grown by QED Inc.) (1995).

[157] van de Ven, J., Rutten, G. M. J., Raaijmakers, M. J., and Giling, L. J., "Gas phase depletion and flow dynamics in horizontal MOCVD reactors," *J. Cryst. Growth*, **76**, 352–72 (1986).

[158] Jensen, K. F., "Transport phenomena and chemical reaction issues in OMVPE of compound semiconductors," *J. Cryst. Growth*, **98**, 148–66 (1989).

[159] Moffat, H. and Jensen, K. F., "Complex flow phenomena in MOCVD reactors. I. Horizontal reactors," *J. Cryst. Growth*, **77**, 108–19 (1986).

[160] Mason, N. J. and Walker, P. J., "Influence of gas mixing and expansion in horizontal MOVPE reactors," *J. Cryst. Growth*, **107**, 181–7 (1991).

[161] Moerman, I., Coudenys, G., Demeester, P., Turner, B., and Crawley, J., "Influence of gas mixing on the lateral uniformity in horizontal MOVPE reactors," *J. Cryst. Growth*, **107**, 175–80 (1991).

[162] Frijlink, P. M. J., "A new versatile, large size MOVPE reactor," *J. Cryst. Growth*, **93**, 207 (1988).

[163] Frijlink, P. M., Nicolas, J. L., Ambrosius, H. P. M. M., Linders, R. W. M., Waucquez, C., and Marchal, J. M., "The radial flow planetary reactor: low pressure versus atmospheric pressure MOVPE," *J. Cryst. Growth*, **115**, 203–10 (1991).

[164] Frijlink, P. M., Nicolas, J. L., and Suchet, P., "Layer Uniformity in a multiwafer MOVPE reactor for III-V compounds," *J. Cryst. Growth*, **107**, 166 (1991).

[165] Bergunde, T., Dauelsberg, M., Kadinski, L., Makarov, Yu. N., Weyers, M., Schmitz, D., Strauch, G., and Juergensen, H., "Heat transfer and mass transport in a multiwafer MOVPE reactor: modeling and experimental studies," *J. Cryst. Growth*, **170**, 66–71 (1996).

[166] Keppler, J., Private communication (1998).

[167] Fotiadis, D. I., Kieda, S., and Jensen, K. F., "Transport phenomena in vertical reactors for metalorganic vapor phase epitaxy. I. Effects of heat transfer characteristics, reactor geometry, and operating conditions," *J. Cryst. Growth*, **102**, 441–70 (1990).

[168] Breiland, W. G. and Evans, G. H., "Design and verification of nearly ideal flow and heat transfer in a rotating disk chemical vapor deposition reactor," *J. Electrochem. Soc.*, **138**, 1806–16 (1991).

[169] Biber, C. R., Wang, C. A., and Motakef, S., "Flow regime map and deposition rate uniformity in vertical rotating-disk OMVPE reactors," *J. Cryst. Growth*, **123**, 545–54 (1992).

[170] Wang, C. A., Groves, S. H., Palmateer, S. C., Weyburne, D. W., and Brown, R. A., "Flow visualization studies for optimization of OMVPE reactor design," *J. Cryst. Growth*, **77**, 136–43 (1986).

[171] Fotiadis, D. I., Kremer, A. M., McKenna, D. R., and Jensen, K. F., "Complex flow phenomena in vertical MOCVD reactors: effects on deposition uniformity and interface abruptness," *J. Cryst. Growth*, **85**, 154–64 (1987).

[172] Wang, C. A., Patnaik, S., Caunt, J. W., and Brown, R. A., "Growth characteristics of a vertical rotating-disk OMVPE reactor," *J. Cryst. Growth*, **93**, 228–34 (1988).

[173] Killeen, K. P., Schneider, R. P., and Figiel, J. J., Unpublished data (1994).

[174] Hou, H. Q., Chui, H. C., Choquette, K. D., Hammons, B. E., Breiland, W. G., and Geib, K. M., "Highly uniform and reproducible vertical-cavity surface-emitting lasers grown by metalorganic vapor phase epitaxy with in situ reflectometry," *IEEE Photonics Technol. Lett.*, **8**(10), 1285–7 (1996).

[175] Tompa, G. S., Zawadzki, P. A., Moy, K., McKee, M., Thompson, A. G., Gurary, A. I., Wolak, E., Esherick, P., Breiland, W. G., Evans, G. H., Bulitka, N., Hennessy, J., and Moore, C. J. L., "Design and operating characteristics of a metalorganic vapor phase epitaxy production scale, vertical, high speed, rotating disk reactor," *J. Cryst. Growth*, **145**, 655–61 (1994).

[176] Gurary, A. I., Tompa, G. S., Thompson, A. G., Stall, R. A., Zawadzki, P. A., and Schumaker, N. E., "Thermal and flow issues in the design of metalorganic chemical vapor deposition reactors," *J. Cryst. Growth*, **145**, 642–9 (1994).

[177] Lee, P., McKenna, D., Kapur, D., and Jensen, K. F., "MOCVD in inverted stagnation point flow. I. Deposition of GaAs from TMAs and TMGa," *J. Cryst. Growth*, **77**, 120–7 (1986).

[178] Hummel, S., Joh, S., and Bhat, R., "Theoretical and experimental evaluations of a vertical reactor with stagnation flow geometry," Presented at *International Conf. on Metalorganic Vapor Phase Epitaxy, Wales, UK*. Unpublished data (1996).

[179] Hummel, S., Unpublished data (1997).

[180] Zhang, X., Moerman, I., Sys, C., Demeester, P., Crawley, J., and Thrush, E., "Highly uniform AlGaAs/GaAs and InGaAs(P)/InP structures grown in a multiwafer vertical rotating susceptor MOVPE reactor," *J. Cryst. Growth*, **170**, 83–87 (1997).

[181] Breiland, W. G., Hou, H. Q., Chui, H. C., and Hammons, B. E., "In situ pre-growth calibration using reflectance as a control strategy for MOCVD fabrication of device structures," *J. Cryst. Growth*, **174**, 564–571 (1997).

[182] Aspnes, D. E., "Real-time optical diagnostics for epitaxial growth," *Surf. Sci.*, **307–309**, 1017–27 (1980).

[183] Aspnes, D. E., "Observation and analysis of epitaxial growth with reflectance-difference spectroscopy," *Mater. Sci. Eng. B, Solid-State Mater. Adv. Technol.*, **30**, 109–19 (1995).

[184] Chalmers, S. A. and Killeen, K. P., "Method for accurate growth of vertical-cavity surface-emitting lasers," *Appl. Phys. Lett.*, **62**(11), 1182–1184 (1993a).

[185] Lum, R. M., McDonald, M. L., Bean, J. C., Vandenberg, J., Pernell, T. L., Chu, S. N. G., Robertson, A., and Karp, A., "In situ reflectance monitoring of InP/lnGaAsP films grown by metalorganic vapor phase epitaxy," *Appl. Phys. Lett.*, **69**, 928–30 (1996).

[186] Bajaj, J., Irvine, S. J. C., Sankur, H. O., and Svoronos, S. A., "Modeling of in situ monitored laser reflectance during MOCVD growth of HgCdTe," *J. Electron. Mater.*, **22**, 899–906 (1993).

[187] Holleman, J., Hasper, A., and Middelhoek, J., "In situ growth rate measurement of selective LPCVD of tungsten," *J. Electrochem. Soc.*, **138**, 989–93 (1991).

[188] Takano, A., Kawasaki, M., and Koinuma, H., "In situ determination of optical constants of growing hydrogenated amorphous silicon film by p-polarized light reflectance measurement on the surface," *J. Appl. Phys.*, **73**, 7987–9 (1993).

[189] Zuiker, C. D., Gruen, D. M., and Krauss, A. R., "Laser-reflectance interferometry measurements of diamond-film growth," *MRS Bull.*, **20**, 29–31 (1995).

[190] Vawter, G. A., Klem, J. F., Hadley, G. R., and Kravitz, S. H., "Highly accurate etching of ridge-waveguide directional couplers using in situ reflectance monitoring and periodic multilayers," *Appl. Phys. Lett.*, **62**, 1–3 (1993).

[191] Breiland, W. G. and Killeen, K. P., "A virtual interface method for extracting growth rates and high temperature optical constants from thin semiconductor films using in situ normal incidence reflectance," *J. Appl. Phys.*, **78**, 6726–36 (1995).

[192] Killeen, K. P. and Breiland, W. G., "In situ spectral reflectance monitoring of III-V epitaxy," *J. Electron. Mater.*, **23**, 179–83 (1994).

[193] Frateschi, N. C., Hummel, S. G., and Dapkus, P. D., "In situ laser reflectometry applied to the growth of $Al_xGa_{1-x}As$ Bragg reflectors by metalorganic chemical vapour deposition," *Electron. Lett.* (UK), **27**(2), 155–157 (1991).

[194] Hummel, S. G. and Dapkus, P. D., "A fully automated optical growth rate calibration

method," Presented at the *Sixth Biennial Workshop on Organometallic Vapor Phase Epitaxy, Palm Springs, CA.* Unpublished data (1993).

[195] Houng, Y. M., Tan, M. R. T., Liang, B. W., Wang, S. Y., and Mars, D. E., "In-situ thickness monitoring and control for highly reproducible growth of distributed Bragg reflectors," *J. Vac. Sci. Technol.*, **B12**, 1221–23 (1994).

[196] Liu, X., Ranalli, E., Sato, D. L., Li, Y., and Lee, H. P., "In situ pyrometric interferometry monitoring and control of Ill-V layered structures during molecular-beam epitaxy growth," *J. Vac. Sci. Technol. B*, **13**, 742–5 (1995).

[197] Jackson, A. W., Pinsukanjana, P. R., Gossard, A. C., and Coldren, L. A., "In situ monitoring and control for MBE growth of optoelectronic devices," *IEEE J. Sel. Topics Quantum Electron.*, **3**, 836–44 (1997).

[198] SpringThorpe, A. J. and Majeed, A., "Epitaxial growth rate measurements during molecular beam epitaxy," *J. Vac. Sci. Technol.*, **B8**, 266–70 (1990).

[199] Grothe, H. and Boebel, F. G., "In-situ control of Ga(Al)As MBE layers by pyrometric interferometry," *J. Cryst. Growth*, **127**, 1010–13 (1993).

[200] Lee, H. P., Ranalli, E., and Liu, X., "Physical modeling of pyrometric interferometry during molecular beam epitaxial growth of III-V layred structures," *Appl. Phys. Lett.*, **67**, 1824–26 (1995).

[201] Zhou, J., Li, Y., Thompson, P., Chu, R., Lee, H., Kao, Y., and Celii, F., "Continuous in-situ growth rate extraction using pyrometric interferometry and laser reflectance measurement during molecular beam epitaxy," *J. Electron. Mater.*, **26**, 1083 (1997).

[202] Walker, J. D., Malloy, K., Wang, S., and Smith, J. S., "Precision AlGaAs Bragg reflectors fabricated by phase-locked epitaxy," *Appl. Phys. Lett.*, **56**, 2493–95 (1990).

[203] Walker, J. D., Kuchta, D. M., and Smith, J. S., "Wafer-scale uniformity of vertical-cavity lasers grown by modified phase-locked epitaxy technique," *Electron. Lett. (UK)*, **29**, 239–40 (1993).

[204] Chalmers, S. A. and Killeen, K. P., "Real-time control of molecular beam epitaxy by optical-based flux monitoring," *Appl. Phys. Lett.*, **63**(23), 3131–3133 (1993).

[205] Chalmers, S. A., Killeen, K. P., and Jones, E. D., "Accurate multiple-quantum-well growth using real-time optical flux monitoring," *Appl. Phys. Lett.*, **65**, 4–6 (1994).

[206] Bour, D. P., "AlGaInP quantum well lasers," Chap. 9 in *Quantum Well Lasers*, P. S. Zory, Jr., ed., Academic Press, San Diego (1993).

[207] Schneider, R. P., Jr., Bryan, R. P., Lott, J. A., and Olbright, G. R., "Visible (657 nm) InGaP/InAlGaP strained quantum well vertical-cavity surface-emitting laser," *Appl. Phys. Lett.*, **60**, 1830–2 (1992).

[208] Lott, J. A. and Schneider, R. P., Jr., "Electrically injected visible (639–661 nm) vertical cavity surface emitting laser," *Electron. Lett.*, **29**, 830–2 (1993).

[209] Schneider, R. P., Jr. and Lott, J. A., "Cavity design for improved electrical injection in InAlGaP/AlGaAs visible (639–661 nm) vertical-cavity surface-emitting laser diodes," *Appl. Phys. Lett.*, **63**, 917–19 (1993).

[210] Lott, J. A., Schneider, R. P., Jr., Choquette, K. D., Kilcoyne, S. P., and Figiel, J. J., "Room temperature continuous wave operation of red vertical cavity surface emitting laser diodes," *Electron. Lett.*, **29**, 1693–4 (1993).

[211] Schneider, R. P., Jr., Choquette, K. D., Lott, J. A., Lear, K. L., Figiel, J. J., and Malloy, K. J., "Efficient room-temperature continuous-wave AlGaInP/AlGaAs visible (670 nm) vertical-cavity surface-emitting laser diodes," *IEEE Photonics Technol. Lett.*, **6**, 313–16 (1994).

[212] Schneider, R. P., Crawford, M. H., Choquette, K. D., Lear, K. L., Kilcoyne, S. P., and Figiel, J. J., "Improved AlGaInP-based red (670–690 nm) surface-emitting lasers with novel C-doped short-cavity epitaxial design," *Appl. Phys. Lett.*, **67**, 329–331 (1995).

[213] Crawford, M. H., Schneider, R. P., Jr., Choquette, K. D., and Lear, K. L., "Temperature-dependent characteristics and single-mode performance of AlGaInP-based 670-690-nm vertical-cavity surface-emitting lasers," *IEEE Photon. Technol. Lett.*, **31**, 724–726 (1995).

[214] Waters, R. G., Bour, D. P., Yellen, S. L., and Ruggieri, N. F., "Inhibited dark-line defect formation in strained InGaAs/AlGaAs quantum well lasers," *IEEE Photonics Technol. Lett.*, **2**, 531–533 (1990).

[215] Garbuzov, D. Z., Antonishkis, N. Y., Bondarev, A. D.,Gulakov, A. B., Zhigulin, S. N., Katsavets, N. I., Kochergin, A. V., and Rafailov, E. V., "High-power 0.8 μm InGaAsP-GaAs SCH SQW lasers," *IEEE J. Quantum Electron.*, **27**, 1531–6 (1991).

[216] Botez, D., Mawst, L. J., Bhattacharya, A., Lopez, J., Li, J., Kuech, T. F., Iakovlev, V. P., Suruceanu, G. I., Caliman, and A., Syrbu, A. V., "66% CW wallplug efficiency from Al-free 0.98 mu m-emitting diode lasers," *Electron. Lett.*, **32**, 2012–13 (1996).

[217] Yellen, S. L., Shepard, A. H., Harding, C. M., Baumann, J. A., Waters, R. G., Garbuzov, D. Z., Pjataev, V., Kochergin, V., and Zory, P. S., "Dark-line-resistant, aluminum-free diode laser at 0.8 μm," *IEEE Photonics Technol. Lett.*, **4**, 1328–30 (1992).

[218] Schneider, R. P., Jr. and Hagerott-Crawford, M., "GaInAsP/AlGaInP-based near-IR (780 nm) vertical-cavity surface-emitting lasers," *Electron. Lett.*, **31**, 554–6 (1995).

[219] Ko, J., Hegblom, E. R., Akulova, Y., Margalit, N. M., and Coldren, L. A., "AlInGaAs/AlGaAs strained-layer 850 nm vertical-cavity lasers with very low thresholds," *Electron. Lett.*, **33**, 1550–1 (1997).

[220] Ko, J., Hegblom, E. R., Akulova, Y., Thibeault, B. J., and Coldren, L. A., "Low-threshold 840-nm laterally oxidized vertical-cavity lasers using AlInGaAs-AlGaAs strained active layers," *IEEE Photonics Technol. Lett.*, **9**, 863–5 (1997).

[221] Kondow, M., Uomi, K., Niwa, A., Kitatani, T., Watahiki, S., and Yazawa, Y., "GaInNAs: A novel material for long-wavelength-range laser diodes with excellent high-temperature performance," *Jpn. J. Appl. Phys.*, **35**, 1273–5 (1996).

[222] Kondow, M., Uomi, K., Kitatani, T., Watahiki, S., and Yazawa, Y., "Extremely large *N* content (up to 10%) in GaNAs grown by gas-source molecular beam epitaxy," *J. Cryst. Growth*, **164**, 175–9 (1996).

[223] Kondow, M., Nakatsuka, S., Kitatani, T., Yazawa, Y., and Okai, M., "Room-temperature pulsed operation of GaInNAs laser diodes with excellent high-temperature performance," *Jpn. J. Appl. Phys.*, **35**, 5711–13 (1996).

[224] Nakahara, K., Kondow, K., Kitatani, T., Yazawa, Y., and Uomi, K., "Continuous-wave operation of long-wavelength GaInNAs/GaAs quantum well laser," *Electron. Lett.*, **32**, 1585–6 (1996).

[225] Sato, S., Osawa, Y., Saitoh, T., and Fujimura, I., "Room-temperature pulsed operation of 1.3 μm GaInNAs/GaAs laser diode," *Electron. Lett.*, **33**, 1386–7 (1997).

[226] Larson, M. C., Kondow, M., Kitatani, T., Yazawa, Y., and Okai, M., "Room temperature continuous-wave photopumped operation of 1.22 μm GaInNAs/GaAs single quantum well vertical cavity surface-emitting laser," *Electron. Lett.*, **33**, 959–60 (1997).

[227] Larson, M. C., Kondow, M., Kitatani, T., Tamura, K., and Okai, M., "Photopumped lasing at 1.25 μm of GaInNAs-GaAs multiple-quantum-well vertical-cavity surface-emitting lasers," *IEEE Photonics Technol. Lett.*, **9**, 1549–51 (1997).

[228] Larson, M. C., Kondow, M., Kitatani, T., Nakahara, K., Tamura, K., Inoue, H., and Okai, M., "Room temperature pulsed operation of GaInNAs/GaAs long-wavelength vertical cavity lasers," *Postdeadline Abstracts from the 1997 Meeting of the Lasers and Electro-Optics Society (LEOS), San Francisco, CA*, Paper PD1.3 (1997).

[229] Wang, C. A. and Choi, H. K., "Lattice-matched GaSb/AlGaAsSb double-heterostructure diode lasers grown by MOVPE," *Electron. Lett.*, **32**, 1779–81 (1996).

[230] Wang, C. A. and Choi, H. K., "GaInAsSb/AlGaAsSb multiple-quantum-well diode lasers grown by organometallic vapor phase epitaxy," *Appl. Phys. Lett.*, **70**, 802–4 (1997).

[231] Kurtz, S.R., Biefeld, R.M., Allerman, A. A., Howard, A. J., Crawford, M. H., and Pelczynski, M. W., "Pseudomorphic InAsSb multiple quantum well injection laser emitting at 3.5 μm," *Appl. Phys. Lett.*, **68**, 1332–4 (1997).

[232] Anan, T., Shimomura, H., and Sugou, S., "Improved reflectivity of AlPSb/GaPSb Bragg reflector for 1.55 μm wavelength," *Electron. Lett.*, **30**, 2138–9 (1994).

[233] Shimomura, H., Anan, T., and Sugou, S., "Growth of AlPSb and GaPSb on InP by gas-source molecular beam epitaxy," *J. Cryst. Growth*, **162**, 121–5 (1996).

[234] Almuneau, G., Genty, F., Chusseau, L., Bertru, N., Fraisse, B., and Jacquet, J., "Molecular beam epitaxy growth of 1.3 μm high-reflectivity AlGaAsSb/AlAsSb Bragg mirror," *Electron. Lett.*, **33**, 1227–8 (1996).

[235] Blum, O., Fritz, I. J., Dawson, L. R., and Drummond, T. J., "Digital alloy AlAsSb/AlGaAsSb distributed Bragg reflectors lattice matched to InP for 1.3–1.55 μm wavelength range," *Electron. Lett.*, **31**, 1247–8 (1995).

[236] Lambert, B., Toudic, Y., Rouillard, Y., Gauneau, M., Baudet, M., Alard, F., Valiente, I., and Simon, J. C., "High reflectivity 1.55 μm (Al)GaAsSb/AlAsSb Bragg reflector lattice matched on InP substrates," *Appl. Phys. Lett.*, **66**, 442–4 (1995).

[237] Blum, O., Klem, J. F., Lear, K. L., Vawter, G. A., and Kurtz, S. R., "Optically pumped, monolithic, all-epitaxial 1.56 μm vertical cavity surface emitting laser using Sb-based reflectors," *Electron. Lett.*, **33**, 1878–80 (1997b).

[238] Genty, F., Almuneau, G., Chusseau, L., Boissier, G., Malzac, J.-P., Salet, P., and Jacquet, J., "High reflectivity Te-doped GaAsSb/AlAsSb Bragg mirror for 1.5 μm surface emitting lasers," *Electron. Lett.*, **33**, 140–2 (1997).

[239] Dias, I. F. L., Nabet, B., Kohl, A., and Harmand, J. C., "High reflectivity, low resistance Te doped AlGaAsSb/AlAsSb Bragg mirror," *Electron. Lett.*, **33**, 716–17 (1997).

[240] Blum, O., Geib, K. M., Hafich, M. J., Klem, J. F., and Ashby, C. I. H., "Wet thermal oxidation of AlAsSb lattice matched to InP for optoelectronic applications," *Appl. Phys. Lett.*, **68**, 3129–31 (1996).

[241] Blum, O., Hafich, M. J., Klem, J. F., Baucom, K., and Allennan, A., "Wet thermal oxidation of AlAsSb against As/Sb ratio," *Electron. Lett.*, **33**, 1097–9 (1997a).

[242] Legay, P., Petit, P., Le Roux, G., Kohl, A., Dias, I. F. L., Juhel, M., and Quillec, M., "Wet thermal oxidation of AlAsSb alloys lattice matched to InP," *J. Appl. Phys.*, **81**, 7600–3 (1997).

[243] Peter, M., Forker, J., Winkler, K., Bachem, K. H., and Wagner, J., "A novel pseudomorphic $(GaAs_{1-x}Sb_xIn_yGa_{1-y}As)/GaAs$ bilayer-quantum-well structure lattice-matched to GaAs for long-wavelength optoelectronics," *J. Electron. Mater.*, **24**, 1551–5 (1995).

[244] Anan, T., Nishi, K., Sugou, S., and Kasahra, K., "Novel GaAsSb quantum well for 1.3 μm VCSELs on GaAs substrates," *Postdeadline Abstracts for the 1997 Annual Meeting of the Lasers and Electro-Optics Society (LEOS)* (1997).

[245] Davies, G. J., Duncan, W. J., Skevington, P. J., French, C. L., and Foord, J. S., "Selective area growth for opto-electronic integrated circuits (OEICs)," *Mater. Sci. Eng. B, Solid-State Mater. Adv. Technol.*, **B9**, 93–100 (1991).

[246] Bhat, R., "Current status of selective area epitaxy by OMCVD," *J. Cryst. Growth*, **120**, 362–8 (1992).

[247] Thrush, E. J., Stagg, J. P., Gibbon, M. A., Mallard, R. E., Hamilton, B., Jowett, J. M., and Allen, E. M., "Selective and non-planar epitaxy of InP/GaInAs(P) by MOCVD," *Mater. Sci. Eng. B, Solid-State Mater. Adv. Technol.*, **B21**, 130–46 (1993).

[248] Caneau, C., Bhat, R., Frei, M. R., Chang, C. C., Deri, R. J., and Koza, M. A., "Studies on the selective OMVPE of (Ga, In)/(As, P)," *J. Cryst. Growth*, **124**, 243–8 (1992).

[249] Suzuki, M., Aoki, M., Tsuchiya, T., and Taniwatari, T., "1.24–1.66 μm quantum energy tuning for simultaneously grown InGaAs/InP quantum wells by selective-area metalorganic vapor phase epitaxy," *J. Cryst. Growth*, **145**, 249–255 (1994).

[250] Sasaki, T., Yamaguchi, M., and Kitamura, M., "10 wavelength MQW-DBR lasers fabricated by selective MOVPE growth," *Electron. Lett.*, **30**, 785–786 (1994).

[251] Bhat, R., Kapon, E., Simhony, S., Colas, E., Hwang, D. M., Stoffel, N. G., and Koza, M. A., "Quantum wire lasers by OMCVD growth on nonplanar substrates," *J. Cryst. Growth*, **107**, 716–23 (1991).

[252] Zhao, H., MacDougal, M. H., Frateschi, N. C., Siala, S., Dapkus, P. D., and Nottenburg, R. N., "High efficiency InGaAs/GaAs single-quantum-well lasers using single-step metalorganic chemical vapor deposition," *IEEE Photonics Technol. Lett.*, **6**, 468–70 (1994).

[253] Zhao, H., Uppal, K., MacDougal, M. H., Dapkus, P. D., Lin, H., and Rich, D. H., "Growth and doping properties of AlGaAs/GaAs/InGaAs structures on nonplanar substrates for applications to low threshold lasers," *J. Cryst. Growth*, **145**, 824–31 (1994).

[254] Zhao, H., MacDougal, M. H., Dapkus, P. D., Uppal, K., Cheng, Y., Yang, G.-M., "Submilliampere threshold current InGaAs-GaAs-AlGaAs lasers and laser arrays grown on nonplanar substrates," *IEEE J. Sel. Topics Quantum Electron.*, **1**, 196–202 (1995).

[255] Buydens, L., Demeester, P., van Ackere, M., Ackaert, A., and van Daele, P., "Thickness variations during MOVPE growth on patterned substrates," *J. Electron. Mater.*, **19**, 317–321 (1990).

[256] Koyama, F., Mukaihara, T., Hayashi, Y., Ohnoki, N., Hatori, N., and Iga, K., "Two-dimensional multiwavelength surface emitting laser arrays fabricated by nonplanar MOCVD," *Electron. Lett.*, **30**, 1947 (1994).

[257] Koyama, F., Mukaihara, T., Hayashi, Y., Ohnoki, N., Hatori, N., and Iga, K., "Wavelength control of vertical cavity surface-emitting lasers by using nonplanar MOCVD," *IEEE Photonics Technol. Lett.*, **7**, 10–12 (1995).

[258] Ortiz, G. G., Luong, S. Q., Sun, S. Z., Cheng, J., Hou, H. Q., Vawter, G. A., and Hammons, B. E., "Monolithic, multiple-wavelength vertical-cavity surface-emitting laser arrays by surface-controlled MOCVD growth rate enhancement and reduction," *IEEE Photon. Technol. Lett.*, **9**, 1069–1071 (1997).

[259] Eng, L. E., Bacher, K., Yuen, W., Larson, M., Ding, G., Harris, J. S., Jr., and Chang-Hasnain, C. J., "Wavelength shift in vertical cavity laser arrays on a patterned substrate," *Electron. Lett.*, **31**, 562–3 (1995).

[260] Eng, L. E., Bacher, K., Wupen, Y., Harris, J. S., Jr., and Chang-Hasnain, C. J., "Multiple-wavelength vertical cavity laser arrays on patterned substrates," *IEEE J. Sel. Top. Quantum Electron.*, **1**, 624–8 (1995).

[261] Wupen Y., Li, G. S., and Chang-Hasnain, C. J., "Multiple-wavelength vertical-cavity surface-emitting laser arrays with a record wavelength span," *IEEE Photonics Technol. Lett.*, **8**, 4–6 (1996).

[262] Wupen Y., Li, G. S., and Chang-Hasnain, C. J., "Multiple-wavelength vertical-cavity surface-emitting laser arrays," *IEEE J. Sel. Topics Quantum Electron.*, **3**, 422–8 (1997).

[263] Larson, M. C., Sugihwo, F., Massengale, A. R., and Harris, J. S., Jr., "Micromachined tunable vertical-cavity surface-emitting lasers," *International Electron Devices Meeting. Technical Digest*, 405–8 (1996).

[264] Sugihwo, F., Larson, M. C., and Harris, J. S., Jr., "Low threshold continuously tunable vertical-cavity surface-emitting lasers with 19.1 nm wavelength range," *Appl. Phys. Lett.*, **70**, 547–9 (1997).

[265] Sugihwo, F., Larson, M. C., and Harris, J. S., Jr., "Simultaneous optimization of membrane reflectance and tuning voltage for tunable vertical cavity lasers," *Appl. Phys. Lett.*, **72**, 10–12 (1998).

[266] Vail, E. C., Li, G. S., Wupen, Y., and Chang-Hasnain, C. J., "High performance and novel effects of micromechanical tunable vertical-cavity lasers," *IEEE J. Sel. Topics Quantum Electron.*, **3**, 691–7 (1997).

[267] Li, Y. M., Yuen, W., Li, G. S., and Chang-Hasnain, C. J., "Top-emitting micromechanical VCSEL with a 31.6-nm tuning range," *IEEE Photonics Technol. Lett.*, **10**, 18–20 (1998).

[268] Nakamura, S., "InGaN-based blue laser diodes," *IEEE J. Sel. Topics Quantum Electron.*, **3**, 712–18 (1997).

[269] Fritz, I. J. and Drummond, T. J., "AlN-GaN quarter-wave reflector stack grown by gas-source MBE on (100) GaAs," *Electron. Lett.*, **31**, 68–9 (1995).

[270] Redwing, J. M., Loeber, D. A. S., Anderson, N. G., Tischler, M. A., and Flynn, J. S., "An optically pumped GaN-AlGaN vertical cavity surface emitting laser," *Appl. Phys. Lett.*, **69**, 1–3 (1996).

5

Fabrication and Performance of Vertical-Cavity Surface-Emitting Lasers

Kent D. Choquette and Kent M. Geib

5.1 Introduction

Vertical-cavity surface-emitting lasers were invented at the Tokyo Institute of Technology in 1978 [1, 2], and since the late 1980s they have been the subject of research in the United States [3, 4] and throughout the world. The development of VCSELs has stimulated a wide range of research, from basic studies of microcavity and semiconductor heterostructure physics, to advances in epitaxial growth and device fabrication technologies. Since the mid 1990s, several companies have launched into VCSEL manufacture and numerous applications have begun to penetrate the marketplace. Thus after more than a decade of research, VCSELs are transitioning from research laboratories toward the manufacturing arena, making VCSEL diode structures and their fabrication technologies of notable consequence [5].

The topological distinction of emitting light perpendicular, rather than parallel, to the wafer surface of a VCSEL is enabled through the use of distributed Bragg reflector (DBR) mirrors, to form the longitudinal laser cavity [6, 7]. Hence a fundamental difference between VCSELs and conventional edge-emitting lasers is that the need to fabricate a facet mirror by either cleaving or dry etching is eliminated [8]. Because of the perpendicular emission, VCSELs have several advantages over edge-emitting lasers, such as a low divergence circular (rather than an elliptical) output beam, wafer-level device testing before packaging, and the feasibility of dense two-dimensional laser arrays. However, similar to edge-emitting lasers, the requirement of transverse confinement of photons and charge carriers within the cavity of the VCSEL still persists and is a primary concern in VCSEL device fabrication.

The performance characteristics of VCSELs are strongly influenced by the epitaxial structure and fabrication processes that are employed. For example, one focus of VCSEL development has been toward reduced threshold current; therefore, processing technologies producing laser structures designed for enhanced transverse confinement have had significant impact. Figure 5.1 shows schematically the primary VCSEL structures that have been demonstrated to provide transverse optical and/or electrical confinement: etched air–post, ion-implanted, etched/regrown, and selectively oxidized VCSELs. These various laser structures require disparate fabrication technologies and exhibit differing electrical, optical, and thermal characteristics [9].

The fabrication of VCSEL diode structures, the requisite processing technologies, and the impact on the laser performance are described in this chapter. Top-surface-emitting VCSEL structures possessing semiconductor DBR mirrors are predominately discussed, since monolithic VCSELs are presently the primary focus of manufacturing endeavors. In Section 5.2, epitaxial issues related to VCSELs are briefly reviewed, including the effects

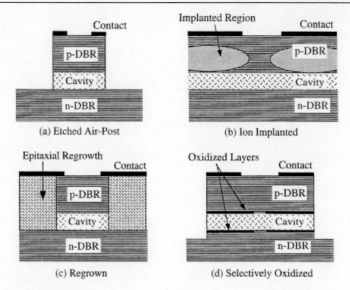

Figure 5.1. VCSEL device structures used for transverse optical and/or electrical confinement: (a) etched air–post; (b) ion implanted; (c) buried heterostructure; and (d) selectively oxidized [9].

of the interplay between the cavity resonance and laser gain bandwidth. The fabrication and characteristics of ion-implanted VCSELs are described in Section 5.3. In Section 5.4 the etching techniques and characteristics of air–post VCSELs are examined. Section 5.5 reviews regrowth techniques used to fabricate buried heterostructure VCSELs. Wet oxidation of AlGaAs alloys and its application to VCSEL fabrication are considered in Section 5.6. In addition, the impact of the electrical and optical confinement afforded by the oxide apertures on VCSEL performance is presented. Finally, Section 5.7 concludes with a comparison of the fabrication technologies with regard to VCSEL performance, application requirements, and their potential for high volume and low cost manufacture.

5.2 Epitaxial Issues

5.2.1 Growth Methods

The components of a VCSEL are two high-reflectivity DBR mirrors separated by an optical cavity [10]. The mirrors and optical cavity consist of multiple semiconductor layers with stringent growth tolerances. The growth of VCSEL wafers is very demanding since accurate thickness, composition, and doping concentration are all simultaneously necessary for each of the possibly hundreds of layers that compose the VCSEL structure. For example, the reflectance wavelengths of the top and bottom DBR mirrors (determined by the composition and thickness of the DBR layers), the cavity resonance wavelength (determined by the composition and thickness of the optical cavity and the DBR layers), and the laser gain spectrum (determined by the composition and thickness of the quantum wells) must all be aligned within a specific wavelength tolerance for viable VCSEL operation. Moreover, sophisticated composition and doping profiles are employed across the interfaces of the high- and low-index DBR layers to reduce the diode series resistance. Due to this epitaxial complexity,

VCSEL development has depended on the continuing progress of compound semiconductor growth and processing technologies. In situ diagnostics during growth combined with a stable and reproducible growth platform are therefore advantageous for epitaxial VCSEL growth.

Extensive VCSEL growth using molecular beam epitaxy (MBE) has been demonstrated [3, 11–13]. MBE is an evaporation process occurring under ultrahigh vacuum conditions involving the reaction between atomic (or molecular) beams with a single crystal substrate surface. This is a layer by layer growth process and thus stringent control of thickness, composition, and doping is inherent to MBE growth. Techniques for monitoring during VCSEL growth by MBE have included in situ reflectance [14–16], electron diffraction oscillations [17], and thermal emissions [18].

Metalorganic vapor-phase epitaxy (MOVPE) was initially employed to grow the first VCSELs, which were hybrid structures (dielectric mirrors surrounding a semiconductor active medium) [1, 19], and it has recently proved to be superior for monolithic VCSEL growth [20–24]. MOVPE uses the reaction between metalorganic and hydride precursor gases at or near atmospheric pressure to epitaxially deposit compound semiconductor materials on a crystalline substrate at elevated temperature. The rapid growth rate, high wafer throughput, run-to-run stability, and wafer uniformity make MOVPE an appropriate VCSEL manufacturing platform [23]. Furthermore, the broad flexibility of materials and dopants that can be employed, combined with the ease of continuous composition grading, are specific advantages of MOVPE. As an example of the epitaxial complexity that can be realized, visible VCSELs can employ a phosphorus-based optical cavity (AlInP/GaInP) and arsenide-based DBRs (AlGaAs/AlAs), implying two complete changes of the group V element during the growth sequence [22]. Other advantages of MOVPE growth include the ease of doping with C (which is a more readily activated p-type dopant than Be) combined with continuous alloy grading at the heterointerfaces of the DBR mirror to enable low series resistance. Reflectance techniques have also been established for in situ MOVPE monitoring [24]. Finally, as described in Section 5.6, the accessibility of the complete alloy range using MOVPE is a significant advantage for selectively oxidized VCSELs.

5.2.2 Distributed Bragg Reflectors

The DBR mirrors surrounding the optical cavity are required for the longitudinal confinement of light. The mirror reflectivity must be greater than 99% because of the low round-trip gain typically found in a VCSEL. As depicted in Figure 5.2(a), the mirrors consist of alternating pairs of quarter-wavelength thick high and low refractive index layers, composed of either monolithically grown semiconductor layers [6, 7] or dielectric materials deposited after epitaxial growth [1, 25, 26]. By spacing multiple high-to-low index interfaces a distance of $\lambda/2$ apart, the reflectivity of each interface adds constructively to produce mirrors with a maximum reflectance of >99%. The calculated reflectance spectrum for a monolithic infrared VCSEL is shown in Figure 5.2(b). The transmission "dip" in the center of the mirror stopband at 850 nm in Figure 5.2(b) corresponds to the cavity resonance and thus the VCSEL emission wavelength. The reflectivity and width of the mirror stopband are proportional to the refractive index difference between the high and low index layers in the DBR mirrors.

An advantage of hybrid dielectric DBR mirrors is that a large refractive index difference between the layers can be achieved. The high index contrast produces high-reflectivity mirrors with a wide mirror stopband using only a few periods. For example, five periods of

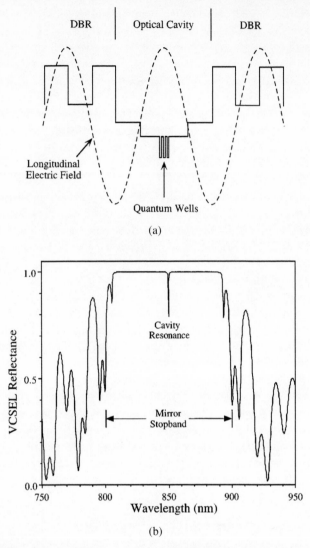

Figure 5.2. (a) The refractive index profile and longitudinal electric field in the distributed Bragg reflector and optical cavity within a VCSEL. (b) The calculated VCSEL reflectance spectrum showing the mirror stopband and the cavity resonance at 850 nm [5].

ZnSe/CaF$_2$ produces greater than 99% reflectivity [27]. In spite of the greater fabrication complexity, depositing a top and/or bottom dielectric DBR mirror during device fabrication can increase the versatility of the VCSEL design. Extremely high contrast mirrors have been demonstrated using Al$_2$O$_3$/GaAs mirrors [28] fabricated with selective oxidation of AlAs as described in Section 5.6. For these mirrors greater than 99% reflectivity can be achieved with a wide mirror stopband of 460 nm employing only four mirror periods [29]. Note that intracavity electrical contacts that bypass the insulating mirrors are required for current injection into the active medium resulting in an increase of fabrication complexity.

Monolithic semiconductor DBR mirrors were a key enabling innovation in the development of VCSELs [6, 7]. Monolithic mirrors are fashioned during the epitaxial growth, which simplifies the VCSEL fabrication and allows current injection through the mirrors into the active medium. Monolithic DBR mirrors typically have lower index contrast and thus have a narrower mirror stopband requiring a greater number of periods for high reflectivity. For example, using GaAs/AlAs layers, 20 DBR periods produce a calculated reflectivity of 99.86% at 850 nm with a 100 nm mirror stopband as shown in Figure 5.2(b). The composition of the DBR layers are chosen to maximize their index contrast and yet be transparent to the laser emission. Hence VCSELs emitting at 980 or 650 nm could use pairs of GaAs/AlAs or $Al_{0.5}Ga_{0.5}As$/AlAs, respectively, for the high/low index layers in the DBR mirrors. In semiconductor systems that lack high index contrast alloys (such as the InGaAsP/InP system) nonlattice-matched DBRs can be grown separately and then wafer bonded to the optical cavity. For example, DBR mirrors composed of GaAs/AlAs layers have been grown separately and then fused to InGaAsP active regions to fabricate VCSELs emitting at long wavelengths such as 1.3 μm [30, 31] and 1.55 μm [32].

Current transport through the semiconductor DBR mirrors is an important monolithic VCSEL issue [9]. Although many heterointerfaces providing a large index difference are necessary for high DBR reflectivity, they also imply that many large energy band offsets are present. These discontinuities in the conduction and valence energy bands form potential barriers that impede current flow and can lead to high series resistance. To overcome this problem, the heterointerfaces between the DBR layers can be modified using composition grading along with specific doping profiles. A variety of compositional profiles between the DBR layers have been reported, including staircase [33], linear [34, 35], and parabolic [22, 36] heterointerface grading. Unfortunately, the doping level within the DBR (and the optical cavity) cannot be arbitrarily increased, owing to the induced free-carrier optical absorption. To reduce optical absorption, the doping profile within monolithic VCSELs is carefully tailored to be minimized in regions where the optical field is at a maximum, such as at high/low refractive index interfaces [37, 38] (see Figure 5.2(a)) and in the vicinity of the optical cavity. Reduction of the series resistance using sophisticated heterointerface designs and doping profiles within the DBR mirrors is important to reduce the mirror series resistance and concomitant parasitic ohmic heating, which can lead to cavity resonance/laser gain spectral mismatch, as discussed in Section 5.2.4.

5.2.3 Optical Cavity

The high-reflectivity DBR mirrors are separated by a thickness of a multiple of $\lambda/2$ (where λ is the wavelength of light in the semiconductor) to form a high-finesse Fabry–Perot cavity. The optical cavity of the VCSEL is typically a single wavelength (λ) thick and contains the active medium of the laser as sketched in Figure 5.2(a). The DBRs and optical cavity are doped such that the active medium is located at the p-n junction of the laser diode. The active medium within the optical cavity is commonly one or more very thin (≤ 10 nm) semiconductor layers surrounded by barrier layers with a greater bandgap. The resulting offset in the conduction (see Figure 5.2(a)) and valence bands between the lower bandgap semiconductor and the barrier material confine the charge carriers and form quantum wells for optical recombination. Compressive or tensile strain can be introduced into the quantum wells if there is a mismatch between the lattice constant of the quantum well and the surrounding barrier material. As shown in Figure 5.2(a), the quantum wells are positioned within the optical cavity to optimally overlap the antinodes of the electromagnetic field to provide optical gain [39]. The quantum well materials are selected to provide gain for the desired VCSEL emission wavelength. For example InGaP, AlGaAs, GaAs, InGaAs, and InGaAsP are appropriate quantum well materials for VCSELs emitting near 650, 780, 850,

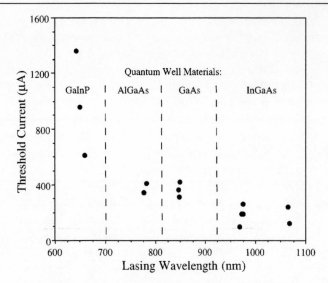

Figure 5.3. Minimum threshold currents for selectively oxidized VCSELs emitting at various wavelengths and employing different quantum-well materials.

980, and 1,300 nm, respectively. Figure 5.3 shows the minimum threshold current achieved from selectively oxidized VCSELs fabricated at Sandia National Laboratories containing various quantum well materials. For some material systems, lower threshold currents have been observed by other research groups [40–43]. The trend of decreasing threshold current with increasing wavelength is due to the incorporation of strain into the quantum wells (producing greater optical gain) and/or greater quantum well confinement (reducing leakage current) obtained from the different semiconductors used in the active region [9].

5.2.4 Resonance/Gain Alignment

Because of their shorter longitudinal cavity length, VCSELs are fundamentally different from their edge-emitting laser counterparts. The separation between the resonances of an optical cavity (free spectral range) is inversely proportional to the length of the optical cavity. Edge-emitting lasers with relatively long cavity lengths (100s of λ) have many longitudinal cavity modes that overlap their gain bandwidth. In contrast, the short optical cavity length of a VCSEL (typically a single or few λ) implies that only one resonance will overlap with the laser gain bandwidth. However, a VCSEL will typically have a relatively large cavity diameter ($\approx$10–20 μm), which may permit numerous transverse spatial modes. Therefore, the VCSEL emission consists of a single longitudinal mode, but with possibly multiple transverse optical modes.

The spectral alignment between the single longitudinal resonance and the laser gain as depicted in Figure 5.4(a) will profoundly influence the performance of the VCSEL [44]. For example, plotted in Figure 5.4(b) is the threshold current density versus radial position for implanted 980 nm VCSELs from a wafer grown with intentional thickness nonuniformity. The resulting variation in the thickness of the optical cavity significantly changes the cavity resonance wavelength, but it has a relatively small effect on the peak laser gain wavelength. Since the gain bandwidth is relatively unaffected, the changes in the threshold current (and threshold voltage) arise in Figure 5.4(b) from the spectral misalignment of the

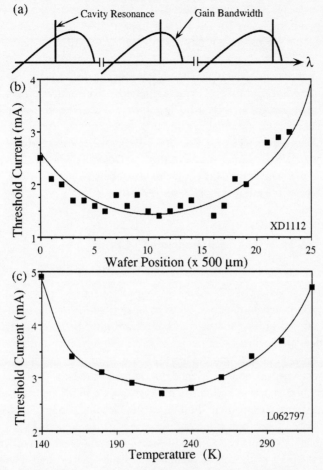

Figure 5.4. (a) Schematic of the spectral alignment between the cavity resonance and the laser gain bandwidth in a VCSEL, where the alignment can be influenced by epitaxial layer thickness and operating temperature; (b) the variation of ion implanted VCSEL threshold current arising from thickness nonuniformity (producing different lasing wavelength) across a wafer; and (c) variation of threshold current for an 850 nm ion implanted VCSEL as a function of operating temperature [5].

cavity resonance relative to the laser gain profile [45]. Hence the minimum threshold current occurs approximately at the resonance wavelength that corresponds to the peak of the laser gain. As the cavity resonance shifts to longer or shorter wavelengths away from the peak gain, the lasing threshold increases. In fact, by spectrally scanning the single longitudinal cavity resonance through the laser gain bandwidth (through layer thickness or temperature variations), the quantum-well gain profile can be analyzed through measurements of threshold voltage or current as a function of lasing wavelength [46].

The cavity resonance/laser gain alignment also dominates the temperature dependence of VCSEL operation [47]. Note that as temperature increases, both the cavity resonance and laser gain shift to longer wavelengths owing to the refractive index and bandgap temperature dependence, respectively. However, the laser gain shifts to longer wavelengths faster than the cavity resonance, causing spectral misalignment between the cavity resonance and peak gain, leading to degradation of the laser performance. The laser gain profile and amplitude also change with temperature [48], which dominates the temperature dependence of edge-emitting lasers but has a lesser impact on VCSEL operation. Figure 5.4(c) shows the threshold current for an implanted 850 nm VCSEL as a function of substrate temperature [49]. The minimum threshold occurs approximately at the temperature where the cavity resonance overlaps the peak laser gain. Because of this, the cavity resonance is often intentionally designed to be at a slightly longer wavelength relative to the peak laser gain at room temperature, so that at higher current injection and thus, higher operating temperature, the peak laser gain shifts into alignment with the cavity resonance to yield optimum VCSEL performance [50]. Therefore during the epitaxial growth of the VCSEL, the cavity resonance/gain alignment can be designed to obtain low threshold or high output power at a particular temperature [51] or to produce relatively invariant threshold properties over a wide range of operating temperatures [52–54].

5.3 Implanted VCSELs

5.3.1 Ion Implantation

VCSEL fabrication generically involves establishing electrical contact to the anode and cathode of the diode and defining the transverse extent of the optical cavity. For the latter, a means of confinement for photons and/or electrons must be implemented. A method to achieve electrical confinement in a planar VCSEL topology is to utilize ion implantation [55–57]. Implantation of ions into the top DBR mirror can be used to render the material around the laser cavity nonconductive and thus to concentrate the injected current into the active medium as depicted in Figure 5.1(b). Damage, primarily crystal vacancies created by the implanted ions, compensate the free carriers leading to regions of high resistivity. Hence the ion dose is chosen to sufficiently compensate the dopant impurities in the DBR.

The ion implantation energy required to achieve current confinement within a VCSEL depends upon the mass of the ion used and the implant depth desired. Thus the maximum vacancy concentration can be tailored to a specific depth within the DBR mirror. Various ion species have been employed (H^+, O^+, N^+, F^+), although proton implants are the most common. The peak implant damage is usually designed to occur somewhat above the quantum wells to avoid excessive damage to the active medium [58]. Figure 5.5(a) shows the vacancy distribution into $Al_{0.5}Ga_{0.5}As$ (appropriate for a GaAs/AlAs DBR) that results from a proton implant done at a 300 keV implant energy as calculated using TRIM-1992 [59]. In Figure 5.5(a) we have also included a 100 nm thick Au contact on top of the semiconductor. In spite of the vacancy spike created at the metal semiconductor interface, formation of the electrical contact prior to implantation can be advantageous, as described below. The maximum implant damage in Figure 5.5(a) occurs at a depth of 2.44 μm. Figure 5.5(b) shows the depth of the maximum vacancy concentration (projected range) into $Al_{0.5}Ga_{0.5}As$ as a function of the proton implantation energy. A substantial density of vacancies are present above this projected range extending to the wafer surface, as shown in Figure 5.5(a). This damage will induce greater unintended lateral resistance in the DBR mirror, which is particularly significant for top-emitting VCSEL structures with annular contacts. Rapid thermal annealing (RTA) can significantly reduce the damage near the wafer surface and increase the lateral conductivity.

A detailed sketch of a planar implanted VCSEL is presented in Figure 5.6(a). Fabrication of implanted VCSELs will usually begin with the deposition of electrical contacts. It is

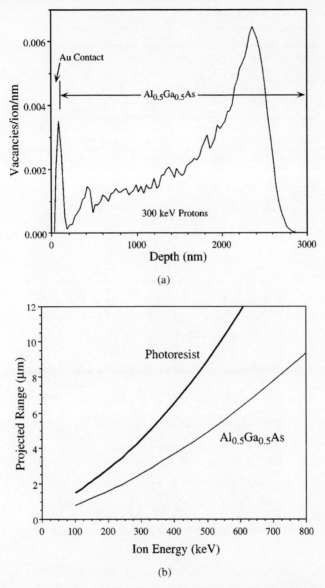

(a)

(b)

Figure 5.5. (a) Simulated vacancy density versus depth into $Al_{0.5}Ga_{0.5}As$ showing the damage profile from 300 keV implanted protons; (b) the depth of maximum implant damage versus proton implantation energy into $Al_{0.5}Ga_{0.5}As$ and photoresist.

advantageous to deposit the top contact as the first processing step to avoid surface damage (such as unintentional removal of a GaAs cap layer protecting underlying AlGaAs DBR layers) during subsequent steps, which may degrade the ohmic contact. Next an implantation mask of photoresist [57] or plated metal can be used to block the ions and thus define the laser cavity. The simplicity and versatility of photoresist implant masks usually make them more favorable as compared to metal masks. It is also beneficial to have the implant mask extend over the metal contact as depicted in Figure 5.6(a) to ensure that electrical contact to an unimplanted region of the VCSEL is maintained [44].

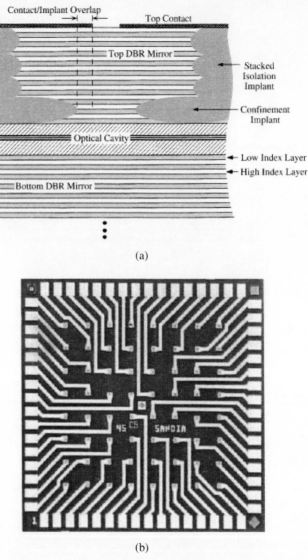

(a)

(b)

Figure 5.6. (a) Cross-section sketch of a planar ion-implanted VCSEL with stacked implantations used for electrical isolation; (b) top view of an 8×8 individually addressable VCSEL array showing the interconnect metal and bonding pads around the periphery.

Deep proton implants near the active region are used to define the transverse extent of the laser cavity. As a generic example, a 300 keV proton implant with a dose of 4×10^{14} cm^{-3} could be employed for an 850 nm VCSEL with a 20-period Al$_{0.16}$Ga$_{0.84}$As/AlAs DBR (2.6 μm thick). As apparent from Figure 5.5(b), a 300 keV proton implant would require a photoresist implant mask thicker than 6 μm. To avoid ion channeling during implantation, the samples are typically inclined by $7°$ from normal. The photoresist mask can be angle etched to match the trajectory of the ion beam [60] and avoid "shadowing" of the ions from the inclined mask, which will produce thin regions of highly damaged material at shallow depths around the cavity perimeter.

To electrically isolate the VCSEL from neighboring devices, multiple stacked implants with successively decreased energy can be used [56]. Surrounding the periphery of the VCSEL contact in Figure 5.6(a) are stacked implanted regions defined by a larger implant mask to fashion a highly resistive region extending from the optical cavity up to the wafer surface. A combination of deep proton and shallow O^+ implantation is effective to laterally isolate the VCSELs; the latter implant is beneficial for the near surface region because of the high dopant concentration typically present. Finally, an interconnect metal layer can be defined to link the VCSELs with contact pads such as shown in Figure 5.6(b). Therefore, a completely planar VCSEL structure can be implemented with a combination of implantation steps, which is amenable to low cost and high volume manufacture [61].

5.3.2 Electrical Characteristics of Implanted VCSELs

For any VCSEL structure, ohmic contacts to the anode and cathode of the laser diode must be made. The majority of VCSELs to date are grown on an n-doped substrate with a lower n-type DBR and a p-type top DBR, although reverse configurations have also been provided [62]. The ohmic contacts are deposited films of metal alloys appropriate for the particular doped semiconductor. For example, Ge/Au/Ni/Au and AuBe/Ti/Au metal films can be used for ohmic contacts to n- and p-type GaAs, respectively. In many cases, a broad-area contact deposited on the backside of the wafer is a common contact for all VCSELs in an array, while a top contact is defined for each individual laser. Figure 5.6(b) shows an individually addressable 8×8 VCSEL array with metal runners connecting each VCSEL to a bonding pad located around the periphery of the array. Alternatively, by etching down to the appropriate layers, contacts to both the n- and p-type DBRs can be fashioned from the top surface of the wafer.

An important concern for efficient current injection through the semiconductor DBR mirror is electrical resistance as described in Section 5.2.2. Two contributions to the series resistance arising from fabrication are the contact resistance and lateral resistance. The latter can be particularly important for top-emitting lasers with annular contacts [63]. To minimize the contact resistance, the deposited metal films are typically subjected to rapid thermal annealing. This process causes the metal and semiconductor to intermix in a region under the contact, enabling the formation of a low resistance ohmic contact. In addition, it can be beneficial to use a heavily doped top layer with a thickness of 0.75λ (a half-wave thick layer can be added to any quarter-wave thick layer in the DBR and not effect its optical reflectivity) to ensure that contact intermixing with the semiconductor does not penetrate through to a high bandgap layer in the DBR. The RTA step can also heal implant damage in the upper portion of the DBR mirror (see Figure 5.5(a)) under the contact, increasing the lateral conductivity. Without this RTA step, "self-annealing" effects manifest as changing light–voltage versus current (L–I–V) characteristics during initial laser operation can be apparent. Typical self-annealing effects observed from implanted VCSELs include increasing threshold current, lower voltage, and greater maximum output power during the first few minutes of continuous wave (CW) operation or the first few repeated scans of the injection current. The threshold current and output power increase are caused by an increase in the effective cross-section area of the implanted VCSEL as the damage at the periphery of the implanted region is locally annealed. Thus RTA schedules or burn-in operation can be necessary to achieve uniform and constant characteristics from implanted VCSELs.

5.3.3 Optical Characteristics of Implanted VCSELs

The transverse optical confinement afforded by a particular VCSEL structure will affect many of its optical properties. Specifically, the thermally induced optical confinement of an implanted VCSEL influences its transverse optical modes, threshold characteristics under pulsed versus CW operation, and modulation response. In addition, the spectral alignment between the cavity resonance and laser gain can impact other optical properties. This latter influence on the maximum output power and emission polarization are described below for implanted VCSELs but are representative of VCSELs in general.

Although the damage from ion implantation defines an electrical current path into the optical cavity as depicted in Figure 5.6(a), no transverse optical confinement is manifest. Instead, implanted VCSELs require the creation of a "thermal lens" at its center, which is a thermally induced refractive index gradient that will support the lasing optical mode [64, 65]. Hence the formation of the thermal lens determines how various transverse modes arise during VCSEL operation. As described in Section 5.2.4, VCSELs will typically exhibit multiple transverse optical modes. The evolution of the transverse modes that arise in a broad-area implanted VCSEL with increasing injection current is shown in Figure 5.7. As commonly observed for implanted VCSELs, the fundamental Gaussian mode arises initially at threshold, followed sequentially by higher order transverse optical modes with increasing injection current [66, 67]. Although the thermal index profile tends to promote the Gaussian mode, higher order modes will subsequently appear with increasing injection current due to spatial and spectral hole burning effects [60, 68]. Also notice in Figure 5.7 that the transverse modes monotonically shift to longer wavelength with increasing current and the concomitant increasing junction temperature as a result of the temperature dependence of the refractive indices of the constituent VCSEL layers.

Because the thermal lens aids the formation of the transverse optical mode, threshold currents of implanted VCSELs under pulsed and CW operation can vary significantly [44, 69]. If the carriers are injected into the active region faster than the formation of the thermal lens, then the lasing threshold will be significantly higher due to greater diffraction loss in the cavity. Under constant current injection, the induced thermal lens from the parasitic ohmic heating will foster a lasing mode at a lower carrier density because of reduced optical loss. Furthermore, owing to the relatively long time required to form a thermal lens

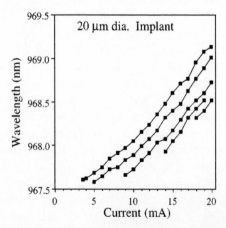

Figure 5.7. Transverse optical mode evolution with increasing operating current for a broad-area (20 μm diameter) proton-implanted 850 nm VCSEL.

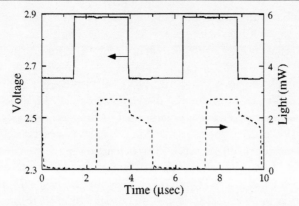

Figure 5.8. Temporal response from a 10 μm diameter implanted 980 nm VCSEL to square wave modulation about lasing threshold [69].

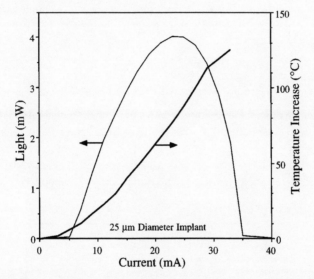

Figure 5.9. The output light intensity and corresponding junction temperature increase versus injection current for a 25 μm diameter implanted 850 nm VCSEL showing the "roll over" behavior of the output light due to the high junction temperature.

($\approx$1 μsec), switching turn-on delays of the VCSEL output can be apparent under pulsed operation. This is explicitly demonstrated in Figure 5.8, which shows the temporal offset between the input signal and the output response of an implanted VCSEL. Note that the VCSEL studied in Figure 5.8 exhibited low series resistance, which exaggerates this effect. Nevertheless, the lack of index confinement ultimately limits the small signal modulation bandwidth of an implanted VCSEL to $\leq$15 GHz [70, 71].

The generic optical properties that are affected by the spectral alignment between the cavity resonance and the laser gain bandwidth include the high power operation [44] and the emission polarization of the VCSEL [72]. Plotted in Figure 5.9 is the light output power and change in the junction temperature as a function of injection current in an implanted VCSEL. The cavity temperature increase is measured by the spectral shift of the lasing emission; the proportionality constant is approximately 0.06 nm/°C. As demonstrated in Figure 5.9, with increasing operating current the VCSEL output increases to a maximum, then decreases with stimulated emission eventually ceasing for junction temperatures in excess of 100°C. This "rollover" behavior of the VCSEL output in Figure 5.9 arises from the increase of the cavity temperature and the thermally induced misalignment of the cavity resonance and laser gain during operation. Thus for any VCSEL structure, low series resistance to limit parasitic heating and sufficient heat dissipation are important to achieve high-power operation. The thermal characteristics of implanted VCSELs are relatively good, since their planar configuration provides good heat sinking. However, specific steps to increase the thermal dissipation will enable higher VCSEL output power.

Implanted VCSELs exhibit unique polarization properties which, unlike their edge-emitting laser counterparts, can also be affected by the resonance/gain spectral overlap [72, 73]. The linear transverse electric (TE) polarization of a VCSEL is, in general, free to be randomly oriented in the plane of the active medium. However, the lasing emission is distributed between two orthogonal TE polarization eigenstates. Similar to the threshold behavior discussed in Section 5.2.4, the dominant polarization eigenstate has been found to vary depending upon the relative spectral alignment between the polarization cavity resonance and the laser gain profile induced from wafer nonuniformity [72] or cavity temperature variations [49].

5.3.4 Implanted VCSEL Reliability

The crystal damage inherent to an implanted VCSEL could potentially nucleate defects in the crystal or cause degradation of the electrical contacts. Nevertheless, excellent VCSEL lifetime has been observed by numerous groups [74, 75], including several VCSEL manufacturers [62, 76]. Figure 5.10 shows the constant power aging characteristics of implanted 980 nm VCSELs grown and fabricated at Sandia National Laboratories [77]. The active

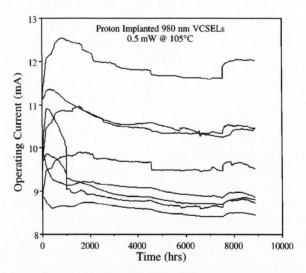

Figure 5.10. Operating current required to achieve 0.5 mW output at an ambient temperature of 105°C for 15 μm diameter ion-implanted 980 nm VCSELs. Prior to the 105°C experiment, these VCSELs were operated to emit 1 mW at 50–70°C for >2,000 hours. Therefore, these devices have demonstrated >10^4 hours of total lifetime at elevated temperature.

region of these devices contain three $In_{0.2}Ga_{0.8}As$ quantum wells. The 15 μm diameter implanted VCSELs were fabricated using 330 keV protons with a dose of 1×10^{14} cm^{-2}. The drive current required to produce a given output power (0.5 mW) at 105°C is plotted in Figure 5.10. Note that prior to the data shown in the figure, these VCSELs were operated at 1 mW output at 50°C ambient temperature for 650 hrs, followed by 70°C operation for 1,900 hrs. Hence, each of the VCSELs have operated for $> 10^4$ total hours. During the initial operation of these lasers the required injection current monotonically decreased, indicative of the self-annealing processes that can occur with implanted VCSELs.

More extensive lifetime experiments have produced similar results for VCSELs operating at other wavelengths. For example, 850 nm VCSELs have exhibited greater than 2.4 million cumulative hours for operation at 70°C [62]. Using a failure criteria defined by the particular VCSEL application considered (e.g., optical fiber data link), a mean time to failure of 3×10^7 hrs for implanted 850 nm VCSELs has been predicted by several vendors [62, 76]. Although VCSELs deployed in actual systems will likely not operate for thousands of years, it does seem likely that they can function reliably until system obsolescence.

In summary, ion-implanted VCSELs have the advantages of a planar device geometry with enhanced heat sinking and excellent reliability. Their disadvantages include a lack of inherent optical confinement to support the lasing modes, which can lead to varying thresholds and modulation limitations. Nevertheless, the performance of implanted VCSELs is appropriate for many emerging applications, and the manufacture of ion-implanted VCSELs is presently being pursued by several companies.

5.4 Etched Air–Post VCSELs

5.4.1 Etch Methods

The simplest means to transversely define the laser cavity is to etch a pillar or post as illustrated in Figure 5.1(a). In fact, the first demonstration of monolithic VCSEL diodes were etched air–post lasers [11]. Various wet and dry etching techniques have been employed to remove the epitaxial material around the laser cavity. Nonselective wet etches of GaAs/AlAs materials, such as a 1:1:10 solution of phosphoric acid, hydrogen peroxide, and deionized water, will produce an isotropic etching profile. Isotropic wet etches have several limitations, such as the undesirable "undercut" edge profile and the lack of process control necessary for achieving small feature size. Anisotropic dry etching techniques such as chemically assisted ion beam etching (CAIBE) [78] or reactive ion etching (RIE) [12, 79] are able to accomplish small diameter air–posts with smooth vertical sidewalls. The small cavity diameter is desirable to achieve a low active volume and thus a low threshold current, while smooth sidewalls are necessary for mitigation of optical loss. Examples of air–post VCSELs fabricated using a $SiCl_4$ RIE process are shown in Figure 5.11 [79]. Important process issues for dry etching air–post VCSELs include the etch uniformity as well as the process reproducibility.

To fabricate air–post VCSELs, contacts can be initially deposited. Following this an etch mask is defined to delineate the laser cavity. During the subsequent etching, a means of monitoring the VCSEL etch depth (e.g., using optical reflectometry) [80] is desirable. By measuring the intensity of a laser beam reflected off of the wafer surface during etching, the etch depth can be determined. Shown in Figure 5.12 is the reflected intensity from a 670 nm laser obtained during etching of an unpatterned 850 nm VCSEL sample. The minima and maxima apparent in Figure 5.12 correspond to etching through the low-index (AlAs) and high-index ($Al_{0.2}Ga_{0.8}As$) DBR layers, respectively, within the VCSEL. In general,

(a) 2 μm

Figure 5.11. Scanning electron microscopy (SEM) views of air–post VCSELs etched through both mirrors for (a) top-emitting and (b) bottom-emitting lasers. For the top-emitting lasers, SiO$_2$ covers the region within the metal contact annulus to protect the top laser facet during the reactive ion etching process [79].

(b) 2 μm

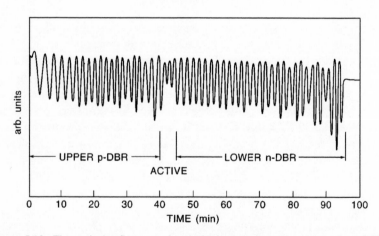

Figure 5.12. The optical reflectance signal obtained during the etching of a VCSEL wafer. The signal maxima (minima) correspond to etching through the high (low) refractive index layers of the distributed Bragg reflector mirror.

interference due to reflections from deeper (unetched) interfaces will also modulate the reflected signal and can be taken into account to accurately model the reflectometry. Nevertheless, it is very typical to observe clear features due to etching through each quarterwave layer in the DBR, providing sufficient accuracy (e.g., ±30 nm) for air–post VCSEL fabrication.

5.4.2 Air–Post VCSEL Performance

The air–post VCSEL structure influences several electrical and optical properties. First, the ability to define a pillar with a small diameter allows the fabrication of high-density ultralow threshold current VCSELs. For example, millions of VCSELs with diameters as small as 1.5 μm have been produced [81], and submilliamp threshold currents have been reported [12, 82]. Secondly, as a consequence of the strong index-guiding present in air–post VCSELs (due to the air/semiconductor interface surrounding the laser cavity), small cavity diameters (<5 μm) are necessary to ensure a single transverse mode laser waveguide [66]. Thus air–post VCSELs with diameters ≥ 10 μm typically exhibit several transverse optical modes above threshold, with wavelengths typically separated by ≈ 0.1 nm. The induced index guiding can also enable polarization control by introducing an anisotropic cavity geometry, which is easily implemented in an air–post structure [83, 84].

The etch depth in a large measure determines the trade-offs between the electrical and optical loss experienced by an air–post VCSEL [85, 86]. For example, loss in an air–post VCSEL that has been etched only into the top DBR mirror (shallow air–post) arises from diffraction caused by the longitudinal index variation and from current spreading away from the active region. Loss in air–post VCSELs etched through the active region into the lower DBR mirror (deep air–post) will suffer less from optical diffraction since the VCSEL is primarily surrounded by air (see Figure 5.11), but will have a contribution from nonradiative recombination at the exposed sidewalls of the active region. For either etch depth, optical scattering from sidewall roughness will be significant as the cavity radius decreases.

To examine the consequences of etch depth and the various loss mechanisms, Figure 5.13 shows the pulsed threshold current, I_{th}, for shallow and deep air–post VCSELs etched using a SiCl$_4$ RIE process [79]. The 850 nm VCSELs contain five GaAs quantum wells in their active region. The data in Figure 5.13 are from a cluster of neighboring VCSELs at each etch depth from nearby pieces of the same wafer. Deep and shallow air–posts with equal

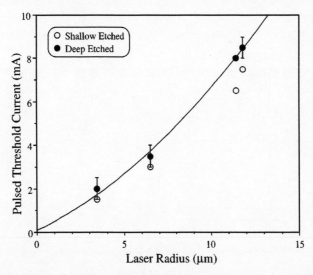

Figure 5.13. Pulsed threshold current versus radius of air–post 850 nm VCSELs etched through the top mirror (shallow etched) or the bottom mirror (deep etched). The fit to the deep etched laser data yields a surface recombination velocity of 7×10^5 cm/sec for the GaAs quantum wells [79].

radius have nearly equal I_{th}. Sidewall scattering from deep air–post VCSELs should be greater than that from shallow etched lasers; the approximately equal I_{th} in Figure 5.13 implies that other loss mechanisms are more significant. Loss in the deep air–post VCSELs is dominated by nonradiative recombination in the active region due to the high surface recombination velocity, S, of the exposed GaAs quantum wells. Under pulsed excitation

$$I_{th} = qd\pi n_{th}\left(\frac{r^2}{\tau_r} + 2Sr\right) \tag{5.1}$$

where q is the electronic charge, d is the total thickness of the GaAs quantum wells, n_{th} is the threshold carrier density, τ_r is the radiative recombination lifetime (assumed below to be 1 nsec), and r is the laser radius [79]. Fitting the deep air–post VCSEL data in Figure 5.13 we find that S ranges from 2×10^5 to 7×10^5 cm/sec, which is comparable to the relatively high value inherent to unetched GaAs surfaces. (Note that for InGaAs surfaces, $S \approx 1 \times 10^3$ cm/sec, which indicates that air–post 980 nm VCSELs are inherently more viable.) Loss due to diffraction and current spreading in the shallow etched lasers have roughly the same impact on laser operation as the nonradiative recombination of the deep air–post VCSELs.

The thermal characteristics are also dramatically impacted by the air–post VCSEL structure. An air–post VCSEL is particularly subject to the output power rollover as described in Section 5.3.3, since by its very nature, the heat sink material is removed from around the laser cavity. One means to increase the heat sinking is to encapsulate the air–post VCSELs with a material with higher thermal conductivity [87, 88]. In Figure 5.14 we compare the $L–I–V$ characteristics of an implanted VCSEL to a shallow etched air–post laser covered with Au on its sidewalls where both VCSELs were fabricated from the same wafer [87]. The encapsulating Au film was evaporated in situ immediately after etching, ensuring adhesion and intimate electrical contact to the exposed AlGaAs/AlAs layers of the upper DBR. Figure 5.14 shows that the operating voltage of the encapsulated air–post VCSEL is reduced and the maximum power is increased as compared to the implanted VCSEL. The reduced current path through the DBR mirror due to the side injection from the sidewall Au contact produces the lower voltage characteristic in Figure 5.14. The greater output power is

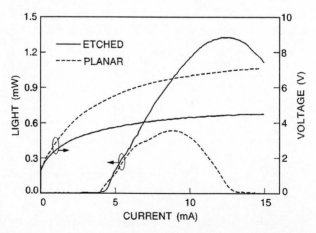

Figure 5.14. Light output and voltage versus injection current characteristics for metal-encapsulated etched air–post (solid curve) and planar ion-implanted (dashed curve) 20 μm diameter 850 nm VCSELs fabricated from the same epitaxial wafer [87].

a manifestation of the reduced thermal impedance of the encapsulated VCSEL, resulting from the enhanced thermal conduction of the sidewall Au contact.

In summary, etched air–post VCSELs have the advantages of ease of fabrication, strong index-guiding, polarization control, and potentially low threshold current. The drawbacks of air–post VCSELs include large nonradiative recombination, greater diffraction and scattering losses, and reduced thermal dissipation leading to a high thermal impedance.

5.5 Regrown VCSELs

5.5.1 Regrowth on AlGaAs

To achieve index-guiding in a planar VCSEL topology requires some manner of lateral variation of the refractive index around the laser. One method to this end is an etch/regrowth process to deposit a semiconductor with a different composition (and thus refractive index) around the laser cavity such as sketched in Figure 5.1(c). This can be accomplished by depositing a robust etch mask (e.g., SiO_2 or SiN_x), etching a pillar, and then regrowing new material around the etched cavity. Regrowth around the laser cavity can also provide a means for efficient current confinement or lateral current injection, passivation of the active region sidewall, and good heat sinking. However, epitaxial regrowth around VCSELs is particularly challenging since it requires growth on high Al content AlGaAs surfaces, inherent to monolithic DBR mirrors. The highly reactive AlGaAs surfaces may be subjected to chemical processes, ion bombardment, and atmospheric exposure, all of which may preclude the possibility of epitaxial regrowth. In particular, the native oxides that form on AlGaAs surfaces can be very tenacious to remove. Thus special cleaning/etching techniques before growth or avoidance of air exposure is necessary for epitaxial regrowth.

Three fabrication techniques have demonstrated successful regrowth on high Al composition layers within a VCSEL. The first buried heterostructure VCSELs were dry etched followed by liquid phase epitaxy (LPE) utilizing melt-back cleaning [89, 90]. The melt-back step employed in LPE regrowth can be difficult to control, compromising the fidelity of small features. Furthermore, LPE regrowth will initiate only on the GaAs substrate, requiring very deep etching ($\geq 8\ \mu$m) and regrowth of several microns of semiconductor material to bury the cavity.

A second method used in situ dry etching and MBE regrowth [91–93]. The vacuum integrated etching and MBE process chambers were linked by ultrahigh vacuum transfer to avoid atmospheric exposure of the highly reactive AlGaAs surfaces. A transmission electron microscope (TEM) image of such a VCSEL, etched through the top DBR mirror using electron cyclotron resonance plasma etching, followed by MBE regrowth of GaAs around the cavity is shown in Figure 5.15 [94]. As apparent in Figure 5.15, excellent quality regrowth can be achieved, although polycrystalline deposits occur on the mask layer, which must later be removed for top-emitting lasers. Unfortunately, the usefulness of etching and epitaxial regrowth on arbitrary reactive surfaces using vacuum integrated processing is somewhat abated by the inherent complexity and expense of this technology.

The third technique used a combination of dry and chemical etching before MOVPE regrowth [95, 96]. Using MOVPE provides the unique advantage of selective area regrowth (i.e., regrowth is inhibited on dielectric masking materials) and it is considered a desirable manufacturing platform. However, successful MOVPE regrowth on high Al content AlGaAs is critically dependent upon developing a nonselective and controllable etching pretreatment.

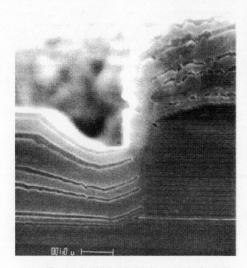

Figure 5.15. Dark field transmission electron microscopy (TEM) images of the molecular beam epitaxial regrowth on in situ plasma-etched VCSELs. Note that amorphous material is deposited on the SiO$_2$ dielectric etch mask, but epitaxial material is observed on exposed semiconductors, including AlAs surfaces.

Successful demonstrations of MOVPE regrowth on dry-etched VCSELs have used brief wet etches of the native oxides (e.g., with bromine:ethanol and/or hydrofloric acid solutions) immediately before loading into the MOVPE reactor.

5.5.2 Regrown VCSEL Performance

The sophisticated processing techniques outlined above have demonstrated buried heterostructure VCSELs with a variety of attributes. Using LPE regrowth of AlGaAs around VCSELs with sufficiently small cavity diameters (<5 μm), index-guided VCSELs exhibiting single-mode operation have been demonstrated [97]. VCSELs with higher index material, GaAs, regrown around the cavity have been demonstrated using vacuum integrated MBE regrowth and MOVPE regrowth. Figure 5.16 shows a comparison of the $L-I-V$ characteristics of a broad-area (20 μm diameter) air–post and etched/regrown VCSEL using vacuum integrated MBE regrowth [92]. In Figure 5.16 similar electrical characteristics are found for both VCSEL structures as expected since the current path is identical for the two VCSELs. The equivalent threshold current implies effective current blocking in the regrown material with no current leakage path at the etched/regrown interface. However, the etched/regrown VCSEL has a greater operating current range and exhibits a higher output power owing to a lower thermal impedance provided by the heat sinking of the regrown material. In addition, antiguided VCSELs have been demonstrated using MOVPE regrowth [95, 96]. The higher index material surrounding 8 μm diameter VCSELs leads to greater optical loss and thus discrimination against higher order transverse modes, yielding single-mode VCSEL operation. Amorphous GaAs layers deposited around etched pillar VCSELs have also provided antiguiding behavior [98].

Regrown VCSELs have the advantages of engineered index-guiding, transverse mode control, and restored heat sinking. In addition, the epitaxially regrown semiconductor material could allow monolithic integration of other optoelectronic or microelectronic devices

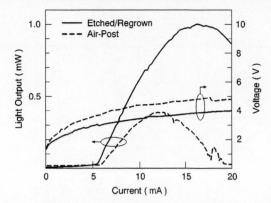

Figure 5.16. Light output and voltage versus injection current characteristics for etched air–post (dashed) and etched/regrown (solid) 20 μm diameter 850 nm VCSELs fabricated from the same epitaxial wafer [87].

[99]. The disadvantages are associated with the challenges, complexity, and expense involved in the epitaxial regrowth on reactive high Al containing alloys used in VCSELs.

5.6 Selectively Oxidized VCSELs

5.6.1 Oxidation of AlGaAs Alloys

A recent innovation is to provide lateral material variation surrounding a semiconductor laser diode through the selective wet oxidation of buried AlGaAs layers [100, 101]. Exposing AlGaAs alloys to temperatures from 350 to 500°C in a steam environment converts the semiconductor into a mechanically robust, chemically inert, insulating, and low refractive index oxide [102]. Wet oxidation of AlGaAs has been successfully employed in the fabrication of edge-emitting lasers [103–105] and has recently been applied to VCSEL fabrication. Oxide layers converted from AlAs or AlGaAs have been incorporated into hybrid VCSELs, which use a dielectric top DBR mirror [106], and into monolithic VCSELs, which use a semiconductor top DBR mirror [107] as illustrated in Figure 5.1(d). The oxide layers within selectively oxidized VCSELs effectively confine both electrons and photons and thus serve to define the transverse optical cavity.

A stable and reproducible oxidation technology requires control of carrier gas flow and composition, water vapor content, and the oxidation furnace temperature [108, 109]. These criteria can be accomplished by employing mass flow controllers to regulate the steam flow, supplied by bubbling an inert gas (e.g., N_2) through water at a constant temperature into a temperature-controlled furnace. The lateral oxidation of AlGaAs has a temperature-dependent linear oxidation rate as shown in Figure 5.17(a) [110, 111] and as reported by other researchers [100, 112–114]. Notice that oxidation occurs without an induction time preceding the onset of oxidation, even when the AlGaAs layers to be oxidized are etched and exposed to atmospheric conditions for several days before oxidation [109]. From the Deal and Grove model [115], the oxide thickness, d_{ox}, achieved in a time t is expressed as

$$d_{\text{ox}}^2 + A d_{\text{ox}} = Bt, \qquad (5.2)$$

where B is related to the diffusion of the reactants through the oxide and B/A is proportional to the oxidation rate constant or the rate of reactant supply at the oxide boundary. For thin

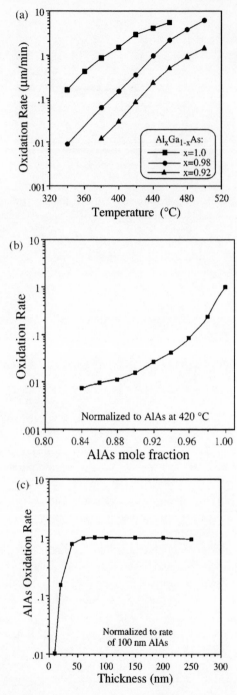

Figure 5.17. Wet oxidation rate of $Al_xGa_{1-x}As$ versus: (a) furnace temperature for $x = 1.0$, 0.98, and 0.92; (b) AlAs mole fraction, x, at 420°C; and (c) AlAs layer thickness at 400°C [109].

native oxides of Si, diffusion of reactants through the oxide layer can be neglected, and Eq. (5.2) reduces to

$$d_{ox} = (B/A)t. \tag{5.3}$$

Based on this model, the linear oxidation rates apparent in Figure 5.17(a) indicate that the lateral oxidation of AlGaAs is reaction rate limited rather than determined by the diffusion of reactant species through the oxide to the reaction front.

The temperature dependence of oxidation follows an Arrhenius behavior, where the oxidation activation energy depends on the AlGaAs composition [109]. Hence the oxidation rate is very sensitive to the composition, as depicted in Figure 5.17(b). For example, the oxidation rate of $Al_xGa_{1-x}As$, for x varying from 1 to 0.84, changes by more than two orders of magnitude [107]. Thus, a high degree of oxidation selectivity between AlGaAs layers can be obtained with only a small change in Al concentration. The thickness of the semiconductor layer will also affect its oxidation rate [116, 117]. Figure 5.17(c) illustrates the oxidation rate of AlAs layers of varying thickness [109]. For layer thickness ≥ 60 nm, a relatively constant lateral oxidation rate is observed; for thinner layers the oxidation rate dramatically decreases. Notice that the oxidation rate dependence on thickness can compensate for the compositional dependence: Thin layers of AlAs may oxidize slower than thick layers of AlGaAs. The reduced oxidation rate of thin layers is not due to decreased reactant transport but likely is influenced by strain and the surface energy at the oxide terminus [117]. Therefore stringent control of Al content and thickness of the AlGaAs layer as well as the oxidation temperature is necessary to ensure a reproducible oxidation technology.

The microstructure of selectively oxidized AlGaAs layers under different process conditions are shown in the TEM images in Figure 5.18. Immediately after oxidation the oxide layer shown in Figure 5.18(a) is an amorphous solid solution of $(Al_xGa_{1-x})_2O_3$, as verified by electron diffraction [118]. This amorphous microstructure has been identified in oxidized $Al_xGa_{1-x}As$ layers with $1 \leq x \leq 0.8$. Annealing the oxide layer will produce the polycrystalline $\gamma - (Al_xGa_{1-x})_2O_3$ microstructure apparent in Figure 5.18(b) [119]. Furthermore, the amorphous oxide in Figure 5.18(a) can also be transformed into the polycrystalline γ-phase by electron beam induced crystallization during TEM examination [118, 120]. The electron beam can also initiate the formation of voids at the oxide/semiconductor interface [118]. For either the amorphous or polycrystalline microstructure in Figure 5.18 (see also Figure 5.19(b)), no extended defects are found along the oxide/semiconductor interfaces. The thickness of the oxide formed at temperatures above 300°C will be reduced compared to the original semiconductor layer. Specifically, the linear shrinkage is measured to be $\approx 13\%$ in oxide layers formed from AlAs [28, 121], and 7% in oxide layers formed from $Al_{0.92}Ga_{0.08}As$ layers [119]. The reduced thickness of the oxide is thus dependent upon the composition of the converted semiconductor and can result in strain fields at the oxide terminus as discussed further in Section 5.6.5.

5.6.2 Selectively Oxidized VCSEL Fabrication

Fabrication of selectively oxidized monolithic VCSELs is predicated on designing the layer compositions, and thus the oxidation rates, such that specific oxide layers near the laser cavity have a greater lateral extent to define an oxide aperture, as depicted in Figure 5.19(a) [107, 110]. The oxidation selectivity to Al composition is exploited to fabricate the buried

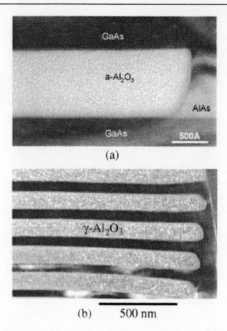

Figure 5.18. Cross-section transmission electron microscopy bright field images of a selectively oxidized layer showing (a) amorphous microstructure after oxidation and (b) polycrystalline γ-phase microstructure after oxidation and rapid thermal annealing [118].

oxide layers within a VCSEL: With minute changes of Ga concentration, one or more AlGaAs layers are induced to oxidize more rapidly (see Figure 5.17(b)) and thus form buried oxide layers for electrical and optical confinement. The selectively oxidized VCSELs described in the following are grown by MOVPE, which has the advantages of accessibility of the complete AlGaAs alloy range and stringent compositional uniformity and control.

Fabrication of selectively oxidized VCSELs begins with the deposition of electrical contacts. A silicon nitride mask is deposited and patterned on the wafer top to encapsulate the metal contact and to form an etch mask. Dry etching, such as reactive ion etching, is used to define mesas or holes to expose the oxidation layers. For mesa structures [110], the oxide aperture is formed by oxide layer(s) extending from the edge into the center of the mesa, as shown in Figure 5.19(a). For etched hole planar structures [122], the aperture is formed by the overlap of oxides that extend outward from each hole. The lateral oxidation extent of a layer within the etched mesa or surrounding the etched hole is controlled by the composition of the layer and the oxidation time. For VCSELs fabricated at Sandia National Laboratories, the low-index layers of the DBR mirrors are typically $Al_{0.92}Ga_{0.08}As$, whereas the low-index layers intended for oxidation are adjusted to $Al_{0.98}Ga_{0.02}As$ to increase their oxidation rate [107]. Typically, an oxidation temperature of 440°C produces a oxidation rate of approximately 1 μm/min for the $Al_{0.98}Ga_{0.02}As$ layers. After oxidation the nitride mask is removed to permit laser testing.

The outline of an oxide aperture within a VCSEL mesa is shown in Figure 5.19(b). For a given oxidation time and thus lateral extent of oxidation, variation of the oxide aperture area within a sample can be obtained using differing mesa sizes or separation between the etched holes. Figure 5.19(c) shows a cross-section TEM image of a selectively oxidized

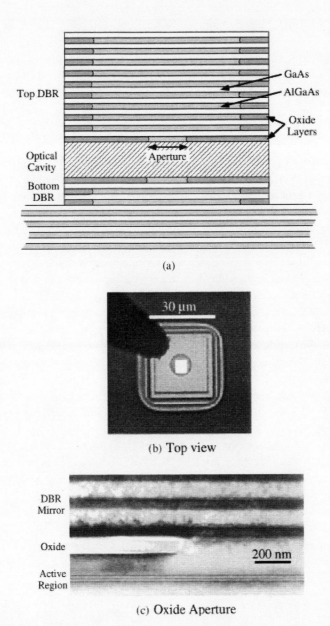

(a)

(b) Top view

(c) Oxide Aperture

Figure 5.19. (a) Cross-section sketch of a monolithic selectively oxidized VCSEL with an oxide aperture on each side of the optical cavity; (b) top view of a VCSEL mesa where the square inner region corresponds to the laser aperture; and (c) transmission electron micrograph of a VCSEL oxide aperture terminus showing no defects or strain apparent along the oxide/semiconductor interface located near the 3 quantum well active region [5, 110].

VCSEL with a single quarter-wave thick oxide aperture next to the optical cavity [110]. Note the absence of extended defects along the oxide/semiconductor interface. For mesa-etched oxide structures, planarization can be accomplished using polyimide as a backfill, or an airbridge technology can be use to allow deposition of metal interconnects. A planar VCSEL configuration conducive to the deposition of bonding pads results using the etched-hole oxide structures, but electrical isolation between VCSELs must still be accomplished by some technique such as stacked ion implantation (see Section 5.3.1).

5.6.3 Electrical Characteristics of Selectively Oxidized VCSELs

Selectively oxidized VCSELs have several advantages over other VCSEL structures arising from the enhanced electrical confinement. In a monolithic selectively oxidized VCSEL, low-resistance DBR mirror designs employing heterointerface grading and doping profiles can be fully exploited by utilizing the entire top mirror to conduct current into the active region. Thus current crowding effects of thin intracavity contact layers [106] or ion implantation damage in the DBR are avoided. Positioning the current apertures immediately adjacent to the optical cavity also eliminates the sidewall nonradiative recombination present in etched air–post VCSELs and minimizes lateral current spreading outside of the laser cavity, which is endemic to ion-implanted VCSELs.

The improved electrical confinement produces reduced threshold voltage and reduced threshold current as compared to other VCSEL structures [123]. Figure 5.20 shows comparison plots of the threshold current and voltage of selectively oxidized and ion-implanted visible VCSELs fabricated from the same epitaxial wafer [124]. Since the insulating oxide layers are located immediately adjacent to the active region, the charge carriers are efficiently confined and injected into the quantum wells contributing to the low threshold current in Figure 5.20(a). By comparison, for implanted VCSELs the maximum implant dosage is necessarily located $\approx$0.5 μm above the quantum wells, allowing significant current spreading outside of the laser cavity and thus the higher threshold current in Figure 20(a). Note that a selectively oxidized infrared VCSEL has exhibited the lowest threshold current observed to date ($<$10 μA) [40]. The lower threshold voltages in Figure 5.20(b) arise from the higher conductivity of the DBR mirrors, which is due to the lack of ion implantation damage. Threshold voltages within 50 meV of the photon energy have been obtained [107]. The combination of lower threshold current and voltage in selectively oxidized VCSELs has enabled record power conversion efficiencies of $>$10% for visible VCSELs [124] and $>$50% for infrared VCSELs [125, 126].

5.6.4 Optical Characteristics of Selectively Oxidized VCSELs

Selectively oxidized VCSELs are strongly index-guided, which profoundly influences the optical characteristics such as threshold current, transverse modes, and output beam stability. The refractive index of the buried oxide layer changes from 3.0 for the original AlGaAs layer to 1.6 for the oxidized layer [104], which induces a significant index difference between the laser cavity and the region surrounding the cavity [127, 128]. Thus selectively oxidized VCSELs are strongly index-guided, but in a quasi-planar configuration amenable to efficient current flow and heat extraction. The index-guiding in selectively oxidized VCSELs contributes to the reduced threshold current in Figure 5.20(a) because of the reduced diffraction loss since formation of a thermal lens is not required [69]. Moreover,

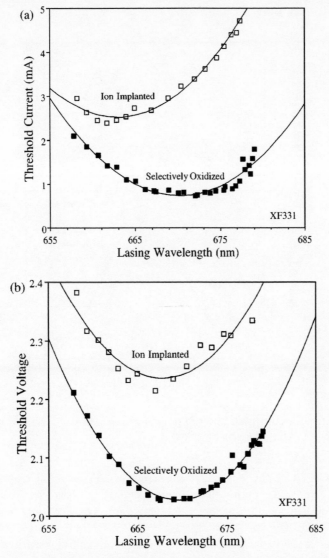

Figure 5.20. Threshold characteristics of ion-implanted and selectively oxidized visible VCSELs fabricated from the same wafer: (a) threshold current and (b) threshold voltage versus lasing wavelength obtained at different wafer locations.

the reduced threshold current and voltage of selectively oxidized VCSELs reduces parasitic ohmic heating, leading to a wider range of current operation and higher output power [123].

The strong optical confinement is also manifest in the transverse mode evolution. In contrast to implanted VCSELs (see Figure 5.7), for selectively oxidized VCSELs several transverse modes can simultaneously arise at threshold since the threshold modal gain is often nearly identical for numerous lower-order transverse modes. The index-guiding of selectively oxidized VCSELs also influences the output beam stability [69]. Figure 5.21(a) shows the beam profile from an implanted VCSEL exhibiting a 3 mA threshold and single-mode operation up to 14 mA. The beam divergence and pointing direction varies with increasing current as evident in Figure 5.21(a). Figure 5.21(b) depicts the beam profile for a

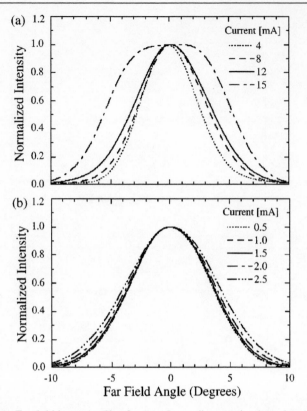

Figure 5.21. Far-field beam profiles for (a) 15 μm diameter implanted and (b) 9 × 9 μm selectively oxidized 980 nm VCSELs at various drive levels. Both lasers are single mode at the first three currents listed [69].

selectively oxidized VCSEL with a threshold current of 0.33 mA and single-mode operation up to 1.8 mA. Notice that during single-mode operation, the selectively oxidized VCSEL maintains a constant pointing direction and beam divergence, which only slightly increases during multimode operation.

A novel optical effect arising from the buried oxide layer is illustrated in Figure 5.22 [123]. The longitudinal cavity resonance is strongly modified under the oxide aperture relative to the as-grown cavity resonance; this phenomenon can lead to lasing in multiple widely separated longitudinal modes. For example, Figure 5.22 shows the observed lasing spectra from an infrared selectively oxidized VCSEL that exhibits two simultaneous lasing emissions. The optical mode observed around the perimeter of the oxide aperture has a wavelength emission 17 nm shorter than the optical mode observed within the aperture at the designed VCSEL wavelength. In this case, a single quarterwave thick oxide layer induces a 17 nm shift in the cavity resonance as shown by the calculated reflectance spectra in Figure 5.22. Note that the oxide mode cavity resonance wavelength can be designed through the control of the thickness and placement of the oxide layer(s). For example, single quarterwave thick oxide layers on each side of the optical cavity shift the oxide mode out of the mirror stopband, thus inhibiting this optical mode. Independent control of the oxide cavity resonance may therefore prove valuable to manipulate the optical cavity surrounding the VCSEL for inhibition of specific lasing modes and/or spontaneous emission, or to tailor the VCSEL emission wavelength.

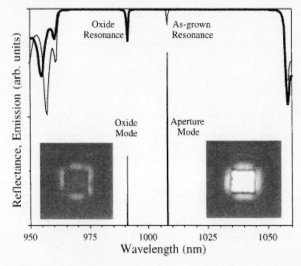

Figure 5.22. The observed lasing emission and calculated reflectance spectra from a 980 nm selectively oxidized VCSEL. The insets show the near-field profiles of aperture and oxide optical modes that originate in the center or at the periphery, respectively, of the oxide apertures [134].

Although the strong optical confinement leads to superior performance for selectively oxidized VCSELs as compared to other device structures, it can negatively impact the optical loss for lasers of small cross-sectional area [129–131]. In Figure 5.23 the threshold current of selectively oxidized VCSELs monotonically decreases with size; however, the threshold current density begins to increase when the oxide aperture area approaches the size of the optical mode. The increase of the threshold current density, and thus the required threshold gain, corresponds to an increase in optical loss originating from the oxide aperture. The optical loss and confinement of selectively oxidized VCSELs can be mediated by reducing the interaction between the optical field and the oxide layer through the design of the longitudinal position [131], thickness [129, 132], and terminus profile [133, 134] of the oxide layer within the cavity. For example, a relatively thin oxide inserted three DBR periods away from the optical cavity has enabled lasing in 850 nm VCSELs with an oxide aperture as small 0.5×0.5 μm [131].

5.6.5 Selectively Oxidized VCSEL Reliability

Wet thermally oxidatized AlAs and AlGaAs have both been employed in the fabrication of selectively oxidized VCSELs. The use of the binary AlAs rather than AlGaAs alloys as the oxidation layer obviously relaxes the compositional control required during growth. However, the properties of the oxides formed from AlAs versus AlGaAs are found to significantly differ and influence the mechanical stability and VCSEL reliability [135]. The stability to thermal shock of selectively oxidized VCSELs depends upon the composition of the original semiconductor layer. Figures 5.24(a) and 5.24(b) show VCSEL mesas containing oxidized $Al_{0.98}Ga_{0.02}As$ and AlAs, respectively, which after oxidation have been subjected to rapid thermal annealing to 350°C for 30 seconds. Note that the mesas containing AlAs (Figure 5.24(b)) have delaminated along the oxide/semiconductor interfaces. This sensitivity to thermal cycling is also apparent during other postoxidation fabrication steps that require

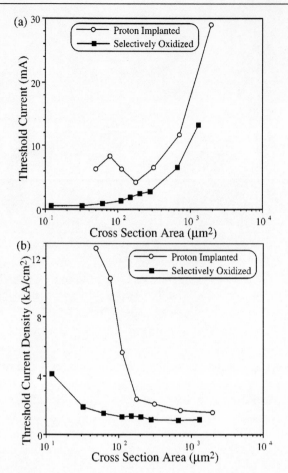

Figure 5.23. (a) Threshold current and (b) threshold current density of proton-implanted and selectively oxidized 850 nm VCSELs.

temperatures of $>100°C$. By comparison, VCSEL mesas containing $Al_{0.98}Ga_{0.02}As$ in the oxide layer are impervious to anneals as demonstrated in Figure 5.24(a).

The apparent difference in mechanical stability is consistent with the degree of strain observed by TEM analysis [135]. The TEM image of an oxide aperture converted from AlAs in Figure 5.24(c) exhibits strain at the terminus of the oxide. By comparison, Figure 5.19(c) of a VCSEL aperture formed from $Al_{0.98}Ga_{0.02}As$ exhibits significantly less strain. Further, the strain apparent in Figure 5.24(c) is in accordance with the greater volume shrinkage of the oxide, as discussed in Section 5.6.1. Thus, the addition of a small amount of Ga to the oxidation layer is found to mitigate the mechanical instability and residual strain.

In spite of their superior performance, the reliability of selectively oxidized VCSELs is perhaps the most pertinent issue for VCSEL applications [77]. The composition of the oxidized layer has been shown to impact the reliability of selectively oxidized VCSELs. VCSELs employing oxidized AlAs layers have exhibited rapid device degradation [135, 136]. The aging characteristics of selectively oxidized VCSELs containing oxidized $Al_{0.98}Ga_{0.02}As$ current apertures with different areas are presented in Figure 5.25 [109]. At an ambient temperature of $80°C$, the current required to maintain 1 mW output power

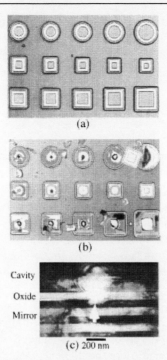

Figure 5.24. Top view after oxidation and rapid thermal annealing (350°C for 30 sec) of VCSEL mesas that contain buried oxide layers converted from (a) $Al_{0.98}Ga_{0.02}As$ and (b) AlAs. (c) Cross-section transmission electron microscopy image of a VCSEL oxide aperture formed from AlAs showing evidence of strain at the oxide terminus (see arrow) [135].

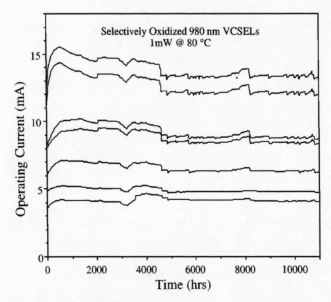

Figure 5.25. Aging characteristics of selectively oxidized VCSELs containing $Al_{0.98}Ga_{0.02}As$ oxide apertures. The drive current required to achieve 1 mW output at 80°C for VCSELs of various sizes is shown.

remains relatively unchanged for $> 10^4$ hrs, which is comparable to the reliability described in Section 5.3.4 and similar to reports for commercial ion-implanted VCSELs. Although these VCSEL lifetime characteristics are promising, further study of the reliability of selectively oxidized VCSELs and their degradation mechanisms is warranted.

5.7 Conclusions

Ultimately, commercial VCSEL development will be driven by the requirements of their applications and manufacturing costs. Present VCSEL applications require only modest laser output power (≈ 1 mW) and benefit from the low operating current, low thermal budget, and circular output beam. Examples of applications where these attributes make VCSEL insertion appropriate include sources for optical fiber-based data links, free-space data links, and optical displays. Furthermore, applications for efficient low-power laser sources should increase as VCSEL performance matures. For example, single transverse mode operation to a few milliwatts is desirable for development of optical disk read/write modules, printing heads, optical scanners, and optical sensors. Continued VCSEL development should also create new applications. Examples include long-wavelength (2–10 μm) VCSELs for deployment in environmental monitoring applications or high density two-dimensional VCSEL arrays integrated with microelectronics for smart pixel applications, such as reconfigurable interconnects or high-speed image processing.

Low cost, high volume manufacture should be an inherent VCSEL attribute of benefit for any application. Creating the high-reflectivity DBR mirrors during the material growth step leads to epitaxial complexity on the front end of manufacture but ultimately simplifies the laser diode fabrication because facet mirrors are eliminated. In addition, monolithic mirrors allow on-wafer testing and device qualification at virtually any stage of device fabrication. Finally, the small real estate taken up by a VCSEL will allow hundreds of thousands of devices to be fabricated from a single wafer. Hence, low manufacturing costs could overcome the limitations of VCSEL performance for some applications. For example, VCSELs are inherently low-power lasers; however, arrays of VCSELs producing modest output power (e.g., 1 W), which are produced inexpensively, could penetrate diode pumping applications where the output power/cost is the predominant concern.

In this chapter ion-implanted, etched air–post, etched/regrown, and selectively oxidized VCSEL structures have been described along with overviews of the various fabrication technologies. At the present time, VCSEL manufacture using ion-implantation has been adopted by VCSEL producers. This well-developed VCSEL technology provides laser characteristics that are suitable for the initial VCSEL products being pursued. Specifically, commercial ion-implanted infrared VCSELs have exhibited submilliamp threshold currents, high efficiency, and up to 15 GHz modulation bandwidth. Most important of all, implanted VCSELs have shown reliability characteristics that are superior to any other semiconductor laser technology, and VCSELs can now be purchased as commodity items.

Etched air–post VCSELs certainly represent the simplest laser structure, yet exhibit the poorest performance relative to the other device structures. Buried heterostructure VCSELs requiring epitaxial regrowth have to date not exhibited suitable performance to justify their potential high cost of manufacture. However, an epitaxial regrowth technology provides a viable path toward III–V microelectronic/optoelectronic integration, which may be crucial for advanced systems applications in the future.

The selectively oxidized VCSEL, under development since 1994, continues to receive extensive scrutiny in research and industrial laboratories. The superior performance of selectively oxidized VCSELs makes these devices appropriate for emerging applications and markets under consideration for VCSELs. The exceptional electrical and optical confinement afforded by the buried oxide apertures has enabled benchmark performance for virtually every VCSEL attribute. In fact, the continuing development of wet thermal oxidation of AlGaAs alloys is driven by the rapid and significant performance advances of selectively oxidized VCSELs. However, issues pertaining to the control, reproducibility, and uniformity of oxidation obviously remain for the manufacture of selectively oxidized VCSELs. In addition, their reliability and modes of degradation must be understood before commercial acceptance of this VCSEL structure is secured. Interestingly, although oxidation has not been extensively practiced on compound semiconductors, it is foundational to the Si microelectronics industry, and thus it represents the continued migration of semiconductor laser diode fabrication toward low cost and high volume manufacture of high performance components.

Acknowledgments

The authors have benefited from interactions with many individuals at AT&T Bell Laboratories, Colorado State University, and Sandia National Laboratories. Particular thanks are due to: H. C. Chui, M. Haggerot Crawford, P. Esherick, J. J. Figiel, B. Freund, G. Hasnain, B. E. Hammons, M. Hong, H. Q. Hou, S. Kilcoyne, K. L. Lear, R. Leibenguth, J. Mannerts, J. Nevers, A. Owyoung, R. P. Schneider, Jr., H. Temkin, R. D. Twesten, C. W. Wilmsen, and J. Wynn.

Sandia National Laboratories is a multiprogram laboratory operated by Sandia Corporation, a Lockheed Martin Company, for the United States Department of Energy under contract No. DE-AC04-94AL85000.

References

[1] H. Soda, K. Iga, C. Kitahara, and Y. Suematsu, "GaInAsP/InP surface emitting injection lasers," *Jpn. J. Appl. Phys.*, **18**, 2329–2330 (1979).

[2] K. Iga, S. Ishikawa, S. Ohkouchi, and T. Nishimura, "Room-temperature pulsed oscillation of GaAlAs/GaAs surface-emitting injection laser," *Appl. Phys. Lett.*, **45**, 348–350 (1984).

[3] P. L. Gourley and T. J. Drummond, "Visible room temperature surface emitting laser using an epitaxial Fabry–Perot resonator with AlGaAs/AlAs quarterwave high reflectors and AlGaAs/GaAs multiple quantum wells," *Appl. Phys. Lett.*, **50**, 348–350 (1987).

[4] J. L. Jewell, S. L. McCall, Y. H. Lee, A. Scherer, A. C. Gossard, and J. H. English, "Lasing characteristics of GaAs microresonators," *Appl. Phys. Lett.*, **54**, 1400–1402 (1989).

[5] K. D. Choquette and H. Q. Hou, "Vertical-cavity surface emitting lasers: moving from research to manufacturing," *Proceedings of IEEE*, **85**, 1730–1739 (1997).

[6] P. L. Gourley and T. J. Drummond, "Single crystal, epitaxial multilayers of AlAs, GaAs, and $Al_xGa_{1-x}As$ for use as optical interferometric elements," *Appl. Phys. Lett.*, **49**, 489–491 (1986).

[7] J. L. Jewell, Y. H. Lee, S. L. McCall, J. P. Harbison, and L. T. Florez, "High-finesse (Al,Ga)As interference filters grown by molecular beam epitaxy," *Appl. Phys. Lett.*, **53**, 640–642 (1988).

[8] P. L. Gourley, K. L. Lear, and R. P. Schneider, "A different mirror," *IEEE Spectrum*, **31**, 31–37 (1994).

[9] W. W. Chow, K. D. Choquette, M. H. Crawford, K. L. Lear, and G. R. Hadley, "Design, fabrication, and performance of infrared and visible vertical-cavity surface-emitting lasers," *IEEE J. Quantum. Electron.*, **33**, 1810–1824 (1997).

[10] J. L. Jewell, J. P. Harbison, A. Scherer, Y. H. Lee, and L. T. Florez, "Vertical-cavity surface-emitting lasers: Design, growth, fabrication, characterization," *IEEE Journal of Quantum Electronics*, **27**, 1332–1346 (1991).

[11] J. L. Jewell, A. Sherer, S. L. McCall, Y. H. Lee, S. Walker, J. P. Harbison, and L. T. Florez, "Low-threshold electrically pumped vertical-cavity surface-emitting microlasers," *Electron. Lett.*, **25**, 1123–1124 (1989).

[12] R. S. Geels, S. W. Corzine, J. W. Scott, D. B. Young, and L. A. Coldren, "Low threshold planarized vertical-cavity surface-emitting lasers," *IEEE Photonics Technology Letters*, **2**, 234–236 (1990).

[13] Y. H. Wang, K. Tai, J. D. Wynn, M. Hong, R. J. Fischer, J. P. Mannaerts, and A. Y. Cho, "GaAs AlGaAs multiple quantum well GRINSCH vertical cavity surface-emitting laser diodes," *IEEE Photonics Technology Letters*, **2**, 456–458 (1990).

[14] K. Bacher, B. Pezishhi, S. M. Lord, and J. S. Harris, "Molecular beam epitaxial growth of vertical cavity optical devices with *in situ* corrections," *Appl. Phys. Lett.*, **61**, 1387–1389 (1992).

[15] S. A. Chalmers and K. P. Killeen, "Method for accurate growth of vertical-cavity surface-emitting lasers," *Appl. Phys. Lett.*, **62**, 1182–1184 (1993).

[16] G. S. Li, W. Yuen, K. Toh, L. E. Eng, S. F. Lim, and C. J. Chang-Hasnain, "Accurate molecular beam epitaxial growth of vertical-cavity surface-emitting laser using diode laser reflectometry," *IEEE Photonics Technology Letters*, **7**, 971–973 (1995).

[17] J. D. Walker, D. M. Kuchta, and J. S. Smith, "Vertical-cavity surface-emitting laser diodes fabricated by phase-locked epitaxy," *Appl. Phys. Lett.*, **59**, 2079–2081 (1991).

[18] Y. M. Houng, M. R. T. Tan, B. W. Liang, S. Y. Wang, and D. E. Mars, "*In-situ* thickness monitoring and control for highly reproducible growth of distributed Bragg reflectors," *J. Vac. Sci. Technol.*, **B12**, 1221–1224 (1994).

[19] F. Koyama, H. Uenohara, T. Sakaguchi, and K. Iga, "GaAlAs/GaAs MOCVD growth for surface-emitting laser," *J. Journal of Appl. Phys.*, **26**, 1077–1081 (1987).

[20] P. Zhou, J. Cheng, C. F. Schaus, S. Z. Sun, K. Zheng, E. Armour, C. Hains, W. Hsin, D. R. Myers, and G. A. Vawter, "Low series resistance high-efficiency GaAs/AlGaAs vertical-cavity surface-emitting lasers with continuously graded mirrors grown by MOVPE," *IEEE Photonics Technology Letters*, **3**, 591–593 (1991).

[21] T. Kawakami, Y. Kadota, Y. Kohama, and T. Tadokoro, "Low-threshold current low-voltage vertical-cavity surface-emitting lasers with low Al-content p-type mirrors grown by MOVPE," *IEEE Photonics Technology Letters*, **4**, 1325–1327 (1992).

[22] R. P. Schneider, Jr., J. A. Lott, K. L. Lear, K. D. Choquette, M. H. Crawford, S. P. Kilcoyne, and J. J. Figiel, "Metalorganic vapor phase epitaxial growth of red and infrared vertical-cavity surface-emitting laser diodes," *J. Crystal Growth*, **145**, 838–845 (1994).

[23] M. K. Hibbs-Brenner, R. A. Morgan, R. A. Walterson, J. A. Lehman, E. L. Kalweit, S. Bounnak, T. Marta, and R. Gieske, "Performance, uniformity, and yield of 850 nm VCSELs deposited by MOVPE," *IEEE Photonics Technology Letters*, **8**, 7–9 (1996).

[24] H. Q. Hou, H. C. Chui, K. D. Choquette, B. E. Hammons, W. G. Breiland, and K. M. Geib, "Highly uniform and reproducible vertical-cavity surface-emitting lasers grown by metalorganic vapor phase epitaxy with in situ reflectometry," *IEEE Photonics Technology Letters*, **8**, 1285–1287 (1996).

[25] D. Botez, L. M. Zinkiewicz, T. J. Roth, L. J. Mawst, and G. Peterson, "Low threshold current density vertical cavity surface emitting AlGaAs/GaAs diode lasers," *IEEE Photonics Technology Letters*, **1**, 205–208 (1989).

[26] C. Lei, T. J. Rogers, D. G. Deppe, and B. G. Streetman, "InGaAs-GaAs quantum well vertical-cavity surface-emitting laser using molecular beam epitaxial regrowth," *Appl. Phys. Lett.*, **58**, 1122–1124 (1991).

[27] C. Lei, T. J. Rogers, D. G. Deppe, and B. G. Streetman, "ZnSe/CaF$_2$ quarter-wave Bragg reflector for the vertical-cavity surface-emitting laser," *J. Appl. Phys.*, **69**, 7430–7434 (1991).

[28] M. H. MacDougal, H. Zhao, P. D. Dapkus, M. Ziari, and W. H. Steier, "Wide-bandwidth distributed Bragg reflectors using oxide/GaAs multilayers," *Electron. Lett.*, **30**, 1147–1149 (1994).

[29] M. H. MacDougal, P. D. Dapkus, V. Pudikov, Z. Hanmin, and Y. Gye Mo, "Ultralow threshold current vertical-cavity surface-emitting lasers with AlAs oxide-GaAs distributed Bragg reflectors," *IEEE Photonics Technology Letters*, **7**, 229–231 (1995).

[30] J. J. Dudley, D. I. Babic, R. Mirin, L. Yang, B. I. Miller, R. J. Ram, T. Reynolds, E. L. Hu, and J. E. Bowers, "Low threshold wafer fused long wavelength vertical cavity lasers," *Appl. Phys. Lett.*, **64**, 1463–1465 (1994).

[31] Y. Qian, Z. H. Zhu, Y. H. Lo, H. Q. Ho, M. C. Wang, and W. Lin, "1.3 μm Vertical cavity surface emitting lasers with double bonded GaAs-AlAs Bragg mirrors," *IEEE Photonics Technology Letters*, **9**, 8–10 (1997).

[32] D. I. Babic, K. Streubel, R. P. Mirin, N. M. Maraglit, J. E. Bowers, E. L. Hu, D. E. Mars, L. Yang, and K. Carey, "Room temperature continuous wave operation of 1.54 um vertical cavity lasers," *IEEE Photonics Technology Letters*, **7**, 1225–1227 (1995).

[33] K. Tai, L. Yang, Y. H. Wang, J. D. Wynn, and A. Y. Cho, "Drastic reduction of series resistance in doped semiconductor distributed Bragg reflectors for surface-emitting lasers," *Appl. Phys. Lett.*, **56**, 2496–2498 (1990).

[34] M. Hong, J. P. Mannaerts, J. M. Hong, R. J. Fischer, K. Tai, J. Kwo, J. M. Vanderberg, Y. H. Wang, and J. Gamelin, "A simple way to reduce series resistance in p-doped semiconductor distributed Bragg reflectors," *J. Crystal Growth*, **111**, 1071–1075 (1991).

[35] S. A. Chalmers, K. L. Lear, and K. P. Killeen, "Low resistance wavelength-reproducible p-type (Al,Ga)As distributed Bragg reflectors grown by molecular beam epitaxy," *Appl. Phys. Lett.*, **62**, 1585–1587 (1993).

[36] K. L. Lear and R. P. Schneider, Jr., "Uniparabolic mirror grading for vertical cavity surface-emitting lasers," *Appl. Phys. Lett.*, **68**, 605–607 (1996).

[37] M. Sugimoto, H. Kosaka, K. Kurihara, I. Ogura, T. Numai, and K. Kasahara, "Very low threshold current density in vertical-cavity surface-emitting laser diodes with periodically doped distributed Bragg reflectors," *Electronics Letters*, **28**, 385–387 (1992).

[38] K. Kojima, R. A. Morgan, T. Mullaly, G. D. Guth, M. W. Focht, R. E. Leibenguth, and M. T. Asom, "Reduction of p-doped mirror electrical resistance of GaAs/AlGaAs vertical-cavity surface-emitting lasers by delta doping," *Electronics Letters*, **29**, 1771–1772 (1993).

[39] S. W. Corzine, R. S. Geels, J. W. Scott, R.-H. Yan, and L. A. Coldren, "Design of Fabry–Perot surface-emitting lasers with a periodic gain structure," *IEEE J. Quantum Electron.*, **25**, 1513–1524 (1989).

[40] G. M. Yang, M. H. MacDougal, and P. D. Dapkus, "Ultralow threshold current vertical-cavity surface-emitting lasers obtained with selective oxidation," *Electronics Letters*, **31**, 886–888 (1995).

[41] D. L. Huffaker, L. A. Graham, H. Deng, and D. G. Deppe, "Sub-40 μA continuous-wave lasing in an oxidized vertical-cavity surface-emitting laser with dielectric mirrors," *IEEE Photonics Technology Letters*, **8**, 974–976 (1996).

[42] Y. Hayashi, T. Mukaihara, N. Hatori, N. Ohnoki, A. Matsutani, F. Koyama, and K. Iga, "Record low-threshold index-guided InGaAs/GaAlAs vertical-cavity surface-emitting laser with a native oxide confinement structure," *Electronics Letters*, **31**, 560–561 (1995).

[43] J. Ko, E. R. Hegblom, Y. Akulova, N. M. Margalit, and L. A. Coldren, "AlInGaAs/AlGaAs strained layer 850 nm vertical cavity lasers with very low thresholds," *Electronics Letters*, **33**, 1550–1551 (1997).

[44] G. Hasnain, K. Tai, L. Yang, Y. H. Wang, R. J. Fischer, J. D. Wynn, B. Weir, N. K. Dutta, and A. Y. Cho, "Performance of gain-guided surface emitting lasers with semiconductor distributed Bragg reflectors," *IEEE J. Quantum Electron.*, **27**, 1377–1385 (1991).

[45] K. D. Choquette, R. P. Schneider, Jr., and J. A. Lott, "Lasing characteristics of visible AlGaInP/AlGaAs vertical-cavity lasers," *Optics Lett.*, **19**, 969–971 (1994).

[46] K. D. Choquette, W. W. Chow, M. H. Crawford, K. M. Geib, and R. P. Schneider, Jr., "Threshold investigation of oxide-confined vertical-cavity laser diodes," *Appl. Phys. Lett.*, **68**, 3689–3691 (1996).

[47] B. Tell, K. F. Brown-Goebeler, R. E. Leibenguth, F. M. Baez, and Y. H. Lee, "Temperature dependence of GaAs-AlGaAs vertical cavity surface emitting lasers," *Appl. Phys. Lett.*, **60**, 683–685 (1992).

[48] W. W. Chow, S. W. Koch, and M. Sargent III, *Semiconductor-Laser Physics*, Springer-Verlag, Berlin (1994).

[49] K. D. Choquette, D. A. Richie, and R. E. Leibenguth, "Temperature dependence of gain-guided vertical-cavity surface emitting laser polarization," *Appl. Phys. Lett.*, **64**, 2062–2064 (1994).

[50] D. B. Young, J. W. Scott, F. H. Peters, M. G. Peters, M. L. Majewski, B. J. Thibeault, S. W. Corzine, and L. A. Coldren, "Enhanced performance of offset-gain high-barrier vertical-cavity surface-emitting lasers," *IEEE J. Quantum Electron.*, **29**, 2013–2022 (1993).

[51] D. B. Young, J. W. Scott, F. H. Peters, B. J. Thibeault, S. W. Corzine, M. G. Peters, S. L. Lee, and L. A. Coldren, "High-power temperature-insensitive gain-offset InGaAs/GaAs vertical-cavity surface-emitting lasers," *IEEE Photonics Technology Letters*, **5**, 129–132 (1993).

[52] E. Goobar, G. Peters, G. Fish, and L. A. Coldren, "Highly efficient vertical-cavity surface-emitting lasers optimized for low-temperature operation," *IEEE Photonics Technology Letters*, **7**, 851–853 (1995).

[53] L. A. Hornak, J. C. Barr, W. D. Cox, K. S. Brown, R. A. Morgan, and M. K. Hibbs-Brenner, "Low-temperature (10–300 K) characterization of MOVPE-grown vertical-cavity surface-emitting lasers," *IEEE Photonics Technology Letters*, **7**, 1110–1112 (1995).

[54] G. Goncher, L. Bo, L. Wen-Lin, J. Cheng, S. Hersee, S. Z. Sun, R. P. Schneider, and J. C. Zolper, "Cryogenic operation of AlGaAs-GaAs vertical-cavity surface-emitting lasers at temperatures from 200 K to 6 K," *IEEE Photonics Technology Letters*, **8**, 316–318 (1996).

[55] K. Tai, R. J. Fischer, K. W. Wang, S. N. G. Chu, and A. Y. Cho, "Use of implant isolation for fabrication of vertical-cavity surface-emitting laser diodes," *Electronics Letters*, **25**, 1644–1645 (1989).

[56] M. Orenstein, A. V. Lehmen, C. J. Chang-Hasnain, N. G. Stoffel, J. P. Harbison, L. T. Florez, E. Clausen, and J. L. Jewell, "Vertical-cavity surface-emitting InGaAs/GaAs lasers with planar lateral definition," *Appl. Phys. Lett.*, **56**, 2384–2386 (1990).

[57] Y. H. Lee, B. Tell, K. F. Brown-Goebeler, and J. L. Jewell, "Top-surface-emitting GaAs four-quantum-well lasers emitting at 0.85 μm," *Electronics Letters*, **26**, 710–711 (1990).

[58] W. Jiang, C. Gaw, P. Kiely, B. Lawrence, M. Lebby, and P. R. Claisse, "Effect of proton implantion on the degradation of GaAs/AlGaAs vertical cavity surface emitting lasers," *Electronics Letters*, **33**, 137–139 (1997).

[59] J. F. Ziegler and J. P. Biersack, "The stopping and range of ions in solids," TRIM-1992, International Business Machines Research, Yorktown, NY (1992).

[60] D. Vakhshoori, J. D. Wynn, G. J. Aydzik, R. E. Leibenguth, M. T. Asom, K. Kojima, and R. A. Morgan, "Top-surface-emitting lasers with 1.9 V threshold voltage and the effect of spatial hole burning on their transverse mode operation and efficiencies," *Appl. Phys. Lett.*, **62**, 1448–1450 (1993).

[61] R. A. Morgan, M. K. Hibbs-Brenner, R. A. Walterson, J. A. Lehman, T. M. Marta, S. Bounnak, E. L. Kalweit, T. Akinwande, and J. C. Nohava, "Producible GaAs-based MOVPE-grown vertical-cavity top-surface emitting lasers with record performance," *Electronics Letters*, **31**, 462–464 (1995).

[62] C. Lei, L. A. Hodge, J. J. Dudley, M. R. Keever, B. Liang, J. R. Bhagat, and A. Liao, "High performance vertical cavity surface-emitting lasers for product applications," in *Vertical Cavity Surface Emitting Lasers I*, K. D. Choquette and D. Deppe, eds., **3003**, 28–33, SPIE (1997).

[63] K. L. Lear, S. P. Kilcoyne, and S. A. Chalmers, "High power conversion efficiencies and scaling issues for multimode vertical-cavity top-surface-emitting lasers," *IEEE Photonics Technology Letters*, **6**, 778–781 (1994).

[64] N. K. Dutta, G. H. L. W. Tu, G. Zydzik, Y. H. Wang, and A. Y. Cho, "Anomalous temporal response of gain guided surface-emitting lasers," s *Electronics Letters*, **27**, 208–210 (1991).

[65] G. R. Hadley, K. L. Lear, M. E. Warren, K. D. Choquette, J. W. Scott, and S. W. Corzine, "Comprehensive numerical modeling of vertical-cavity surface-emitting lasers," *IEEE J. Quantum Electron.*, **32**, 607–616 (1996).

[66] C. J. Chang-Hasnain, M. Orenstein, A. Vonlehmen, L. T. Florez, J. P. Harbison, and N. G. Stoffel, "Transverse mode characteristics of vertical cavity surface-emitting lasers," *Appl. Phys. Lett.*, **57**, 218–220 (1990).

[67] K. Tai, Y. Lai, K. F. Huang, T. D. Lee, and C. C. Wu, "Transverse mode emission characteristics of gain-guided surface-emitting lasers," *Appl. Phys. Lett.*, **63**, 2624–2626 (1993).

[68] G. C. Wilson, D. M. Kuchta, J. D. Walker, and J. S. Smith, "Spatial hole burning and self-focusing in vertical-cavity surface-emitting laser diodes," *Appl. Phys. Lett.*, **64**, 542–544 (1994).

[69] K. L. Lear, R. P. Schneider, Jr., K. D. Choquette, and S. P. Kilcoyne, "Index guiding dependent effects in implant and oxide confined vertical-cavity lasers," *IEEE Photonics Technology Letters*, **8**, 740–742 (1996).

[70] G. Shtengel, H. Temkin, P. Brusenbach, T. Uchida, M. Kim, C. Parsons, W. E. Quinn, and S. E. Swirhun, "High-speed vertical-cavity surface emitting laser," *IEEE Photonics Technology Letters*, **5**, 1359–1362 (1993).

[71] R. A. Morgan, "Vertical cavity surface emitting lasers: present and future," in *Vertical Cavity Surface Emitting Lasers I*, K. D. Choquette and D. Deppe, eds., **3003**, 14–26, SPIE (1997).

[72] K. D. Choquette, R. P. Schneider, Jr., K. L. Lear, and R. E. Leibenguth, "Gain-dependent polarization properties of vertical-cavity lasers," *Selected Topics Quantum Electron.*, **1**, 661–666 (1995).

[73] C. J. Chang-Hasnain, J. P. Harbison, G. Hasnain, A. C. Von Lehmen, L. T Florez, and N. G. Stoffel, "Dynamic, polarization, and transverse mode characteristics of vertical cavity surface emitting lasers," *IEEE J. Quantum Electron.*, **27**, 1402–1408 (1991).

[74] C. C. Wu, K. Tai, T. C. Huang, and K. F. Huang, "Reliability studies of gain-guided 0.85 μm GaAs/AlGaAs quantum well surface-emitting lasers," *IEEE Photonoics Technology Letters*, **6**, 37–39 (1994).

[75] D. Vakhshoori, J. D. Wynn, R. E. Leibenguth, and R. A. Novotny, "Long lasting vertical-cavity surface-emitting lasers," *Electronics Letters*, **29**, 2118–2119 (1993).

[76] J. K. Guenter, R. A. Hawthorne, D. N. Granville, M. K. Hibbs-Brenner, and R. A. Morgan, "Reliability of proton-implanted VCSELs for data communications," in *Fabrication, Testing, and Reliability of Semiconductor Lasers*, M. Fallahi and S. C. Wang, eds., **2683**, 102–113, SPIE (1996).

[77] K. L. Lear, S. P. Kilcoyne, R. P. Schneider, Jr., and J. A. Nevers, "Life-testing oxide confined VCSELs: too good to last?," in *Fabrication, Testing, and Reliability of Semiconductor Lasers*, M. Fallahi and S. C. Wang, eds., **2683**, 114–122, SPIE (1996).

[78] A. Sherer, J. L. Jewell, Y. H. Lee, J. P. Harbison, and L. T. Florez, "Fabrication of microlasers and microresonator optical switches," *Appl. Phys. Lett.*, **55**, 2724–2723 (1989).

[79] K. D. Choquette, G. Hasnain, Y. H. Wang, J. D. Wynn, R. S. Freund, A. Y. Cho, and R. E. Leibenguth, "GaAs vertical-cavity surface emitting lasers fabricated by reactive ion etching," *IEEE Photonics Technology Letters*, **3**, 859–862 (1991).

[80] G. A. Vawter, J. F. Klem, and R. E. Leibenguth, "Improved epitaxial layer design for real-time monitoring of dry etching in III–V compound semiconductor heterostructures with depth accuracy of 8 nm," *J. Vac. Sci. Technol.*, **A12**, 1973–1977 (1994).

[81] J. L. Jewell, Y. H. Lee, A. Scherer, S. L. McCall, N. A. Olsson, J. P. Harbison, and

L. T. Florez, "Surface-emitting microlasers for photonic switching and interchip connections," *Optical Engineering*, **29**, 210–214 (1990).

[82] Y. J. Yang, T. G. Dziura, S. C. Wang, W. Hsin, and S. Wang, "Submilliamp continuous wave room temperature lasing operation of a GaAs mushroom structure surface-emitting laser," *Appl. Phys. Lett.*, **56**, 1839–1840 (1990).

[83] K. D. Choquette and R. E. Leibenguth, "Control of vertical-cavity laser polarization with anisotropic transverse cavity geometries," *IEEE Photonics Technology Letters*, **6**, 40–42 (1994).

[84] T. Yoshikawa, H. Kosaka, K. Kurihara, M. Kajita, Y. Sugimoto, and K. Kasahara, "Complete polarization control of 8 × 8 vertical-cavity surface-emitting laser matrix arrays," *Appl. Phys. Lett.*, **66**, 908–910 (1995).

[85] Y. H. Lee, J. L. Jewell, B. Tell, K. F. Brown-Goebeler, A. Sherer, J. P. Harbison, and L. T. Florez, "Effects of etch depth and ion-implantation on surface emitting microlasers," *Electronics Letters*, **26**, 225–227 (1990).

[86] B. J. Thibeault, T. A. Strand, T. Wipiejewski, M. G. Peters, D. B. Young, S. W. Corzine, L. A. Coldren, and J. W. Scott, "Evaluating the effects of optical and carrier losses in etched-post vertical cavity lasers," *J. Appl. Phys.*, **78**, 5871–5875 (1995).

[87] K. D. Choquette, G. Hasnain, J. P. Mannaerts, J. D. Wynn, R. C. Wetzel, M. Hong, R. S. Freund, and R. E. Leibenguth, "Vertical-cavity surface-emitting lasers fabricated by vacuum integrated processing," *IEEE Photonics Technology Letters*, **4**, 951–954 (1992).

[88] T. Wipiejewski, D. B. Young, M. G. Peters, B. J. Thibeault, and L. A. Coldren, "Improved performance of vertical-cavity surface-emitting laser diodes with Au-plated heat spreading layer," *Electronics Letters*, **31**, 279–281 (1995).

[89] S. Kinoshita and K. Iga, "Circular buried heterostructure (CBH) GaAlAs/GaAs surface emitting lasers," *IEEE J. Quantum Electron.*, **23**, 882–888 (1987).

[90] M. Ogura, W. Hsin, M. Wu, S. Wang, J. R. Whinnery, S. C. Wang, and J. J. Yang, "Surface emitting laser diode with vertical GaAs/GaAlAs quarter-wavelength multilayers and lateral buried heterostructure," *Appl. Phys. Lett.*, **51**, 1655–1657 (1987).

[91] K. D. Choquette, M. Hong, R. S. Freund, S. N. G. Chu, J. P. Mannaerts, R. C. Wetzel, and R. E. Leibenguth, "Molecular beam epitaxial regrowth on in situ plasma-etched AlAs/AlGaAs heterostructures," *Appl. Phys. Lett.*, **60**, 1738–1740 (1992).

[92] K. D. Choquette, M. Hong, R. S. Freund, J. P. Mannaerts, R. C. Wetzel, and R. E. Leibenguth, "Vertical-cavity surface-emitting laser diodes fabricated by in situ dry etching and molecular beam epitaxial regrowth," *IEEE Photonics Technology Letters*, **5**, 284–287 (1993).

[93] M. Hong, D. Vakhshoori, L. H. Grober, J. P. Mannaerts, M. T. Asom, J. D. Wynn, F. A. Thiel, and R. S. Freund, "Buried heterostructure laser diodes fabricated by in situ processing," *J. Vac. Sci. Technol.*, **B12**, 1258–1261 (1994).

[94] K. D. Choquette, M. Hong, R. S. Freund, S. N. G. Chu, R. C. Wetzel, J. P. Mannaerts, and R. E. Leibenguth, "Vacuum integrated fabrication of vertical-cavity surface-emitting lasers," *J. Vac. Sci. Technol.*, **B11**, 1844–1849 (1993).

[95] C. J. Chang-Hasnain, Y. A. Wu, G. S. Li, G. Hasnain, K. D. Choquette, C. Caneau, and L. T. Florez, "Low threshold buried heterostructure vertical cavity surface-emitting laser," *Appl. Phys. Lett.*, **63**, 1307–1309 (1993).

[96] Y. A. Wu, G. S. Li, R. F. Nabiev, K. D. Choquette, C. Caneau, and C. J. Chang-Hasnain, "Single-mode, passive antiguide vertical cavity surface emitting laser," *IEEE J. Selected Topics Quantum Electron.*, **1**, 629–637 (1995).

[97] K. Mori, T. Asaka, H. Iwano, M. Ogura, S. Fujii, T. Okada, and S. Mukai, "Effect of cavity size on lasing characteristics of a distributed Bragg reflector-surface-emitting laser with buried heterostructure," *Appl. Phys. Lett.*, **60**, 21–22 (1992).

[98] B. S. Yoo, H. Y. Chu, M. S. Park, H. H. Park, and E. H. Lee, "Stable transverse mode emission in vertical-cavity surface-emitting lasers antiguided by amorphous GaAs layer," *Electronics Letters*, **32**, 116–117 (1996).

[99] P. Zhou, J. Cheng, J. C. Zolper, K. L. Lear, S. A. Chalmers, G. A. Vawter, R. E. Leibenguth,

and A. C. Adams, "Monolithic optoelectronic switch based on the integration of a GaAs/AlGaAs heterojunction bipolar transistor and a GaAs vertical-cavity surface-emitting laser," *IEEE Photonics Technology Letters*, **5**, 1035–1038 (1993).

[100] W. T. Tsang, "Self-terminating thermal oxidation of AlAs epilayers grown on GaAs by molecular beam epitaxy," *Appl. Phys. Lett.*, **33**, 426–429 (1978).

[101] J. M. Dallesasse, N. Holonyak, Jr., A. R. Sugg, T. A. Richard, and N. El-Zein, "Hydrolyzation oxidation of $Al_xGa_{1-x}As$-AlAs-GaAs quantum well heterostructures and superlattices," *Appl. Phys. Lett.*, **57**, 2844–2846 (1990).

[102] N. Holonyak, Jr., and J. M. Dallesasse, USA Patent #5,262,360 (1993).

[103] J. M. Dallesasse and N. Holonyak, Jr., "Native-oxide stripe-geometry $Al_xGa_{1-x}As$-GaAs quantum well heterostructure lasers," *Appl. Phys. Lett.*, **58**, 394–396 (1991).

[104] F. A. Kish, S. J. Caracci, N. Holonyak, Jr., J. M. Dallesasse, K. C. Hsieh, M. J. Ries, S. C. Smith, and R. D. Burnham, "Planar native-oxide index-guided $Al_xGa_{1-x}As$-GaAs quantum well heterostructure lasers," *Appl. Phys. Lett.*, **59**, 1755–1757 (1991).

[105] S. A. Maranowski, F. A. Kish, S. J. Caracci, N. Holonyak, Jr., J. M. Dallesasse, D. P. Bour, and D. W. Treat, "Native-oxide defined $In_{0.5}(Al_xGa_{1-x})_{0.5}P$ quantum well heterostructure window lasers (660 nm)," *Appl. Phys. Lett.*, **61**, 1688–1690 (1992).

[106] D. L. Huffaker, D. G. Deppe, K. Kumar, and T. J. Rogers, "Native-oxide defined ring contact for low threshold vertical-cavity lasers," *Appl. Phys. Lett.*, **65**, 97–99 (1994).

[107] K. D. Choquette, R. P. Schneider, Jr., K. L. Lear, and K. M. Geib, "Low threshold voltage vertical-cavity lasers fabricated by selective oxidation," *Electronics Letters*, **30**, 2043–2044 (1994).

[108] K. M. Geib, K. D. Choquette, H. Q. Hou, and B. E. Hammons, "Fabrication issues of oxide-confined VCSELs," in *Vertical-Cavity Surface-Emitting Lasers*, K. D. Choquette and D. Deppe, eds., **3003**, 69–74, SPIE (1997).

[109] K. D. Choquette, K. M. Geib, C. I. H. Ashby, R. D. Twesten, O. Blum, H. Q. Hou, D. M. Follstaedt, B. E. Hammons, D. Mathes, and R. Hull, "Advances in selective oxidation of AlGaAs Alloys," *J. Special Topics Quantum Electron.*, **3**, 916–926 (1997).

[110] K. D. Choquette, K. L. Lear, R. P. Schneider, Jr., K. M. Geib, J. J. Figiel, and R. Hull, "Fabrication and performance of selectively oxidized vertical-cavity lasers," *IEEE Photonics Technology Letters*, **7**, 1237–1239 (1995).

[111] K. D. Choquette, K. M. Geib, H. C. Chui, H. Q. Hou, and R. Hull, "Selective oxidation of buried AlGaAs for fabrication of vertical-cavity lasers," in *Mat. Res. Soc. Symp. Proc.*, **421**, 53–61 (1996).

[112] R. S. Burton and T. E. Schlesinger, "Wet thermal oxidation of $Al_xGa_{1-x}As$ compounds," *J. Appl. Phys.*, **76**, 5503–5507 (1994).

[113] H. Nickel, "A detailed experimental study of the wet oxidation kinetics of $Al_xGa_{1-x}As$ layers," *J. Appl. Phys.*, **78**, 5201–5203 (1995).

[114] M. Ochiai, G. E. Giudice, H. Temkin, J. W. Scott, and T. M. Cockerill, "Kinetics of thermal oxidation of AlAs in water vapor," *Appl. Phys. Lett.*, **68**, 1898–1900 (1996).

[115] D. E. Deal and A. S. Grove, "General relationship for the oxidation of silicon," *J. Appl. Phys.*, **36**, 3770 (1965).

[116] J. H. Kim, D. H. Lim, K. S. Kim, G. M. Yang, K. Y. Lim, and H. J. Lee, "Lateral wet oxidation of $Al_xGa_{1-x}As$-GaAs depending on its structure," *Appl. Phys. Lett.*, **69**, 3357 (1996).

[117] R. L. Naone and L. A. Coldren, "Surface energy model for the thickness dependence of the lateral oxidation of AlAs," *J. Appl. Phys.*, **82**, 2277–2280 (1997).

[118] R. D. Twesten, D. M. Follstaedt, and K. D. Choquette, "Microstructure and interface properties of laterally oxidized $Al_xGa_{1-x}As$," in *Vertical-Cavity Surface-Emitting Lasers*, K. D. Choquette and D. G. Deppe, eds., **3003**, 56–62, SPIE (1997).

[119] R. D. Twesten, D. M. Follstaedt, K. D. Choquette, and R. P. Schneider, Jr., "Microstructure of laterally oxidized $Al_xGa_{1-x}As$ layers in vertical-cavity lasers," *Appl. Phys. Lett.*, **69**, 19 (1996).

[120] F. A. Kish, S. J. Caracci, N. Holonyak, Jr., K. C. Hsieh, J. E. Baker, S. A. Maranowski,

A. R. Sugg, J. M. Dallesasse, R. M. Fletcher, C. P. Kuo, T. D. Osentowski, and M. G. Craford, "Properties and use of $In_{0.5}(Al_xGa_{1-x})_{0.5}P$ and $Al_xGa_{1-x}As$ native oxides in heterostructure lasers," *J. Electronic Materials*, **21**, 1133–1139 (1992).

[121] T. Takamori, K. Takemasa, and T. Kamijoh, "Interface structure of selectively oxidized AlAs/GaAs," *Appl. Phys. Lett.*, **69**, 659 (1996).

[122] C. L. Chua, R. L. Thornton, and D. W. Treat, "Planar laterally oxidized vertical-cavity lasers for low threshold high density top surface emitting arrays," *IEEE Photonics Technology Letters*, **9**, 1060–1061 (1997).

[123] K. D. Choquette, K. L. Lear, R. P. Schneider, and K. M. Geib, "Cavity characteristics of selectively oxidized vertical-cavity lasers," *Appl. Phys. Lett.*, **66**, 3413–3415 (1995).

[124] K. D. Choquette, R. P. Schneider, M. H. Crawford, K. M. Geib, and J. J. Figiel, "Continuous wave operation of 640–660 nm selectively oxidized AlGaInP vertical-cavity lasers," *Electronics Letters*, **31**, 1145–1146 (1995).

[125] K. L. Lear, K. D. Choquette, R. P. Schneider, Jr., S. P. Kilcoyne, and K. M. Geib, "Selectively oxidized vertical cavity surface emitting lasers with 50% power conversion efficiency," *Electronics Letters*, **31**, 208–209 (1995).

[126] B. Weigl, M. Grabherr, C. Jung, R. Jager, G. Reiner, R. Michalzik, D. Sowada, and K. Ebeling, "High performance oxide-confined GaAs VCSELs," *Selected Topics Quantum Electron.*, **3**, 409–414 (1997).

[127] D. L. Huffaker, J. Shin, and D. G. Deppe, "Lasing characteristics of low threshold microcavity lasers using half-wave spacer layers and lateral index confinement," *Appl. Phys. Lett.*, **66**, 1723–1725 (1995).

[128] K. L. Lear, K. D. Choquette, R. P. Schneider, Jr., and S. P. Kilcoyne, "Modal analysis of a small surface emitting laser with a selectively oxidized waveguide," *Appl. Phys. Lett.*, **66**, 2616–2618 (1995).

[129] T. H. Oh, D. L. Huffaker, and D. G. Deppe, "Size effects in small oxide confined vertical-cavity surface emitting lasers," *Appl. Phys. Lett.*, **69**, 3152–3154 (1996).

[130] E. R. Hegblom, D. I. Babic, B. J. Thibeault, and L. A. Coldren, "Estimation of scattering losses in dielectrically apertured vertical cavity lasers," *Appl. Phys. Lett.*, **68**, 1757–1759 (1996).

[131] K. D. Choquette, W. W. Chow, G. R. Hadley, H. Q. Hou, and K. M. Geib, "Scalability of small-aperture selectively oxidized vertical-cavity lasers," *Appl. Phys. Lett.*, **70**, 823–825 (1997).

[132] B. J. Thibeault, E. R. Hegbloom, P. D. Floyd, R. Naone, Y. Akulova, and L. A. Coldren, "Reduced optical scattering loss in vertical cavity lasers using a thin (300 Å) oxide aperture," *IEEE Photonics Technology Letters*, **8**, 593–595 (1996).

[133] E. R. Hegblom, B. J. Thibeault, R. L. Naone, and L. A. Coldren, "Vertical cavity lasers with tapered oxide apertures for low scattering loss," *Electronics Letters*, **33**, 869–871 (1997).

[134] K. D. Choquette, W. W. Chow, G. R. Hadley, H. Q. Hou, K. M. Geib, and B. E. Hammons, "Engineering the optical properties of selectively oxidized vertical cavity lasers," in *OSA Trends in Optics and Photonics: Advances in Vertical Cavity Surface Emitting Lasers*, C. J. Chang-Hasnain, ed., **15**, 125–129, Optical Society of America (1997).

[135] K. D. Choquette, K. M. Geib, H. C. Chui, B. E. Hammons, H. Q. Hou, T. J. Drummond, and R. Hull, "Selective oxidation of buried AlGaAs versus AlAs layers," *Appl. Phys. Lett.*, **69**, 1385–1387 (1996).

[136] D. L. Huffaker, J. Shin, and D. G. Deppe, "Low threshold half-wave vertical-cavity lasers," *Electronics Letters*, **30**, 1946–1947 (1994).

6

Polarization Related Properties of Vertical-Cavity Lasers

Dmitri Kuksenkov and Henryk Temkin

6.1 Introduction

The development of vertical-cavity surface-emitting lasers has brought renewed attention to the possibility of using optical devices in high-speed switching and computing [1–3]. Features such as large bandwidth, freedom from electromagnetic interference, and the possibility of free-space propagation with a large degree of parallelism have been cited as advantages of optics in communications between digital processors [4], scalable switches, interconnects for optical backplanes [5, 6], and many similar applications. In many of these situations free-space optical interconnects appear preferable to the fiber- or waveguide-based alternatives, primarily because of a higher physical channel density [7] and the resulting higher data throughput. The conceptually straightforward possibilities of large fan-out and scalability, and very low latency, are also attractive in simplifying system design. However, in the absence of suitable laser sources these ideas could not begin to be implemented.

The VCSEL has made the idea of a high performance optical interconnect feasible. Its structure lends itself naturally to large-scale integration, and low threshold currents permit operation of high density arrays. The surface-normal direction of light emission makes for relatively simple integration with the drive and control electronics. Some of the optical properties of VCSELs relevant to free-space interconnects, such as the circular and narrow angle far-field pattern or the single transverse mode operation, can be optimized simply through adjustment in the active diameter of the laser. VCSELs have demonstrated high longitudinal mode purity and the subsequent long coherence length, properties important in holographic interconnects. Finally, the small cavity volume and large carrier densities characteristic of VCSELs result in high modulation bandwidth [8, 9], sufficient for most of the free-space systems being proposed. Initial demonstrations of VCSEL-based systems, such as crossbar interconnects designed for large number of nodes [10] and switched networks for massively parallel computers [3], confirm that performance competitive with that of purely electronic implementations is possible. Most system architectures incorporating optical devices [11] rely today on one- or two-dimensional arrays of VCSELs or VCSEL-based smart pixels.

Some of the system experiments point out problems associated with the variation in the polarization state of the VCSEL. The polarization instability is the result of circularly symmetric design of the VCSEL cavity, and the difficulties encountered in controlling polarization under CW operation are well known [12]. Since free-space systems frequently rely on polarization-sensitive, diffractive [13] optical elements, such as holograms [14], for splitting and routing of various optical information channels, the source polarization must be also stable, or controllable, on the time scale dictated by the data rate used. Surprisingly, in a number of experiments using circularly symmetric VCSELs, operated in the lowest transverse mode, the laser emission was found to switch rapidly between the two polarization eigenstates.

This chapter discusses properties of VCSELs we consider important in applications to free-space systems with an emphasis on properties related to the polarization. We first review optical properties of VCSELs operated under static bias as well as dynamic modulation. We then describe the polarization instability in VCSELs and quantify its effects on the operation of optical transmission links. This is done in terms of a model transmission system, which tends to exaggerate the problem but allows for quantitative analysis of the signal to noise ratio in the link and the resulting bit error rates. Some of the parameters of VCSELs, such as the polarization resolved internal loss and modal gain, are measured independently in order to describe the temporal evolution of the polarization state. The last section discusses performance of VCSELs specifically designed for improved dynamic polarization stability and determines the difference in the modal loss or gain needed to ensure a complete stability of the polarization state.

6.2 Static and Dynamic Spectral Properties of VCSELs

To be useful in optical interconnect applications, fiber-based and free-space alike, laser sources should have narrow linewidth and low values of chirp under high-speed current modulation. Both the static linewidth and the dynamic chirp in semiconductor lasers are dependent on the coupling strength between the amplitude and phase noise. This is usually expressed as the linewidth enhancement factor α [15,16], given as a ratio of the refractive index and gain derivatives with respect to the carrier density,

$$\alpha = \frac{4\pi}{\lambda} \cdot \frac{dn/dN}{dg/dN}, \tag{6.1}$$

where N is the carrier density, n is the refractive index, g is the gain, and λ is the emission wavelength.

At the early stages of VCSEL development, it was believed that the linewidth of vertical-cavity lasers would be much larger than that of edge-emitters, an increase due to large mirror losses, a short laser cavity, and increased coupling of the spontaneous emission to the lasing mode [17]. First measurements, reported as early as 1989 [18], have shown that in fact the VCSEL linewidth can be as narrow as 50 MHz. Later corrections to the theory [19] have also confirmed that there is no reason to expect the VCSEL linewidth to be significantly broader than the linewidth of edge-emitting lasers.

The measurements of VCSEL linewidth were reported in a number of studies [20–27]. These experiments typically use homodyne or delayed self-heterodyne techniques or scanning Fabry–Perot interferometric measurements to determine the linewidth as a function of the output power. It is known that back reflections can change the linewidth [28] and it is important to optically isolate the VCSEL from the measurement apparatus. It is also important to note that the absence of compensation for self-heating of the active region results in a noticeable increase in the linewidth and, therefore, of the linewidth–output power product, at higher output powers [23, 27]. It is thus necessary to maintain the active region at a constant temperature.

The linewidth of a GaAs/AlGaAs VCSEL, measured with a confocal scanning Fabry–Perot interferometer (with a resolution of 15 MHz) is plotted in Figure 6.1 as a function

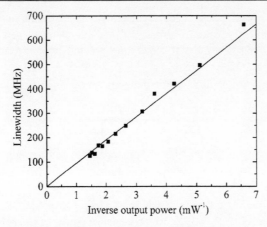

Figure 6.1. The linewidth plotted as a function of the inverse power output [24].

of the inverse output power. Back-reflections were suppressed using 50 dB of optical isolation between the laser diode and the interferometer. The resulting linewidth–output power product is $\Delta \nu \times P = 95$ MHz $\cdot$ mW. This result appears typical of gain-guided VCSELs.

Lasers used in the linewidth measurement illustrated in Figure 6.1 were planar top emitters, with current confinement obtained by double proton implantation, as described in detail in ref. 20. The active region consists of three GaAs quantum wells, each 8 nm thick, and the device is designed to operate at the wavelength of 850 nm. Lasers with 8 μm diameter opening in the top contact and 6 μm diameter proton implant had room-temperature threshold current I_{th} of 2.8 mA (CW) and threshold voltage of 4.7 V. The maximum power output of about 1.1 mW was obtained at the drive current of 7.2 mA.

Large serial resistance of the Bragg mirror results in high thermal resistance of the VCSEL. In the measurements, we compensate for self-heating in the following manner. For any value of the injection current I the temperature of the active region is $T_{ac} = T_{hs} + R_t I V$, where T_{hs} is the heat-sink temperature, V is the voltage drop across the laser, and $R_t = 1.75°C/mW$ is the measured thermal resistance of the VCSEL. Adjusting the T_{hs} for every value of I and V used in the measurement, we can maintain a constant temperature for the active region. In the measurements shown in Figure 6.1 the active region temperature T_{ac} was kept at 52°C.

The measurement of linewidth carried out as a function of the output power can be used to estimate the α-factor. The α-factor is obtained from a linear fit of the data of Figure 6.1 to the classic expression of Henry [15] for the linewidth of a semiconductor laser:

$$\Delta \nu = \frac{R_{sp} \cdot (1 + \alpha^2)}{4\pi \cdot P_0}, \tag{6.2}$$

where R_{sp} is the spontaneous emission rate and P_0 is the average number of photons inside the cavity. The spontaneous emission rate R_{sp} is found from

$$R_{sp} = \beta \cdot \Gamma \cdot \frac{\eta_i \cdot I_{th}}{e}, \tag{6.3}$$

where β is the spontaneous emission factor, Γ is the optical confinement factor, η_i is the internal quantum efficiency, and e is the elementary charge. Taking into account the energy penetration depth into the Bragg mirrors [29] and the resonant gain [30], we calculate the optical confinement factor $\Gamma = 0.045$. Using previously measured values [31, 32] of $\beta = 2 \times 10^{-3}$ and $\eta_i = 0.6$ we obtain $R_{sp} = 1.0 \times 10^{12}$ s^{-1}.

The average number of photons inside the laser cavity is given by [15]:

$$P_0 = \frac{P}{h\nu \cdot \upsilon_g \cdot \alpha_m}, \tag{6.4}$$

where P is the output power, $h\nu$ is the photon energy, υ_g is the group velocity, and α_m are the mirror losses, and we assume for simplicity that all the power is emitted through the top mirror.

Using the technique suggested by Babic et al. [33], we calculate $\alpha_m = 42$ cm^{-1} for our device. Then from Eq. (6.4) $P_0 = 1.2 \times 10^4$ at the power output of 1 mW ($\upsilon_g = 8.5 \times 10^9$ cm/s, $h\nu = 2.334 \times 10^{-19}$ J), and finally from (6.2) the linewidth enhancement factor $\alpha = 3.7$ is obtained. It is in a good agreement with the value of the α-factor reported in ref. 23 for InGaAs/GaAs VCSELs.

It is also possible to estimate the α-factor directly from Eq. (6.1), as a ratio of the refractive index and gain derivatives with respect to the carrier density. The differential gain $A = dg/dN$ can be obtained from small-signal modulation measurements of the resonance frequency f_r plotted as a square root of the average output power [34]. The $f_r(\sqrt{P})$ dependence in gain-guided VCSELs is noticeably superlinear and, in turn, the output power dependence on current is sublinear. These effects persist even after careful compensation for self-heating. There are two likely causes for the sublinearity of the light–current dependence. First is the variation of internal losses with current inherent to the gain-guided device design. The second possibility is a reduction of the modal area, and consequently η_i, due to thermal lensing [35]. This difficulty can be removed by rewriting the usual relation for $f_r(\sqrt{P})$ in the following form:

$$f_r = D \cdot \sqrt{\frac{P}{dP/dI}}, \qquad D = \frac{1}{2\pi} \cdot \sqrt{\frac{\eta_i \cdot \Gamma \cdot \upsilon_g \cdot A}{e \cdot V_a}}, \tag{6.5}$$

where dP/dI is the slope efficiency and V_a is the volume of the quantum wells forming the active layer. In the expression for f_r the coefficient D is independent of cavity losses. Moreover, any changes in the mode volume $V = V_a/\Gamma$ and η_i are expected to cancel each other. Thus the dependence of f_r on $[P/(dP/dI)]^{1/2}$ should be linear. Indeed, the measured dependence, plotted in Figure 6.2, is fairly linear and the measured slope is equal to $D = 3.4$ GHz/mA$^{1/2}$. Then, from Eq. (6.5) and for $V_a = 1.2$ μm^3, one obtains $A = 3.7 \times 10^{-16}$ cm^2.

For the measured differential gain A and the $\lambda = 850$ nm, Eq. (6.1) yields $\alpha = 3.7$, provided we assume the differential refractive index of $dn/dN = 1.0 \times 10^{-20}$cm^3, which is in good agreement with the data reported in the literature [36]. The differential index dn/dN can be measured directly by using, for instance, a method suggested by K. Kishino et al. [37]. The VCSEL wavelength measured as a function of the DC injection current is plotted in Figure 6.3. The absence of any wavelength shift with the above-threshold current is taken as a demonstration of the efficiency of the heat-compensation technique used. The below-threshold data can be well fitted by the square-root current dependence in the form of $\Delta\lambda = -k \cdot ((I/I_{th})^{1/2} - 1)$, where k represents the total wavelength shift from zero to

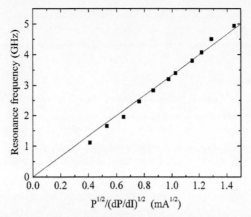

Figure 6.2. The resonance frequency measured under small-signal modulation and plotted as a function of the square root of the output power divided by the derivative of the output power with respect to the injection current. The temperature of the active region was compensated for self-heating as discussed in the text [24].

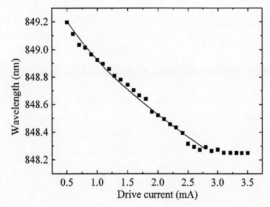

Figure 6.3. The laser wavelength dependence on injection current. The temperature of the active region was kept constant to avoid self-heating related wavelength shift [24].

threshold. The data for the gain-guided, 850 nm VCSEL of Figure 6.3 yields $k = 1.6$ nm. The expression for dn/dN can be written as

$$\frac{dn}{dN} \approx \frac{k \cdot n_{\mathrm{g}} \cdot L_{\mathrm{eff}}}{d_{\mathrm{a}} \cdot \lambda \cdot N_{\mathrm{th}}}, \tag{6.6}$$

where n_{g} is the group refractive index, d_{a} is the total thickness of the quantum wells, N_{th} is the threshold carrier density, and L_{eff} is the effective cavity length equal to the physical length of the cavity augmented by the phase penetration depths for the top and bottom Bragg mirrors. Using $d_{\mathrm{a}} = 24$ nm, $n_{\mathrm{g}} = 3.53$, and $N_{\mathrm{th}} = 7 \times 10^{18}$ cm^{-3} estimated from the turn-on delay measurements [38] and $L_{\mathrm{eff}} = 903$ nm calculated according to ref. 29, we obtain $dn/dN = 3.6 \times 10^{-20}$ cm^3. This is an unreasonably high value, which, if substituted in Eq. (6.1), would produce an unphysically large $\alpha = 13$. The anomalously strong dependence of the lasing wavelength on injection current appears to be caused by the change in the waveguide parameters caused by thermal lensing [39]. Even when the temperature of the active region is kept constant, as it was in the experiment illustrated in Figure 6.3, the heat dissipated and the temperature gradient in the plane of the active region increase with the drive current.

The reported values of the α-factor range from the low of 0.7 (InGaAs/GaAs 980 nm VCSELs, below threshold [40]) to a high of 5.7 (in AlGaAs/GaAs 850 nm VCSELs [21]). Since the differential refractive index is mainly a material parameter, the variation of α from structure to structure has its source in the variation of differential gain. The differential gain has been measured in bulk [41], single quantum well [42], and multiquantum well [43] GaAs lasers. The typical values are 3×10^{-16} cm^2, 7×10^{-16} cm^2, and $(10–12) \times 10^{-16}$ cm^2, respectively. In quantum-well lasers the gain is a logarithmic function of the carrier density and differential gain is expected to vary strongly with the threshold carrier density. In VCSELs, the volume of the active layer and the cavity length are very small. Thus one would expect to find a relatively high value of the threshold carrier density and, therefore, a relatively small value of the differential gain. The differential gain measured in gain-guided VCSELs [24] is comparable to that found in bulk GaAs lasers. Apparently, at the high carrier densities typical of VCSELs the differential gain advantage of quantum well–based structures becomes negligible. Higher values of differential gain can possibly be obtained in VCSELs based on strained or highly doped quantum wells, or in larger diameter VCSELs with lower threshold carrier density. Differential gain can also be increased by the introduction of short-wavelength detuning [44], but this will lead to higher threshold currents.

Under large-signal current modulation the lasing wavelength of VCSELs shifts owing to the change in the carrier density in the active layer resulting in the variation of the refractive index. This effect can be utilized in optical transmission links based on frequency modulation (FM) [45]. However, in the intensity modulation (IM) links the resulting dynamic spectral broadening imposes a limit on the transmission rate and/or system geometry [46]. Dynamic wavelength shift (chirp) under large-signal sinusoidal modulation was investigated by T. Mukaihara et al. [47]. We have also studied the frequency-modulation characteristics of AlGaAs/GaAs VCSELs, of the type described above, both under small- and large-signal current modulation [48].

Optical spectra of a gain-guided VCSEL under current modulation are shown in Figure 6.4. The spectra were measured with a Fabry–Perot interferometer. Figure 6.4(a) shows the spectrum of a DC-biased VCSEL. Figure 6.4(b) shows the spectrum after a small-signal sinusoidal modulation was superimposed on the DC bias. The amplitude modulation depth m was measured with a high-speed InGaAs photodiode and a sampling oscilloscope. In the small-signal experiment m was varied between 0.1 and 0.3 and the modulation frequency f_m was varied between 0.5 and 2 GHz. At a DC bias of 3.5 mA this constituted between 0.15 to 0.6 of the resonance frequency. The spectrum of Figure 6.4(b) was obtained for $m = 0.15$ and $f_m = 2.0$ GHz.

The spectra illustrated in Figure 6.4 can be used to estimate the linewidth enhancement factor α, as shown by Harder et al. [49]. A field $E(t)$ describing a sinusoidally modulated cavity mode can be written as

$$E(t) = E_0\left(1 + \frac{m}{2}\cos(2\pi f_m t)\right)\cos(2\pi f_0 t + b\cos(2\pi f_m t)), \qquad b = \frac{\Delta f}{f_m},$$

(6.7)

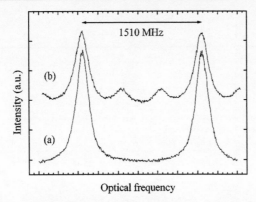

Figure 6.4. VCSEL spectra measured with the Fabry–Perot interferometer: (a) DC bias only with $I_{DC} = 3.5$ mA and (b) DC bias and sinusoidal modulation at a frequency $f_m = 1.0$ GHz and a depth $m = 0.15$. At 850 nm the interferometer has a free spectral range of 1,510 MHz and a resolution of 15 MHz [48].

where f_0 is the lasing frequency, f_m is the modulation frequency, Δf is the maximum frequency deviation, and b is the frequency modulation index. The spectral density of the field is then given by [50]

$$E_0^2\left[J_0^2(b) + m^2 J_1^2(b)\right] \qquad \text{(center line at } f_0\text{)},$$

$$E_0^2\left[J_1^2(b) + \left(\frac{m}{2}(J_2(b) - J_0(b))\right)^2\right] \qquad \text{(first sideband at } f_0 \pm f_m\text{)},$$

(6.8)

where $J_i(b)$ are Bessel functions of the ith order. Under small-signal modulation (linear case) the linewidth enhancement factor is $\alpha = 2b/m$, and the modulation index b can be found from (6.8) by using the measured relative intensity of the modulation sideband. For $f_m = 0.5$ GHz we obtain $|\alpha| = 3.7$, in good agreement with the value obtained from DC linewidth measurements. However, the sideband intensity measurements yield an α-factor that increases with the modulation frequency. For instance, for $f_m = 2.0$ GHz we obtain $|\alpha| = 4.5$. The frequency dependence of the α-factor estimated in this way is discussed below.

Although the small-signal measurements provide a good estimate of the high-frequency modulation characteristics of the laser, we are interested in detailed spectral parameters under conditions of data transmission, that is, under large-signal modulation. In large-signal experiments the VCSEL is DC-biased slightly below threshold and a square-wave modulation is provided by a return-to-zero (RZ) pulse sequence from a bit-error rate tester (BERT).

A pulse response obtained under representative large-signal modulation conditions is shown in Figure 6.5. The VCSEL was driven by a pulse sequence with 700 MHz clock frequency, with each pulse corresponding to the logic "one" (0.7 ns in length and with a rise time of 0.2 ns) followed by 23 "zeros." The low duty cycle pattern was chosen to minimize spectral changes caused by heating. The pulse current was varied from ~2.0 to 6.0 mA. For each pulse current the DC prebias was adjusted to produce a single relaxation peak in the laser output. The laser is thus operated in the "gain-switching" regime in which the dynamic chirp is at a maximum.

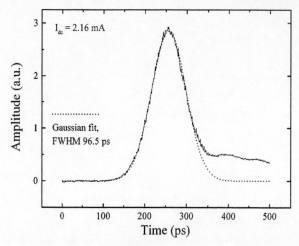

Figure 6.5. Optical response of a VCSEL driven by 0.7 ns long electrical pulse. The pulse amplitude is 3.0 V. Dashed line represents a Gaussian fit [48].

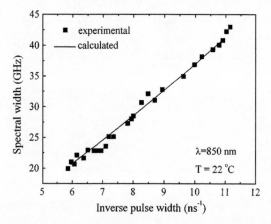

Figure 6.6. Measured and calculated dependence of pulse spectral FWHM on inverse temporal FWHM in gain-switching regime [48].

The output waveform shown in Figure 6.5, obtained for a VCSEL driven by a current pulse of 6 mA, is faithfully approximated by a Gaussian. The fall-time deviation seen in Figure 6.5 is an artifact of the photodiode bias network. The corresponding optical spectra were measured with a grating monochromator and corrected for the instrumental response by assuming a Gaussian instrumental function with a width of 0.05 nm. Under these drive conditions the largest chirped linewidth observed in our measurement was ≈0.1 nm.

The full-width-half-maximum (FWHM) linewidth obtained in large signal measurements is plotted against the inverse temporal FWHM in Figure 6.6. A fit to the data of Figure 6.6, based on a simple theory of linear chirp of instantaneous frequency [16], gives the product of $\Delta\nu \times \Delta\tau = 3.637$, where $\Delta\tau$ is the temporal width (FWHM) of the pulse, implying $\alpha = 8.18$. This is in disagreement with the DC linewidth and the small-signal measurements discussed above.

Table 6.1. *VCSEL parameters used in the rate equation simulation.*

Parameter	Value
Internal quantum efficiency, η_i	0.6
Differential gain, dg/dN	3.7×10^{-16} cm^2
Transparency carrier density, N_0	3.5×10^{18} cm^{-3}
Bimolecular recombination coefficient, B	1.8×10^{-10} cm^3/s
Total losses, $\alpha_m + \alpha_i$	52 cm^{-1}
Spontaneous emission factor, β	2.0×10^{-3}
Optical confinement factor, Γ	0.045
Saturation photon density, S_s	2.5×10^{16} cm^{-3}

The data of Figure 6.6 is modeled by numerically solving rate equations. We use conventional equations for the carrier and photon density [51]:

$$\frac{dN}{dt} = \frac{\eta_i I}{eV_a} - BN^2 - v_g A \frac{(N - N_0)S}{\sqrt{1 + S/S_s}},$$

$$\frac{dS}{dt} = \Gamma v_g A \frac{(N - N_0)S}{\sqrt{1 + S/S_s}} - \frac{S}{\tau_p} + \Gamma \beta B N^2, \tag{6.9}$$

where S is the photon density, I is the injection current, N_0 is the carrier density at transparency, B is the bimolecular recombination coefficient, S_s is the saturation photon density, and τ_p is the photon lifetime in the laser cavity: $\tau_p^{-1} = v_g(\alpha_m + \alpha_i)$, where α_i is the internal loss. The phase equation is in the form

$$\frac{d\phi}{dt} = \frac{2\pi}{\lambda} \cdot v_g \cdot \frac{dn_{eff}}{dN} \cdot (N - N_{th}), \tag{6.10}$$

where ϕ is the phase, N_{th} is the carrier density at threshold and n_{eff} is the effective refractive index of the waveguide. The parameters used in the calculation were measured [24, 31, 32, 52] or calculated according to the present state-of-the-art VCSEL theory [33, 53] and are summarized in Table 6.1.

Assuming a trapezoidal current pulse we calculate the time dependence of the photon number $P(t)$. The optical spectrum is then calculated as a Fourier transform of $P(t) \exp(-i\phi(t))$. The differential effective refractive index dn_{eff}/dN is used as a fitting parameter. The calculated dependence of Δv on the inverse of $\Delta \tau$ for $dn_{eff}/dN = -9 \times 10^{-22}$ cm^3 is shown in Figure 6.6 by the solid line.

With the differential gain $dg/dN = 3.7 \times 10^{-16}$ cm^2 and $\alpha = 3.7$ measured previously, the differential refractive index of the active layer should be $dn_a/dN = -1.0 \times 10^{-20}$ cm^3. Multiplying by the confinement factor of $\Gamma = 0.045$ we obtain $dn_{eff}/dN = -4.5 \times 10^{-22}$ cm^3, which is only half of the value obtained from fitting the data of Figure 6.6. Apparently the dynamic change in the effective refractive index is larger than one would expect from the change of the refractive index of the active layer alone.

For edge-emitting lasers [54], it was shown that a nonconstant spatial carrier distribution or carrier diffusion may result in enhanced frequency modulation. This effect manifests itself in the frequency dependence of the α-factor extracted from small-signal modulation measurements. For low modulation frequencies α is close to its "static" value; it then increases with the modulation frequency. This is the behavior observed in gain-guided VCSELs. We postulate that the anomalously large decrease in the effective refractive index with carrier density arises from a nonuniform carrier distribution in the active layer.

Below threshold, the injected carrier density has its maximum at the center of the active region. Stimulated emission depletes the carrier density in the center of the active region while the carrier density at the periphery, not clamped by stimulated emission, continues to rise. This effect, known as spatial hole burning, causes dynamic change in the laser waveguide parameters, field distribution, effective refractive index, and therefore the lasing wavelength. Under large-signal modulation this may lead to a dynamic formation of higher-order transverse modes. We indeed observe the appearance of higher-order transverse modes when the amplitude of the modulation pulse exceeds 5.0 mA. The second mode appears as a shoulder in the spectrum and does not contribute to the deconvolved FWHM of the spectrum. At the same time the abrupt appearance of the shoulder clearly indicates a significant change in the radial distribution of the carrier density and therefore gain and refractive index. We believe that this effect is responsible for the anomalously large decrease in the effective refractive index with carrier density and the observed difference between the frequency modulation characteristics under small- and large-signal modulation. Frequency modulation characteristics of VCSELs can also be strongly influenced by nonlinear gain, as was shown previously for edge-emitters by Li [55].

6.3 Polarization Instability and Performance of VCSELs in Optical Data Links

6.3.1 Polarization Instability and Relative Intensity Noise

The output of semiconductor lasers exhibits fluctuations in intensity, phase, and frequency even when the laser is DC biased. The main source of this noise is spontaneous emission [56]. Each spontaneously emitted photon adds a small component with a random phase to the coherent field of stimulated emission and thus causes a small perturbation in its amplitude and phase. The random variation in the light output intensity is usually described in terms of relative intensity noise (RIN), defined as the mean square spectral density of the noise power $p^2(f)$ normalized to the average value of the total emitted power P:

$$\text{RIN} = \frac{p^2(f)}{P^2}. \tag{6.11}$$

Intensity fluctuations limit the signal to noise ratio (SNR) and, as a consequence, receiver sensitivity and bit-error rate (BER) of optical data transmission. The knowledge of RIN characteristics of VCSELs is therefore important in optical interconnects and networks. The total intensity RIN measurements were reported for single-mode [57, 58], and multimode [59, 60], devices.

In multimode VCSELs, low-frequency RIN of the dominant mode can be considerably enhanced by the energy exchange with other cavity modes. This phenomenon is known as mode-partition noise and has been extensively studied in edge-emitting lasers [61]. Investigations of transverse mode partition in VCSELs and its influence on the optical transmission system performance were reported in a number of studies [62–64]. In general, the causes and consequences of mode partition in VCSELs and edge-emitters are similar. There is, however, one very important difference.

The optical gain for transverse-electric (TE) modes in VCSELs is polarization degenerate, not dependent on the direction of the electric field vector in the plane of the active layer.

Owing to the intentional, or unintentional, asymmetry of the waveguide characteristics, VCSEL cavities usually have two orthogonal polarization eigenstates, with the electric field vector aligned along the $\langle 011 \rangle$ and $\langle 01\bar{1} \rangle$ crystallographic directions (these are denoted as the $0°$ and $90°$ states, respectively, in the rest of this chapter). The absence of a well-defined polarization selection mechanism leads to the coexistence, switching [65], or bistability [66] of lasing modes and even to the rotation of the polarization eigenstates [67] with a change in the drive current or temperature. When the dominant polarization is different for different transverse modes [68], mode partition will cause fluctuations in the polarization state of the VCSEL output. Increased low-frequency relative intensity noise [69] and pulse distortion under modulation [70] were observed in polarization-resolved measurements.

Polarization instability and mode partition noise have been studied in AlGaAs/GaAs gain-guided VCSELs of the design described above [71]. These lasers had 10 μm diameter contact windows and 8 μm diameter proton implants.

In our experience, in heatsink-mounted and wire-bonded lasers, the alignment of polarization eigenstates along crystallographic directions does not deviate with current, temperature, or modulation, by more than $5°$. Immediately above threshold (under DC bias) lasers operate in a single transverse mode with a Gaussian near-field profile, and the $0°$ state is dominant (Figure 6.7(a)). At a bias of $(1.7\text{–}1.8) \times I_{th}$ (about 7.2–7.3 mA) a second, higher-order transverse mode, blue shifted by ≈ 5Å, appears in the lasing spectrum. For this

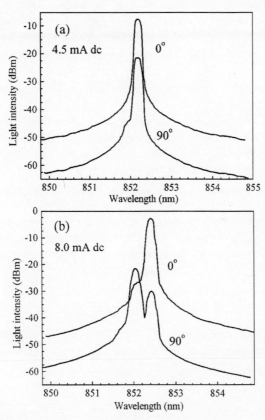

Figure 6.7. Polarization-resolved spectra of a VCSEL biased just above the threshold for: (a) zero-order transverse mode ($I = 4.5$ mA) and (b) higher-order transverse mode ($I = 8$ mA) [71].

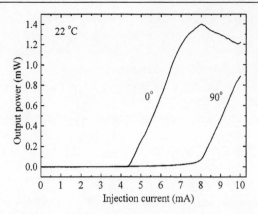

Figure 6.8. Polarization-resolved light–current characteristic for a gain-guided VCSEL.

second mode, the dominant polarization state is 90°(Figure 6.7(b)). The intensity of the higher-order mode increases gradually with the drive current and eventually reaches about 1/6 of the intensity of the Gaussian mode, as illustrated by the polarization-resolved L–I characteristic in Figure 6.8. This allows for RIN measurements for different values of the side-mode suppression ratio (SMSR) in a single laser.

Relative intensity noise measurements were carried out using a surface illuminated InGaAs p-i-n photodiode with a bandwidth of 2.5 GHz connected to a microwave spectrum analyzer. The photodiode signal was amplified by a low-noise three-stage amplifier with a total gain of 62 dB. In polarization-resolved measurements, the collimated laser beam was passed through a 30 dB optical isolator with the input polarizer axis aligned along the $\langle 011 \rangle$ or $\langle 01\bar{1} \rangle$ crystallographic directions of the VCSEL wafer. Here, the isolator serves only as a polarization selective element.

The total light intensity RIN was measured first. The optical isolator was removed and the photodiode was slightly misaligned with respect to the optical axis to avoid back-reflections. The resulting RIN dependence on the relative pumping level $R_p = \log(I/I_{th} - 1)$ is shown in Figure 6.9. It is very similar to that reported in ref. 58.

The quantum mechanical theory of noise in semiconductor lasers was developed by Yamamoto [56] and later expanded by Henry and Kazarinov [72]. Simple analytical expressions for RIN dependence on the output power can be obtained from the analysis of rate equations with Langevin noise sources [73]. Immediately above threshold (low power regime), RIN originates mainly in the beat noise between spontaneous and stimulated emission and is expected to follow the expression

$$\text{RIN} = \frac{2N\beta V^3 (h\nu\eta_{ex})^3}{\tau_p \tau_n^3 A P^3},\tag{6.12}$$

where η_{ex} is the differential quantum efficiency and τ_n is the carrier lifetime. With increasing output power, RIN decreases and approaches the shot noise limit expressed by

$$\text{RIN} = \frac{2h\nu\eta_{ex}}{P}.\tag{6.13}$$

The measured RIN of Figure 6.9, in agreement with (6.12), decreases just above threshold with the slope approximately equal to −30 dB/decade (inversely proportional to the cube of the output power P), then flattens out with increased current, and finally reaches its

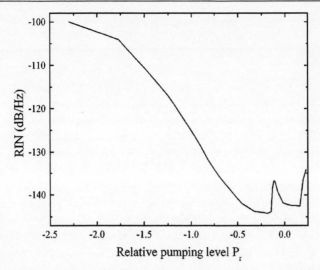

Figure 6.9. Total intensity RIN dependence on the relative pumping level $R_p = \log(I/I_{th} - 1)$.

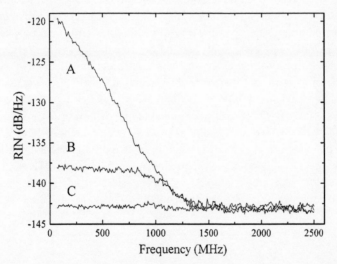

Figure 6.10. Polarization-resolved RIN spectra for different values of the side-mode suppression ratio: curve A – $SMSR = 8$ dB; curve B – $SMSR = 16$ dB; curve C – $SMSR = 20$ dB or higher [71].

minimum value of about -143 dB/Hz. The RIN level shows a jump, of about 7 dB, at a current corresponding to the threshold of the higher-order transverse mode and then decreases again to the minimum level. This temporary increase in the RIN level is caused by an interaction between spontaneous and stimulated emission at the threshold of the new mode, rather than the mode hopping known in edge-emitting lasers [74].

Having established the baseline RIN dependence on the drive current we then measure RIN spectra for the dominant mode in the frequency range of 0.07–2.5 GHz. The results obtained for different values of SMSR are presented in Figure 6.10. For SMSR > 20 dB the measured RIN spectrum (curve A) is flat and coincides with the minimum level of the total intensity RIN. For lower SMSR (curve B) we observe an increase in RIN below 1 GHz. This increase is attributed to mode partition noise. For SMSR = 8 dB (curve C) we observe a 20 dB increase in RIN at 100 MHz. This is significantly higher than reported elsewhere [73, 75] but still much lower than observed experimentally or predicted by theory for edge-emitting lasers [76]. The reason for the relatively small increase in low-frequency RIN may be the high photon density inside the VCSEL cavity. It was noted in ref. 58 that in spite of a short cavity length VCSELs reach the same, or even lower, RIN levels as edge-emitters. The effect of very high resonator quality can be illustrated by a simple example. The mirror loss in VCSELs is estimated at $\alpha_m = 42$ cm^{-1}. An edge-emitting laser with the same mirror loss (and therefore, for the same output power, the same number of photons inside the cavity) would have a resonator length of $L = 264 \, \mu$m. Assuming a stripe width of 2.5 μm this would correspond to a mode volume of $\approx 450 \, \mu$m^3. For an 8 μm diameter VCSEL the mode volume is $\approx 45 \, \mu$m^3. Therefore, for a given output power, the photon density in a VCSEL is at least an order of magnitude higher. The higher photon density leads to higher nonlinear gain, which is known to cause general damping of photon density oscillations [76], including mode partition fluctuations.

Polarization-resolved spectral measurements taken with a higher resolution monochromator reveal that each transverse mode of the VCSEL consists of a doublet of longitudinal modes, polarized orthogonally to each other. Spectra for a zero-order transverse mode are presented in Figure 6.11. A negligibly small difference in the effective refractive index for the two eigen polarization states results in a very small spectral separation of the longitudinal modes. In the case shown in Figure 6.11 the separation is ≈ 0.15 Å. A similar observation was reported in ref. 67.

Under a DC bias, the 90° polarized longitudinal mode of the VCSEL is suppressed by a factor of at least 100. The resulting polarization extinction ratio is 20–22 dB. We define the polarization extinction ratio as a power ratio of the two orthogonal polarizations, which is in this case is equal to the side-mode suppression ratio. When the SMSR is high enough, no low-frequency RIN increase is observed for a zero-order transverse mode in CW operation (Figure 6.10, curve C). This changes dramatically under large-signal current modulation,

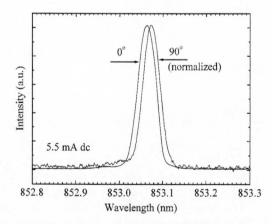

Figure 6.11. Polarization-resolved spectra of a zero-order transverse mode of the VCSEL. 90° polarized light intensity is multiplied by a factor of 100.

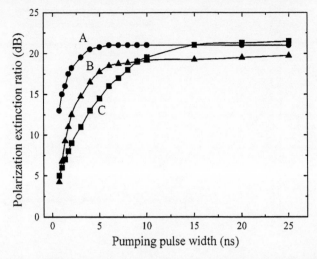

Figure 6.12. Polarization extinction ratio plotted as a function of the half period (pulse width) of the square-wave modulation for different VCSELs: A – 850 nm emission wavelength, 8 μm diameter window; B – 960 nm, 12 μm window; C – 850 nm, 10 μm window [71].

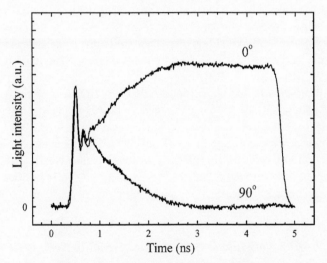

Figure 6.13. Polarization-resolved time response of a VCSEL driven with 10 ns wide rectangular current pulses. The VCSEL is DC prebiased at 4 mA [71].

as shown in a measurement in which variable frequency square wave, 4 mA peak-to-peak, was superimposed on the DC bias adjusted to assure VCSEL operation in a single transverse mode. The change in the polarization extinction ratio, under these bias conditions, is plotted in Figure 6.12 as a function of the half-period of modulation. We also show the data obtained on 8 μm diameter AlGaAs/GaAs and 12 μm diameter InGaAs/GaAs VCSELs, all of the same circular, double-proton implanted design. For all lasers, the polarization extinction ratio degrades significantly for modulation frequencies higher than 100 MHz. Under a steplike excitation it takes a relatively long time to establish a dominant polarization state of a VCSEL. Figure 6.13 shows the averaged polarization-resolved response of a VCSEL to a

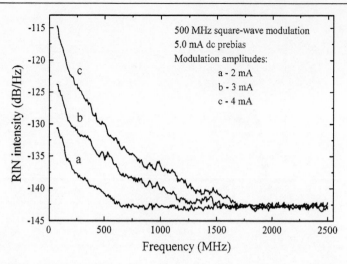

Figure 6.14. RIN spectra for the dominant polarization under square-wave modulation [71].

rectangular current pulse, measured by a fast (20 GHz) photodiode. The polarization extinction ratio increases from ≈ 0 dB at the beginning of the pulse to its maximum value of ≈ 22 dB as lasing develops. The time constant for this process is 2–5 ns, in different devices, and it varies slightly with operating conditions. The data of Figure 6.13 are clear evidence for a very small difference in the modal gain or loss for the two orthogonally polarized longitudinal modes of the VCSEL. This was also illustrated by the rate equation–based model of ref. 71.

Figure 6.14 presents the dominant polarization ($0°$) RIN spectra for a VCSEL under square-wave modulation (the DC prebias is again adjusted so that the VCSEL operates in a single transverse mode). As the modulation frequency increases we observe an increase in the low-frequency RIN, as high as 25 dB for 500 MHz modulation. It is interesting to note that the "cutoff" point for the RIN enhancement is also clearly correlated with the modulation frequency. We believe the observed RIN increase is in fact due to fluctuations in the maximum polarization extinction ratio reached during each current pulse.

6.3.2 Polarization Instability and Signal To Noise Ratio

Under digital modulation in the return-to-zero (RZ) format, the laser is driven with current pulses of the width equal to one half of the inverse clock frequency. Each logic "one" state in the data stream corresponds to the presence of a current pulse, and a logic "zero" state corresponds to the absence of the pulse. In polarization-unstable circular-symmetric devices the polarization extinction ratio reached during each current pulse fluctuates randomly, as was shown in the previous paragraph. This fluctuation is more pronounced at higher modulation frequencies. A number of components of optical interconnects and networks, such as polarizers, quarter- and half-wavelength plates, or large deflection angle holograms are, to some extent, polarization selective. When the light output of a VCSEL is passed through a polarizer, or another polarization-selective element, fluctuations in the polarization extinction ratio are translated into fluctuations in the light pulse amplitude. This is equivalent to degradation of the SNR in the link.

We have suggested and demonstrated a simple technique for estimating the SNR from the measurement of the relative intensity noise spectrum under modulation [77]. The technique is based on the following consideration. When a typical VCSEL is biased below threshold and driven by a rectangular current pulse, with the duration of 5 ns or less, the light output pulse shape is well approximated by a Gaussian (see Figure 6.5). Under current modulation with a square wave at a repetition frequency $f > 100$ MHz (analogous to the "111111" RZ pattern), the light output $L(t)$ can be described as

$$L(t) = \sum_{i=-\infty}^{\infty} P_i h(t - iT), \qquad h(t') = e^{-\frac{t'^2}{w^2}}, \qquad T = \frac{1}{f}, \tag{6.14}$$

where w is the width parameter of the Gaussian pulse.

If we neglect both the phase noise arising from the turn-on jitter and the pulse width fluctuations and consider only the amplitude fluctuations caused by variation in the polarization extinction ratio, then each pulse amplitude can be written as $P_i = P_{\text{peak}} + N_i$, where the noise component N_i is an infinite set of independent random variables with zero mean. A noise signal in the form

$$N(t) = \sum_{i=-\infty}^{\infty} N_i h(t - iT) \tag{6.15}$$

complies with the definition of a cyclostationary random process [78] and therefore has a power spectrum given by

$$S(\omega) = R(0) \cdot \frac{|H(\omega)|^2}{T}, \tag{6.16}$$

where $H(\omega)$ is the Fourier transform of the pulse shape $h(t)$ and $R(0) = E\{N_i^2\}$ is the mean square value of the noise pulse amplitude. For Gaussian-shaped pulses the power spectrum can be written more explicitly as

$$S(f) = \frac{1}{4} \frac{w^2 \pi}{T} e^{-2\pi^2 w^2 f^2} \cdot R(0). \tag{6.17}$$

Assuming a normal distribution, $R(0)$ can be expressed in terms of SNR and the average light output power $\bar{P}$:

$$\bar{P} = \frac{w\sqrt{\pi}}{T} P_{\text{peak}}, \qquad R(0) = \frac{T^2}{w^2\pi} \frac{\bar{P}^2}{(\text{SNR})^2}. \tag{6.18}$$

A final expression for the RIN spectrum is written in the form

$$\text{RIN}(f) = \frac{1}{4} \frac{T}{(\text{SNR})^2} e^{-2\pi^2 w^2 f^2}. \tag{6.19}$$

We can now estimate SNR from the measured RIN, written as $\text{RIN}(0) = T/[4(\text{SNR})^2]$, or by fitting the RIN data with (6.19) and using SNR and w as fitting parameters.

Figure 6.15 shows the measured polarization-resolved RIN spectrum and the corresponding fit for a 4 mA peak-to-peak square-wave modulation at a frequency of 500 MHz. A gain-guided VCSEL with 8 μm diameter contact window is used in the experiment. The laser is prebiased at $I = 2.9$ mA, so that the current sweeps from 2.9 mA (slightly below threshold) to $I = 7$ mA. The estimated SNR is 5.4. Without a polarization-selective element

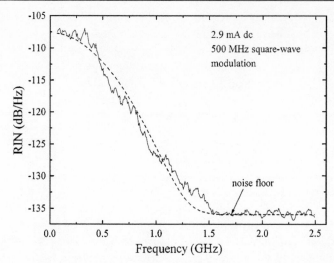

Figure 6.15. Polarization-resolved relative intensity noise spectrum of a VCSEL under square-wave modulation. Dashed curve shows theoretical fit for SNR = 5.4 [77].

between the source and the detector the entire noise spectrum would lie below the noise floor of the spectrum analyzer.

The analysis presented here does not depend on the physical source of noise. It can therefore be used to estimate SNR in any laser transmission system with amplitude noise, as well as for any clock frequency, as long as the light pulse shape can be approximated by a function with a known Fourier transform. The mathematics can also be modified to account for the pulse width and pulse-to-pulse period fluctuations. However, in its present form the analysis does not take into account any pattern dependence of the SNR; that is, it is assumed that the SNR is the same for the "111111" pattern and for pseudo-random bit sequence (PRBS).

We then carry out quantitative measurements of VCSEL performance in a free-space optical link. The experimental arrangement is presented in Figure 6.16. In this experiment VCSELs are biased below threshold and modulated with a $2^7 - 1$ word PRBS with rates between 100 and 500 Mbit/s. The light output is attenuated and detected by an optical receiver. The receiver, consisting of a Si avalanche photodiode (APD) followed by a transimpedance amplifier, has a bandwidth of 1.5 GHz. The link can be made polarization selective by inserting a Glan–Thompson prism in the light path.

The measured bit-error-rate (BER) dependence on the average received power, without and with the polarization-selective element in the link, is shown in Figures 6.17(a) and (b), respectively. Without the polarizer, the receiver sensitivity scales from -39 dBm at 100 Mbit/s to -36 dBm at 500 Mbit/s. With the polarizer inserted into the link the 100 Mbit/s curve is practically unchanged. However, at 300 Mbit/s the power penalty due to the polarization instability is as high as 6 dB. At 500 Mbit/s we observe a bit error floor at BER $\sim 10^{-6}$.

To fit the experimental data, we use the theoretical approach suggested by Agrawal et al. [79]. In this model, the detected signal S_1 or S_0, in "1" or "0" states, is the average current given by

$$S_i = R \langle M \rangle \bar{b}_i \quad (i = 0 \text{ or } 1), \tag{6.20}$$

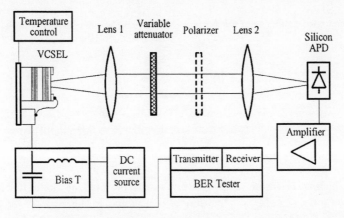

Figure 6.16. Schematic representation of the free-space transmission experiment.

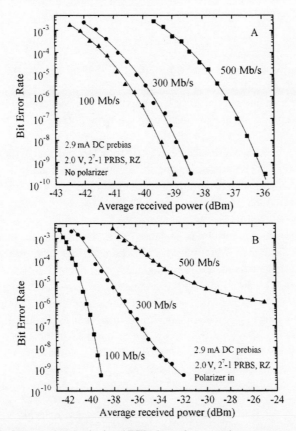

Figure 6.17. Measured and calculated BER dependence on the average received power for an $8\,\mu$m diameter VCSEL. A – nonpolarization-selective link, B – polarization-selective link. Results shown are for prebias of 2.9 mA, 2.0 V modulation, and a $2^7 - 1$ bit RZ word. BER curves do not change with the word length [77].

where $R = (\eta_d e / h\nu)$ is the responsivity of a detector with the quantum efficiency η_d at the incident photon energy of $h\nu$, $\langle M \rangle$ is the average APD gain, e is the elementary charge, and $\bar{b}_i$ is the received optical power. The noise current σ_i is associated with the signal S_i. The probability of an error, or BER, is given by

$$\text{BER} = \frac{1}{2}\text{erfc}\left(\frac{Q}{\sqrt{2}}\right), \tag{6.21}$$

where erfc(x) denotes the complimentary error function and $Q = (S_1 - S_0)/(\sigma_1 + \sigma_0)$ for the optimized decision level. The noise current σ_i can be written in the form

$$\sigma_i^2 = \sigma_c^2 + R\langle M^2 \rangle eB\bar{b}_i + R^2\langle M \rangle^2 \sigma_{Li}^2, \tag{6.22}$$

where B is the bit rate. In Eq. (6.22), the first term represents the circuit noise, the second term is due to the shot noise, and the third term accounts for laser power fluctuations. We assume $\sigma_{Li} = \bar{b}_i/\text{SNR}$.

We fit our experimental data with Eqs. (6.20)–(6.22), using $R = 0.6$ and $\langle M \rangle = 50$. The circuit noise σ_c and the signal to noise ratio SNR are used as fitting parameters. The second moment $\langle M^2 \rangle$ is approximated by [80]

$$\langle M^2 \rangle \approx \langle M \rangle^2 F(\langle M \rangle), \qquad F(\langle M \rangle) = k\langle M \rangle + (2 - \langle M \rangle^{-1})(1 - k), \tag{6.23}$$

where the ionization coefficient ratio k for our Si APD is taken to be 0.05.

Without the polarizer all three curves of Figure 6.17(a) are best fit with SNR = 30. For the polarization-selective link of Figure 6.17(b), the fitted SNR decreases to 7 at 300 Mbit/s and 4.9 at 500 Mbit/s, in excellent agreement with the value obtained from the noise spectrum measurement.

For comparison, we have also taken BER data for a larger device with a contact window diameter of 10 μm. The measurements on this device, without and with the polarizer, are shown in Figures 6.18(a) and (b). In the absence of a polarizer we estimate SNR $\sim$50. With the polarizer inserted into the link we obtain SNRs of 20, 6, and 4.6 at 100, 300, and 500 Mbit/s, respectively. The measured time constant for the stabilization of the polarization state after injection current turn-on is 3 ns in a 10 μm device and 2 ns in an 8 μm diameter device. A small difference in the polarization characteristics can probably be explained by temperature gradients being more pronounced in smaller VCSELs. As can be seen by comparing Figure 6.17 and Figure 6.18 even this small change in the polarization stability has a large effect on VCSEL performance in an optical data transmission link.

Our experiments suggest that without any measures taken to ensure polarization stability, practical data transmission rates in strongly polarization-selective links based on VCSELs are limited to $\sim$300 Mbit/s. The consequences of polarization instability will be of course less pronounced in links with weaker selectivity.

6.3.3 VCSELs with Improved Polarization Stability

The guarantee of polarization stability represents a challenge to the VCSELs design and fabrication technology, and a number of approaches to improving polarization stability have been attempted. These include using anisotropic stress from a hole in the substrate [81], utilizing asymmetric cavities [82, 83], inscribing subwavelength [84] or metal-interlaced [85] gratings on the top mirror surface, using an asymmetric superlattice in the active region

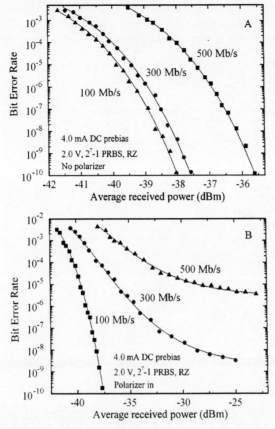

Figure 6.18.   Measured and calculated BER dependence on the average received power for a $10\,\mu$m diameter VCSEL. A – nonpolarization-selective link, B – polarization-selective link. The laser is prebiased at 4.0 mA; modulation conditions are same as those of Figure 6.17 [77].

[86], and growing on $\langle n11 \rangle$ (where $n \leq 3$) oriented substrates [87]. Most of these measures are effective in ensuring stable polarization in CW operation.

A complete polarization control of 8×8 arrays of index-guided VCSELs was reported in ref. 83. The main design feature of these transversely injected, InGaAs/GaAs devices was an asymmetric rectangular air–post mesa etched in the upper Bragg mirror. The improvement in the polarization stability is achieved by the introduction of a small difference in diffraction losses for differently polarized modes. We have studied the dynamic properties of such asymmetric air–post devices and analyzed their performance in polarization-selective interconnects [88]. It is very instructive to compare results obtained with these lasers with the measurements carried out on circularly symmetric, gain-guided lasers.

VCSELs with $6 \times 4\,\mu$m air–post mesa size have threshold current of $\sim$1.5 mA, and maximum output power of $\sim$2 mW and operate in the zero-order transverse mode under DC drive currents up to about 6 mA. Figure 6.19(a) shows the polarization-resolved spectra for a DC current bias of $I_{DC} = 5$ mA. Similarly to circularly symmetric VCSELs, the lasing spectrum consists of two longitudinal modes polarized orthogonally to each other, with the spectral separation of less than 0.3 Å (not resolved by the spectrometer in Figure 6.19).

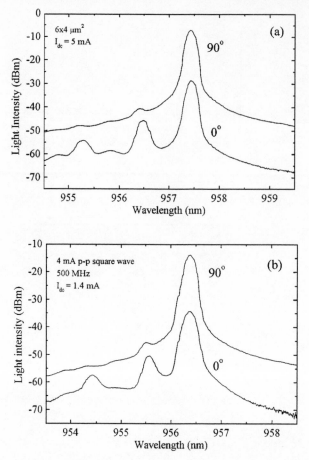

Figure 6.19. Polarization-resolved spectra of rectangular air–post VCSELs: (a) under a DC bias of 5 mA, and (b) under a DC prebias of 1.4 mA combined with a 4 mA p–p square wave at 500 MHz [88].

However, unlike the circularly symmetric VCSELs, all the asymmetric mesa devices in the 8×8 array operate with the dominant mode having the electric field vector parallel to the long side of the mesa (the $\langle 01\bar{1} \rangle$ direction, 90° eigenstate). The CW polarization extinction ratio reaches 24 dB, about 2 dB higher than that observed in circularly symmetric VCSELs.

Under modulation, the polarization extinction ratio decreases. Figure 6.19(b) shows the polarization-resolved VCSEL spectra under 500 MHz square-wave modulation. A square wave with a 4 mA peak-to-peak current swing is superimposed on a DC prebias set at 95% of the threshold current. Under these conditions the polarization extinction ratio is reduced to 20 dB, only 4 dB less than under the DC bias alone. As was shown above, the extinction ratio in polarization-unstable VCSELs decreases from 20–22 dB under DC bias to only 6–9 dB under similar modulation conditions. This is a significant improvement, and we will show below that it is sufficient to strongly enhance performance of VCSELs in polarization-selective optical data links. However, a complete polarization stability is not yet achieved.

Figure 6.20 presents the polarization-resolved time response of a VCSEL to a 1 ns long rectangular current pulse. The response was again measured with a high speed (20 GHz)

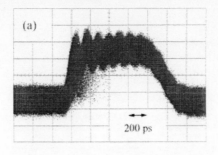

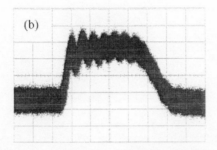

Figure 6.20. Response of a polarization-controlled VCSEL to a short current pulse. The laser is operated without DC prebias; (a) dominant polarization response and (b) total intensity response. The high-speed digital oscilloscope is operated in the infinite persistence mode to allow for observation of low probability events [88].

photodiode connected to a digitizing oscilloscope. The top curve shows the dominant polarization only; the bottom curve displays the total light intensity emitted by the VCSEL. Both waveforms were taken with the oscilloscope left in the infinite persistence mode for 20 minutes, so that thousands of single sweeps overlap in each image of Figure 6.20. This allows for the observation of even very rare signal deviations. In the absence of polarization selectivity, Figure 6.20(b), the response of a VCSEL is consistent from pulse to pulse, that is, there are no unusual low probability events. This is clearly not the case of the data of Figure 6.20(a), which displays a randomly varying delay, or a variable rise time in the optical response, with some of the delay times as long as 400 ps.

The data of Figure 6.20 can be explained as follows. Due to the asymmetry of the etched-mesa geometry one of the two orthogonally polarized lasing modes of the VCSEL experiences larger diffraction losses. Under a DC bias, the difference is sufficient to ensure a large (24 dB) suppression ratio for this mode. The situation is different under pulse excitation. It is well known that in VCSELs the spontaneous emission coupling to the lasing mode is large compared to edge-emitters [31]. After the population inversion is reached, lasing starts as amplification of the spontaneous emission. Since the spontaneous emission is a random process, there is a finite probability that a spontaneous emission fluctuation will result in lasing in a mode with a greater loss. This mode will therefore dominate for a short time, or at least have an intensity comparable to that of a more probable mode.

The transient mode partition was studied before in nearly single-mode edge-emitting lasers [89–91]. In VCSELs this effect manifests itself in the polarization-resolved optical response as a randomly varying delay or a variable rise time of the optical pulse. Typical

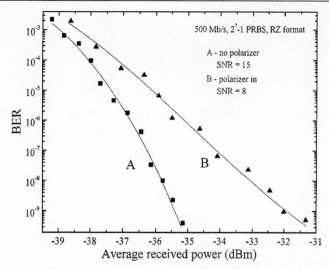

Figure 6.21. BER plotted as a function of average received optical power for a link based on polarization-stabilized VCSELs. Curve (a) illustrates performance of the link without a polarizer; curve (b) shows the effect of a polarizer inserted into the link. Bit rate is 500 Mbit/s. Signal to noise ratio is calculated from the fit to BER data shown by solid lines [88].

transient polarization partition time for circularly symmetric devices is 2–5 ns, as follows from the data of Figure 6.12 and Figure 6.13. A maximum delay seen in the waveform of Figure 6.20(a) is only 400 ps. The asymmetric air–post VCSELs can be thus operated even in strongly polarization-selective links at data rates approaching 1 Gbit/s.

Figure 6.21 shows the results of BER measurement for air–post VCSELs in the experimental arrangement presented above in Figure 6.16. The 500 Mbit/s data are taken (a) without and (b) with a polarization-selective element (polarizer) inserted between the laser and the receiver. The packaged laser is modulated directly with the signal of a BER tester in RZ format, with the amplitude of current modulation of 4 mA peak-to-peak and DC prebias set at 2.0 mA, or 1.3 times threshold (the current signal sweeps from 2.0 to 6.0 mA). The solid line shows the fit to the measured BER data from which the signal to noise ratio is estimated. In the absence of a polarizing element we obtain a SNR of 15. With the polarizer present, the SNR is reduced to 8. The insertion of a polarizing element results in a power penalty of 4 dB for BER $= 10^{-9}$ and of approximately 8 dB for BER $= 10^{-12}$. At 500 Mbit/s and with BER $= 10^{-12}$ the receiver sensitivity is -27 dBm. For the average emitted power of -6 dBm this gives a system margin of 21 dB.

With a circularly symmetric VCSEL the signal to noise ratio in a polarization-selective link operating at 500 Mbit/s was roughly ten times lower than that in a nonpolarization-selective link. With the rectangular air–post VCSELs the relative decrease in SNR, also at 500 Mbit/s, is only about 45%.

The signal to noise ratio of an optical link is also strongly dependent on the DC prebias level relative to threshold. A similar dependence was found for edge-emitting lasers [92]. Figure 6.22 plots the SNR for rectangular air–post VCSELs, found from the fit to BER measurement, as a function of the DC prebias for polarization selective and nonselective links. In this experiment, the modulation amplitude was kept constant and the DC offset

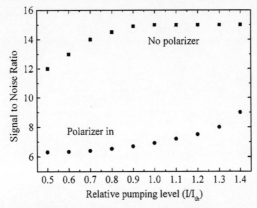

Figure 6.22. Signal to noise ratio dependence on the DC bias relative to threshold for a polarization selective and nonselective links. In the polarization selective link, higher SNR is obtained for VCSELs prebiased above threshold to reduce the contribution of the unpolarized spontaneous emission [88].

level was varied. Without the polarizer in the link, the SNR does not change with the DC prebias above threshold. A slight decrease with the prebias reduced below threshold can be attributed to the appearance of relaxation oscillations in the optical response of the laser. In a polarization-selective link, the SNR is found to increase with the increasing prebias. The relatively low SNR at low prebias levels is consistent with the high relative intensity of unpolarized spontaneous emission in the laser cavity prior to the onset of a pump current pulse. Because of self-heating and the resulting decrease in the quantum efficiency, rectangular air–post VCSELs were not tested at DC offset levels higher than ~ 1.3 times the threshold current.

6.3.4 Conditions of Complete Polarization Stability

During the current turn-on process, even VCSELs with improved polarization stability still exhibit a fair amount of polarization partition. In order to completely suppress stimulated emission in one of the polarization eigenstates, a certain degree of polarization anisotropy of gain or a certain amount of polarization-selective loss should be provided. The necessary loss difference could be provided, for instance, by polarization-selective optical feedback. Using feedback in a T-shaped external cavity [93], we have previously demonstrated a VCSEL with complete polarization stability [94]. The same experimental arrangement also allows for a polarization-resolved measurement of internal loss and modal gain in a VCSEL. In Section 6.3.4.1, we discuss the general approach to the measurement of internal loss in VCSELs. Section 6.3.4.2 describes the technique used for the polarization control and measurement of internal loss and modal gain in external cavity experiments. In Section 6.3.4.3, we discuss a data transmission experiment using a polarization-stable VCSEL.

6.3.4.1 Measurement of Internal Loss in VCSELs

For a known layer composition in the quarter-wavelength pairs of the VCSEL Bragg mirrors, the modal reflection, the field penetration depth, and therefore the device mirror loss α_m can be accurately calculated using, for example, the approach suggested in ref. 29 and ref. 33. If the internal differential quantum efficiency η_i is also known, then the internal loss α_i can be found from the measured external differential quantum efficiency η_{ex} using a standard formula:

$$\eta_{ex} = \eta_i \frac{\alpha_m}{\alpha_m + \alpha_i}. \tag{6.24}$$

This approach is widely used for the estimation of the optical loss in VCSELs and its dependence on the device size for different laser structures [95]. However, the internal quantum efficiency can vary from 0.5 to almost 1 for different device designs, and it is difficult to estimate independently.

In edge-emitting lasers, the internal loss can be found from the measurement of amplified spontaneous emission (ASE) spectra below threshold [96, 97]. Essentially, this technique is based on the determination of modal gain at the transparency point, where it is exactly equal to the total optical loss. Unfortunately, it is not possible to determine the modal gain of the VCSEL in this way, because the VCSEL cavity is too short to provide the necessary "ripples" in the ASE spectra. Instead, modal gain spectrum can be found from the true spontaneous emission spectrum (TSE) recorded from the side of the chip using the technique suggested by Henry et al. [98]. However, in the case of VCSELs this approach has serious disadvantages. First, to record the TSE spectrum it is necessary to cleave the VCSEL structure close to the active region. Second, an additional reference point is still needed to obtain an absolute value of the modal gain at the transparency point.

Both the internal quantum efficiency and internal losses in edge-emitting semiconductor lasers are commonly determined from measurements of the external efficiency dependence on the cavity length [99]. This assumes that these quantities are independent, to the first order, of the carrier density. Rewriting (6.24) as

$$\frac{1}{\eta_{ex}} = \frac{1}{\eta_i} \cdot \left(1 + \alpha_i \cdot \frac{1}{\alpha_m} \right), \tag{6.25}$$

the inverse internal efficiency is obtained as the y-axis intercept of the linear fit to the measured $1/\eta_{ex}(1/\alpha_m)$. The internal loss is obtained from the slope of this line. The mirror loss is varied by cleaving samples of different cavity length from the same wafer. This method is not directly applicable to VCSELs, where the resonator length is fixed and usually equal to the lasing wavelength. However, it is still possible to vary the mirror losses in VCSEL, for example, removing a certain number of layers in the top mirror by wet chemical etching, as was suggested by Yang et al. [100].

In ref. 32, we have suggested using a mirror of an external cavity to return a part of the emitted light back into the VCSEL. In the experiment (see Figure 6.24) the light emitted from a VCSEL is collected and collimated by a lens and then reflected back by a highly reflective dielectric mirror. A variable attenuator/beamsplitter placed inside the external cavity is used to control the amount of light returned to the VCSEL and, at the same time, to couple out a small fraction of light for the measurement. In the experiment, light–current curves for the light reflected off the beamsplitter are taken for different values of the attenuator transmittance t_{bs}. For each value of t_{bs}, the threshold current I_{th} and the "output" slope efficiency $(dP/dI)_{out}$ just above threshold are determined.

In the presence of an external reflection the VCSEL can be characterized by "effective" values of the mirror loss (α_m^{eff}) and external efficiency (η_{ex}^{eff}), dependent on the quality of the

a

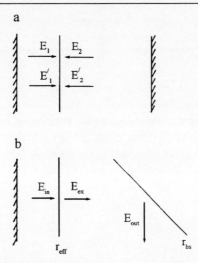

b

Figure 6.23. Schematic depiction of (a) the scattering matrix approach and (b) the effective efficiency calculation [32].

external cavity. Assuming that these effective values are still related through the expression (6.25), the internal loss and efficiency are obtained as from the linear fit to the $1/\eta_{ex}^{eff}(1/\alpha_m^{eff})$ dependence. The effective mirror loss of a VCSEL can be expressed in terms of the effective mirror reflection coefficient r_{eff}:

$$\frac{1}{\alpha_m^{eff}} = \frac{L_{eff}}{\ln(1/r_{eff})}, \tag{6.26}$$

where L_{eff}, the effective length of the VCSEL cavity, is equal to the physical length plus the penetration depth into Bragg mirrors. The effective reflection coefficient r_{eff} can be found using the scattering matrix approach [101], as illustrated in Figure 6.23(a). A 2×2 matrix S relates the electric field components in the two coupled cavities, in this case the VCSEL itself and the external cavity:

$$\begin{pmatrix} E_1' \\ E_2' \end{pmatrix} = \begin{pmatrix} S_{11} & S_{12} \\ S_{21} & S_{22} \end{pmatrix} \begin{pmatrix} E_1 \\ E_2 \end{pmatrix}. \tag{6.27}$$

The coupling between cavities is governed by the top Bragg mirror of a VCSEL, and the exact values of the S coefficients can be calculated for plane waves by transfer matrix theory [102]. Another approach is to model the Bragg reflector as a hard mirror with a reflection coefficient r, located at a distance equal to the phase penetration depth from the top boundary of the VCSEL cavity [29]. For this mirror, $S_{11} = r$, $S_{22} = -r$, and $S_{12} = S_{21} = \sqrt{1 - r^2}$. If we now define the effective reflection coefficient in the presence of external cavity as $r_{eff} = E_1'/E_1$, we have:

$$r_{eff} = S_{11} + \frac{S_{12} \cdot S_{21} \cdot t}{1 - S_{22} \cdot t} = r + \frac{t \cdot (1 - r^2)}{1 + t \cdot r}, \tag{6.28}$$

where $t = E_2/E_2'$ is the transmittance of the external resonator and $t = t_c \cdot t_r \cdot t_a$, where t_c is the coupling efficiency (estimated at $t_c = 0.75$), $t_r = (1 - r_{bs})^2$ accounts for the losses caused by reflection off the beamsplitter r_{bs}, and t_a is the attenuation of the beamsplitter.

In the general case the expression for t is complex. However, for a long enough external cavity it can be assumed that the lasing wavelength of a VCSEL is always tuned to the nearest external cavity resonance and consequently the phase change after a round trip in the external cavity is a multiple of 2π.

The value actually measured in the experiment is the slope efficiency of the output light power reflected off the beamsplitter $(dP/dI)_{\text{out}}$. To estimate the internal efficiency and loss it is then convenient to calculate the values of the "effective" slope efficiency $(dP/dI)_{\text{eff}}$, the efficiency the VCSEL would have if the reflection coefficient of its top mirror was indeed equal to r_{eff}. In this case, illustrated in Figure 6.23(b), the electric fields inside and outside the laser cavity are connected through

$$E_{\text{ex}} = \sqrt{1 - r_{\text{eff}}^2} \cdot E_{\text{in}},\qquad (6.29)$$

where the electric field inside the laser cavity is $E_{\text{in}} = E_1$ and the electric field coupled out by the beamsplitter is $E_{\text{out}} = \sqrt{t_c} \cdot r_{\text{bs}} \cdot E_2'$. Therefore,

$$E_{\text{out}} = \sqrt{t_c} \cdot r_{\text{bs}} \cdot \frac{\sqrt{1 - r^2}}{1 + t \cdot r} \cdot \frac{E_{\text{ex}}}{\sqrt{1 - r_{\text{eff}}^2}}.\qquad (6.30)$$

The measured and "effective" slope efficiencies are then related by

$$\left(\frac{dP}{dI}\right)_{\text{eff}} = \frac{E_{\text{ex}}^2}{E_{\text{out}}^2} \cdot \left(\frac{dP}{dI}\right)_{\text{out}} = \frac{\left(1 - r_{\text{eff}}^2\right) \cdot (1 + t \cdot r)^2}{t_c \cdot r_{\text{bs}}^2 \cdot (1 - r^2)} \cdot \left(\frac{dP}{dI}\right)_{\text{out}}.\qquad (6.31)$$

Assuming for simplicity that the reflection of the bottom Bragg mirror is unity and even in the presence of external cavity all output power is emitted through the top mirror, the inverse effective quantum efficiency is given by

$$\frac{1}{\eta_{\text{ex}}^{\text{eff}}} = \frac{h\nu/e}{(dP/dI)_{\text{eff}}},\qquad (6.32)$$

where $h\nu$ is the photon energy and e is the elementary charge.

Using this technique, values of $\eta_i = 0.6$ and $\alpha_i = 10\,\text{cm}^{-1}$ were obtained in ref. 32 for 8 μm diameter AlGaAs/GaAs gain-guided VCSELs.

6.3.4.2 Polarization-Resolved Measurements of Internal Loss and Modal Gain

Only a slight modification of the experimental setup is needed to make the internal loss measurement for VCSEL polarization resolved. We add a polarization beamsplitting cube and a second highly reflective mirror to the arrangement suggested in ref. 32, as shown in Figure 6.24. The optical axis of the polarization beamsplitter is aligned with the 0° direction, so that each branch of the T-shaped external cavity is returning light into its own polarization eigenstate of the VCSEL. Using two branches separately, we are able to measure internal loss and internal quantum efficiency for each polarization. Using both branches, we can set the effective loss difference between the two polarization eigenstates to a fixed value, which can be calculated from expressions (6.26) and (6.28).

In the experiments we use gain-guided GaAs/AlGaAs 850 nm VCSELs of the design discussed above with a 10 μm diameter window in the top contact. The device cavity is axially symmetric, and no polarization-selective features are introduced intentionally. The upper Bragg mirror consists of 18.5 quarter-wavelength pairs of AlGaAs/GaAs and

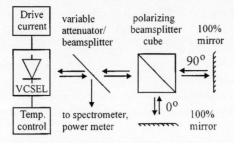

Figure 6.24. Schematic diagram of the T-shaped external cavity experiment [93].

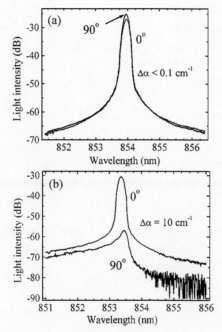

Figure 6.25. Optical spectra of a VCSEL in a T-shaped external cavity; (a) nearly equal effective cavity loss for two orthogonal polarization states; (b) effective loss difference of about $10\,\text{cm}^{-1}$ [93].

is designed to provide the field reflection coefficient of $r = 0.9965$. Without the external cavity, we calculate the mirror loss for these devices of $\alpha_m \approx 39\,\text{cm}^{-1}$ (from (6.26), for the effective cavity length of $L_{\text{eff}} = 903$ nm). For one of the branches of external cavity perfectly aligned and the attenuation of the beamsplitter set to minimum, we calculate an effective mirror loss for the corresponding polarization state of $\alpha_m^{\text{eff}} \approx 8.8\,\text{cm}^{-1}$. Thus, the maximum loss difference between the two polarization eigenstates that can be produced in the experiment is about $30\,\text{cm}^{-1}$.

Figure 6.25 shows polarization-resolved spectra of a VCSEL placed in the T-shaped external cavity. When the effective loss difference between the two polarization eigenstates is negligibly small (Figure 6.25(a)), the two longitudinal modes are almost completely degenerate and have equal intensity. When the effective loss difference is $\approx 10\,\text{cm}^{-1}$ (Figure 6.25(b)), stimulated emission is observed in only one of the polarization states ($0°$)

and the other longitudinal mode is completely suppressed. In this case, the polarization extinction ratio, defined as the power ratio of the two orthogonal polarizations, is about 28 dB (630:1). In comparison, in single-mode edge-emitting lasers (distributed feedback or distributed Bragg reflector) the side-mode suppression ratio can be as high as 40–45 dB (10,000:1). The reason for the relatively low polarization extinction ratio in VCSELs is the relatively low output power and high spontaneous emission coupling to the lasing mode [31]. While, strictly speaking, the SMSR for the case illustrated in Figure 6.25(b) is infinity (the side mode does not lase), the measured polarization extinction ratio is the ratio of stimulated emission power in one of the polarization states to the spontaneous emission power in the other polarization state, and it only reaches the value of 28 dB.

The spectra of Figure 6.25 were measured for the drive current equal to 1.3 times the threshold value. Measuring the polarization-resolved $L–I$ curves, we found that for the effective loss difference of 10 cm^{-1} the stimulated emission in the 90° polarization state is suppressed over the entire current range of the measurement ($0–4 \times I_{\text{th}}$). In contrast, in case of mode degeneracy, random abrupt switching between the two polarization states is observed with increasing drive current.

Without the external cavity the 0° polarization state is dominant, with the polarization extinction ratio of about 60:1 measured at the drive current of $\approx 1.5 \times I_{\text{th}}$. Figure 6.26 presents the dependence of the inverse effective quantum efficiency on the inverse effective mirror loss, measured for the two branches of the T-shaped external cavity. For each measurement point, we measure the $L–I$ curve for the light power reflected off the beamsplitter and extract values of the threshold current I_{th} and slope efficiency $(dP/dI)_{\text{out}}$ immediately above threshold. The inverse effective quantum efficiency is then calculated according to (6.31)–(6.32), and the inverse effective mirror loss is determined from the parameters of the external cavity (the reflection coefficient and attenuation of the beamsplitter) using relations (6.26) and (6.28). We estimate the coupling coefficient between the VCSEL and external cavity $t_c \approx 0.8$ from the measurement of the output power with a large-area photodetector, with and without the collimating lens. The data of Figure 6.26 are well fit with the straight line for both polarizations, except for the two points corresponding to the highest quality of the external cavity. These points are off the line because of the slight

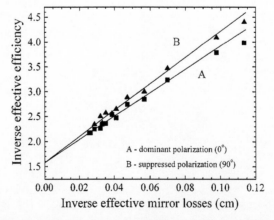

Figure 6.26. Inverse effective efficiency plotted as a function of the inverse effective mirror loss for two polarization eigenstates [93].

dependence of the internal loss on the drive current attributable, most likely, to free-carrier absorption.

The internal loss is estimated at 15 cm^{-1} for the 0° polarization state and 16.9 cm^{-1} for the 90° polarization state. The internal quantum efficiency is estimated at 0.64, for both states. The relatively low internal quantum efficiency of gain-guided VCSELs is attributed to current leakage, carrier diffusion, and barrier recombination [43, 52]. The difference of approximately 2 cm^{-1} in the internal loss for the two polarization states is responsible for the dominance of the 0° mode under CW operation. The reason for this difference is not clear for nominally axially symmetric devices. It is known that waveguide properties of gain-guided VCSELs are strongly influenced by thermal lensing effects [39]. Thus the small difference in diffraction loss can be caused by asymmetric temperature gradients caused by the self-heating, as well as possible asymmetry caused by processing imperfections.

It is interesting to compare the values of material gain for the two longitudinal modes of the laser. For a given mode, at threshold, its modal gain is equal to the sum of all losses $\Gamma g = \alpha_m^{\text{eff}} + \alpha_i$, where g is the material gain and Γ is the optical confinement factor ($\Gamma = 0.045$ for VCSELs under study). With the increasing quality of the external cavity, the effective mirror loss α_m^{eff} and, consequently, the threshold gain g_{th} decrease. In the experiment we are measuring the threshold current I_{th} and calculating the corresponding effective mirror loss α_m^{eff} for varying quality of the external cavity. Assuming that the estimated value of internal loss α_i is not dependent on the drive current, we can plot the dependence of material gain on the injection current for both polarizations, as shown in Figure 6.27. Both data sets of Figure 6.27 are well fit with the simple logarithmic gain formula of the form $g = g_0 \ln(I/I_0)$, with the gain constant of $g_0 = 1{,}125$ cm^{-1} and the transparency current of $I_0 = 1.66$ mA. We take this as evidence of quantum-well material gain being indeed independent of the polarization of light.

6.3.4.3 Transmission Experiments with VCSELs in the T-Shaped External Cavity

As was shown above, for the effective loss difference of 10 cm^{-1} or higher, stimulated emission in one of the polarization states in CW operation is completely suppressed. To

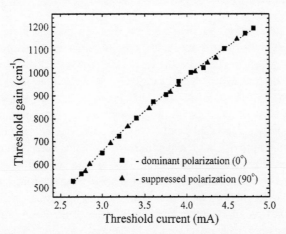

Figure 6.27. Threshold gain dependence on injection current for the two polarization eigenstates [93].

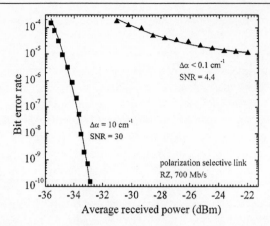

Figure 6.28. The measured BER dependence on the average received power and the corresponding theoretical fit for a polarization selective optical transmission link [93].

ascertain the effect of the difference in losses on the dynamic polarization stability, we have performed free-space transmission experiments with VCSELs placed in the T-shaped external cavity, using once more the experimental setup depicted in Figure 6.16. The amplitude of the RZ current modulation is 4 mA peak-to peak and DC prebias in each experiment is set approximately at threshold for the longitudinal mode with the lower loss. As we already know, under these conditions any transient polarization partition caused by incomplete polarization stability of the device results in considerable power penalty for a given BER value.

Figure 6.28 shows the difference in the measured BER dependence on the average received power when the two longitudinal modes of the laser are almost completely degenerate ($\Delta\alpha_m^{\text{eff}} < 0.1\,\text{cm}^{-1}$) and when one of the modes is suppressed ($\Delta\alpha_m^{\text{eff}} = 10\,\text{cm}^{-1}$), for the bit rate of 700 Mbit/s. In the case of mode degeneracy, polarization instability is so strong that we observe a BER floor at 10^{-5}, that is, a BER lower than 10^{-5} cannot be achieved for any value of the average received power. When one of the polarization states is suppressed ($\Delta\alpha_m^{\text{eff}} = 10\,\text{cm}^{-1}$ or higher), we do not observe any power penalty owing to the polarization instability.

For complete polarization stability ($\Delta\alpha_m^{\text{eff}} = 10\,\text{cm}^{-1}$), the best fit to the measured BER data gives SNR = 30. We believe this value to be limited only by the turn-on jitter and an imperfect impedance match between the BER test set and the laser. For the polarization-degenerate case, the calculated fit shows a decrease in SNR to approximately 4.4. The decrease in SNR is attributed to the transient polarization partition. It might seem that polarization-selective feedback from the long external cavity (the total length of each branch of the T-shaped cavity in our experiment is about 30 cm, which corresponds to a round-trip time of $\approx$2 ns) cannot provide sufficient polarization stability for a laser driven with current pulses shorter than 1 ns. However, our experimental data, as well as some previously reported observations on nearly single longitudinal mode edge-emitting lasers [103], suggest that transient polarization partition is not observed when there is no stimulated emission associated with the side mode in the laser cavity prior to the onset of the driving current pulse. Therefore, a reasonable prediction of the degree of dynamic polarization stability for a laser of any design can be made simply by measuring polarization-resolved spectra above threshold.

In conclusion, our experimental results suggest that complete polarization stability for any VCSEL design requires a difference of at least 10 cm^{-1} in the modal gain or loss between the two polarization eigenstates of the device. It is doubtful that such selectivity can be provided by the diffraction loss difference in asymmetric cavity designs. Therefore, we believe that the most promising way of achieving complete polarization stability is to use VCSEL structures with anisotropic gain or integrated polarization-selective reflectors.

References

[1] J. W. Goodman, F. I. Leonberger, S. Y. Kung, and R. A. Athale, *Proc. IEEE*, **72**, 850 (1984).

[2] M. R. Feldman, S. C. Esner, C. C. Guest, and S. H. Lee, *Appl. Optics*, **27**, 1742 (1988).

[3] F. E. Kiamilev, P. Marchand, A. V. Krishnamoorthy, S. C. Esner, and S. H. Lee, *J. Lightwave Technol.*, **9**(12), 1674 (1991).

[4] S. Araki, M. Kajita, K. Kasahara, K. Kubota, K. Kurihara, I. Redmond, E. Schenfeld, and T. Suzaki, *Appl. Optics*, **35**, 1269 (1996).

[5] R. K. Kostuk, D. L. Ramsey, and T.-J. Kim, *Appl. Optics*, **36**, 4722 (1997).

[6] H. J. Zhou, V. N. Morozov, J. A. Neff, and A. S. Fedor, *Appl. Optics*, **36**, 3835 (1997).

[7] M. W. Haney and M. P. Christensen, in *Massively Parallel Processing Using Optical Interconnections*, Maui, Hawaii, p.16 (1996).

[8] G. Shtengel, H. Temkin, P. Brusenbach, T. Uchida, M. Kim, C. Parsons, W. E. Quinn, and S. Swirhun, *Photonic Technol. Lett.*, **5**, 1359 (1993).

[9] K. L. Lear, H. Q. Hou, J. J. Banas, and B. E. Hammons, *Electron. Lett.*, **33**, 783–4, 1997, to be published.

[10] Y. Li, T. Wang, and R. A. Linke, *Appl. Optics*, **35**, 1282 (1996).

[11] T. Sakano, K. Noguchi, and T. Matsumoto, *Appl. Optics*, **29**, 1094 (1990).

[12] J. Martin-Regalado, F. Prati, M. San Miguel, and N. B. Abraham, *IEEE J. Quantum Electron.*, **33**(5), 765 (1997).

[13] K. S. Urquhart, P. Marchand, Y. Fainman, and S. H. Lee, *Appl. Optics*, **33**, 3670 (1994).

[14] V. N. Morozov, J. A. Neff, H. Temkin, and A. S. Fedor, *Optical Engineering*, **34**, 523 (1995).

[15] C. H. Henry, *IEEE J. Quantum Electron.*, **QE-18**, 259 (1982).

[16] T. L. Koch and J. E. Bowers, *Electron. Lett.*, **20**, 1028 (1984).

[17] G. P. Agrawal and G. R. Gray, *Appl. Phys. Lett.*, **59**, 399 (1991).

[18] H. Tanobe, F. Koyama, and K. Iga, *Electron. Lett.*, **25**, 1444 (1989).

[19] G. Bjork, A. Karlsson, and Y. Yamamoto, *Appl. Phys. Lett.*, **60**, 304 (1992).

[20] Y. H. Lee, B. Tell, K. Brown-Goebler, J. L. Jewell, C. A. Burrus, and J. M. V. Howe, *IEEE Photon. Technol. Lett.*, **2**, 686 (1990).

[21] G. R. Olbright, R. P. Bryan, W. S. Fu, R. Apte, D. M. Bloom, and Y. H. Lee, *IEEE Photon. Technol. Lett.*, **3**, 779 (1991).

[22] R. S. Geels, S. W. Corzine, and L. A. Coldren, *IEEE J. Quantum Electron.*, **27**, 1359 (1991).

[23] B. Moller, E. Zeeb, U. Fiedler, T. Hackbarth, and K. J. Ebeling, *IEEE Photon. Technol. Lett.*, **6**, 921 (1994).

[24] D. Kuksenkov, S. Feld, C. Wilmsen, H. Temkin, S. Swirhun, and R. Leibenguth, *Appl. Phys. Lett.*, **66**, 277 (1995).

[25] J. L. A. Chilla, B. Benware, M. E. Watson, P. Stanko, J. J. Rocca, C. Wilmsen, S. Feld, and R. Leibenguth, *IEEE Photon. Technol. Lett.*, **7**, 449 (1995).

[26] G. Reiner, E. Zeeb, B. Moller, M. Ries, and K. J. Ebeling, *IEEE Photon. Technol. Lett.*, **7**, 730 (1995).

[27] W. Shmid, C. Jung, B. Weigl, G. Reiner, R. Michalzik, and K. J. Ebeling, *IEEE Photon. Technol. Lett.*, **8**, 1288 (1996).

[28] G. P. Agrawal, *IEEE J. Quantum Electron.*, **QE-20**, 468 (1984).

[29] D. I. Babic and S. W. Corzine, *IEEE J. Quantum Electron.*, **28**, 514 (1992).

[30] S. W. Corzine, R. S. Geels, J. W. Scott, R. H. Yan, and L. A. Coldren, *IEEE J. Quantum Electron.*, **25**, 1513 (1989).

[31] G. Shtengel, H. Temkin, T. Uchida, M. Kim, P. Brusenbach, and C. Parsons, *Appl. Phys. Lett.*, **64**, 1062 (1994).

[32] D. V. Kuksenkov, H. Temkin, and S. Swirhun, *Appl. Phys. Lett.*, **66**, 1720 (1995).

[33] D. I. Babic, Y. Chung, N. Dagli, and J. E. Bowers, *IEEE J. Quantum Electron.*, **29**, 1950 (1993).

[34] K. Y. Lau and A. Yariv, *IEEE J. Quantum Electron.*, **QE-21**, 121 (1985).

[35] D. Vakshoori, J. D. Wynn, G. J. Zydzik, R. E. Leibenguth, M. T. Asom, K. Kojima, and R. A. Morgan, *Appl. Phys. Lett.*, **62**, 1448 (1993).

[36] J. Manning, R. Olshansky, and C. B. Su, *IEEE J. Quantum Electron.*, **QE-19**, 1525 (1983).

[37] K. Kishino, S. Aoki, and Y. Suematsu, *IEEE J. Quantum Electron.*, **QE-18**, 343 (1982).

[38] R. W. Dixon and W. B. Joyce, *J. Appl. Phys.*, **50**, 4591 (1979).

[39] K. L. Lear, R. P. Schneider, Jr., K. D. Choquette, and S. P. Kilcoyne, *IEEE Photon. Technol. Lett.*, **8**, 740 (1996).

[40] P. Dowd, H. D. Summers, I. H. White, M. R. T. Tan, Y. M. Houng, and S. Y. Wang, *Electron. Lett.*, **31**, 557 (1995).

[41] T. Takahashi, M. Nishioka, and Y. Arakawa, *Appl. Phys. Lett.*, **58**, 4 (1991).

[42] C. A. Zmudzinski, P. S. Zory, G. G. Lim, L. M. Miller, K. J. Beernink, T. L. Cockerill, J. J. Coleman, C. S. Hong, and L. Figueroa, *IEEE Photon. Technol. Lett.*, **3**, 1057 (1991).

[43] J. D. Ralston, S. Weisser, I. Esquivias, E. C. Larkins, J. Rosenzweig, P. J. Tasker, and J. Fleissner, *IEEE J. Quantum Electron.*, **29**, 1648 (1993).

[44] T. Yamanaka, Y. Yoshikuni, K. Yokoyama, W. Lui, and S. Seki, *IEEE J. Quantum Electron.*, **QE-29**, 1609 (1993).

[45] F. S. Choa, Y. H. Lee, T. L. Koch, C. A. Burrus, B. H. Tell, J. L. Jewell, and R. E. Leibenguth, *IEEE Photon. Technol. Lett.*, **3**, 697 (1991).

[46] V. N. Morozov, *Sov. J. Quantum Electron.*, **3**, 367 (1974).

[47] T. Mukaihara, M. Araki, H. Maekawa, Y. Hayashi, N. Ohnoki, F. Koyama, and K. Iga, *Electron. Lett.*, **30**, 1677 (1994).

[48] D. V. Kuksenkov, H. Temkin, and S. Swirhun, *Appl. Phys. Lett.*, **66**, 3239 (1995).

[49] C. Harder, K. Vahala, and A. Yariv, *Appl. Phys. Lett.*, **42**, 328 (1983).

[50] H. Olesen and G. Jacobsen, *IEEE J. Quantum Electron.*, **QE-18**, 2069 (1982).

[51] G. P. Agrawal, *IEEE J. Quantum Electron.*, **26**, 1901 (1990).

[52] J. W. Bae, G. Shtengel, D. V. Kuksenkov, H. Temkin, and P. Brusenbach, *Appl. Phys. Lett.*, **66**, 2031 (1995).

[53] J. W. Scott, R. S. Geels, S. W. Corzine, and L. A. Coldren, *IEEE J. Quantum Electron.*, **29**, 1295 (1993).

[54] S. Kobayashi, Y. Yamamoto, M. Ito, and T. Kimura, *IEEE J. Quantum Electron.*, **QE-18**, 582 (1982).

[55] H. Li, *IEEE Photon. Technol. Lett.*, **8**, 1594 (1996).

[56] Y. Yamamoto, *IEEE J. Quantum Electron.*, **QE-19**, 34 (1983).

[57] K. Morito, F. Koyama, and K. Iga, *Japan. J. Appl. Phys., Part 1*, **29**, 299 (1990).

[58] D. M. Kuchta, J. Gamelin, J. D. Walker, J. Lin, K. Y. Lau, J. S. Smith, M. Hong, and J. P. Mannerts, *Appl. Phys. Lett.*, **62**, 1194 (1993).

[59] K. H. Hahn, M. R. Tan, and S. Y. Wang, *Electron. Lett.*, **30**, 139 (1994).

[60] L. Raddatz, I. H. White, K. H. Hahn, M. R. Tan, and S. Y. Wang, *Electron. Lett.*, **30**, 1991 (1994).

[61] K. Ogawa, in *Semiconductors and Semimetals*, Vol. 22C, Academic, Orlando, 1985.

[62] D. M. Kuchta and C. J. Mahon, *IEEE Photon. Technol. Lett.*, **6**, 288 (1994).

[63] U. Fiedler, B. Moller, G. Reiner, D. Wiedenmann, and K. J. Ebeling, *IEEE Photon. Technol. Lett.*, **7**, 1116 (1995).

[64] D. G. Cunningham, M. C. Nowell, P. Dowd, L. Raddatz, and I. H. White, *Electron. Lett.*, **31**, 2147 (1995).

[65] K. D. Choquette, R. P. Schneider, Jr., K. L. Lear, and R. E. Leibenguth, *IEEE J. Select. Topics Quantum Electron.*, **1**, 661 (1995).

[66] Z. G. Pan, S. Jiang, M. Dagenais, R. A. Morgan, K. Kojima, M. T. Asom, R. E. Leibenguth, G. D. Guth, and M. W. Focht, *Appl. Phys. Lett.*, **63**, 2999 (1993).

[67] K. D. Choquette, D. A. Richie, and R. E. Leibenguth, *Appl. Phys. Lett.*, **64**, 2062 (1994).

[68] C. J. Chang-Hasnain, J. P Harbison, G. Hasnain, A. C. Von Lehmen, L. T. Florez, and N. G. Stoffel, *IEEE J. Quantum Electron.*, **27**, 1402 (1991).

[69] T. Mukaihara, N. Ohnoki, Y. Hayashi, N. Hatori, F. Koyama, and K. Iga, *IEEE Photon. Technol. Lett.*, **7**, 1113 (1995).

[70] M. S. Wu, L. A. Buckman, G. S. Li, K. Y. Lau, and C. J. Chang-Hasnain, *14th IEEE Semicond. Laser Conf.*, Paper P32, Conf. Digest, p. 145 (1994).

[71] D. V. Kuksenkov, H. Temkin, and S. Swirhun, *Appl. Phys. Lett.*, **67**, 2141 (1995).

[72] C. H. Henry and R. F. Kazarinov, *Rev. Modern Phys.*, **68**, 801 (1996).

[73] F. Koyama, K. Morito, and K. Iga, *IEEE J. Quantum Electron.*, **27**, 1410 (1991).

[74] M. Ohtsu and Y. Otsuka, *Appl. Phys. Lett.*, **46**, 108 (1985).

[75] K. P. Ho, J. D. Walker, and J. M. Kahn, *IEEE Photon. Technol. Lett.*, **5**, 892 (1993).

[76] G. P. Agrawal, *Phys. Rev. A.*, **37**, 2488 (1988).

[77] D. V. Kuksenkov, H. Temkin, and S. Swirhun, *IEEE Photon. Technol. Lett.*, **8**, 703 (1996).

[78] W. A. Gardner and L. E. Franks, *IEEE Trans. Inform. Theory*, **IT-21**, 4 (1975).

[79] G. P. Agrawal and T. M. Shen, *IEEE J. Lightwave Technol.*, **LT-4**, 58 (1986).

[80] R. G. Smith and S. D. Personick, in *Semiconductor Devices for Optical Communication*, Springer-Verlag, New York, 1982.

[81] T. Mukaihara, F. Koyama, and K. Iga, *IEEE Photon. Technol. Lett.*, **5**, 133 (1993).

[82] K. D. Choquette and R. E. Leibenguth, *IEEE Photon. Technol. Lett.*, **6**, 40 (1994).

[83] T. Yoshikawa, H. Kosaka, K. Kurihara, M. Kajita, Y. Sugimoto, and K. Kasahara, *Appl. Phys. Lett.*, **66**, 908 (1995).

[84] S. J. Schablitsky, L. Zhuang, R. C. Shi, and S. Y. Chou, *Appl. Phys. Lett.*, **69**, 7 (1996).

[85] J. H. Ser, Y. G. Ju, J. H. Shin, and Y. H. Lee, *Appl. Phys. Lett.*, **66**, 2769 (1995).

[86] D. Vakhshoori and R. E. Leibenguth, *Appl. Phys. Lett.*, **67**, 1045 (1995).

[87] M. Takahashi, P. Vaccaro, K. Fujita, T. Watanabe, T. Mukaihara, F. Koyama, and K. Iga, *IEEE Photon. Technol. Lett.*, **8**, 737 (1994).

[88] D. V. Kuksenkov, H. Temkin, and T. Yoshikawa, *IEEE Photon. Technol. Lett.*, **8**, 977 (1996).

[89] S. E. Miller, *IEEE J. Quantum Electron.*, **QE-22**, 16 (1986).

[90] P.-L. Liu and M. M. Choy, *IEEE J. Quantum Electron.*, **25**, 1767 (1989).

[91] J. C. Cartledge, *IEEE J. Quantum Electron.*, **26**, 2046 (1989).

[92] P.-L. Liu and M. M. Choy, *IEEE J. Quantum Electron.*, **25**, 854 (1989).

[93] D. V. Kuksenkov and H. Temkin, *IEEE J. Sel. Topics in Quant. Electron.*, **3**(2), 390 (1997).

[94] H. Kawaguchi, T. Irie, and M. Murakami, *IEEE J. Quantum Electron.*, **31**, 447 (1995).

[95] B. Thibeault, T. A. Strand, T. Wipiejewski, M. G. Peters, D. B. Young, S. W. Corzine, L. A. Coldren, and J. W. Scott, *J. Appl. Phys.*, **78**, 10 (1995).

[96] P. A. Andrekson, N. A. Olsson, T. Tanbun-Ek, R. A. Logan, D. L. Coblentz, and H. Temkin, *Electron. Lett.*, **28**, 171 (1992).

[97] G. E. Shtengel and D. A. Ackerman, *Electron. Lett.*, **31**, 1157 (1995).

[98] C. H. Henry, R. A. Logan, and F. R. Merrit, *J. Appl. Phys.*, **51**, 3042 (1980).

[99] H. C. Casey, M. B. Panish, W. O. Schlosser, and T. L. Paoli, *J. Appl. Phys.*, **45**, 322 (1974).

[100] G. M. Yang, M. H. MacDougal, V. Pudikov, and P. D. Dapkus, *IEEE Photon. Technol. Lett.*, **7**, 1228 (1995).

[101] H. K. Choi, K. L. Chen, and S. Wang, *IEEE J. Quantum Electron.*, **QE-20**, 385 (1984).

[102] G. Bjork and O. Nilsson, *J. Lightwave Technol.*, **LT-5**, 140 (1987).

[103] P. L. Liu, in *Coherence, Amplification and Quantum Effects in Semiconductor Lasers*, Wiley, New York, 1991.

7

Visible Light Emitting Vertical-Cavity Lasers

Robert L. Thornton

7.1 Background

7.1.1 Introduction

This chapter describes progress toward the realization of vertical-cavity lasers emitting visible light. Research into the demonstration of vertical-cavity lasers emitting over a broad range of wavelengths in the visible portion of the spectrum has accelerated since the early demonstration of such devices. This chapter first provides perspective on the wide range of applications for visible light emitting VCSELs. It then focuses on specific wavelength regions within the visible, starting at the long wavelength red visible and progressing through the spectrum to blue and into the ultraviolet. In each range we will present materials issues as well as device design issues and, where available, describe device results. The reader will hopefully be left with a sense of the state of current research as well as performance capabilities and potentials of lasers within this wavelength range.

7.1.2 Applications for Visible Light Emitting Lasers

There are many applications that will be enabled or greatly enhanced by the availability of VCSELs in the visible. Visible light emitting semiconductor lasers generally provide three types of utility relative to their infrared counterparts. Firstly, their visibility provides improved performance in applications where beam alignment, pointing, scanning, or display are important. In Figure 7.1 we show a diagram of the eye sensitivity as a function of photon wavelength. Also indicated are the perceived colors corresponding to the various wavelengths.

An intriguing potential application of visible light emitting lasers in general, and VCSELs in particular, is the retinal display, which would provide for full color imaging by directly scanning an image from three lasers representing the primary colors of red, blue, and green directly onto the retina. This architecture would provide for very efficient optical collection and an extremely compact display, particularly when implemented using VCSEL devices combined with micro-electromechanical systems (MEMS) based scanning elements [1, 2]. One embodiment of such a system is shown in Figure 7.2.

Secondly, there are those applications where some important system parameter scales with the wavelength used. For example, in optical imaging applications such as optical recording and laser printing, the shorter wavelength allows for a smaller spot size, resulting in increased density of imaging for a given size of optic or numerical aperture. As a result, higher resolution, more compact systems are possible.

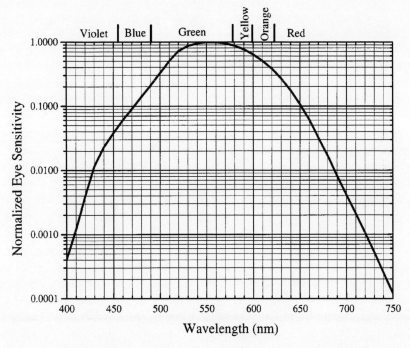

Figure 7.1. Human eye response as a function of emission wavelength, defining the visible portion of the spectrum. The ranges of perception of various colors are also indicated.

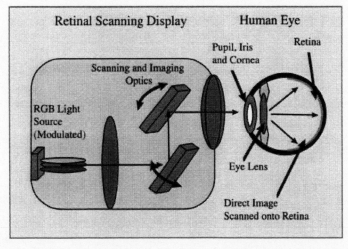

Figure 7.2. Schematic diagram of a retinal scanning display, which utilizes coherent red, green, and blue light sources to project full color images directly into the human eye.

Finally, there are those applications that are enabled by the availability of higher energy photons. In some applications, such as, again, laser printing, photoactivated processes can be made to happen more quickly, efficiently, or reliably with higher energy visible light photons. In addition, as emission wavelengths increase to the blue end of the visible, coherent high energy photons may be used in frequency down-conversion processes, such as phosphor excitation. This results in efficient light emission for use in creating visible light sources spanning the entire visible wavelength range.

Generally, target applications for VCSELs are also candidates for edge-emitting lasers of the same visible wavelength. It has to date proven to be true that the advance of VCSEL devices has followed on the heels of advances in the field of edge-emitting laser device performance at a given wavelength. It is therefore currently the case that the greatest opportunities for visible VCSEL device applications will be those in which the unique potential advantages of VCSELs, (i.e., lower cost at moderate power, and relative ease of fabrication into arrays) can be brought to bear.

Certainly prominent among the potential applications of visible VCSELs is that of high density optical storage, where the recording density available scales inverse quadratically with the optical wavelength used. The current predominant wavelength of emission for optical storage CD lasers is 780 nm. The transition to higher density storage using devices emitting in the 630–680 nm range is well underway. The benefits in this regard to short wavelength green, blue, or even UV semiconductor lasers is clearly recognized. This market is currently dominated by edge-emitting lasers; however, even the venerable 780 nm CD laser market may experience VCSEL replacement, driven by reduced cost of packaging [3]. In addition to reduced cost, the two-dimensional array ability may offer opportunities for VCSEL devices to increase system throughput. For the visible VCSEL device to enter this market, strategies for parallel optical scanning for increased performance would be a key enabler [4].

In the area of laser printing, VCSELs offer the general advantage of greatly facilitating the use of arrays of lasers rather than single lasers for writing multiple scan lines simultaneously. The impact of such sources is greatest when one considers high speed, high volume printers for which the great need for durability and longevity places stringent demands on the light-sensitive photoreceptor materials. These demands are currently best satisfied by visible light sensitive materials. As an example, current commercially available 120+ page per minute laser printers will typically use a photosensitive material with sensitivity in the visible range to about 680 nm. Many such devices, which initially used bulky He–Ne gas lasers coupled with beamsplitters and acousto-optic modulators [5] are prime candidates for insertion of arrays of semiconductor lasers. These systems will initially adopt red edge-emitting lasers. As throughput and resolution requirements increase, VCSEL devices become an increasingly attractive solution owing to their ability to be fabricated into arrays, thus multiplexing several scan lines simultaneously through the optical system. In Figure 7.3 we show a schematic diagram of the scanning subsystem in a laser printer. In Figure 7.4 we show the dependence of system throughput on the polygon rotation speed and the number of laser beams in a system having 600 spi resolution. The leverage obtained by the use of multiple beams is quite clear. Not only is throughput increased, but cost is reduced because of the use of a less costly rotating polygon mirror.

In the area of data communications, solutions are still in a state of flux for low cost, high data rate fiber insertion into local area networks and high volume applications such as fiber to home. For cost of connectorization considerations, plastic optical fiber is a strong candidate

Rotating Polygon Laser Printer

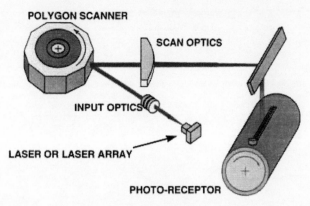

Figure 7.3. Schematic diagram of the laser subsystem of a laser printer, showing the rotating polygon, semiconductor laser, and other key components.

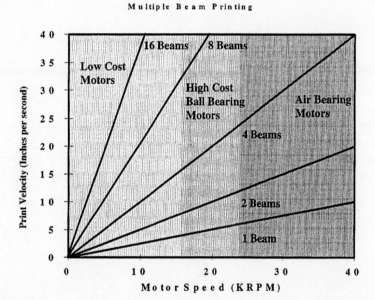

Figure 7.4. The dependence of laser printing velocity on the number of beams used, and its impact on the speed and cost of the rotating polygon mirror.

for this application [6–10]. In order to effectively utilize plastic optical fiber, low cost emitters that operate at the transmission minimum wavelength for plastic optical fiber are needed. The most widely implemented plastic optical fiber has optical transmission minima at 570 nm and 650 nm, offering opportunities for low cost emitters at these wavelengths. The transmission spectrum through conventional plastic optical fiber is shown in Figure 7.5. The minima in absorption losses in the visible for this fiber may be a driver for the application of visible light emitting VCSELs, as the cost of packaging for the devices as well as its improved stability of wavelength with temperature, enable superior performance in this application.

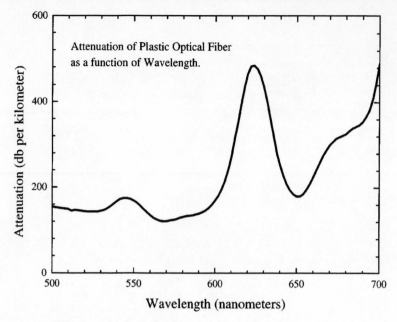

Figure 7.5. Optical transmission curves for plastic optical fiber, showing minima in propagation loss at 570 nm and 650 nm.

7.1.3 Summary

In conclusion, visible light emitting VCSELs are of interest for a wide variety of applications. Many of these applications are driven by the rapidly expanding knowledge economy, where faster data communications, higher density storage, higher quality visual displays, and higher quality printed output of digital information are all powerful moving forces. Throughout most of these applications, the availability of cost effective, arrayable, efficient, and coherent light sources offers great opportunities.

7.2 Visible Laser Materials Considerations

7.2.1 Overview

Many of the fundamental issues regarding the feasibility and fabrication of VCSEL devices have been established using VCSEL devices fabricated on a GaAs substrate, utilizing an InGaAs quantum well active region, and emitting in the 980 nm wavelength range [11–12]. This is in large part because VCSELs in this wavelength range are able to utilize the maximum refractive index contrast available between AlAs and GaAs and because the emission is not absorbed by the GaAs substrate, enabling bottom surface emitting VCSELs for which electrical injection and optical extraction are simplified considerably. With shorter wavelengths, successive challenges are encountered. As photon energy increases, the GaAs substrate first becomes absorbing as one goes into the near infrared to red; the situation then becomes even more problematic as one abandons the III–V Zincblende material system to access the green and blue emission regions. Within the III–V compounds, as wavelength decreases, the range of available refractive indices decreases, as a result of increasing the Al content of the low Al content mirror layer. This is necessary to avoid optical absorption in the higher refractive index layer. This necessitates an increase in the number of layer pairs to achieve high reflectivity. With wider bandgap materials, the availability of suitable index contrast, as well as the increased difficulty of p-type doping in epitaxial mirrors, often dictates dielectric rather than epitaxial mirrors.

As of this writing, vertical-cavity lasers in the red are rapidly becoming an established technology, while encouraging demonstrations of photopumped [13–16] and electrically injected [17] lasers at shorter wavelengths have also been reported.

In the push to realize VCSEL devices at shorter wavelengths, one must consider the options for materials that emit at the appropriate wavelength. One encounters increasing levels of challenge from the materials systems that are required. This is perhaps best seen by examining the bandgap versus lattice parameter diagram in Figure 7.6. Materials in this figure are all currently the subject of active research for their application toward the fabrication of lasers within their respective wavelength regions. The challenges associated with the various wavelengths are varied and in some cases substantial. We shall next discuss the issues for specific wavelengths starting from red and progressing through the near UV.

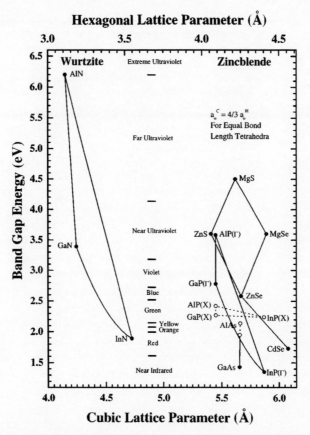

Figure 7.6. Bandgap versus lattice parameter for several materials systems used to make devices that emit visible light. The direct and indirect bands are indicated for AlGaInP, and the direct to indirect transition for AlGaAs is indicated by the transition from solid to dashed lines. Since the definitions of lattice parameter for hexagonal and cubic crystals are not consistent, these curves are plotted to reconcile the bond length for the tetrahedral bonds, which is the common measure for lattice mismatch between the hexagonal Wurtzite and cubic Zincblende systems. Also indicated on the figure are the visual color ranges for these materials, ranging from the near infrared to the extreme ultraviolet.

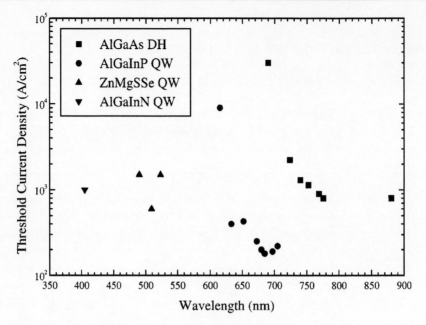

Figure 7.7. Threshold current density versus wavelength for edge-emitting lasers in the AlGaAs, AlGaInP, MgZnSSe, and AlGaInN materials systems. AlGaAs data are for double heterostructure lasers; all other data are for quantum well lasers.

7.2.2 Short-Wavelength AlGaAs Lasers

There have been numerous studies of the capability of AlGaAs material to produce efficient light emission and lasing in the red portion of the spectrum [18–21]. As the wavelength decreases from 780 nm down into the deep red region around 700 nm, lasing operation becomes increasingly difficult to achieve in this material system. This is due to reduction of band offsets and increasing proximity to the direct–indirect transition as wavelength is reduced. Further, oxygen and other impurities introduce nonradiative deep levels that are increasingly adverse to laser performance as the wavelength decreases [22]. In Figure 7.7 we show the threshold current density versus wavelengths for edge-emitting lasers fabricated in AlGaAs [20], AlGaInP [23–25], ZnMgSSe [26, 27], and AlGaInN [28] as a function of wavelength. It is noted that for edge-emitting lasers in AlGaAs, threshold current densities increase rapidly at wavelengths shorter than 690 nm. Therefore, although laboratory demonstrations of edge-emitters as short as 690 nm and VCSELs as short as 685 nm have been reported, practical, commercially viable laser devices have not been established substantially shorter than 780 nm in the AlGaAs system.

7.2.3 AlGaInP Red Lasers

The most established viable strategy for true red light emission is the AlGaInP material system. In Figure 7.6 we show both the Γ and X conduction band minima as a function of lattice parameter. By appropriate balancing of the In content of this material to approximately 50% of the column III species, this material can be grown lattice-matched to a GaAs substrate with a direct bandgap ranging from 1.92 to 2.3 eV. The fundamental reason for the improvement of devices in this material system relative to the AlGaAs-based material system is the fact that the energy level of the indirect X-band energy is increased from its value of 1.9–2.15 eV in the AlGaAs system to 2.3 eV in the AlGaInP system when lattice-matched to a GaAs substrate. In AlInGaP, this value varies only slightly with composition. In fact, the direct gap energy of AlAs, at 3.0 eV is considerably greater than the direct gap of AlInP lattice-matched to GaAs.

Availability of an optimal substrate for growth can have a major impact on the development of semiconductor laser light sources, as we will see further when the yet shorter wavelength ranges are addressed. For the case of AlGaInP lattice-matched to a GaAs substrate, extensive research and development through the mid to late 1980s has resulted in the commercial availability of edge-emitting lasers in this material system spanning the wavelength range from 690 to 630 nm. There is perhaps some room to reach yet shorter wavelengths into the 600 nm region in AlInGaP.

7.2.4 ZnSe-Based Blue-Green Lasers

As one targets laser emission wavelengths shorter than 600 nm, the ZincBlende III–V compounds become impractical for the fabrication of room temperature CW lasers. In AlGaInP, the X-minimum transition energy occurs at about 2.3 eV, with only a very weak dependence on the Al composition. As the laser photon energy approaches this value, carrier loss into the indirect minimum becomes increasingly problematic, resulting in reduced efficiency of carrier injection and laser emission. To avoid this loss in efficiency, both conduction and valence band offsets must be sufficient to provide a barrier to carrier diffusion well above the thermal energy of the carriers. This results in a maximum practical lasing photon energy on the order of 2.1 eV, or a wavelength of 590 nm.

As can be seen from Figure 7.6, the primary candidate for light emission in the shorter wavelength region are composed of the II–VI compound MgZnSSe and related compounds, which potentially span the emission range from red through the near ultraviolet. One attractive feature of these materials is that they can also be grown lattice-matched to a GaAs substrate, and in fact, the first blue-green semiconductor edge-emitting laser was demonstrated in this material system in 1991 [29]. Extensive effort in this material system to date has resulted in dramatic improvements in device performance [30]. However, there are particularly difficult reliability issues with lasers fabricated in this material system. Lifetimes at present are measured in hundreds of minutes rather than thousands of hours [31]. Nevertheless, with continuing work on reliability issues this material system may yet prove suitable for the fabrication of edge-emitting laser and VCSEL devices approaching the blue portion of the spectrum.

7.2.5 AlGaInN Lasers

As the desired wavelength of emission extends into the blue and ultraviolet, the AlGaInN system becomes of increasing relevance. This material system is based on the GaN material with a bandgap of 3.3 eV in the ultraviolet. Major breakthroughs have occurred in the advance of materials quality in this system, and blue light emitting diodes were introduced commercially in the mid 1990s, spurring extensive research worldwide on GaN-based materials. A milestone in this effort was reached when the first AlGaInN-based electrically injected laser was reported [32]. One striking feature of this material system appears to be the relative insensitivity of the efficiency of light emission to defects [33]. Commercial LEDs contain defect levels previously considered prohibitive to efficient light emission, and early laser demonstrations involved similarly heavily defected materials. New epitaxial growth processes, involving dramatic substrate defect reduction by epitaxial lateral overgrowth to create very low defect density nitrides [34], further promise to improve the reliability and manufacturability of nitride-based devices. As a result, we currently operate in a climate of great optimism regarding the potential for this material system to demonstrate reliable devices at the short end of the visible emission spectrum. Further intensive investigation in this material system is underway.

7.2.6 Summary

In this introduction, we have attempted to give the reader a broad overview of the materials issues governing current research into the development of VCSEL devices emitting in the visible portion of the spectrum. By and large, these materials issues mirror those that are relevant in the realization of edge-emitting lasers. We have referred extensively to the edge-emitting laser data as a guide to potential device capability. For the remainder of this chapter we will delve deeper into the issues governing development of visible VCSELs in specific materials systems and in specific wavelength ranges.

7.3 AlGaAs Deep Red VCSELs

7.3.1 Growth Issues

The AlGaAs material system is, without a doubt, the most thoroughly studied and understood of all the III–V compound semiconductor alloy systems. There is considerable leverage to be obtained by utilizing this material system for device architectures, and extensive exploration of the short wavelength limits of emission has taken place. Although the epitaxial techniques that enable these laser structures are well developed, pushing these devices into the visible tests the limits of the capabilities of this material system. This can be understood by referring to Figure 7.8, which shows the conduction and valence band potentials of AlGaAs as a function of alloy composition, as well as the direct and indirect energy bandgap values as a function of composition. The band potentials determine the magnitude of the potential steps in the conduction and valence bands between two compositions of AlGaAs. These dependences are also shown in Figure 7.8 for the AlGaInP material system, which will be discussed in Section 7.4. Note that for AlGaAs, the conduction band offset has a maximum value of 0.33 eV at the direct–indirect band transition energy of 1.95 eV.

In Figure 7.9 we have calculated the maximum available electron confinement barrier as a function of wavelength, again for both AlGaAs and AlGaInP systems. These curves are calculated assuming the compositions of maximum electron potential for the confining layer and the composition with bandgap corresponding to the photon emission energy for the active layer. Due to the high electron mobility relative to the hole mobility, electron confinement is typically most problematic. We therefore consider electron confinement only. It is reasonable to expect that this confinement potential must be well above the thermal excitation energy of the electrons to provide effective carrier confinement for laser emission. At room temperature, where $kT = 25.6$ meV, one can see by reference to Figure 7.7 that

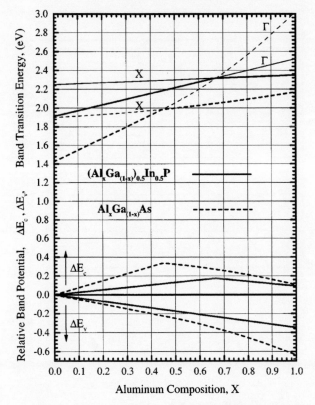

Figure 7.8. Bandgaps and band offsets for the AlGaAs and AlGaInP material systems as a function of Al composition. ΔE_c and ΔE_v are the conduction and valence band offsets, respectively, and E_Γ and E_x are the gaps relative to the Γ direct and X indirect conduction band minima, respectively.

for edge-emitting lasers both in AlGaAs and in AlInGaP, performance is largely unaffected by wavelength until the confinement potential reaches $\sim 4.5\ kT$. Beyond this point, laser performance degrades quite rapidly.

It is often the case, however, that performance degrades substantially long before the wavelengths indicated in Figure 7.9. This is a result in part of the adverse effect of oxygen and its associated nonradiative deep levels on the performance of these devices [22]. Without extreme care in elimination of this contamination, laser performance will typically be quite poor at wavelengths below 750 nm in AlGaAs.

7.3.2 Device Structures

Despite the limitations of the AlGaAs material system, the maturity of growth technology in this system is a strong justification for efforts to extend the wavelength of accessibility of VCSELs in this system as far as possible. This strategy has been extensively explored. The typical result of effort in this direction is a structure as depicted in Figure 7.10.

Quantum-well active regions with substantial aluminum content are grown, and the aluminum content of the low refractive index layer in the epitaxial mirror is increased to avoid absorption in this layer at the shorter emission wavelength. As a result, a larger number of

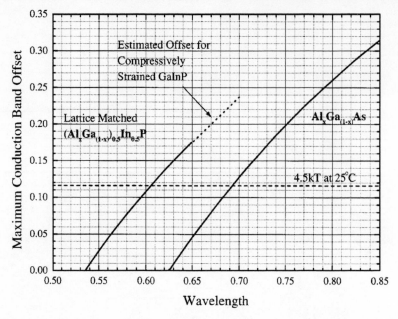

Figure 7.9. Maximum available conduction band offsets for lasers fabricated in the AlGaAs and AlGaInP materials systems, as a function of emission wavelength. Offset value at which room temperature laser performance tends to degrade rapidly is also indicated, as 4.5 kT.

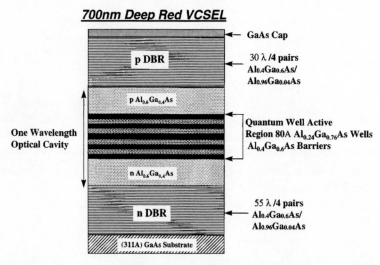

Figure 7.10. Structure of AlGaAs short-wavelength VCSEL structure emitting at 700 nm wavelength.

mirror layer pairs is required to achieve laser emission when compared to devices emitting in the infrared.

Within the limitations of quantum-well injection and radiative efficiency, however, this structure is quite tenable and has enabled the demonstration of VCSEL devices at wavelengths as short as 685 nm under pulsed conditions and 690 nm under room-temperature CW

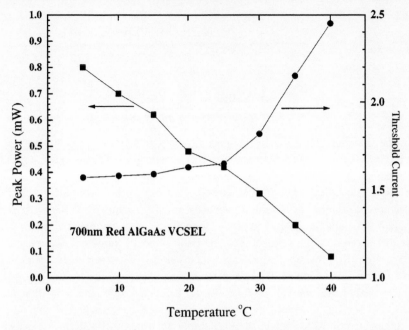

Figure 7.11. Threshold current and peak power versus temperature for AlGaAs VCSELs emitting at 700 nm.

conditions, which are sufficiently short to access several important applications for visible light emission. In addition, although the eye is approximately 300 times less sensitive to this wavelength as it is to wavelengths at the sensitivity peak in the green, these devices can in fact be easily seen with the human eye. Hence, these devices are often referred to as emitting in the "deep red" wavelength range.

7.3.3 Device Performance

There have been extensive studies of AlGaAs VCSEL device performance into the 680 nm region [35–39]. It has been found that room-temperature CW operation is indeed possible at wavelengths as short as 685 nm. However, as the wavelength approaches the potential barrier limitations as shown in Figure 7.9, the temperature sensitivity of the laser is greatly increased. In Figure 7.11 we show the temperature dependence of a 700 nm VCSEL fabricated in the AlGaAs system. For devices at this extreme end of the AlGaAs wavelength range, temperature sensitivity remains a substantial issue. For most applications, adequate operation over an extended temperature range is required. This temperature range may typically be from 0°C to 70 or 85°C.

7.3.4 Summary

We come to the conclusion that there is a region of potential utility for AlGaAs-based devices in the visible. For wavelengths even as short as 690 nm, encouraging results have been obtained for VCSEL devices. This is largely a result of refining the material to greatly reduce the negative impact of the impurities on device efficiency. Current devices, however, are rapidly approaching the hard limits imposed by the direct–indirect transition. Without the implementation of more exotic bandgap engineering techniques such as, perhaps, multiple quantum barriers used as electron Bragg reflectors [40–42], practical devices will be limited to the wavelength range longer than 700 nm.

7.4 AlGaInP Red VCSELs

7.4.1 Lattice Match and Strain

As has been mentioned previously, the most effective strategy for obtaining laser emission in the red is by the use of InGaAlP material lattice-matched to a GaAs substrate. This system has had great success in producing commercial edge-emitting lasers in the wavelength range from 690 to 630 nm [43–47]. Two key advances in materials capability that enable this advance are the addition of phosphorous-based growth capability and the ability to grow layers with sufficient compositional precision to maintain lattice matching when AlGaInP material is grown on a GaAs substrate. The issues here are similar to those encountered when attempting to grow long-wavelength InGaAsP lasers lattice-matched to an InP substrate. This implies a compositional precision during growth on the order of 0.3% in order to avoid exceeding the critical layer thickness for defect formation given the total thickness of several microns for the epitaxial layers. This precision can be achieved with reasonable ease with either the metal-organic chemical vapor deposition (MOCVD) [48–54] or MBE [55–60] growth methods. It is true, however, that the ability to incorporate the P material by MBE is somewhat less developed than by MOCVD, and many of the key advances in this material system have been accomplished by MOCVD.

Within the AlGaInP materials system, it is possible to intentionally introduce strain into the quantum well, which can be either compressive or tensile. The potential benefits of strained quantum wells have been explored extensively [46, 61–68]. Owing to the resolution of the degeneracy between the light hole and heavy hole bands, there is a benefit to the addition of either compressive or tensile strain. Adding tensile strain by reducing In composition also shortens the wavelength. This may be one part of an effective strategy for pushing the wavelength limits in AlGaInP.

7.4.2 Band Offsets and Carrier Confinement

As in the AlGaAs materials system, the wavelength range accessible to this material system is determined in large part by the band offsets and carrier confinement that are available. Referring to Figure 7.8, we see the dependence of the relative conduction and valence band energies on composition for the direct and indirect energy transitions in AlGaInP. Although the bandgap is significantly greater in AlGaInP than in AlGaAs, the available conduction band offset is considerably reduced. The net impact on wavelength is shown in Figure 7.9. If we again compare this curve to the edge-emitting laser threshold data in Figure 7.7, it can be seen that, as was the case in the AlGaAs system, laser performance is largely wavelength insensitive until the photon energy corresponds to an offset barrier potential of $\sim4.5\,kT$ in the conduction band.

7.4.3 Spontaneous Ordering and Suppression

An important property of the AlGaInP system that has had a substantial effect on the development of laser devices within this material system is the phenomenon of spontaneous

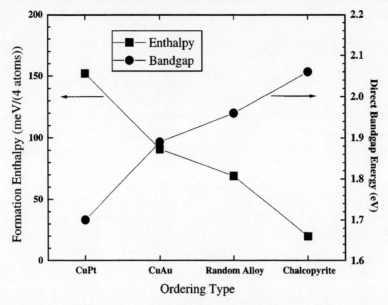

Figure 7.12. Theoretical values of the bandgap [74] and formation enthalpy [73] for a random $Ga_{0.5}In_{0.5}P$ alloy as well as for three ordered variants.

ordering. Under appropriate growth conditions, it has been observed that the In distribution on the column III sublattice is not random but is ordered in a way such that a higher order periodicity is established in the epitaxial crystal [69–72]. The type of ordering observed corresponds to a superlattice in the ⟨111⟩ direction. This type of ordering is commonly referred to as CuPt ordering.

As a result of this higher-order periodicity, the band structure of the crystal is modified. Theoretical calculations of the bulk formation enthalpy [73] and the bandgap [74] resulting from several different ordered structures have been performed, and these are shown in Figure 7.12. In addition to the disordered alloy, three types of ordered structure are considered: chalcopyrite, CuAu-like, and CuPt-like. The structure for these ordered alloys can be described as binary superlattices in specific crystallographic directions with specific repeat periods [74]. If we call the crystallographic direction G and the repeat period p (i.e, $(GaP)_p(InP)_p$ superlattice) then CuAu is $p = 1$, $G = ⟨001⟩$, chalcopyrite is $p = 2$, $G = ⟨201⟩$, and the observed CuPt is $p = 1$, $G = ⟨111⟩$. From Figure 7.12 we see that CuPt ordering results in a bandgap for the crystal that is narrower than the bandgap of the fully random alloy, resulting in a longer emission wavelength at a given quantum well composition. Thus, if the goal is to produce shorter wavelengths of emission, the effect of spontaneous ordering is an undesirable one. Figure 7.12 also indicates that the observed CuPt ordered phase is least favored energetically in the bulk. It is therefore believed that the driving force for CuPt ordering is surface energy minimization during growth. There is theoretical support for this conclusion [75]. Further, it has been shown that the ordering process is not homogeneous, and inhomogeneous broadening of the gain spectrum results, reducing the available gain from the quantum well [76].

For optimal laser performance, it is therefore generally desirable to suppress the spontaneous ordering to as great an extent as possible. The two methods for accomplishing this that have had the greatest success are modification of the growth temperature and growing on other than ⟨100⟩ oriented substrates [77]. Spontaneous ordering is the result of a balance between surface mobility and surface binding energy on a reconstructed surface [75]. Ordering is therefore most pronounced within a limited temperature range. For lower growth temperatures, column III species lack sufficient surface mobility to find their lowest energy ordered lattice sites. For higher growth temperatures, these species have sufficient thermal energy to overcome the energy barriers associated with the ordered sites and again become randomly ordered. The other approach to suppression of spontaneous ordering is the growth of structures on substrate orientations other than ⟨100⟩. This is consistent with the picture that the CuPt ordered phase is not most stable in the bulk, but that its presence is driven by reconstruction of the ⟨100⟩ growth surface. For surface orientations away from ⟨100⟩, surface reconstructions on the available free surface are modified such that the spontaneous ordering properties of the ⟨100⟩ surface are suppressed. The substrate orientation approach may in some ways be superior to the growth temperature approach, as it allows growth temperature optimization for other optoelectronic properties without the constraint of temperature optimization for minimum ordering.

7.4.4 Device Architecture

For the fabrication of red VCSELs, it has been established that InGaAlP active region layers provide the highest quantum efficiencies at the high carrier densities required for laser operation. This is a result of the higher bandgap and greater Γ and X band separation relative to AlGaAs material and the resulting improved carrier confinement at a given wavelength as in Figure 7.9. However, the realization of epitaxial mirrors in the AlGaInP system is quite challenging [78]. The reason for this can be seen in Figure 7.13, where we show the dependence of refractive index on Al composition for AlGaInP [79] and AlGaAs [80] as a function of photon energy. It can be seen that the refractive index contrast available in the AlGaInP system is substantially less than that available in the AlGaAs system. This reduced contrast has a direct impact on the maximum reflectivity and the reflectivity passband of the mirrors, resulting in more layer pairs being required to reach suitable high reflectivity mirrors in the AlGaInP system. The narrower high reflectivity passbands of these mirrors increases the amount of precision required in their growth. A specific comparison is given in Figure 7.14, where we show the reflectivity spectrum of two mirrors, both designed to have 55 layer pairs and to have high refractive index layers of bandgap 2.0 eV to avoid absorption in the high refractive index layer at an emission wavelength of 680 nm. For a required reflectivity of 99.9%, we see that the AlGaAs mirror has a bandwidth of 34 nm, while the AlInGaP has a bandwidth of 16 nm. The greater bandwidth of the AlGaAs mirror significantly reduces growth tolerances for the laser structure. In addition, the thermal properties of AlGaInP-based mirrors are significantly poorer than those of AlGaAs mirrors, resulting in greater sensitivity to heat dissipation with an AlGaInP-based mirror structure. Consequently, the most efficient red VCSEL devices have been fabricated using AlGaInP active region layers combined with AlGaAs multilayers for the mirrors.

Because of the combination of P-based active layers with As-based mirrors, great demands are placed on the epitaxial growth technology in order to accomplish effective switching between As and P materials in close proximity to the active region with high quality transitional interfaces. Indeed, early AlInGaP red VCSEL devices extended the volume of the optical cavity to 8 wavelengths to remove the As/P transition interfaces from close proximity to the active region quantum wells [81, 82]. Although this strategy allowed

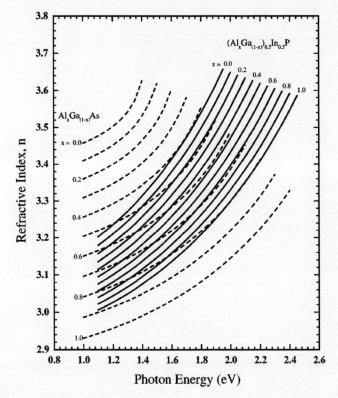

Figure 7.13. Refractive indices for AlGaAs [80] and AlGaInP [79] for varying Al compositions, as a function of photon energy.

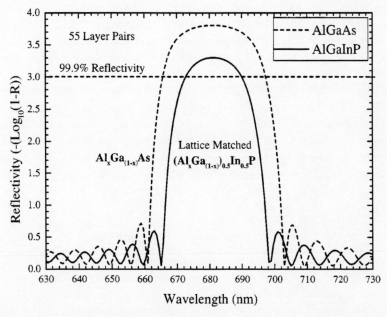

Figure 7.14. Comparison of the reflectivity spectra of 55 layer pair, 680 nm mirrors fabricated in AlGaAs and AlGaInP.

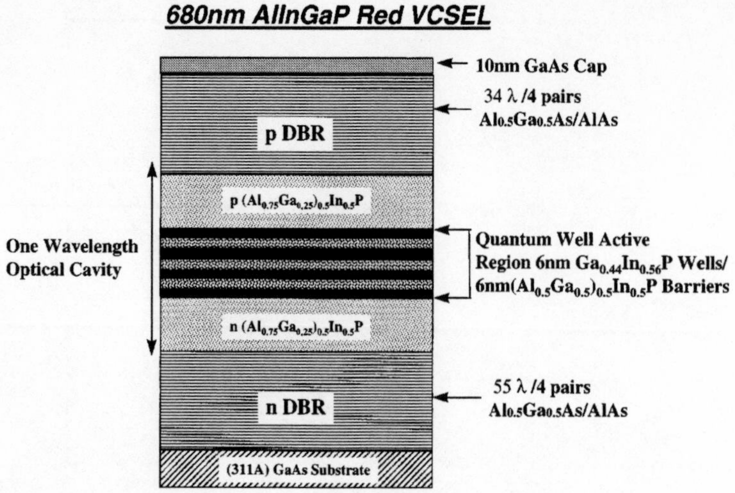

680nm AlInGaP Red VCSEL

Figure 7.15. Structure for AlGaInP-based red VCSEL device emitting at 680 nm [83].

the first demonstrations of room-temperature CW laser operation in the red, the resulting increase in free-carrier absorption provided the impetus to improve the optical and electronic quality of the transition interfaces, allowing a single-wavelength optical cavity to be implemented [83]. To date, the highest output powers and highest operating temperatures have been achieved in single-wavelength cavity devices [84–87].

The epitaxial structure for a typical red VCSEL is shown in Figure 7.15. To avoid absorption of the laser emission in the high refractive index layer of the mirror, the compositions chosen for the mirrors are $Al_{0.5}Ga_{0.5}As$ and AlAs. Fifty-five layer pairs are included in the lower mirror, while thirty-three layer pairs are included in the upper mirror. Issues of mirror conductivity are relaxed in one sense, in that the heterointerfaces that must be graded for low mirror voltages are much lower in this mirror structure than in the infrared mirror structure. However, the number of layer pairs has increased substantially, increasing both electrical and thermal concerns.

Doping for the mirror layers is an important issue for good room-temperature CW performance of red VCSEL devices. Although n-type doping appears reasonably effective with a variety of n-dopants including Si, Se, and Te, p-type doping is more critical, owing to the higher current densities present in the p-layer of the typical n-substrate VCSEL. Although Mg, Zn, and Be have all shown reasonable effectiveness as p-dopants, the addition of C doping in appropriate layers can have a substantial impact on reducing diode voltages and thereby increasing device operating temperature and efficiency [83].

7.4.5 Device Confinement Strategies

Because of the similarity of the layer structures in the design of the red VCSEL to those in the infrared AlGaAs VCSEL, many of the same processing techniques can be applied to this device structure. In particular, good performance has been achieved in red VCSELs using the planar ion implantation approach, which has been demonstrated to be commercially viable for infrared VCSELs [88]. Figure 7.16 shows performance data for a typical AlGaInP-based, ion-implanted red VCSEL emitting at 685 nm, demonstrating high single-mode

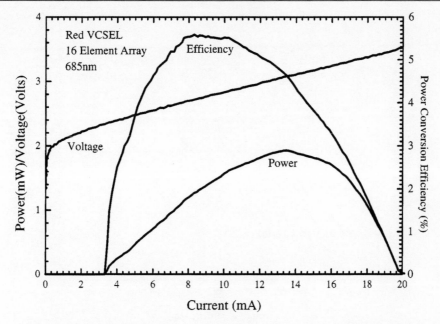

Figure 7.16. Typical optical emission characteristics for an ion implant apertured red VCSEL operating at 685 nm. These results are for a device with an implant diameter of 15 microns and an aperture diameter of 11 microns.

output power in excess of 1 mW and a peak efficiency of 5.5%. At the relatively long red wavelengths, record high output powers of 8 mW and total efficiencies in excess of 10% have been reported [86]. Owing to the simplified structure of the ion-implanted VCSEL, this device may also prove commercially viable in red VCSELs.

The ion-implanted structure, however, does have its drawbacks. The ion implant energy required is optimally on the order of 350 keV for the structure shown in Figure 7.15. The high energy is required to reach the optimal penetration depth in the upper mirror layer, which is approximately 4 μm thick. This requires a very thick mask layer technology to pattern this high energy implant, and this fact, combined with increased lateral straggle of the implant, limits the minimum aperture size practical with ion-implanted devices. Consequently, alternate means for achieving current and optical confinement in visible VCSELs are very desirable.

To find an improved method of confinement in visible VCSELs, oxide confinement based VCSELs have been explored. Device performance improvements achieved by this process have been substantial. This approach has been driven by the great benefits established by lateral oxidation of infrared VCSELs [89]. The oxide confinement process has enabled threshold currents for red VCSELs as low as 660 μA as well as emission wavelengths as short as 642 nm for VCSEL devices [90].

7.4.6 Device Performance

One of the ongoing challenges in the operation of red VCSEL devices is the high temperature sensitivity of light output power. In Figure 7.17 we plot typical maximum output powers and threshold currents versus temperature for a 680 nm red AlInGaP VCSEL [86]. Temperature

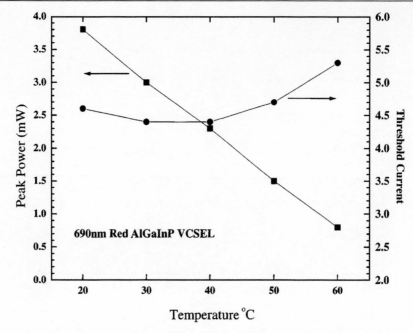

Figure 7.17. Temperature dependence of threshold current and output power for an AlGaInP red VCSEL emitting at 680 nm [85].

sensitivity becomes an increasing problem as the wavelength decreases from the 680 nm region through the 650 nm region. As in edge-emitting red lasers, this is due to the poorer carrier confinement resulting from the lower heterojunction band offsets available in this material system. In particular, electron leakage into the p-mirror layer poses a significant problem. Edge-emitting lasers counter this problem by increasing the p-type doping of the cladding layer [91]. In the VCSEL, this strategy must be weighed against the resulting increase in free-carrier absorption. Many of the advanced techniques for improving carrier confinement in visible edge-emitters, such as electron Bragg reflectors [92–94] and tensile strained quantum wells [24, 95], may prove useful here as well.

Owing to the high temperature sensitivity of these devices, the increased thickness of the p-type layers, and the difficulty of doping the high Al content layers of p-type, reduction of electrical resistance and the resulting ohmic heating are particularly beneficial to the performance of red VCSEL devices. This is illustrated by the positive impact of the application of indium tin oxide (ITO) as a transparent top contact to red VCSELs, which lowers the series resistance and drive voltage for the device. In Figure 7.18 we show the peak emission power for two nominally identical devices, one with an ITO transparent top contact and the other with a conventional metal top contact. Operating voltage, and therefore operating temperature, are reduced for the ITO contacted device, resulting in a significant extension of high temperature operation.

Because of the interest in red light emitting VCSELs for applications in high performance printing, high density 16-element array lasers have been fabricated using ion implantation with good performance. Figure 7.19 shows a 4×4 array on 50 μm centers of 680 nm VCSELs, with eight of the individual elements in the array illuminated, demonstrating

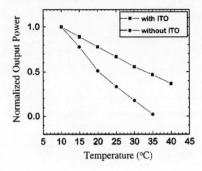

Figure 7.18. Improvement in the maximum output power of a red VCSEL as a result of indium tin oxide (ITO) transparent top contacts.

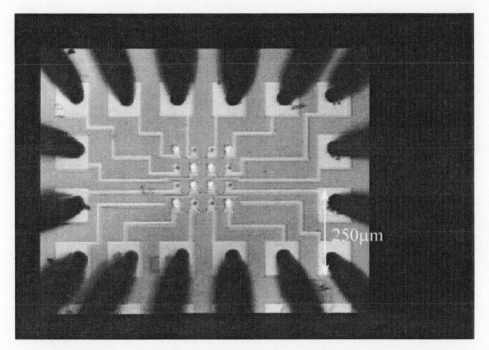

Figure 7.19. 4 × 4 Array of red VCSEL devices for printing applications, with eight elements illuminated. Center to center spacing is 50 μm.

good isolation between elements via ion implantation. The greater temperature sensitivity of the red VCSEL device results in an increased level of crosstalk between devices when array packing density is high. Such issues must be resolved in order to design high density array systems based on red VCSELs.

In general, the transverse mode behavior of red VCSELs is similar to that of infrared VCSELs of comparable aperture sizes. Single-mode emission up to output powers on the order of 1.9 mW have been reported [86]. In contrast to IR VCSELs, however, the red VCSELs demonstrate considerably more stable polarization properties [96]. Polarization oriented parallel to the ⟨011⟩ direction is believed to be related to the orientation of the substrate in the 10° off ⟨100⟩ toward the ⟨110⟩ direction. It is known that the gain is orientation dependent in the plane. Substrate orientation away from exact ⟨100⟩ is preferred for growth as a result of its effectiveness in the suppression of spontaneous ordering, as previously mentioned. Similar polarization control has been observed in IR VCSELs when fabricated on a ⟨311⟩ substrate [97].

Red VCSELs have been shown to be capable of modulation at high data rates in excess of 1.5 GB/s [98]. Small-signal modulation data showing 2.25 GHz bandwidths and 1.5 GB/s modulation rates with bit error rates of 10^{-10} have been demonstrated. As threshold current for these devices are reduced, yet higher modulation rates will be expected.

7.4.7 Reliability

The red VCSEL has yet to be subjected to thorough reliability evaluation. This is in part due to the difficulty in maintaining reasonable output powers when operated at the elevated temperatures that are generally required for accelerated aging studies. Extended operating life has, however, been demonstrated for red VCSEL devices operated at room temperature. Figure 7.20 indicates that the required drive current for devices operated at constant power at room temperature was approximately constant up to 2000 hours. These results provide encouragement that, as device efficiencies and elevated temperature operation improve, reliability issues will prove resolvable. Red VCSEL devices fabricated on (311) orientation substrates have, however, been shown to display a reliability challenge that is not related to device operation. In infrared VCSEL devices, which are generally grown on a substrate orientation close to (100), it has been found that mirror layers containing AlAs are stable with respect to extended oxidation decomposition at room temperature. However, when the substrate orientation is changed to (311), the AlAs mirror layer may become unstable with respect to room temperature oxidative decomposition throughout the mirror layer. This is thought to be due to the higher reactivity of the exposed high Al content surfaces to oxygen. This process has been observed, over a period of several months, to result in oxide encroachment into the lasing aperture, converting the electrically conducting mirror layers into electrically insulating layers.

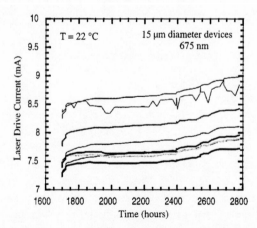

Figure 7.20. Variation of current required to maintain 1 mW output power over time for red VCSELs operating at room temperature. (Courtesy Sandia National Laboratories.)

7.4.8 Summary

It appears that red VCSELs are near to being a practical reality in the 670–690 nm wavelength range. As in their edge-emitting counterparts, the remaining challenge is the extension of the wavelength of accessibility into the 630–650 nm region where substantial volume applications appear in optical data storage and plastic fiber data communications.

7.5 ZnSe-Based Blue-Green Visible VCSELs

7.5.1 Overview

As wavelengths shorter than 620 nm become of interest, the quantum efficiency of devices fabricated in the AlGaInP/AlGaAs system becomes inadequate for reliable room-temperature CW laser operation. This is again due to the encroachment of the lasing photon energy on the indirect bandgap energy of AlGaInP and the resulting loss of electron confinement. Further progress toward shorter wavelengths requires the adoption of wider bandgap materials. The most thoroughly researched materials system for shorter than red wavelength lasers is the ZnMgSSe-based II–IV materials system. Considerable progress has been made in the area of edge-emitting lasers in this materials system [30, 31]. In this section we will explore the status of vertical-cavity laser development. As a result of the particularly difficult issues concerning reliability of II–VI lasers, these materials have currently fallen out of favor for short-wavelength coherent light emission relative to the even wider bandgap nitride materials. This is true despite the early spectacular advances in blue-green light emission that were demonstrated in the II–Vs. Nevertheless, for wavelengths in the orange (620 nm) to green (520 nm), there exists a window that may prove very difficult to access reliably in any other material system. Strained CdZnSe quantum wells allow the II–VI materials to access most of this wavelength range, in a manner similar to strained InGaAs quantum wells enabling access to longer than GaAs wavelength devices on a GaAs substrate. There is, consequently, substantial device capability to be gained by resolving the reliability issues and fabricating laser devices in the MgZnSSe material system.

7.5.2 Materials and Device Strategies

One of the advantages to lasers fabricated in the MgZnSSe materials system is that they may be grown lattice-matched to a GaAs substrate. However, as is increasingly the case with wider bandgap semiconductor materials, effective p-type doping offered a long-standing barrier to the advancement of photonic devices. In 1990, however, it was demonstrated that the use of a remote nitrogen plasma source was effective for achieving high p-type doping when using MBE to grow ZnSe-based materials [99, 100]. This achievement precipitated rapid progress in the development of ZnSe p-n junction devices. A major milestone of this development was the 1991 demonstration of an edge-emitting quantum-well laser in ZnSSe on GaAs [29]. Following this initial demonstration, improvement of device performance was rapid [30]. However, at this point, reliability is the key area of focus for visible II–VI laser research. There is evidence that the reliability of II–VI visible lasers may be particularly problematic owing to the strongly ionic bonding in these materials, and the resulting ease of defect propagation, which compromises device life [101, 102]. Extensive investigation of defect mechanisms and defect reduction strategies has brought this field to the

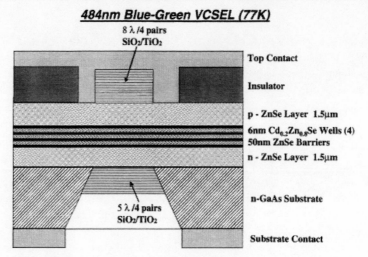

Figure 7.21. Structural diagram of the first electrically injected vertical-cavity laser in the ZnMgSSe system, emitting at 484 nm at 77 K [17].

current state of the art, where room-temperature operating lifetimes of several tens of hours have been achieved [31].

7.5.3 High-Reflectivity Mirrors

For the fabrication of high reflectivity mirrors, the compound MgS provides the widest bandgap and therefore lowest refractive index in the MgZnSSe system. Although demonstrations have been made of epitaxial mirrors [103], the difficulties achieving the required level of growth precision are substantial. Dielectric deposited mirrors have also been demonstrated [104] and have been the route by which laser emission results have been reported to date.

Optically pumped vertical-cavity lasing has been observed in ZnCdSSe using dielectric mirrors, first at low temperature [13, 14, 105] and subsequently at room temperature [15]. Incorporating electric contacting into dielectric mirror devices, however, presents additional challenges. In Figure 7.21 we show the structure for an electrically injected ZnCdSSe laser [17] emitting at 484 nm at 77 K. Electrical injection was facilitated by the combination of a patterned top surface dielectric mirror and an etched aperture in the substrate allowing bottom mirror deposition and bottom surface light emission.

7.5.4 Summary

The MgZnSSe system has demonstrated the shortest wavelength electrically injected VCSEL to date. Despite the very encouraging current status, however, it appears that progress in this material system has slowed somewhat. This is in part due to the particularly challenging reliability issues that have surfaced in the effort to develop commercial edge-emitters in this system. Another substantial contributing factor, however, is the astonishing rate at which the III–V nitrides have advanced in their level of performance in the past several years. As a result, much of the effort that had been devoted to the II–VIs has been redirected to the nitrides. Nonetheless, it remains to be seen whether the nitrides will be able to fully access the entire visible spectrum, obviating the need for II–VI-based devices.

7.6 AlGaInN Blue/UV Devices

7.6.1 Introduction

At the time of this writing, we are in the midst of rapid development of semiconductor lasers within the AlGaInN materials system. In 1989, p-type doping of GaN was demonstrated [106–110], eliminating one of the major hurdles to the realization of efficient p-n junction based devices. Soon thereafter, very efficient emission was reported from LEDs in this material system [107]. The efficiencies reported were astounding in that they were for devices fabricated on sapphire substrates with 14% lattice-mismatch, and at defect densities so high that prior experience in III–V materials would indicate that light emission would be negligible. This was the first clear indication that GaN light emission would be a major departure from what has been observed in the As- and P-based III–Vs, as well as the II–VIs, where defect density reduction and precise lattice-matching of substrates are essential to efficient radiative recombination. Following these reports, research on AlGaN and related materials has expanded dramatically, with edge-emitting lasers having been demonstrated [32, 111] and with these devices appearing well on their way to becoming commercially available light sources.

7.6.2 Growth Techniques

By far the predominant method for the growth of AlGaInN has been MOCVD [112–114]. There has also been work by MBE that has demonstrated laser devices [115]. This was mainly a result of the development of techniques for effective p-doping by postgrowth treatment of the annealed GaN p-type material in order to activate the typically Mg dopants [116, 117]. C-plane sapphire substrates have been used most often for AlGaN epitaxy to date, although lasers have been reported on SiC substrates as well [115]. Substrate selection for AlGaN growth is a topic of ongoing research. This is in large part a result of the fact that the GaN binary alloy crystal is extremely difficult to grow and is at present not available in substrate size or quantity large enough to support epitaxial device development. The sapphire substrate has reached its current level of prominence primarily because of its hexagonal structure, which is compatible with wurtzite GaN, as well as its high quality and low cost. Although SiC offers a greatly reduced lattice-mismatch, the cost of SiC has to date hampered its widespread use for the development of GaN-based devices.

7.6.3 Band Offsets, Carrier Confinement, and Gain

There has been considerable theoretical work on the calculation of the gain in bulk GaN [118–120] as well as in GaN quantum wells [121]. Experimental measurements of the gain have also been performed [122, 123]. Experimentally, it has been repeatedly observed that strong stimulated emission and lasing can be achieved by photopumping in GaN vertical structures that have relatively low reflectivities. This observation indicates that GaN is capable of very high gain at sufficient carrier density. However, it has also been concluded that the transparency carrier density is quite high in GaN. The introduction of strain will reduce transparent carrier density to some degree [119, 121]. As a result, GaN-based laser devices have tended to operate with extremely thin quantum well structures of 2–6 nm thickness. Despite the thin quantum wells, the very high unsaturated gain in GaN at high carrier densities [118] may enable vertical-cavity laser operation at reasonable threshold currents.

7.6.4 Structural Issues

There are substantial difficulties associated with the growth of materials in the AlGaInN materials system over the entire composition range. The most desirable strategy would be to grow AlN low refractive index mirror layers along with GaN high index layers and InGaN quantum wells, to maximize the available refractive index contrast and carrier confinement and minimize required thickness of the epitaxial mirror layers. However, the optimal temperatures for growth of AlN, GaN, and InN are sufficiently different as to introduce substantial difficulty in the growth of AlN/GaN multilayer dielectric films along with InGaN quantum well structures. Consequently, the maximum Al composition grown to date for the purpose of mirror fabrication is on the order of 40% [16]. With advances in growth capabilities, the ability to grow high quality AlN material will provide significant advantage because of the large refractive index difference that would then be achievable.

7.6.5 Laser Device Considerations

Photopumped laser operation has been achieved in nitride-based VCSEL structures with metal mirrors [124] and epitaxial mirrors [16]. The structure used for the latter device, depicted in Figure 7.22, used $Al_{0.4}Ga_{0.6}N$ as the low refractive index layer in the mirrors. Laser action in this case is facilitated by the use of a bulk GaN active layer of 10 μm

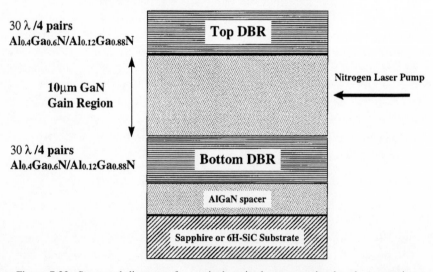

Figure 7.22. Structural diagram of a vertical-cavity laser operating by photopumping in AlGaN, at a wavelength of 364 nm [16].

thickness, thereby providing for a large amount of gain to overcome the high mirror loss and potential scattering losses. These devices have been fabricated on both GaN and SiC substrates, with the SiC substrate device providing somewhat improved performance.

7.6.6 Electrically Injected Laser Operation

Electrical injection of a GaN-based VCSEL has not to date been achieved. For such a demonstration, bulk GaN active layers require a very high transparent carrier density of $(6-9) \times 10^{18}/cm^3$ [118, 120]. It is essential to incorporate well-designed quantum well active regions into these structures to lower transparency and threshold current densities. However, with the use of quantum well active regions, the total available gain will also be reduced due to the reduced layer thickness. Therefore, cavity losses will need to be reduced as well through the use of high reflectivity mirrors. Surface roughness must also be minimized, a particular challenge in nitrides where high defect density tends to enhance surface topology. The resulting scattering may set a limit on the maximum reflectivities achievable. Due to the high gain above transparency available in the nitrides, however, it appears reasonable that, with a suitable substrate, electrically injected vertical-cavity lasers will be achievable in the nitride materials.

As a conceptual example, the structure shown in Figure 7.23 embodies the key design elements of an electrically injected blue VCSEL. The structure consists of multilayer epitaxial mirrors of GaN/Al$_{0.4}$Ga$_{0.6}$N and an active region containing four 4 nm thick InGaN quantum wells separated by InGaN spacers. The low refractive index contrast necessitates 80 layer pairs in the upper mirror and 43 layer pairs in the lower mirror of the optical cavity. Figure 7.24 shows the reflectivity spectrum of the resultant laser structure, along with the assumed refractive index values used for the various layers within the structure. The 14 nm (3%) bandwidth of the mirror indicates that the growth precision for this structure will be a particular challenge. In Figure 7.25 we show the electric field distribution within this structure at resonance, from a one-dimensional calculation. We also show critical

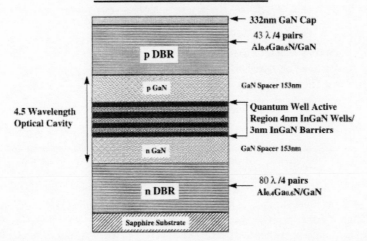

Figure 7.23. Structural diagram of an electrically injected 415 nm AlInGaN violet light emitting vertical-cavity laser.

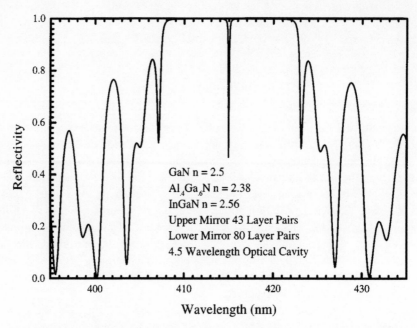

Figure 7.24. Reflectivity spectrum for the AlGaInN VCSEL structure. Assumed refractive indices for the various layers are shown.

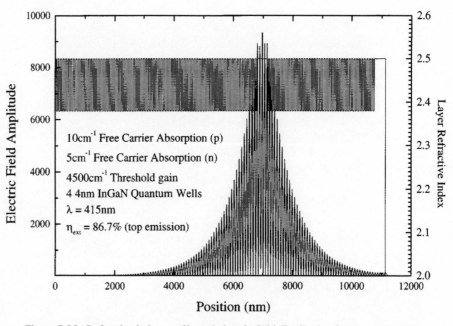

Figure 7.25. Refractive index profile and electric field distribution for a prototype blue AlGaInN VCSEL. Inset shows assumed values of free carrier absorption, calculated threshold gain, and calculated external differential quantum efficiency assuming 100% internal quantum efficiency.

performance parameters for this device to achieve laser operation. Note the threshold gain of 4500 cm^{-1} per quantum well. This value, although high, has been calculated as being obtainable in a 4 nm thick quantum well [125]. The growth challenges with a structure such as this should be evident as well from Figure 7.25. The total epitaxy thickness of 11 μm must be grown within a precision of 1%. The epitaxial layers must provide for current conduction, implying optimized grading of the heterointerfaces. Processing techniques for defining laser apertures must be established. Clearly there is much interesting work yet to be done toward the realization of electrically injected VCSELs in the nitride system.

7.6.7 Conclusions

AlGaInN-based VCSELs appear to be fraught with substantial challenges. However, it appears that many of the fundamental requirements for constructing an AlGaInN VCSEL have, to date, already been demonstrated. What remains is the work of bringing together optimized strategies for materials growth, device architectures, and processing technologies to make further progress in this direction.

7.7 Technology Directions and Challenges

Vertical-cavity lasers in the visible are significantly more difficult to fabricate than their edge-emitting counterparts at a given wavelength. This is generally the result of an increased impact of the higher relative current densities required and the impact of these higher densities on carrier confinement and device voltage. Further development of visible VCSEL devices will be application driven. Focus will therefore be on those applications for which VCSELs are an enabling technology, and therefore for technologies that are cost sensitive or require significant levels of parallelism.

For red lasers, the near-term outlook is quite encouraging. Although all issues of reliability have not yet been resolved, important milestones have been met in this direction. With the advent of 630 nm compact disk lasers, a major market is open to these devices; however, the challenge of reliability at this relatively short wavelength for the AlGaInP VCSEL is all the greater.

As wavelengths move progressively shorter, one moves into the dangerous territory of predicting the future. At this time, the technological challenges associated with blue VCSELs appear quite formidable. However, the potential benefits to the availability of such components are equally powerful. The question to ask is if the physics of the wide bandgap materials prohibits the fabrication of these devices. The answer to this question appears to be no. Cognizant of this fact, and with blue edge-emitting device technology reasonably well in hand, we see that the forerunners in GaN laser technology have already begun to take on the VCSEL challenge.

The author recently (August 1997) attended a conference, where there were topical meetings on both vertical cavity lasers and gallium nitride materials and devices. After the conference, while outside the hotel waiting for an airport shuttle, the author encountered a participant in the GaN conference and made the mistake of expressing enthusiasm over the prospects of blue VCSELs. The GaN expert muttered, almost to himself, "that does not seem very likely." To this, the author felt obligated to respond that, three years prior, even with LEDs in GaN edge-emitting lasers in this material were also thought by many to be "not very likely." The exciting thing about research in this area is that our intuitions about the feasible and the likely have often been shown wrong. This is a great thing, because it opens up new worlds of possibilities.

References

[1] De Wit, G. C., "Resolution matching in a retinal scanning display," *Applied Optics*, **36**, 5587–93 (1997).

[2] Kollin, J. S. and Tidwell, M. *Novel Optical Systems Design and Optimization*, San Diego, CA, USA, **2537**, 48–60 (1995).

[3] Hyun-Kuk, S., Il, K., Eui-Joong, K., Jin-Hwan, K., Eun-Kyung, L., Moon-Kyu, L., Jong-Kuk, M., Choon-Seong, P., and You-Shin, Y., "Vertical-cavity surface-emitting lasers for optical data storage," *Japanese Journal of Applied Physics, Part 1 (Regular Papers & Short Notes)*, **35**, 506–7 (1996).

[4] Goto, K.,*Conference Digest (Cat. No. 97TH8273) 1997 Optical Data Storage Topical Meeting ODS Conference Digest*, Tucson, AZ, USA, IEEE, pp. 64–5.

[5] Guerin, J.-M., "Dual-beam modulation subsystem for high speed digital copier," *SPIE Proceedings*, **1987**, 341–353 (1993).

[6] Nihei, E., Ishigure, T., Tanio, N., and Koike, Y., "Present prospect of graded-index plastic optical fiber in telecommunication," *IEICE Transactions on Electronics*, **E80-C**, 117–22 (1997).

[7] Ishigure, T., Nihei, E., and Koike, Y., "Optimum refractive-index profile of the graded-index polymer optical fiber, toward gigabit data links," *Applied Optics*, **35**, 2048–53 (1996).

[8] Ishigure, T., Nihei, E., Koike, Y., Forbes, C. E., LaNieve, L., Straff, R., and Deckers, H. A., "Large-core, high-bandwidth polymer optical fiber for near infrared use," *IEEE Photonics Technology Letters*, **7**, 403–5 (1995).

[9] Ishigure, T., Nihei, E., Yamazaki, S., Kobayashi, K., and Koike, Y., "2.5 Gbit/s 100 m data transmission using graded-index polymer optical fibre and high-speed laser diode at 650 nm wavelength," *Electronics Letters*, **31**, 467–9 (1995).

[10] Ishigure, T., Nihei, E., and Koike, Y., "Graded-index polymer optical fiber for high-speed data communication," *Applied Optics*, **33**, 4261–6 (1994).

[11] Geels, R. S., Corzine, S. W., and Coldren, L. A., "InGaAs vertical-cavity surface-emitting lasers," *IEEE Journal of Quantum Electronics*, **27**, 1359–67 (1991).

[12] Young, D. B., Scott, J. W., Peters, F. H., Peters, M. G., Majewski, M. L., Thibeault, B. J., Corzine, S. W., and Coldren, L. A., "Enhanced performance of offset-gain high-barrier vertical-cavity surface-emitting lasers," *IEEE Journal of Quantum Electronics*, **29**, 2013–22 (1993).

[13] Floyd, P. D., Merz, J. L., Luo, H., Furdyna, J. K., Yokogawa, T., and Yamada, Y., "Optically pumped CdZnSe/ZnSe blue-green vertical cavity surface emitting lasers," *Applied Physics Letters*, **66**, 2929–31 (1995).

[14] Jeon, H., Kozlov, V., Kelkar, P., Nurmikko, A. V., Grillo, D. C., Han, J., Ringle, M., and Gunshor, R. L., "Optically pumped blue-green vertical cavity surface emitting lasers," *Electronics Letters*, **31**, 106–8 (1995).

[15] Jeon, H., Kozlov, V., Kelkar, P., Nurmikko, A. V., Chu, C. C., Grillo, D. C., Han, J., Hua, G. C., and Gunshor, R. L., "Room-temperature optically pumped blue-green vertical cavity surface emitting laser," *Applied Physics Letters*, **67**, 1668–70 (1995).

[16] Redwing, J. M., Loeber, D. A. S., Anderson, N. G., Tischler, M. A., and Flynn, J. S., "An optically pumped GaN-AlGaN vertical cavity surface emitting laser," *Applied Physics Letters*, **69**, 1–3 (1996).

[17] Yokogawa, T., Yoshii, S., Tsujimura, A., Sasai, Y., and Merz, J., "Electrically pumped CdZnSe/ZnSe blue-green vertical-cavity surface-emitting lasers," *Japanese Journal of Applied Physics, Part 2 (Letters)*, **34**, L751–3 (1995).

[18] Miller, B. I., Ripper, J. E., Dyment, J. C., Pinkas, E., and Panish, M. B., "Semiconductor lasers operating continuously in the 'visible' at room temperature," *Applied Physics Letters*, **18**, 403–5 (1971).

[19] Alferov, Z. I., Andreev, V. M., Belousova, T. Y., Borodulin, V. I., Gorbylev, V. A., Pak, G. T., Petrov, A. I., Portnoi, E. L., Chernousov, N. P., Shveikin, V. I., and Yashchumov, I. V., [Efficient heterojunction injection lasers operating in the wavelength range 7400–9000 Å]. *Fizika i Tekhnika Poluprovodnikov*, **6**, 568–9 (1972).

[20] Kressel, H. and Hawrylo, F. Z., "Red-light-emitting laser diodes operating CW at room temperature," *Applied Physics Letters*, **28**, 598–600 (1976).

[21] Ettenberg, M., "Very low-threshold double-heterojunction $Al_xGa_{1-x}As$ injection lasers," *Applied Physics Letters*, **27**, 652–4 (1975).

[22] Foxon, C. T., Clegg, J. B., Woodbridge, K., Hilton, D., Dawson, P., and Blood, P., *Proceedings of the Third International Conference on Molecular Beam Epitaxy*, San Francisco, CA, USA, **3**, 703 (1984).

[23] Hamada, H., Tominaga, K., Shono, M., Honda, S., Yodoshi, K., and Yamaguchi, T., "Room-temperature CW operation of 610 nm band AlGaInP strained multiquantum well laser diodes with multiquantum barrier," *Electronics Letters*, **28**, 1834–6 (1992).

[24] Bour, D. P., Treat, D. W., Thornton, R. L., Paoli, T. L., Bringans, R. D., Krusor, B. S., Geels, R. S., Welch, D. F., and Wang, T. Y., "Low threshold, 633 nm, single tensile-strained quantum well $Ga_{0.6}In_{0.4}P/(Al_xGa_{1-x})_{0.5}In_{0.5}P$ laser," *Applied Physics Letters*, **60**, 1927–9 (1992).

[25] Rennie, J., Watanabe, M., Okajima, M., and Hatakoshi, G., "High temperature (90 degrees C) CW operation of 646 nm InGaAlP laser containing multiquantum barrier," *Electronics Letters*, **28**, 150–1 (1992).

[26] Okuyama, H., Miyajima, T., Morinaga, Y., Hiei, F., Ozawa, M., and Akimoto, K., "ZnSe/ZnMgSSe blue laser diode," *Electronics Letters*, **28**, 1798–9 (1992).

[27] Nakayama, N., Itoh, S., Ohata, T., Nakano, K., Okuyama, H., Ozawa, M., Ishibashi, A., Ikeda, M., and Mori, Y., "Room temperature continuous operation of blue-green laser diodes," *Electronics Letters*, **29**, 1488–9 (1993).

[28] Nakamura, S., Senoh, M., Nagahama, S. I., Iwasa, N., Yamada, T., Matsushita, T., Sugimoto, Y., and Kiyoku, H., "High-power, long-lifetime InGaN multi-quantum-well-structure laser diodes," *Japanese Journal of Applied Physics, Part 2 (Letters)*, **36**, L1059–61 (1997).

[29] Haase, M. A., Qiu, J., DePuydt, J. M., and Cheng, H., "Blue-green laser diodes," *Applied Physics Letters*, **59**, 1272–4 (1991).

[30] Ishibashi, A., "II-VI Blue-green laser diodes," *IEEE Journal on Selected Topics in Quantum Electronics*, **1**, 741–8 (1995).

[31] Taniguchi, S., Hino, T., Itoh, S., Nakano, K., Nakayama, N., Ishibashi, A., and Ikeda, M., "100h II-VI blue-green laser diode," *Electronics Letters*, **32**, 552–3 (1996).

[32] Nakamura, S., Senoh, M., Nagahama, S., Iwasa, N., Yamada, I., Matsushita, T., Kiyoku, H., and Sugimoto, Y., "Characteristics of InGaN multi-quantum-well-structure laser diodes," *Applied Physics Letters*, **68**, 3269–71 (1996).

[33] Lester, S. D., Ponce, F. A., Craford, M. G., and Steigerwald, D. A., "High dislocation densities in high efficiency GaN-based light-emitting diodes," *Applied Physics Letters*, **66**, 1249–51 (1995).

[34] Nakamura, S., Senoh, M., Nagahama, S., Iwasa, N., Yamada, I., Matsushita, T., Kiyoku, H., Sugimoto, Y., Kozaki, T., Umemoto, H., Sano, M., and Chocho, K., "InGaN/GaN/AlGaN-based laser diodes with modulation-doped strained-layer superlattices grown on an epitaxially laterally overgrown GaN substrate," *Applied Physics Letters*, **72**, 211–13 (1998).

[35] Gourley, P. L. and Drummond, T. J., "Visible, room-temperature, surface-emitting laser using an epitaxial Fabry–Perot resonator with AlGaAs/AlAs quarter-wave high reflectors and AlGaAs/GaAs multiple quantum wells," *Applied Physics Letters*, **50**, 1225–7 (1987).

[36] Lee, Y. H., Tell, B., Brown-Goebeler, K. F., Leibenguth, R. E., and Mattera, V. D., "Deep-red continuous wave top-surface-emitting vertical-cavity AlGaAs superlattice lasers," *IEEE Photonics Technology Letters*, **3**, 108–9 (1991).

[37] Tell, B., Brown-Goebeler, K. F., and Leibenguth, R. E., "Low temperature continuous operation of vertical-cavity surface-emitting lasers with wavelength below 700 nm," *IEEE Photonics Technology Letters*, **5**, 637–9 (1993).

[38] Tell, B., Leibenguth, R. E., Brown-Goebeler, K. F., and Livescu, G., "Short wavelength (699 nm) electrically pumped vertical-cavity surface-emitting lasers," *IEEE Photonics Technology Letters*, **4**, 1195–6 (1992).

[39] Sale, T. E., Roberts, J. S., Woodhead, J., David, J. P. R., and Robson, P. N., "Room temperature visible (683–713 nm) all-AlGaAs vertical-cavity surface-emitting lasers VCSELs," *IEEE Photonics Technology Letters*, **8**, 473–5 (1996).

[40] Loh, T., Miyamoto, T., Koyama, F., and Iga, K., "Strain-compensated multi-quantum barriers for reduction of electron leakages in long-wavelength semiconductor lasers," *Japanese Journal of Applied Physics, Part 1 (Regular Papers & Short Notes)*, **34**, 1504–5 (1995).

[41] Kishino, K., Kikuchi, A., Kaneko, Y., and Nomura, I., "Enhanced carrier confinement effect by the multiquantum barrier in 660 nm GaInP/AlInP visible lasers," *Applied Physics Letters*, **58**, 1822–4 (1991).

[42] Morrison, A. P., Considine, L., Walsh, S., Cordero, N., Lambkin, J. D., O'Connor, G. M., Daly, E. M., Glynn, T. J., and van der Poel, C. J., "Photoluminescence investigation of the carrier confining properties of multiquantum barriers," *IEEE Journal of Quantum Electronics*, **33**, 1338–44 (1997).

[43] Hatakoshi, G., Itaya, K., Ishikawa, M., Okajima, M., and Uematsu, Y., "Short-wavelength InGaAlP visible laser diodes," *IEEE Journal of Quantum Electronics*, **27**, 1476–82 (1991).

[44] Endo, K., Kobayashi, K., Fujii, H., and Ueno, Y., "Accelerated aging for AlGaInP visible laser diodes," *Applied Physics Letters*, **64**, 146–8 (1994).

[45] Ou, S. S., Jansen, M., Yang, J. J., Fu, R. J., and Hwang, C. J., "Reliable high-power, singlemode, 630–640 nm $Ga_{0.5}In_{0.5}P$/GaAlInP ridge waveguide laser diodes," *Electronics Letters*, **29**, 233–4 (1993).

[46] Hashimoto, J., Katsuyama, T., Shinkai, J., Yoshida, I., and Hayashi, H., "Highly stable operation of AlGaInP/GaInP strained multiquantum well visible laser diodes," *Electronics Letters*, **28**, 1329–30 (1992).

[47] Hino, I., Kogure, N., Kanno, K., Morihisa, T., Unozawa, K., Kishi, T., Furuse, T., and Suzuki, T., "670 nm-Wavelength visible light semiconductor laser diode," *NEC Technical Journal*, **41**, 235–8 (1988).

[48] Kim, T. J., Lee, S. W., Kang, B. K., Kim, D. S., Kim, A. S., Shin, K. H., Park, M. K., and Sin, Y. K., "LP MOVPE growth of AlGaInP/GaInP and its application to visible laser diodes," *Optical and Quantum Electronics*, **27**, 465–71 (1995).

[49] Valster, A., van der Heijden, J., Boermans, M., and Finke, M., "GaInP/AlGaInP visible-light emitting laser diodes grown by metal organic vapour phase epitaxy," *Philips Journal of Research*, **45**, 267–77 (1990).

[50] Yuan, J. S., Hsu, C. C., Cohen, R. M., and Stringfellow, G. B., "Organometallic vapor phase epitaxial growth of AlGaInP," *Journal of Applied Physics*, **57**, 1380–3 (1985).

[51] Hsu, C. C., Cohen, R. M., and Stringfellow, G. B., "OMVPE growth of GaInP," *Journal of Crystal Growth*, **62**, 648–50 (1983).

[52] Bour, D. P. and Shealy, J. R., "Organometallic vapor phase epitaxial growth of $(Al_xGa_{1-x})_{0.5}In_{0.5}P$ and its heterostructures," *IEEE Journal of Quantum Electronics*, **24**, 1856–63 (1988).

[53] Morita, E., Ikeda, M., Inoue, M., and Kaneko, K., "Epitaxial growth of GaInP on (111)A and (111)B surfaces by metalorganic chemical vapor deposition," *Journal of Crystal Growth*, **106**, 197–207 (1990).

[54] Garcia, J. C., Maurel, P., Bove, P., and Hirtz, J. P., "Metal organic molecular beam epitaxy growth of $Ga_{0.5}In_{0.5}P$/GaAs quantum well structures," *Japanese Journal of Applied Physics, Part 1 (Regular Papers & Short Notes)*, **30**, 1186–9 (1991).

[55] Blood, P., Roberts, J. S., and Stagg, J. P., "GaInP grown by molecular beam epitaxy doped with Be and Sn," *Journal of Applied Physics*, **53**, 3145–9 (1982).

[56] Wicks, G. W., Koch, M. W., Varriano, J. A., Johnson, F. G., Wie, C. R., Kim, H. M., and Colombo, P., "Use of a valved, solid phosphorus source for the growth of $Ga_{0.5}In_{0.5}P$ and $Al_{0.5}In_{0.5}P$ by molecular beam epitaxy," *Applied Physics Letters*, **59**, 342–4 (1991).

[57] Hayakawa, T., Takahashi, K., Hosoda, M., Yamamoto, S., and Hijikata, T., "GaInP/AlInP quantum well structures and double heterostructure lasers grown by molecular beam epitaxy on (100) GaAs," *Japanese Journal of Applied Physics, Part 2 (Letters)*, **27**, L1553–5 (1988).

[58] Kikuchi, A., Kishino, K., and Kaneko, Y., "High-optical-quality GaInP and GaInP/AlInP double heterostructure lasers grown on GaAs substrates by gas-source molecular-beam epitaxy," *Journal of Applied Physics*, **66**, 4557–9 (1989).

[59] Tappura, K., Aarik, J., and Pessa, M., "High-power GaInP-AlGaInP quantum-well lasers grown by solid source molecular beam epitaxy," *IEEE Photonics Technology Letters*, **8**, 319–21 (1996).

[60] Kishino, K., Kikuchi, A., Nomura, I., and Kaneko, Y., "Gas source molecular beam epitaxial growth and characterization of 600–660 nm GaInP/AlInP double-heterostructure lasers," *Thin Solid Films* **231**, 173–89 (1993).

[61] Blood, P. and Smowton, P., "Strain dependence of threshold current in fixed-wavelength GaInP laser diodes," *IEEE Journal on Selected Topics in Quantum Electronics*, **1**, 707–11 (1995).

[62] Bour, D. P., Geels, R. S., Treat, D. W., Paoli, T. L., Ponce, F., Thornton, R. L., Krusor, B. S., Bringans, R. D., and Welch, D. F., "Strained $Ga_xIn_{1-x}P/(AlGa)_{0.5}In_{0.5}P$ heterostructures and quantum-well laser diodes," *IEEE Journal of Quantum Electronics*, **30**, 593–607 (1994).

[63] Dawson, M. D. and Duggan, G., "Band-offset determination for GaInP-AlGaInP structures with compressively strained quantum well active layers," *Applied Physics Letters*, **64**, 892–4 (1994).

[64] Kamiyama, S., Uenoyama, T., Mannoh, M., and Ohnaka, K., "Theoretical studies of GaInP-AlGaInP strained quantum-well lasers including spin-orbit split-off band effect," *IEEE Journal of Quantum Electronics*, **31**, 1409–17 (1995).

[65] Kamiyama, S., Uenoyama, T., Mannoh, M., and Ohnaka, K., "Strain effect on 630 nm GaInP/AlGaInP multi-quantum well lasers," *Japanese Journal of Applied Physics, Part 1 (Regular Papers & Short Notes)*, **33**, 2571–8 (1994).

[66] Serreze, H. B. and Chen, Y. C., "Low-threshold, strained-layer, GaInP/AlGaInP GRINSCH visible diode lasers," *IEEE Photonics Technology Letters*, **3**, 397–9 (1991).

[67] Smowton, P. M., Blood, P., Mogensen, P. C., and Bour, D. P., "Role of sublinear gain-current relationship in compressive and tensile strained 630 nm GaInP lasers," *International Journal of Optoelectronics*, **10**, 383–91 (1995).

[68] Yoshida, J., Kishino, K., Kikuchi, A., and Nomura, I., "Continuous-wave (CW) operation of GaInP-AlGaInP visible compressively strained multiple quantum-wire (CS-WQWR) lasers," *IEEE Journal on Selected Topics in Quantum Electronics*, **1**, 173–82 (1995).

[69] Ueda, O., Takikawa, M., Komeno, J., and Umebu, I., "Atomic structure of ordered InGaP crystals grown on (001) GaAs substrates by metalorganic chemical vapor deposition," *Japanese Journal of Applied Physics, Part 1 (Regular Papers & Short Notes)*, **26**, L1824–7 (1987).

[70] Kondow, M., Kakibayashi, H., Minagawa, S., Inoue, Y., Nishino, T., and Hamakawa, Y., *Fourth International Conference on Metalorganic Vapor Phase Epitaxy*, Hakone, Japan, **93**, 412–17 (1988).

[71] Kondow, M., Kakibayashi, H., Tanaka, T., and Minagawa, S., "Ordered structure in $Ga_{0.7}In_{0.3}P$ alloy," *Physical Review Letters*, **63**, 884–6 (1989).

[72] Okuda, H., Anayama, C., Tanahashi, T., and Nakajima, K., "X-ray investigation of the ordered structure in AlGaInP quaternary alloys," *Applied Physics Letters*, **55**, 2190–2 (1989).

[73] Bernard, J. E., Dandrea, R. G., Ferreira, L. G., Froyen, S., Wei, S. H., and Zunger, A., "Ordering in semiconductor alloys," *Applied Physics Letters*, **56**, 731–3 (1990).

[74] Wei, S. H. and Zunger, A., "Band-gap narrowing in ordered and disordered semiconductor alloys," *Applied Physics Letters*, **56**, 662–4 (1990).

[75] Froyen, S. and Zunger, A., "Surface-induced ordering in GaInP," *Physical Review Letters*, **66**, 2132–5 (1991).

[76] Fouquet, J. E., Robbins, V. M., Rosner, S. J., and Blum, O., "Unusual properties of photoluminescence from partially ordered $Ga_{0.5}In_{0.5}P$," *Applied Physics Letters*, **57**, 1566–8 (1990).

[77] Chen, G. S., Jaw, D. H., and Stringfellow, G. B., "Effects of substrate misorientation on ordering in $GaAs_{0.5}P_{0.5}$ grown by organometallic vapor phase epitaxy," *Applied Physics Letters*, **57**, 2475–7 (1990).

[78] Schneider, R. P. and Lott, J. A., Jr., "InAlP/InAlGaP distributed Bragg reflectors for visible vertical cavity surface-emitting lasers," *Applied Physics Letters*, **62**, 2748–50 (1993).

[79] Tanaka, H., Kawamura, Y., and Asahi, H., "Refractive indices of $In_{0.49}Ga_{0.51-x}Al_xP$ lattice matched to GaAs," *Journal of Applied Physics*, **59**, 985–6 (1986).

[80] Adachi, S., "GaAs, AlAs, and $Al_xGa_{1-x}As$: Material parameters for use in research and device applications," *Journal of Applied Physics*, **58**, R1–29 (1985).

[81] Lott, J. A. and Schneider, R. P., Jr., "Electrically injected visible (639–661 nm) vertical cavity surface emitting laser," *Electronics Letters*, **29**, 830–2 (1993).

[82] Schneider, R. P., Jr. and Lott, J. A., "Cavity design for improved electrical injection in InAlGaP/AlGaAs visible (639–661 nm) vertical-cavity surface-emitting laser diodes," *Applied Physics Letters*, **63**, 917–19 (1993).

[83] Schneider, R. P., Jr., Hagerott Crawford, M., Choquette, K. D., Lear, K. L., Kilcoyne, S. P., and Figiel, J. J., "Improved AlGaInP-based red (670–690 nm) surface-emitting lasers with novel C-doped short-cavity epitaxial design," *Applied Physics Letters*, **67**, 329–31 (1995).

[84] Hagerott Crawford, M., Schneider, R. P., Jr., Choquette, K. D., Lear, K. L., Kilcoyne, S. P., and Figiel, J. J., "High efficiency AlGaInP-based 660–680 nm vertical-cavity surface emitting lasers," *Electronics Letters*, **31**, 196–8 (1995).

[85] Crawford, M. H., Schneider, R. P., Jr., Choquette, K. D., and Lear, K. L., "Temperature-dependent characteristics and single-mode performance of AlGaInP-based 670–690-nm vertical-cavity surface-emitting lasers," *IEEE Photonics Technology Letters*, **7**, 724–6 (1995).

[86] Crawford, M. H. and Schneider, R. P., Jr., *Summaries of Papers Presented at the Conference on Lasers and Electro-Optics (IEEE Cat. No. 95CH35800) CLEO '95. Conference on Lasers and Electro-Optics. (Papers in summary form only received)*, Baltimore, MA, USA, Opt. Soc. America, pp. 168–9 (1995).

[87] Schneider, R. P., Jr., Choquette, K. D., Lott, J. A., Lear, K. L., Figiel, J. J., and Malloy, K. J., "Efficient room-temperature continuous-wave AlGaInP/AlGaAs visible (670 nm) vertical-cavity surface-emitting laser diodes," *IEEE Photonics Technology Letters*, **6**, 313–16 (1994).

[88] Morgan, R. A., Hibbs-Brenner, M. K., Walterson, R. A., Lehman, J. A., Marta, T. M., Bounnak, S., Kalweit, E. L., Akinwande, T., and Nohava, J. C., "Producible GaAs-based MOVPE-grown vertical-cavity top-surface emitting lasers with record performance," *Electronics Letters*, **31**, 462–4 (1995).

[89] Jager, R., Grabherr, M., Jung, C., Michalzik, R., Reiner, G., Weigl, B., and Ebeling, K. J., "57% Wallplug efficiency oxide-confined 850 nm wavelength GaAs VCSELs," *Electronics Letters*, **33**, 330–1 (1997).

[90] Choquette, K. D., Schneider, R. P., Crawford, M. H., Geib, K. M., and Figiel, J. J., "Continuous wave operation of 640–660 nm selectively oxidised AlGaInP vertical-cavity lasers," *Electronics Letters*, **31**, 1145–6 (1995).

[91] Bour, D. P., Treat, D. W., Thornton, R. L., Geels, R. S., and Welch, D. F., "Drift leakage current in AlGaInP quantum-well lasers," *IEEE Journal of Quantum Electronics*, **29**, 1337–43 (1993).

[92] Takagi, T., Koyama, F., and Iga, K., "Modified multiquantum barrier for 600 nm range AlGaInP lasers," *Electronics Letters*, **27**, 1081–2 (1991).

[93] Takagi, T., Koyama, F., and Iga, K., "Potential barrier height analysis of AlGaInP

multi-quantum barrier (MQB)," *Japanese Journal of Applied Physics, Part 2 (Letters)*, **29**, L1977–80 (1990).

[94] Furuya, A. and Tanaka, H., "Superposed multiquantum barriers for InGaAlP heterojunctions," *IEEE Journal of Quantum Electronics*, **28**, 1977–82 (1992).

[95] Kamiyama, S., Monnoh, M., Ohnaka, K., and Uenoyama, T., "Studies of threshold current dependence on compressive and tensile strain of 630 nm GaInP/AlGaInP multi-quantum-well lasers," *Journal of Applied Physics*, **75**, 8201–3 (1994).

[96] Chen, Y. H., Wilkinson, C. I., Woodhead, J., Button, C. C., David, J. P. R., Pate, M. A., and Robson, P. N., "Polarisation characteristics of InGaAlP/AlGaAs visible vertical cavity surface emitting lasers," *Electronics Letters*, **32**, 559–60 (1996).

[97] Takahashi, M., Vaccaro, P., Fujita, K., Watanabe, T., Mukaihara, T., Koyama, F., and Iga, K., "An InGaAs-GaAs vertical-cavity surface-emitting laser grown on GaAs(311)A substrate having low threshold and stable polarization," *IEEE Photonics Technology Letters*, **8**, 737–9 (1996).

[98] Kuchta, D. M., Schneider, R. P., Choquette, K. D., and Kilcoyne, S., "Large- and small-signal modulation properties of red (670 nm) VCSELs," *IEEE Photonics Technology Letters*, **8**, 307–9 (1996).

[99] Park, R. M., Troffer, M. B., Rouleau, C. M., DePuydt, J. M., and Haase, M. A., "p-Type ZnSe by nitrogen atom beam doping during molecular beam epitaxial growth," *Applied Physics Letters*, **57**, 2127–9 (1990).

[100] Ohkawa, K. and Mitsuyu, T., "p-Type ZnSe homoepitaxial layers grown by molecular beam epitaxy with nitrogen radical doping," *Journal of Applied Physics*, **70**, 439–42 (1991).

[101] Chu, C. C., Ng, T. B., Han, J., Hua, G. C., Gunshor, R. L., Ho, E., Warlick, E. L., Kolodziejski, L. A., and Nurmikko, A. V., "Reduction of structural defects in II-VI blue green laser diodes," *Applied Physics Letters*, **69**, 602–4 (1996).

[102] Kuo, L. H., Salamanca-Riba, L., Wu, B. J., Haugen, G. M., DePuydt, J. M., Hofler, G., and Cheng, H., *22nd Conference on Physics and Chemistry of Semiconductor Interfaces*, Scottsdale, AZ, USA, **13**, 1694–704 (1995).

[103] Uusimaa, P., Rakennus, K., Salokatve, A., Pessa, M., Aherne, T., Doran, J. P., O'Gorman, J., and Hegarty, J., "Molecular beam epitaxy growth of MgZnSSe/ZnSSe Bragg mirrors controlled by in situ optical reflectometry," *Applied Physics Letters*, **67**, 2197–9 (1995).

[104] Honda, T., Yanashima, K., Yoshino, J., Kukimoto, H., Koyama, F., and Iga, K., "Fabrication of a ZnSe-based vertical Fabry–Perot cavity using SiO_2/TiO_2 multilayer reflectors and resonant emission characteristics," *Japanese Journal of Applied Physics, Part 1 (Regular Papers & Short Notes)*, **33**, 3960–1 (1994).

[105] Honda, T., Sakaguchi, T., Koyama, F., Iga, K., Yanashima, K., Inoue, K., Munekata, H., and Kukimoto, H., "Optically pumped ZnSe-based vertical cavity surface emitter with SiO_2/TiO_2 multilayer reflector," *Journal of Applied Physics*, **78**, 4784–6 (1995).

[106] Amano, H., Kitoh, M., Hiramatsu, K., and Akasaki, I., *Proceedings of the Sixteenth International Symposium*, Karuizawa, Japan, Institute of Physics, pp. 725–30 (1986).

[107] Nakamura, S., Mukai, T., and Senoh, M., "High-power GaN P-N junction blue-light-emitting diodes," *Japanese Journal of Applied Physics, Part 2 (Letters)* **30**, L1998–2001 (1991).

[108] Nakamura, S., Senoh, M., and Mukai, T., "Highly p-typed Mg-doped GaN films grown with GaN buffer layers," *Japanese Journal of Applied Physics, Part 2 (Letters)*, **30**, L1708–11 (1991).

[109] Nakamura, S., Mukai, T., Senoh, M., and Iwasa, N., "Thermal annealing effects on P-type Mg-doped GaN films," *Japanese Journal of Applied Physics, Part 2 (Letters)*, **31**, L139–42 (1992).

[110] Nakamura, S., Iwasa, N., Senoh, M., and Mukai, T., "Hole compensation mechanism of P-Type GaN films," *Japanese Journal of Applied Physics, Part 1 (Regular Papers & Short Notes)*, **31**, 1258–66 (1992).

[111] Nakamura, S., *Physics and Simulation of Optoelectronic Devices IV, San Jose, CA, USA*, **2693**, 43–56 (1996).

[112] Nakamura, S., Harada, Y., and Seno, M., "Novel metalorganic chemical vapor deposition system for GaN growth," *Applied Physics Letters*, **58**, 2021–3 (1991).

[113] Asif Khan, M., Kuznia, J. N., Van Hove, J. M., Olson, D. T., Krishnankutty, S., and Kolbas, R. M. "Growth of high optical and electrical quality GaN layers using low-pressure metalorganic chemical vapor deposition," *Applied Physics Letters*, **58**, 526–7 (1991).

[114] Amano, H., Kitoh, M., Hiramatsu, K., and Akasaki, I., "Growth and luminescence properties of Mg-doped GaN prepared by MOVPE," *Journal of the Electrochemical Society*, **137**, 1639–41 (1990).

[115] Bulman, G. E., Doverspike, K., Sheppard, S. T., Weeks, T. W., Kong, H. S., Dieringer, H. M., Edmond, J. A., Brown, J. D., Swindell, J. T., and Schetzina, J. F., "Pulsed operation lasing in a cleaved-facet InGaN/GaN MQW SCH laser grown on 6H-SiC," *Electronics Letters*, **33**, 1556–7 (1997).

[116] Amano, H., Kito, M., Hiramatsu, K., and Akasaki, I., "p-Type conduction in Mg-doped GaN treated with low-energy electron beam irradiation (LEEBI)," *Japanese Journal of Applied Physics, Part 2 (Letters)*, **28**, L2112–14 (1989a).

[117] Li, X. and Coleman, J. J., "Time-dependent study of low energy electron beam irradiation of Mg-doped GaN grown by metalorganic chemical vapor deposition," *Applied Physics Letters*, **69**, 1605–7 (1996).

[118] Domen, K., Kondo, K., Kuramata, A., and Tanahashi, T., "Gain analysis for surface emission by optical pumping of wurtzite GaN," *Applied Physics Letters*, **69**, 94–6 (1996).

[119] Domen, K., Horino, K., Kuramata, A., and Tanahashi, T., "Optical gain for wurtzite GaN with anisotropic strain in c plane," *Applied Physics Letters*, **70**, 987–9 (1997).

[120] Chow, W. W., Knorr, A., and Koch, S. W., "Theory of laser gain in group-III nitrides," *Applied Physics Letters*, **67**, 754–6 (1995).

[121] Chow, W. W., Wright, A. F., and Nelson, J. S., "Theoretical study of room temperature optical gain in GaN strained quantum wells," *Applied Physics Letters*, **68**, 296–8 (1996).

[122] Nakamura, S., Senoh, M., Nagahama, S., Iwasa, N., Yamada, T., Matsushita, T., Sugimoto, Y., and Kiyoku, H., "Optical gain and carrier lifetime of InGaN multi-quantum well structure laser diodes," *Applied Physics Letters*, **69**, 1568–70 (1996).

[123] Kuball, M., Jeon, E. S., Song, Y. K., Nurmikko, A. V., Kozodoy, P., Abare, A., Keller, S., Coldren, L. A., Mishra, U. K., DenBaars, S. P., and Steigerwald, D. A., "Gain spectroscopy on InGaN/GaN quantum well diodes," *Applied Physics Letters*, **70**, 2580–2 (1997).

[124] Khan, M. A., Kuznia, J. N., Van Hove, J. M., and Olson, D. T., "Reflective filters based on single-crystal GaN/$Al_xGa_{1-x}N$ multilayers deposited using low-pressure metalorganic chemical vapor deposition," *Applied Physics Letters*, **59**, 1449–57 (1991).

[125] Chow, W. W., Wright, A. F., Girndt, A., Jahnke, F., and Koch, S. W., "Microscopic theory of gain for an InGaN/AlGaN quantum well laser," *Applied Physics Letters*, **71**, 24xx–24yy (1997).

8

Long-Wavelength Vertical-Cavity Lasers

Dubravko I. Babić, Joachim Piprek, and John E. Bowers

8.1 Introduction

Vertical-cavity surface-emitting lasers operating at 1300 nm and 1550 nm are potentially low-cost sources for optical communications. The market for multimode 1300 nm VCSELs lies in short-distance data-communication links, which presently utilize 1300 nm light-emitting diodes and 850 nm vertical-cavity lasers as sources (Fibre Channel). Owing to the larger distance-bandwidth product achievable in silica-based fiber at 1300 nm, the maximum data-transmission rate and point-to-point distance in these links is expected to increase more than two times with the introduction of multimode vertical-cavity lasers operating at 1300 nm. Single-mode vertical-cavity lasers operating at both 1300 nm and 1550 nm are also viable choices for sources in telecommunications and wavelength-division multiplexing (WDM) applications due to the expected low cost of VCSELs, the possibility of fabricating arrays [1], and the integrability of the VCSEL structure.

GaAs-based VCSELs operating in the 780–980 nm range have exhibited tremendous progress in their performance over the past few years: Threshold currents of several tens of microamperes and wall-plug efficiencies greater than 50% have been demonstrated [2] and the first 850 nm VCL products for high-speed asynchronous transfer mode (ATM) [3] and gigabit local-area networks (LANs) have been introduced by a number of U. S. manufacturers. Even though the first VCSEL operated at 1300 nm [4], the development of long-wavelength VCSELs has been slower over the past few years in comparison to GaAs-based VCSELs owing to several technological difficulties. The key issues have been the difficult realization of high-reflectivity mirrors in the 1300–1550 nm wavelength range and of active layers with sufficient gain at elevated temperatures. The InGaAsP and AlInGaAs material systems lattice-matched to InP (Figure 8.1) have been the natural choices for the fabrication of long-wavelength lasers [5, 6]. However, neither of these systems offers a large enough range of refractive index values that can produce high-reflectivity quarter-wave mirrors of both p- and n-type at 1300 and 1550 nm. This has prompted a more intense investigation of amorphous-dielectric mirrors and other compound semiconductor-material systems for fabrication of long-wavelength VCSELs. The reports on long-wavelength VCSELs over the past years have been dealing with very few devices at a time and of a variety of different structures. Hence, it has been a difficult task to develop consistent models for these devices and reach a clear understanding on the mechanisms limiting improved device performance. For this reason, effort has concentrated on improving all aspects of the laser structures: developing active layers with higher gain, fabricating quarter-wave mirrors with higher reflectivity, and designing device structures with lower thermal resistance. The device

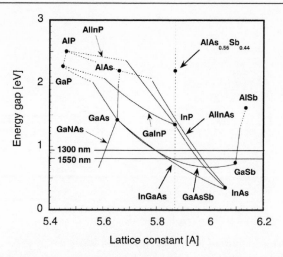

Figure 8.1. Bandgap versus lattice constant for III–V compound semiconductors relevant for use in long-wavelength vertical-cavity lasers. The dots indicate binary compounds, the full lines direct-gap ternary compounds, while the dashed lines indicate indirect-gap ternary compounds.

structures presently investigated for long-wavelength applications can be divided into three groups: 1) etched-well VCSEL structures that use amorphous-dielectric mirrors on both sides of active layer [7], 2) VCSEL structures with one semiconductor and one amorphous-dielectric mirror that utilize ring contacts [8], and 3) all-epitaxial devices that mimic the single-growth GaAs-VCSEL structures [9]. The simplified schematics of these device structures are shown in Figure 8.2. In recent years, significant progress has been reported in all three device structures, owing to improvements in the epitaxial growth techniques and the application of fusion-bonding technology to surface-normal optoelectronic devices.

This chapter reviews the current status of long-wavelength VCSEL development. We start by discussing the active layers and quarter-wave mirrors for long-wavelength applications and then describe their practical implementation in various VCSEL structures. We conclude the chapter with a discussion of design issues for vertical-cavity lasers fabricated by fusion bonding, since this technology has been producing the best performing long-wavelength VCSELs at present time.

8.2 Active Layers

Because of the shorter gain region, higher threshold gain, and typically higher thermal resistance, the requirements on active regions in VCSELs are more stringent than those of edge-emitting lasers. In addition, at long wavelengths (1300 nm and 1550 nm), the performance of semiconductor lasers is plagued by increased Auger recombination, carrier leakage, and carrier-related optical losses [5, 10]. The desired temperature range for operation of long-wavelength VCSELs is 0° to 70°C for multimode data communications and up to 85°C for single-mode optical communications. These requirements place high demands on the design and material development and leave a narrow margin for fabrication error.

Bulk InGaAsP active layers were used during most of the initial development of long-wavelength VCSELs, with quantum wells coming to prevalent use only in recent years [11].

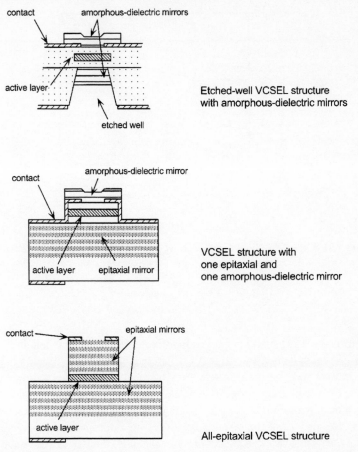

Figure 8.2. Three long-wavelength VCSEL structures: Etched-well structure, VCSEL with one epitaxial and one amorphous-dielectric mirror, and the all-epitaxial VCSEL structure.

It is now clear that the lower transparency and higher differential gain achievable with strained quantum wells is necessary to produce above-room-temperature operating long-wavelength VCSELs. Due to still high (and largely uncertain) values of optical cavity loss and the quantum-well gain saturation, the number of quantum wells used in these devices has always been large (up to 15 wells have been reported). The reason for using a large number of wells is illustrated in Figure 8.3. This figure shows the dependence of VCSEL threshold current on the round-trip loss and the number of quantum wells positioned optimally at the peak of the optical field. The example threshold current density is calculated for a typical logarithmic gain–current relationship (constants taken from ref. 12). It is evident from this relationship that for every cavity loss value there is a certain number of wells that produces lowest threshold current density, and this optimum number of wells increases with the round-trip cavity loss. If the value of loss is uncertain, it is always advantageous to use a larger number of wells because of the smaller sensitivity of threshold current to the variations in the round-trip loss.

Strained quantum well active layers with a large number of wells often require epitaxial growths of epilayers that exceed the critical thickness for stable epitaxial multilayers with no dislocations [13]. In this case the net strain can be reduced or entirely eliminated by utilizing

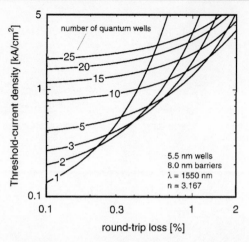

Figure 8.3. VCSEL threshold-current density as a function of number of quantum wells and round-trip cavity loss. The relationship between the material gain and the current density was assumed logarithmic ($g = g_0 \ln(J/J_0)$) with data taken from [12] as an example ($g_0 = 726\,\mathrm{cm}^{-1}$ and $J_0 = 72\,\mathrm{A/cm^2}$). Standing wave effects (gain enhancement) are included. For every loss value there is an optimum number of wells.

strain compensation, originally employed to increase the differential gain in edge-emitting lasers [14]. The first demonstration of VCSELs with strain-compensated multi-quantum-well active layers were reported in refs. 15, 16, and 17. It was the application of strain-compensated and near-strain-compensated quantum wells to long-wavelength VCSELs that has resulted in above-room-temperature continuous-wave operating devices [9, 18, 19].

The use of separate confinement regions is of key importance in the transverse-mode definition in edge-emitting lasers because it provides index-guiding. In vertical-cavity lasers this requirement does not exist and the separate confinement (SCH) regions may substantially be reduced in size [19], step-graded [20], or completely eliminated [12]. In this way, optical losses related to carrier leakage (free-carrier and inter-valence-band absorption) in the SCH layers are greatly reduced. The elimination of wide SCH regions improves the high temperature laser performance, but high carrier-related absorption still remains in the multi-quantum-well regions. The effect of the carrier-related losses on laser performance is further intensified by *loss multiplication*: The presence of carriers in the active region increases the cavity losses and hence more carriers are required to sustain threshold gain. The additional carriers then raise the loss even further, requiring even more carriers. The *positive feedback* realized in this way results in a very strong temperature dependence of the cavity optical losses [21, 22]. Carrier-related optical losses [22, 23, 24] have especially been pronounced in long-wavelength InGaAsP active layers that exhibit high carrier leakage due to poor electron confinement (in respect to the GaAs system). It has been demonstrated that the electron confinement can be improved by using a different material system with higher conduction band offset, such as AlInGaAs grown on InP [6] or on InGaAs substrates [25], and InGaNAs grown on GaAs substrates [26].

8.3 Quarter-Wave Mirrors for Long-Wavelength Applications

The realization of high-reflectivity mirrors in a particular material system critically depends on the availability of two materials, one of high and one of low refractive index, that can be grown or deposited on a specified substrate to form a multilayer structure of alternating quarter-wavelength thick layers. The number of layers necessary to achieve the required reflectivity and bandwidth of such a mirror primarily depends on the refractive index ratio between the high and the low index material. The larger the refractive index ratio, the smaller the number of layers necessary for a given reflectivity and the broader the mirror stop band. In addition, in the presence of material absorption, the material combination with a larger refractive index ratio produces quarter-wave mirrors with higher maximum possible reflectivity. Beside the optical properties, the choice of the material system determines the thermal and electrical resistances of the mirrors. Electrical conduction is possible with epitaxially grown mirrors, but not with amorphous-dielectric mirrors. The practicality of a certain material system depends profoundly on all of these parameters, and they ultimately determine the performance and the structure of these lasers. Even at present time, the choice for the long-wavelength VCSEL mirror materials has not been settled, as it has for the GaAs-based VCSELs.

The peak reflectivity of a quarter-wave mirror depends on the number of quarter-wave layers, the refractive indexes of the incident and the exit media, the high and low index materials, and the material absorption. Figure 8.4 shows a comparison between reflectivity dependence on number of layers for three quarter-wave mirrors used in 1550 nm vertical-cavity lasers: a–Si/SiO$_2$, AlAs/GaAs, and InGaAsP/InP. It is clear that because of a smaller refractive index ratio between InGaAsP and InP, this mirror requires twice as many layers to reach a specified reflectivity as the AlAs/GaAs one at the same wavelength. The a–Si/SiO$_2$ material combination, in contrast, requires even fewer number of layers because the refractive index ratio is much larger between the two materials. In real quarter-wave mirrors the reflectivity is reduced by the presence of material absorption and light scattering, which manifests itself as the saturation in the value of the reflectivity in Figure 8.4. Material absorption has typically been the dominant mechanism determining the peak value of VCSEL mirror reflectivity, owing to the small roughness of state-of-the-art semiconductor

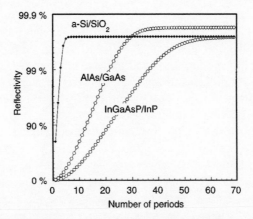

Figure 8.4. Calculated reflectivity of three quarter-wave mirrors used in long-wavelength VCSELs. The refractive index used is taken at 1550 nm. The AlAs/GaAs and InGaAsP/InP mirrors have $10\,\mathrm{cm}^{-1}$ loss uniformly distributed over the layers illustrating that a mirror with lower refractive index ratio saturates at a lower maximum reflectivity. The a–Si/SiO$_2$ mirror requires a substantially smaller number of layers to saturate the reflectivity, but due to high absorption in amorphous silicon the saturate reflectivity value is ~99.6%.

and amorphous-dielectric mirrors used in vertical-cavity lasers [27]. We consider the effect of material absorption in more detail, while we refer the reader to more complete treatments of optical scattering, for example ref. 28. The presence of weak scattering in quarter-wave mirrors can be treated as an effective absorption in a simplified treatment described in ref. 29.

To illustrate the mutual dependence of loss in the layers and the refractive index difference we consider the maximum achievable reflectivity of a quarter-wave mirror in the presence of absorption loss. The *maximum reflectivity* is a hypothetical value of the mirror reflectivity that would be achieved with an infinite number of layers. It is given analytically for the case of small absorption (always the case in VCSEL mirrors) [30] by

$$R_{n_1=n_H} = 1 - \lambda \frac{n_1(\alpha_L + \alpha_H)}{2(n_H^2 - n_L^2)}, \qquad R_{n_1=n_L} = 1 - \lambda \frac{\alpha_L n_H^2 + \alpha_H n_L^2}{2n_1(n_H^2 - n_L^2)}, \qquad (8.1)$$

where λ is the free-space center wavelength. The incident medium refractive index is denoted with n_1 and the absorption coefficient in the high and low index materials are denoted with α_H and α_L. The equations differ depending on whether the first layer index is high (n_H) or low (n_L). In order to observe the effect of loss for all mirrors on equal footing we introduce a normalized relationship for the maximum reflectivity by setting the refractive index of the incident medium equal to that of one of the quarter-wavelength layers. The maximum reflectivity is then written as $R_{max} = 1 - (2\bar{n}/\Delta n - 1)(\bar{\alpha}\lambda/2\bar{n})$, where $\bar{n} = (n_L + n_H)/2$ and $\Delta n = n_H - n_L$ are the average and the difference between the refractive indexes n_H and n_L. Here we define the *normalized absorption coefficient* as $\lambda\bar{\alpha}/\bar{n}$ (dimensionless) and the *fractional refractive index difference* as $\Delta n/\bar{n}$. There are two definitions of $\lambda\bar{\alpha}/\bar{n}$: When the refractive index of the first mirror layer is of high index, we have $\bar{\alpha} = (\alpha_L + \alpha_H)/2$; if the first layer in the mirror is of low index (in which case the incident medium has a high refractive index), we have $\bar{\alpha} = \alpha_L n_H/2n_L + \alpha_H n_L/2n_H$. The fractional refractive index difference is a figure of merit for quarter-wave mirrors and can be used to compare quarter-wave mirrors of different materials in terms of efficiency in realizing high-reflection coefficients. In Figure 8.5 we show a family of constant-reflectivity curves in the $\Delta n/\bar{n} - \lambda\bar{\alpha}/\bar{n}$ coordinate system. The fractional refractive index differences for four epitaxial quarter-wave mirrors and one amorphous-dielectric mirror (a–Si/SiO$_2$) are also given for comparison between the mirrors. The mirrors are tuned to 1550 nm and 1300 nm (as indicated) and use the corresponding values of the refractive indexes. This representation clearly illustrates that the maximum reflectivity of a quarter-wave mirror is more susceptible to the presence of absorption if the refractive index difference is small.

Epitaxial mirrors with small refractive index difference require relatively long growth times, and the specified growth may be difficult to perform repeatedly. The random variation or drift in the layer thicknesses always results in reduced peak reflectivity. However, it is important to note that even with perfect tuning and an infinite number of layers, the material absorption places the ultimate limit on the maximum reflectivity value. The interplay among the refractive indexes, the absorption/scattering, and the peak reflectivity of the quarter-wave mirrors represents the strongest argument in favor of mirrors with large refractive index difference. This is particularly important to consider in extrinsic epitaxial mirrors. A useful rule of thumb for deciding whether a certain epitaxial material combination can be used to achieve a peak reflectivity R is given by $\bar{\alpha}\lambda/\Delta n < 1 - R$, where we have assumed that $n \gg \Delta n$ for epitaxial mirrors. The constant reflectivity curves in Figure 8.5 graphically

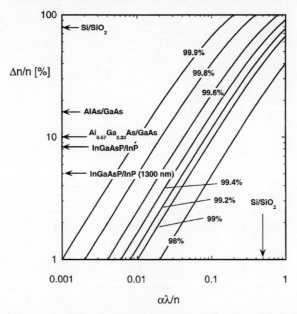

Figure 8.5. A family of constant-reflectivity curves in a $\Delta n/\bar{n} - \lambda\bar{\alpha}/\bar{n}$ coordinate system showing the maximum reflectivity possible with a given refractive index and material absorption combination. The fractional refractive index difference is indicated with horizontal lines for four epitaxial quarter-wave mirrors and one amorphous mirror (a–Si/SiO$_2$). The mirrors are tuned to 1550 nm and 1300 nm (as indicated) and use the appropriate value of the refractive indexes. Note that the same value of normalized absorption coefficient does not correspond to the same value of mirror doping level since we are dealing with a variety of different materials.

illustrate this relationship for all mirrors. The peak reflection coefficient is achievable if the material combination falls to the left of a specified constant-reflectivity curve.

8.4 Etched-Well VCSELs and Amorphous-Dielectric Mirrors

The invention of the VCSEL and the subsequent key development of these lasers have taken place at the Tokyo Institute of Technology, Japan [7]. The first VCSEL, demonstrated by Soda et al. [4], operated at 1300 nm with a 77 K pulsed threshold current of 900 mA. This structure used two AuZn metal mirrors deposited on both sides of a 90 μm thick InP wafer. The high threshold current density was a result of relatively low reflectivity of the AuZn alloyed mirrors ($R \sim 80\%$) and the cavity absorption and diffraction losses. The threshold current of this prototype vertical-cavity laser was subsequently reduced by implementing buried heterostructure active layers, by shortening the cavity length to several micrometers, by etching wells in the substrates, and finally by using amorphous-dielectric mirrors combined with ring electrodes [7]. The resulting structure has been modified and investigated to present date and is commonly known as the *etched-well vertical-cavity laser* because of the necessity of etching vias (wells) in the substrate to access the back of the active layer (shown in Figure 8.2).

In this VCSEL structure, both cavity mirrors are fabricated using amorphous semiconductor and insulator materials that are deposited by low-temperature deposition techniques such

Table 8.1. *Amorphous mirrors tuned to 1550 nm.[a]*

	Low index $\rightarrow$		MgF	CaF$_2$	SiO$_2$	MgO	Al$_2$O$_3$	Si$_3$N$_4$	TiO$_2$
High index	κ [W/cmK] $\rightarrow$		—	0.1	0.012	0.53	0.36	0.16	0.09
$\downarrow$	$\downarrow$	n	1.35	1.43	1.45	1.71	1.74	2.0	2.44
a-Si	0.026	3.6	91%	86%	85%	71%	70%	57%	38%
SiC	2.5	2.57	62%	57%	56%	40%	39%	25%	5%
ZnSe	0.19	2.46	58%	53%	52%	36%	34%	21%	1%
TiO$_2$	0.09	2.44	58%	52%	51%	35%	33%	20%	—
Si$_3$N$_4$	0.16	2.0	39%	33%	32%	16%	14%	—	—
Al$_2$O$_3$	0.36	1.74	25%	20%	18%	2%	—	—	—
MgO	0.53	1.71	24%	18%	16%	—	—	—	—

[a] A summary of thermal conductivity and refractive index at 1550 nm for amorphous mirrors for use in long-wavelength VCSEL fabrication. Each mirror contains one high-index (row) and one low-index (column) material. The percentage shown at the row–column crossing corresponds to the fractional refractive index $\Delta n / \bar{n}$. The table may be also used for 1300 nm mirrors. (The refractive index values at 1300 nm and 1550 nm are very close for insulators in which the bandgap energies are much larger than the photon energy.) The data have been taken from Ref. [138].

as electron-beam evaporation, sputtering, and plasma-enhanced chemical-vapor-deposition (PECVD). All of these materials are electrically insulating but exhibit a relatively large range of thermal conductivities ($0.01 \leq \kappa \leq 2.5$ W/cm K). Thermal conductivities and refractive index values of select amorphous materials used for long-wavelength VCSELs are given in Table 8.1 along with the fractional refractive index ratio. For long-wavelength vertical-cavity-lasers, the most common material combination has been (amorphous) SiO$_2$ as the low index material and amorphous silicon (a–Si) as high index material [31]. Owing to the large refractive index ratio these mirrors only require a few periods to achieve reflectivity above 99.5%. However, in amorphous semiconductors the absorption tail extends deep into the forbidden gap [32], and hence the reflectivity of these mirrors is often limited by high material absorption (typical absorption coefficients for a–Si are $\alpha \approx 100$ cm^{-1} at 1550 nm and $\alpha \approx 1,000$ cm^{-1} at 1300 nm [33]). In addition to the material absorption, the thermal conductivity of both of these materials is quite poor ($\kappa \approx 0.012$ W/K cm for a–SiO$_2$ [34] and $\kappa \approx 0.026$ W/K cm for a–Si [35]). For a long time, this has limited the etched-well devices to pulsed operation at room temperature. The recent implementation of a–Si/MgO [36, 37] and a–Si/Al$_2$O$_3$ [18] mirrors in which the low-index material has higher thermal conductance has produced a dramatic improvement of device performance, despite the fact that the refractive indexes of both MgO and Al$_2$O$_3$ are higher than that of SiO$_2$, and therefore the maximum reflectivity of mirrors employing these two materials (a–Si/MgO and a–Si/Al$_2$O$_3$) is lower than that of the a–Si/SiO$_2$ combination. Other materials may be used to improve the thermal conductivity of amorphous mirrors without substantial reduction in maximum reflectivity [38]. It is important to note that to minimize the thermal resistance between the active layer and the heat sink, etched-well devices are almost always mounted epi-down.

The state of the art in etched-well long-wavelength VCSELs is the structure fabricated by Uchiyama et al. [18] (the device schematic is shown in Figure 8.6(a)). This device employs a partly strain-compensated InGaAsP multi-quantum-well active layer and a current constriction scheme with two InP/InGaAsP regrowths. Figure 8.6(b) shows the scanning electron micrograph of the cross section through the regrown active layer. The 12-quantum-well active layer was formed using metal-organic vapor-phase epitaxy (MOVPE). A diamond-shaped mesa [39]; with sides aligned to $\langle 100 \rangle$ planes were defined by silicon nitride patterning and wet chemical etching. The silicon nitride coating served as a mask for the growth of alternating p-InP and n-InP blocking layers around the mesa. It was subsequently removed and the cavity was completed by adding optical thickness to the p-side and the contact layers during the second regrowth. These devices have been mounted epi-down on diamond heat sinks. The heat was taken out through the a–Si/Al$_2$O$_3$ mirror, which has higher thermal conductivity (as shown in Figure 8.6(a)). This device operated continuous-wave up to 36°C with lasing wavelength of 1,310 nm. The characteristic temperature $T_0 = dT/d \ln J_{\text{th}}(T)|_{T=300 \text{ K}}$ was approximately 50 K.

In summary, the etched-well structure, originally plagued with poor mirror reflectivity, a current constriction scheme, and high thermal resistance, has kept its place as the present day state-of-the-art 1.3 μm VCSEL. This structure has been predominantly used for 1300 nm lasers by many authors [7, 18, 31, 37, 40–43], and also has been used near 1550 nm [44–46].

8.5 VCSELs with Epitaxial Mirrors

A VCSEL structure with one epitaxial and one amorphous-dielectric mirror, shown in Figure 8.2, is attractive because it allows one to use a planar process with a uniform current injection through a conductive mirror. The structure necessarily involves ring contacts and a current/mode confinement scheme. We first discuss the properties of epitaxial quarter-wave mirrors relevant to long-wavelength VCSEL applications and then discuss the investigated device structures.

8.5.1 Properties of Epitaxial Mirrors

Epitaxial mirrors realized with compound III–V semiconductor alloys of interest for long-wavelength applications exhibit a relatively narrow range of refractive index values ($2.9 \leq n \leq 3.5$) [47, 48] and a moderate range of thermal conductivities ($0.02 \leq \kappa \leq 0.9$ W/cm K) [49]. Tables 8.2A and 8.2B show select material combinations that can be used for long-wavelength applications tuned to 1300 nm and 1550 nm wavelengths. The fractional refractive index difference is listed for every possible combinations as a figure of merit. The main advantage of epitaxial mirrors is that they can be made conductive by doping, and, in some cases, they can be realized in situ at the same time as the active layer.

The design of any epitaxial quarter-wave mirror that is also intended for current supply involves the optimization between the reflectivity and the electrical resistance. These two parameters are connected through the doping, which effects the absorption and the mirror conduction. The voltage drop across these mirrors depends on the ohmic resistance of the bulk doped layers and on the potential drop across the large number of heterojunction barriers that are incorporated in a typical quarter-wave mirror. Due to lower mobility, larger valence-band offset, and larger effective mass of holes, p-type mirrors exhibit higher resistances at given doping profile. The reduction of heterojunction barriers involves grading and doping

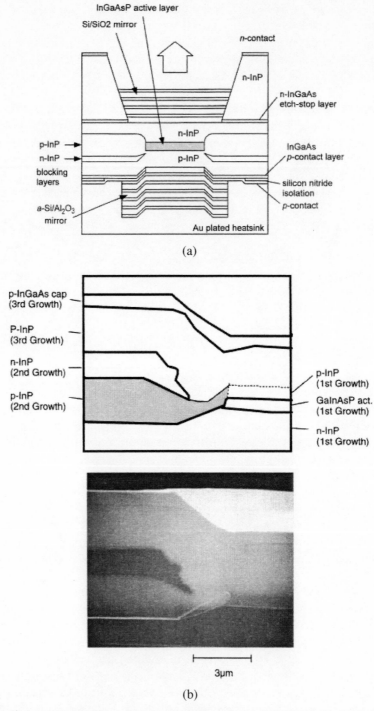

Figure 8.6. (a) The etched-well device of Uchiyama et al. [18]. The device is mounted epi-down on a Au-plated heatsink and the light was taken out through the mirror on the substrate side (a–Si/SiO$_2$). The higher thermal conductivity a–Si/Al$_2$O$_3$ quarter-wave mirror was used on the bottom side of the device. (b) Cross-sectional SEM micrograph showing two regrowths of device shown in (a).

Table 8.2A. *Epitaxial mirrors tuned to 1300 nm.*[a]

High index	κ^c	n	Low index → AlAs	$Al_{0.67}Ga_{0.33}As$	$AlAs_{0.56}Sb_{0.44}$	$AlP_{0.4}Sb_{0.6}$	$Al_{0.48}In_{0.52}As$	InP	$Al_{0.2}In_{0.8}As$
κ^c →			0.91	0.16	0.057	0.039	0.045	0.68	0.07
n →			2.92	3.06	3.15	3.10	3.24	3.21	3.04
GaAs	0.44	3.41	15.5%	10.8%	—	—	—	—	—
InGaAsP[b]	0.043	3.37	—	—	6.7%	8.3%	3.9%	4.9%	—
AlInGaAs[b]	0.042	3.35	—	—	6.2%	7.8%	3.3%	4.3%	—
AlGaAsSb[b]	0.050	3.57	—	—	12.5%	14%	9.7%	10.6%	—
AlGaPSb[b]	0.038	3.57	—	—	12.5%	14%	9.7%	10.6%	—
$In_{0.2}Ga_{0.8}As$	0.06	3.47	—	—	—	—	—	—	13%

[a] A summary of thermal conductivities and refractive index values for epitaxial mirrors used for long-wavelength VCSELs tuned to 1300 nm (Table 8.2A) and 1550 nm (Table 8.2B). Each mirror contains one high-index (row) and one low-index (column) material. The percentage shown at the row–column crossing corresponds to the fractional refractive index $\Delta n/\bar{n}$. The grouped material combinations correspond to three different substrates: GaAs, InP, and $In_{0.2}Ga_{0.8}As$. The data have been taken from Refs. [48, 49, 54, 74, 87, 141, and 142].
[b] Bandgap wavelength $\approx$ 1,150 nm.
[c] Thermal conductivity in [W/cm K].

Table 8.2B. *Epitaxial mirrors tuned to 1550 nm.*[a]

Low index →			AlAs	Al$_{0.67}$Ga$_{0.33}$As	AlAs$_{0.56}$Sb$_{0.44}$	AlP$_{0.4}$Sb$_{0.6}$	Al$_{0.48}$In$_{0.52}$As	InP
High index	κ^c →		0.91	0.16	0.057	0.039	0.045	0.68
→		n	2.89	3.04	3.10	3.05	3.21	3.17
GaAs	0.44	3.37	15.2%	10.3%	—	—	—	—
InGaAsP[b]	0.045	3.45	—	—	10.7%	12.3%	7.2%	8.5%
AlInGaAs[b]	0.045	3.47	—	—	11.3%	12.9%	7.8%	9%
AlGaAsSb[b]	0.062	3.6	—	—	15%	16.5%	—	—
AlGaPSb[b]	0.046	3.55	—	—	13.5%	15.2%	—	—

[a] A summary of thermal conductivities and refractive index values for epitaxial mirrors used for long-wavelength VCSELs tuned to 1300 nm (Table 8.2A) and 1550 nm (Table 8.2B). Each mirror contains one high-index (row) and one low-index (column) material. The percentage shown at the row–column crossing corresponds to the fractional refractive index $\Delta n/\bar{n}$. The grouped material combinations correspond to three different substrates: GaAs, InP, and In$_{0.2}$Ga$_{0.8}$As. The data have been taken from Refs. [48, 49, 54, 74, 87, 141, and 142].
[b] Bandgap wavelength ≈ 1,400 nm.
[c] Thermal conductivity in [W/cm K].

of the interfaces to reduce and/or remove the hole/electron barriers [50, 51]. Most grading schemes involve relatively high levels of doping or charge accumulation at the interfaces [51–53]. However, the interface and bulk doping level have to be carefully selected, because the presence of carriers produces carrier-related material absorption.

The thermal properties of epitaxial mirrors vary with the alloy composition. Because of alloy scattering [54], ternary and quaternary alloys exhibit substantially lower thermal conductivities than the binary alloys (even in binary-alloy layers, a reduction of the thermal conductivity is observed because of interface scattering [55]). This somewhat diminishes the attractiveness of epitaxial mirrors that employ ternary and quaternary alloys, but some of the difficulties may be alleviated using top-down mounting and regrowth to provide the necessary thermal conductance.

There are several material choices for the fabrication of quarter-wave mirrors for the realization of long-wavelength vertical-cavity lasers. The deciding factor on what material combinations can be used is the wavelength of the fundamental absorption and the range of the refractive index that can be realized in the transparent regime. The primary compound semiconductor alloys that are used for 1300 nm and 1550 nm quarter-wave mirrors are grown lattice-matched to GaAs or InP. Two important properties of the semiconductor refractive index are that all semiconductors exhibit a refractive index *decrease* below the fundamental absorption edge and that the refractive index at a given wavelength in the transparent regime generally *reduces* with the larger energy gap of the semiconductor. Therefore, in the epitaxial mirrors, the high-index material has the smaller bandgap of the two.

8.5.2 VCSELs Employing InGaAsP and AlInGaAs Materials

The most common choice for epitaxial quarter-wave mirrors for long-wavelength applications is the InGaAsP/InP system lattice-matched to InP. Numerous researchers have investigated this system for fabrication of mirrors [56–59] and vertical-cavity lasers [8, 45, 60–69]. The main difficulty in realizing these mirrors is their thickness (often more than 10 μm) and their susceptibility to reduction in reflectivity due to the presence of optical loss. Nevertheless, with the progress in MOVPE growth techniques, devices with improved performance are being fabricated. The lowest pulsed threshold current density measured for a device with one InGaAsP/InP and one a–Si/SiO$_2$ mirror is 13 kA/cm^2 reported by Streubel et al. [68]. The structure and the SEM micrograph of this device are shown in Figures 8.7(a) and 8.7(b). The structure consists of an OMVPE-grown 50-period InGaAsP/InP quarter-wave mirror topped with a multi-quantum-well active layer. The active layer consists of nine strain-compensated InGaAsP quantum-well regions grown under a constant As/P ratio [70]. The light output is taken from the 5-period a–Si/SiO$_2$ quarter-wave mirror. Probably the most significant fact exhibited by this laser is that pulsed operation has been demonstrated over a 200°C range (between −160 and 42°C). This is a very encouraging result that implies that with suitable mode-to-gain-peak offset and low thermal resistance, the full temperature range required for optical communications (<85°C) may be obtained using this type of device. The improvements to thermal and diffraction properties of devices with bottom InGaAsP/InP mirrors can be realized by creating a vertical waveguide surrounded by thermally conductive (lower index) InP. This requires etching tall and vertical pillars in InGaAsP/InP mirrors [57, 71] and then regrowing InP selectively around these mesas [72]. A cross-sectional SEM micrograph of such regrown VCSEL structure is shown in

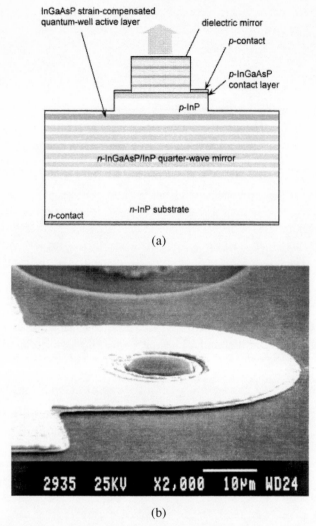

(a)

(b)

Figure 8.7. VCSEL structure with one amorphous-dielectric mirror and one InGaAsP/InP mirror reported by Streubel et al. [68]. (b) SEM micrograph of fabricated device shown in (a).

Figure 8.8 [73]. The a–Si/SiO$_2$ mirror is used as the regrowth mask, while two micrographs show that the planarization is slightly dependent on the crystallographic orientation.

The AlInGaAs/AlInAs and AlInGaAs/InP systems have very similar refractive index differences [74] (the latter having a slightly larger refractive index ratio because the refractive index of InP is ∼0.05 lower than that of AlInAs at 1550 nm). To date a number of researchers have reported such mirrors (AlInAs: [75, 76], AlInGaAs/InP: [77, 78]), but there have not been any reports of vertical-cavity lasers fabricated using this material combination.

In all aforementioned quaternary materials, the bandgap is a continuous function of the composition ending with InGaAs at the long-wavelength side (Figure 8.1). An optimum composition exists for the narrow gap choice: One desires the highest refractive index (smaller gap) with lowest absorption coefficient (larger gap). The smallest bandgap of the

Figure 8.8. Cross-sectional SEM micrograph through regrown 1550 nm VCSEL with one amorphous-dielectric and one InGaAsP/InP mirror [73].

InGaAsP alloy used for mirrors at 1550 nm is in the neighborhood of 0.87 eV [8, 57, 63, 67]. The normalized refractive index difference between this composition and InP is $\Delta n / \bar{n} \approx 8.5\%$ and the mirrors require over 80 layers to achieve reflectivities over 99.5%. At 1300 nm, the smallest bandgap of the quaternary alloy should be in the neighborhood of 1.07 eV, which results in an even smaller refractive index difference ($\Delta n / \bar{n} \approx 5\%$). There have been very few reports of using InGaAsP/InP mirrors at 1.3 μm [58].

Practically all mirrors employing InGaAsP and AlInGaAs materials reported are of the n-type, because of lower optical losses and better conductance: n-type InGaAsP/InP mirrors exhibit low resistance and high reflectivity [57, 68] that are sufficient for fabrication of VCSELs, while there are very few reports of (and no lasers) using p-type InGaAsP/InP mirrors [79]. In n-type epitaxial mirrors, the absorption for energies above the bandgap can also be reduced [80–82] by band-filling using degenerate n-type doping (Burstein shift) [83].

8.5.3 Other Material Combinations

There are other material combinations lattice-matched to InP that promise a large refractive index ratio and can be used for fabricating long-wavelength quarter-wave mirrors [49]. They involve mixed group V elements, such as As/Sb and P/Sb alloys. As listed in Tables 8.2A and 8.2B, these materials can be used to obtain refractive index ratios that are superior to the InGaAsP/InP and AlInGaAs/AlInAs systems. The first reported attempt to grow In-GaAsP/AlAsSb mirrors was reported by Tai et al. [84]. In recent years, the AlAsSb/GaAsSb system lattice-matched to InP has been more thoroughly investigated by Blum et al. [85], who recently demonstrated the photo-pumped operation of a 1,560 nm VCSEL in this material system [86]. For application at 1300 and 1550 nm, one must employ the quaternary AlGaAsSb alloy to bring the absorption edge above the photon energy [87, 88]. Furthermore, the large valence-band offset between AlAsSb and GaAsSb requires bandgap engineering to reduce the electrical resistance of these mirrors [86]. A similar case is that of AlPSb/GaPSb grown by gas-source molecular beam epitaxy, which is also expected to provide a very large refractive index ratio [89]. Long-wavelength AlGaSb/AlSb mirrors (matched to GaSb) have also been reported [90, 91]. However, there have not been any reports of electrically-pumped room-temperature operating vertical-cavity lasers fabricated using any of the antimony-containing alloys.

The outstanding performance of present-day GaAs-based VCSELs is largely a result of the fact that, in the AlGaAs material system, both high-quality active layers and high-reflectivity mirrors can be grown on GaAs in a single epitaxial growth. The AlGaAs alloy on GaAs offers a large range of refractive index values with low extrinsic material losses at <1 μm wavelengths, resulting in high reflectivity mirrors of both polarities. In addition, the AlAs/GaAs alloy combination exhibits the highest thermal conductance of all epitaxial mirrors in this wavelength range. For these reasons, quarter-wave mirrors fabricated in AlGaAs material system would be a good choice for use in long-wavelength devices as well. The fractional refractive index difference between Al(Ga)As/GaAs system is still quite large at 1300 and 1550 nm ($\Delta n/\bar{n} \approx 15\%$) [47]. Unfortunately, because of a large lattice mismatch, long-wavelength InGaAsP or AlInGaAs active-layer materials cannot be incorporated into a GaAs-lattice-matched cavity. This difficulty may be avoided by developing materials that emit at 1300 and 1550 nm while grown directly on GaAs. Self-assembled InGaAs quantum dots [92, 93], and strained InGaNAs active layers grown on GaAs [26] are examples of this approach.

The GaAsN alloy features an interesting peculiarity in the bandgap-lattice constant diagram. Generally, semiconductor-energy bandgap increases with the decreasing lattice constant (Figure 8.1). However, for small amounts of nitrogen, the bandgap of the GaAsN alloy actually decreases, due to a large bowing parameter [26]. As GaAsN grows tensile on GaAs, indium is used to reduce the strain while at the same time lowering the emission energy. In this way, InGaAsN quaternary active layers with emission wavelengths longer than were ever possible with just InGaAs grown on GaAs. The development of this material is centered around determining the growth conditions and corrects amounts of nitrogen to extend the wavelength to 1300 nm or longer: Even though only minute amounts of nitrogen ($\sim 1\%$) are necessary, the optical quality decreases with the introduction of increasing amounts of nitrogen. Interest in this material system has been rapidly growing in recent years, owing to successive demonstrations of edge-emitting lasers with ever increasing wavelength. Room-temperature continuous-wave operation of an InGaAsN ridge laser operating at 1300 nm has been recently demonstrated [94]. The longest room-temperature wavelength reported for electrically pumped InGaNAs VCSELs is 1,220 nm [95]. It is important to note that the compatibility of the InGaAsN active layer with AlGaAs-based mirrors and the high-temperature performance of the InGaAsN strained-quantum-well active layers make this material system highly attractive for the fabrication of 1300 nm vertical-cavity lasers in the future.

Several methods have been investigated in an attempt to evade the lattice-mismatch between GaAs and long-wavelength emitting active materials. Compressively strained InGaAs active layers ($x_{In} \approx 0.4$) on ternary substrates of InGaAs ($x_{In} \approx 0.2$) can be used in 1300 nm VCSELs. The fractional refractive index difference of such AlInAs/InGaAs mirrors is $\Delta n/\bar{n} \approx 13\%$ [96], which is close to that of AlGaAs/GaAs mirrors. In addition, the electron confinement in this material system is expected to improve laser performance at elevated temperatures [25]. These devices involve the development of ternary InGaAs substrates [97] or buffer layers that increase the lattice constant slightly beyond GaAs [96, 98]. The recently demonstrated growth of crystalline InGaAs ($x_{In} \approx 0.2$) on compliant substrates [99] may offer the same possibility for fabricating 1300 nm and 1550 nm lasers on GaAs wafers. Finally, the lattice-matching requirement can be completely eliminated using the technique of *fusion bonding* in which InGaAsP or AlInGaAs active layers are bonded to GaAs-based mirrors. This technique has been used very successfully in recent years to fabricate both 1300 nm and 1550 nm vertical-cavity lasers [9, 19, 69, 100, 101]. In the next section we discuss the fused VCSELs in more detail.

8.6 Fusion Bonding and Devices

8.6.1 Single-Fused Vertical-Cavity Lasers

While the use of InGaAsP/InP mirrors has been almost exclusively applied to 1550 nm VCSELs, alternative solutions have been investigated for fabrication of lasers at 1300 nm with epitaxial mirrors. The need for thermally and electrically conductive mirrors has led to the application of fusion-bonding technology to vertical-cavity lasers. The technique of fusion bonding is technologically similar to silicon-direct bonding [102, 103], but results in fundamentally different bond properties [104–106]. This technique enables the fabrication of optoelectronic (and electronic) devices by bonding of two semiconductors with vastly different lattice constants. Since the fusion-bonded junction between the semiconductors is electrically and thermally conductive, as well as optically transparent, it is ideally suited for realization of many novel surface-normal optoelectronic devices in addition to VCSELs: resonant-cavity photodetectors [107], avalanche photodetectors [108], bistable-optical switches [109], and light-emitting diodes [110, 111]. The application of this technology to vertical-cavity lasers was pioneered by Dudley et al. [100] who used a single fused interface to integrate a GaAs/AlAs mirror with an InGaAsP active region and demonstrated the first fused 1300 nm vertical-cavity laser operating under room-temperature pulsed conditions. The structure of this device is shown in Figure 8.9. The light output was taken from the top 3-period Si/SiO$_2$ mirror, while the back high-reflector was realized using a 28-period n-type AlAs/GaAs mirror. The active layer was realized using a 300 nm thick bulk InGaAsP layer grown by OMVPE. This laser demonstrated the record-low room-temperature pulsed threshold current of 9 mA current density of 9.6 kA/cm^2. The optical losses in this cavity are limited by free-carrier and intervalence-band absorption in the active layer and the insufficient reflectivity of the top mirror (a–Si/SiO$_2$). Further improvement on the single-fused device structure has been demonstrated at 1550 nm by bonding AlAs/GaAs mirrors to InGaAsP/InP mirrors and active layers [69]. This device features a hybrid InGaAsP/InP–Si/SiO$_2$ mirror operated under room-temperature continuous-wave conditions with a 8 mA threshold current and highest operating temperature of 27°C.

The state-of-the-art single-fused device operating at 1300 nm uses an AlInGaAs active layer bonded to a p-AlAs/GaAs mirror [19]. The cross-sectional SEM micrograph of this device is shown in Figure 8.10(a). There are several noteworthy features of this device: The active layer uses strain-compensated AlInGaAs quantum wells, which exhibit better carrier confinement than the InGaAsP system [6, 112], and the current constriction is realized by oxygen implantation performed before fusion bonding. In this way the insulating region is embedded within the fused junction and there is no need to implant through the active layer. The lowest room-temperature (20°C) continuous-wave threshold current of this fused

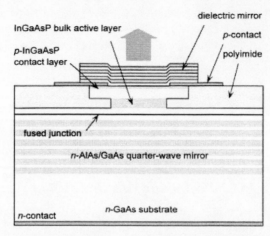

Figure 8.9. The structure of the first vertical-cavity laser fabricated by fusion bonding [100]. This device operated at 1300 nm under room-temperature pulsed conditions.

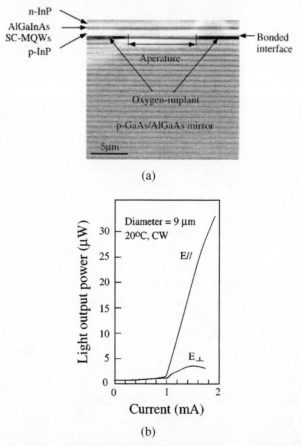

(a)

(b)

Figure 8.10. (a) Cross-sectional SEM micrograph of a single-fused device [19] showing fused p-type AlAs/GaAs mirror and current constriction realized by oxygen implantation. (b) The polarization-resolved continuous-wave 20°C light–current characteristics of a 9 μm diameter device shown in (a).

VCSEL was 1 mA measured on a 9 μm diameter device, which corresponds to 1.57 kA/cm^2. These devices operate continuously up to 40°C. The polarization-resolved light–current characteristic of this device is shown in Figure 8.10(b). The top amorphous-dielectric mirror was realized using ZnSe/MgF quarter-wave layers.

8.6.2 Current- and Mode-Constriction Schemes

Since the amorphous mirrors are insulating, both etched-well VCSELs and devices with one amorphous mirror use ring contacts and various current-constriction schemes to funnel the current into the active layer. In this way the overlap between the current-injected region and the optical mode is maximized. The transverse modes of VCSEL cavities may be defined by the current-constricting scheme or the mirror, depending on which of the two provides a stronger guiding mechanism [113]. In any case, the overlap between the ring contacts and the injected active layer is always minimized, because alloyed contacts do not produce sufficient reflectivity for VCSEL cavities. When the threshold current density is high, current crowding appears near the contacts and at the edges of the active layers [114]. The detrimental effects of nonuniform current injection are higher power dissipation and the reduced overlap between the optical mode and the gain profile. In addition, nonuniform injection is a highly nonlinear process in which the nonuniformity increases with the magnitude of the current. Therefore, the main approach in eliminating nonuniform current injection is to fabricate cavities with low operating current densities. This effect has also been observed in GaAs-based VCSELs, where injection-leveling layers have been investigated for operation at higher current levels [115].

A number of current constricting schemes have been employed in both GaAs- and InP-based VCSELs. The two most widely used schemes used in GaAs-VCSEL technology involve proton implantation and lateral AlGaAs oxidation to provide insulating regions. Indium-phosphide lattice-matched materials do not lend themselves to efficient current constriction to proton bombardment because of the ease of damage and passivation annealing at low temperatures (as low as room temperature). The creation of insulating layers by lateral oxidation may in the future be performed by lateral oxidation of AlInAs [116] or AlAsSb [117] layers. The regrowth technology is more common in InP-matched materials, and current and mode constriction has been very efficiently realized by regrowing insulating/blocking layers around the active layer mesa [37, 118] (as described in Section 8.6.1). The lateral etching of the active layer has also been used but may pose future reliability problems if the active layer is left exposed as in Ref. [41]. The problem may be alleviated by using a separate layer to undercut, as described in Ref. [43].

Wafer-fused VCSELs offer the possibility of using the current-constriction schemes already developed for GaAs-based VCSELs, namely proton implantation and AlGaAs lateral oxide confinement. In addition, a great feature of fusion-bonding technology is that it allows one to bury patterned insulating (or conducting) layers in the fused junction to produce either a current constriction or a current path (Section 8.6.1). The key issue in selecting the implantation species is that the damage and/or the impurity passivation must remain after the high-temperature fusion cycle. Since the insulating properties of proton-implanted GaAs, commonly used in VCSELs, degrade at temperatures as low as 350°C, it is necessary to use heavier atoms that produce insulating or blocking layers that remain after thermal treatments as high as 500–650°C. Blocking and conductive layers can also be realized by embedding implanted or diffused doped layers in the fused junction, whereas the impurities may be redistributed by simultaneous fusion and diffusion. Buried voids that contain insulators or high-temperature metals can be used to produce index-guiding and contacts for future VCSELs. Such novel devices have not yet been explored.

8.6.3 All-Epitaxial VCSEL Structures

Nonuniform injection resulting from ring contacts, poor thermal conductivity and optical absorption in amorphous materials used for mirrors have driven the investigation of device structures in which both mirrors were thermally and electrically conductive. This was pursued in an attempt to mimic the GaAs-based all-epitaxial VCSEL structure (shown for example in ref. 119). As we pointed out in previous sections, for long-wavelength applications, to present date, no material combination has been able to realize a cavity with two epitaxial mirrors in a single growth. This task may in the future become feasible by using any of the techniques described in Section 8.5.3, but presently all-epitaxial long-wavelength VCSELs are exclusively realized by using fusion bonding, and currently the double-fused vertical-cavity laser structure [120] has been among the most successful in achieving pulsed and continuous-wave operation at elevated temperatures [9, 101, 121]. In this structure, the InGaAsP (or AlInGaAs) active layers are sandwiched between two Al(Ga)As/GaAs mirrors. The double-fused laser structure lends itself to a variety of device structures that have already been investigated for GaAs-based VCSELs since both the substrate and the top epitaxial layers contain the AlGaAs alloy. The first room-temperature continuous-wave operation at 1550 nm was realized using the air–post index-guided structures [9] (shown in Figure 8.11). Many other structures employed in the fabrication of GaAs-based VCSELs can be directly applied to the double-fused vertical-cavity laser structure. This fact makes the double-fused vertical-cavity laser structure very attractive in terms of putting the

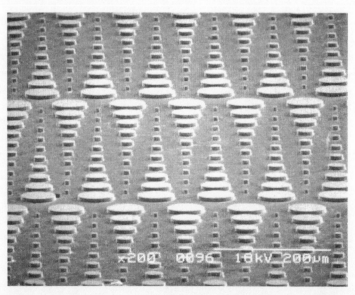

Figure 8.11. SEM micrograph of double-fused air–post index-guided VCSELs [120]. This structure yielded the first above-room-temperature continuous-wave operating long-wavelength VCSELs [9].

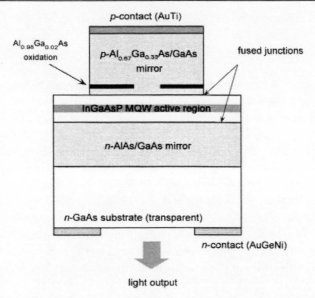

Figure 8.12. Double-fused vertical-cavity laser structure with current constriction realized by lateral oxidation of AlGaAs. Devices reported by Margalit et al. [124].

long-wavelength VCSEL on the same footing as the GaAs-based VCSELs. They both use similar fabrication processes (past wafer fusion), device testing procedures, and packaging.

The state-of-the-art double-fused vertical-cavity laser has been reported by Margalit et al. [121]. This structure, shown in Figure 8.12, utilizes lateral AlGaAs oxidation to produce a current aperture. The top mirror is p-AlGaAs/GaAs with parabolically graded interfaces and seven strain-compensated InGaAsP active layers grown under the constant As/P ratio. The lowest room-temperature (23°) continuous-wave threshold current measured on this device is 0.8 mA, and the highest operating temperature is 64°C. This remarkable result places the fused long-wavelength VCSEL performance alongside the GaAs-based VCSELs and enables practical implementation of these lasers in communication systems.

The availability of room-temperature continuous-wave operating 1550 nm VCSELs has enabled numerous data-transmission experiments. Blixt et al. reported 1 Gbit/s single-mode transmission over 6 km [122] and multimode transmission [123] using air–post index-guided devices reported in ref. 9. The devices fabricated by Margalit et al. [124] have exhibited more output power, and Sheng et al. [125] have reported single-mode transmission over 200 km at 622 Mbit/s. The maximum 3-dB and intrinsic bandwidths on these devices are 4.7 GHz and 24 GHz [124]. A linewidth of 39 MHz and linewidth enhancement factor of 4.0 have been measured on these devices.

8.6.4 Design Issues of Fused VCSELs

The development of fused VCSELs involves the independent design of active region and the Al(Ga)As/GaAs mirrors. There are two primary issues connected with the design of fused lasers: optimizing mirror reflectivity at optical-communication wavelengths and dealing with the standing-wave pattern in a cavity (which involves fused junctions).

As pointed out in Section 8.6.3, the design of epitaxial mirrors requires optimization between the doping level and the mirror resistance because of increased material absorption in extrinsic materials. This has been a challenge for GaAs-based VCSELs in past years, especially with the p-type mirrors [52, 126, 127]. The constraints on the p-type doping level are more stringent at long wavelengths (1300 nm and 1550 nm). Namely, free-carrier absorption and intervalence-band absorption in GaAs both increase dramatically below the bandgap for extrinsic p-type GaAs [128], and, for this reason, the bulk doping of mirror layers used at longer wavelengths has to be lower than what would be used at 980 nm. The p-GaAs absorption coefficient varies approximately linearly with the hole concentration with slopes [12] of $d\alpha/dp = 7$ cm$^{-1}/10^{18}$ cm^{-3} for 980 nm, $d\alpha/dp = 15$ cm$^{-1}/10^{18}$ cm^{-3} for 1300 nm, and $d\alpha/dp = 29$ cm$^{-1}/10^{18}$ cm^{-3} for 1550 nm. It is evident from these data that the absorption in p-GaAs seems to be approximately four times more sensitive to hole concentrations at 1550 nm than at 980 nm. For this reason, the doping levels of long-wavelength GaAs/AlAs mirrors should be reduced with respect to 980 nm devices ($\approx 10^{18}$ cm^{-3}). The absorption in n-type GaAs exhibits a substantially weaker wavelength dependence in this range [129]; therefore the doping level typically used for 980 nm GaAs-based VCSELs is appropriate for long-wavelength applications. In double-fused VCSELs reported in refs. 9, 12, and 124, the average doping level in p-mirrors is $\approx 5.10^{17}$ cm^{-3}. However, lower reflectivity of p-mirrors may be overcome with higher active-layer gain. This is the case with the single-fused device reported by Qian et al. [19] in which the p-type AlAs/GaAs mirrors are doped higher ($\approx 10^{18}$ cm^{-3}).

While the detailed optical properties of fused junctions are still under investigation, the standing-wave pattern of the fused devices has been designed to exhibit a null at the fused junction. There have been a few reports of studies of the electrical properties of fused nn-junctions [130–132]. These reports indicate that the band offset at (001/001) fused junctions is small and produces only small nonlinearities in the nn-junction voltage-current density characteristics. Owing to the lower mobility and larger effective masses of holes, fused pp-junctions are expected to exhibit more nonlinear behavior. To reduce any barriers at the fused junctions, in all double-fused VCSELs, the GaAs fused surfaces were heavily doped. Placing of the fused junction at a null of the standing-wave pattern reduces the free-carrier losses from the high carrier concentrations. Given this phase condition at the edge of the active-layer structure (fused junction), the smallest optical length of the active epilayers can be $\sim\lambda/2$ (~ 250 nm). However, it has been shown by Ram et al. [106] that all dislocations occurring during fusion are localized within 200 nm from the junction. For this reason, the quantum wells have been distanced from both junctions, making the active-epilayer optical length equal to $3\lambda/2$ satisfies this condition. In addition, the fused junctions are located in regions where majority carrier flow dominates. Therefore, no minority carrier recombination is expected at the fused junctions.

The research in the field of fusion bonding still contains many issues that need to be investigated: The electrical, thermal, and optical properties of these junctions are highly dependent on the bonding process conditions and surface preparation. These effects are not completely understood at present day. The most critical unresolved issue in fusion bonding is the reliability. Inasmuch as the active layers of long-wavelength VCSELs are based on InP, the dark-line defect degradation common to the AlGaAs system is not expected to be as pronounced, and this may provide impetus for further development of fused-InP devices.

8.7 Modeling of Long-Wavelength VCSELs

Because VCSELs exhibit a very complex interaction of electronic, optical, and thermal processes, there is considerable interest in advanced VCSEL modeling and simulation to predict and to analyze device performance. There are numerous reports on VCSEL modeling and a good overview is provided in ref. 133. Most of these reports address GaAs-based VCSELs, for which accurate optical modeling has recently become a challenging topic [134]. Motivated by practical device problems, workers simulating long-wavelength VCSELs are so far more concerned with the temperature sensitivity of laser threshold [21,135]. Finite-element analysis of the internal heat dissipation has been applied to long-wavelength VCSELs with various DBR materials to find concepts that promise low self-heating in continuous-wave operation [38, 136–139]. Such investigations pointed toward the need for mirror materials with higher thermal conductivity and for top-down mounting. Wafer-fused devices employing a binary GaAs/AlAs DBR on the side of the heat sink give the lowest thermal resistance [38]. Small active areas in oxide-confined VCSELs help to reduce continuous-wave self-heating considerably [140]. Assuming uniform but variable VCSEL temperature, numerical simulation of pulsed measurements was used to analyze loss mechanisms in fused 1.55 μm VCSELs [21]. This model includes computations of band structure and gain for the strain-compensated multi-quantum well, transfer matrix calculations of the optical field, and two-dimensional finite-element simulation of carrier drift and diffusion. Careful adjustment of material parameters led to agreement with a variety of measurements. Besides spreading-current losses, Auger recombination and intervalence band absorption were found to prevent those VCSELs from high-temperature operation. Main contributions to the temperature sensitivity arise from the offset between the wavelength of maximum gain and the emission wavelength. Simulations predict that VCSELs with intentional negative gain offset give a minimum temperature sensitivity at room temperature [140]. Advanced optical and dynamic VCSEL models are expected to be applied to long-wavelength VCSELs in the near future.

8.8 Conclusion

Long-wavelength vertical-cavity lasers have exhibited tremendous progress since the invention of the vertical-cavity laser in 1979. The progress has been a result of the development in material growth technology, as well as innovation in the device design and fabrication. However, 1300 nm and 1550 nm VCSELs have a long way ahead of them to compete with the performance of present-day 850 nm VCSELs, and especially long-wavelength edge-emitting devices. Nevertheless, the knowledge gained in studying GaAs-based VCSELs continues to greatly contribute to the steeper learning and development curve of long-wavelength VCSELs.

Acknowledgments

The authors would like to thank Near Margalit and Sheng Zhang from U. C. Santa Barbara, California; Dr. Seiji Uchiyama from Furukawa Electric Co., Yokohama, Japan; Dr. Klaus Streubel from Royal Institute of Technology, Stockholm, Sweden; Drs. Yu-Hwa Lo and Yi Qian from Cornell University, New York, for providing photographs and information; and Dr. Lisa Buckman for help with the proofreading of this chapter.

References

[1] M. Orenstein, A. C. Von Lehmen, C. Chang-Hasnain, N. G. Stoffel, J. P. Harbison, and L. T. Florez, "Matrix addressable vertical-cavity surface-emitting laser array," *Electron. Lett.*, **27**(5), 437–438, 1991.

[2] K. Lear, K. D. Choquette, R. P. Schneider, Jr., S. P. Kilcoyne, and K. M. Geib, "Selectively oxidised vertical-cavity surface emitting lasers with 50% power conversion efficiency," *Electron. Lett.*, **31**(3), 208–209, 1995.

[3] D. Smith, "850 nm vertical cavity surface emitting lasers for 622 Mbps multimode fiber interface," Presentation to ATM Forum PHY Working Group, Document 95-420, 1995.

[4] H. Soda, K. Iga, C. Kitahara, and Y. Suematsu, "GaInAsP/InP Surface Emitting Injection Lasers," *Jpn. J. Appl. Phys.*, **18**(12), 2329–2330, 1979.

[5] G. P. Agrawal and N. K. Dutta, *Semiconductor Lasers*, Van Nostrand Reinhold, New York, 1993.

[6] C. E. Zah, R. Bhat, B. N. Pathak, and F. Favire, "High-performance uncooled 1.3-μm $Al_X Ga_y In_{1-x} As_{1-y}$/InP strained-layer quantum-well lasers for subscriber loop applications," *IEEE J. Quantum Electron.*, **30**(2), 511–523, 1994.

[7] K. Iga and F. Koyama, "Vertical-cavity surface-emitting lasers and arrays," in *Surface-Emitting Semiconductor Lasers and Arrays*, G. A. Evans and J. M. Hammer, eds., pp. 71–117, Academic Press, San Diego, 1993.

[8] T. Tadokoro, T., H. Okamoto, Y. Kohama, T. Kawakami, and T. Kurokawa, "Room temperature pulsed operation of 1.5 μm GaInAsP/InP vertical cavity surface emitting lasers," *IEEE Photon. Technol. Lett.*, **4**(5), 409–411, 1992.

[9] D. I. Babić, K. Streubel, R. P. Mirin, N. M. Margalit, J. E. Bowers, E. L. Hu, D. E. Mars, L. Yang, and K. Carey, "Room-temperature continuous-wave operation of 1.54 μm vertical-cavity lasers," *IEEE Photon. Technol. Lett.*, **7**(11), 1025–1027, 1995.

[10] L. A. Coldren and S. W. Corzine, *Diode Lasers and Photonic Integrated Circuits*, Wiley, New York, 1995.

[11] K. Tai, F. S. Choa, W. T. Tsang, S. N. G. Chu, J. D. Wynn, and A. M. Sergent, "Room temperature photopumped 1.5 μm quantum-well surface-emitting lasers with InGaAsP/InP distributed Bragg reflectors," *Electron. Lett.*, **27**(17), 1540–1542, 1991.

[12] D. I. Babić, J. Piprek, K. Streubel, R. P. Mirin, N. M. Margalit, D. E. Mars, J. E. Bowers, and E. L. Hu, "Design and analysis of double-fused 1.55-μm vertical-cavity lasers," *IEEE J. Quantum Electron.*, **33**(8), 1369–1383, 1997.

[13] J. W. Matthews and A. E. Blakeslee, "Defects in epitaxial multilayers," *J. Crystal Growth*, **27** 118–125, 1974.

[14] B. I. Miller, U. Koren, M. G. Young, and M. D. Chien, "Strain-compensated strained-layer superlattices for 1.5 μm wavelength lasers," *Appl. Phys. Lett.*, **58**(18), 1952–1954, 1991.

[15] C. H. Lin, C. L. Chua, Z. H. Zhu, F. E. Ejeckam, T. C. Wu, Y. H. Lo, and R. Bhat, "Photopumped long-wavelength vertical-cavity surface-emitting lasers using strain-compensated multiple quantum-wells," *Appl. Phys. Lett.*, **64**(25), 3395–3397, 1994.

[16] C. L. Chua, C. H. Lin, Z. H. Zhu, Y. H. Lo, M. Hong, J. P. Mannearts, and R. Bhat, "Dielectrically-bonded long wavelength vertical cavity laser using strain-compensated multiple quantum wells," *IEEE Photon. Technol. Lett.*, **6**(12), 1400–1402, 1994.

[17] C. L. Chua, Z. H. Zhu, Y. H. Zhu, Y. H. Lo, R. Bhat, and M. Hong, "Low-threshold 1.57-μm VCSEL's using strain-compensated quantum wells and oxide/metal backmirror," *IEEE Photon. Technol. Lett.*, **7**(5), 444–445, 1995.

[18] S. Uchiyama, N. Yokouchi, T. Ninomiya, "Continuous-wave operation up to 36°C of 1.3-μm GaInAsP/InP strained-layer multiquantum-wells surface-emitting laser," *IEEE Photon. Technol. Lett.*, **9**(1), 8–10, 1997.

[19] Y. Qian, Z. H. Zhu, Y. H. Lo, H. Q. Hou, B. E. Hammons, D. L. Huffaker, D. G. Deppe, W. Lin, and Y. K. Tu, "Low threshold room-temperature CW 1.3 μm single-bonded vertical-cavity surface-emitting lasers using oxygen-implanted confinement," *Proc. 1997 Conf. Optical Fiber Communications*, Dallas, Texas, post-deadline PD14-1.

[20] S. Uchiyama and T. Ninomiya, "1.3-μm GaInAsP/InP multi-quantum-well surface-emitting lasers," *Optical Rev.*, **3**(2), 59–61, 1996.

[21] J. Piprek, D. I. Babić, and J. E. Bowers, "Simulation and analysis of 1.55 μm double-fused vertical-cavity lasers," *J. Appl. Phys.*, **81**(8), 3382–3390, 1997.

[22] S. Seki, H. Oohashi, H. Sugiura, T. Hirono, and K. Yokoyama, "Study of the dominant mechanism for the temperature sensitivity of threshold current in 1.3-μm InP-based strained-layer quantum-well lasers," *IEEE J. Quantum Electron.*, **32**(8), 1478–1485, 1996.

[23] I. Joindot and J. L. Beylat, "Intervalence band absorption coefficient measurements in bulk layer, strained and unstrained multiquantum well 1.55 μm semiconductor lasers," *Electron. Lett.*, **29**(7), 604–606, 1993.

[24] N. Yokouchi, T. Uchida, T. Miyamoto, Y. Inaba, F. Koyama, and K. Iga, "An optical absorption property of highly beryllium-doped GaInAsP grown by chemical beam epitaxy," *Jpn. J. Appl. Phys. Part 1*, **31**(5A), 1255–1257, 1992.

[25] H. Ishikawa and I. Suemune, "Analysis of temperature dependent optical gain of strained quantum well taking account of carriers in the SCH layer," *IEEE Photon. Technol. Lett.*, **6**(3), 344–346, 1994.

[26] M. Kondow, K. Uomi, A. Niwa, T. Kitatani, S. Watahiki, and Y. Yazawa, "GaInNAs: A novel material for long-wavelength-range laser diodes with excellent high-temperature performance," *Jpn. J. Appl. Phys. Part 1*, **35**(2B), 1273–1275, 1996.

[27] P. L. Gourley, L. R. Dawson, T. M. Brennan, B. E. Hammons, J. C. Stover, C. F. Schaus, and S. Sun, "Optical scatter in epitaxial semiconductor multilayers," *Appl. Phys. Lett.*, **58**(13), 1360–1362, 1991.

[28] J. C. Stover, *Optical Scattering*, McGraw-Hill, New York, 1990.

[29] O. Arnon, "Loss mechanisms in dielectric optical interference devices," *Appl. Opt.*, **16**(8), 2147–2151, 1977.

[30] H. A. MacLeod, *Thin-Film Optical Filters*, Adam Hilger Ltd., Bristol, 1986.

[31] M. Oshikiri, F. Koyama, and K. Iga, "Flat surface circular buried heterostructure surface-emitting laser with highly reflective Si/SiO$_2$ mirrors," *Electron. Lett.*, **27**(22), 2038–2039, 1991.

[32] F. Stern, "Band-tail model for optical absorption and for the mobility edge in amorphous silicon," *Phys. Rev. B*, **3**(8), 2636–2645, 1971.

[33] D. I. Babić, T. E. Reynolds, E. L. Hu, and J. E. Bowers, "In situ characterization of thin film optical coatings using a normal incidence laser reflectometer," *J. Vac. Sci. Technol. A*, **10**(2), 514–524, 1992.

[34] D. G. Cahill, S. K. Watson, and R. O. Pohl, "Lower limit to the thermal conductivity of disordered crystals," *Phys. Rev. B*, **46**(10), 6131–6140, 1992.

[35] H. J. Goldsmid, M. M. Kaila, and G. L. Paul, "Thermal conductivity of amorphous silicon," *Phys. Status Solidi A*, **76**, K31, 1983.

[36] H. Tanobe, M. Oshikiri, M. Araki, F. Koyama, and K. Iga, "A preliminary study on MgO/Si multilayer reflectors for improving thermal conductance in surface-emitting lasers," *Proc. IEEE Lasers Electro-Optics Soc. Annu. Mtg.*, Boston, MA, Paper DLTA12.2, 1992.

[37] T. Baba, Y. Yogo, K. Suzuki, F. Koyama, and K. Iga, "Near room-temperature continuous wave lasing characteristics of GaInAsP/InP surface-emitting laser," *Electron. Lett.*, **29**(10), 913–914, 1993.

[38] J. Piprek, "Heat flow analysis of long-wavelength VCSELs with varius DBR materials," Proc. 1994 *IEEE/LEOS Ann. Mtg.*, 286–287, Boston, MA, 1994.

[39] S. Uchiyama and S. Kashiwa, "GaInAsP/InP square buried-heterostructure surface-emitting lasers regrown by MOCVD," *IEICE Trans. on Electronics*, **E78-C**(9), 1311–1314, 1995.

[40] L. Yang, M. C. Wu, K. Tai, T. Tanbun-Ek, and R. A. Logan, "InGaAsP(1.3 μm)/InP vertical-cavity surface-emitting laser grown by metalorganic vapor phase epitaxy," *Appl. Phys. Lett.*, **56**(10), 889–891, 1990.

[41] H. Wada, D. I. Babic, D. L. Crawford, T. E. Reynolds, J. J. Dudley, J. E. Bowers, E. L. Hu, J. L. Merz, B. I. Miller, U. Koren, and M. G. Young, "Low-Threshold, High-Temperature Pulsed Operation of InGaAsP/InP Vertical Cavity Surface Emitting Lasers," *IEEE Phot. Technol. Lett.*, **3**(11), 977–979, 1991.

[42] A. Plais, P. Salet, C. Starck, A. Pinquier, E. Derouin, C. Fortin, J. Jacquet, and F. Brillouet,

"Thermal behavior of 1.3 μm vertical-cavity surface-emitting laser," *Proc. Int. Conf. Indium Phosphide and Related Materials*, Paper ThA 1-3, p. 723, 1996.

[43] P. Salet, A. Plais, E. Derouin, C. Fortin, C. Starck, H. Bissessur, J. Jacquet, and F. Brillouet, "Undercut ridge structures: a novel approach to 1.3/1.5 μm Vertical-Cavity lasers designed for continuous-wave operation," *Special Issue of IEEE J. Quantum Electron.*, 1997.

[44] D. G. Deppe, S. Singh, R. D. Dupuis, N. D. Gerrard, G. J. Zydzik, J. P. van der Ziel, C. A. Green, and C. J. Pinzone, "Room-temperature photopumped operation of an InGaAsP/InP vertical-cavity surface-emitting laser," *Appl. Phys. Lett.*, **56**(22), 2172–2174, 1990.

[45] T. Uchida, T. Miyamoto, N. Yokoushi, Y. Inaba, F. Koyama, and K. Iga, "CBE grown 1.5 μm GaAsInAs-InP surface emitting lasers," *IEEE J. Quantum Electron.*, **29**(6), 1975–1980, 1993.

[46] K. Uomi, S. J. B. Yoo, A. Scherer, R. Bhat, N. C. Andreadakis, C. E. Zah, M. A. Koza, and T. P. Lee, "Low threshold, room temperature pulsed operation of 1.5 μm vertical-cavity surface-emitting lasers with an optimized multi-quantum well active layer," *IEEE Phot. Technol. Lett.*, **6**(3), 317–319, 1994.

[47] M. A. Afromovitz, "Refractive index of $Ga_{1-x}Al_xAs$," *Solid-State Comm.*, **15**, 59–63, 1974.

[48] S. Adachi, "Refractive indices of III-V compounds: Key properties of InGaAsP relevant to device design," *J. Appl. Phys.*, **53**(8), 5863–5869, 1982.

[49] M. Guden and J. Piprek, "Material parameters of quaternary III-V semiconductors for multilayer mirrors at 1.55-μm wavelength," *Modeling and Simulation in Materials Science and Engineering*, **4**(4), 349–357, 1996.

[50] J. R. Hayes, F. Capasso, R. J. Malik, A. C. Gossard, and W. Wiegmann, "Optimium emitter grading for heterojunction bipolar transistors," *Appl. Phys. Lett.*, **43**(10), 949–951, 1983.

[51] M. G. Peters, D. B. Young, F. H. Peters, B. J. Thibeault, J. W. Scott, S. W. Corzine, R. W. Herrick, and L. A. Coldren, "High wall-plug efficiency temperature-insensitive vertical-cavity lasers with low-barrier p-type mirrors," *Proc. SPIE*, **2147**, 1–11, 1994.

[52] K. L. Lear and R. P. Schneider, Jr., "Uniparabolic mirror grading for vertical-cavity surface-emitting lasers," *Appl. Phys. Lett.*, **68**(5), 605–607, 1996.

[53] K. Tai, L. Yang, Y. H. Wang, J. D. Wynn, and A. Y. Cho, "Drastic reduction of series resistance in doped semiconductor distributed Bragg reflectors for surface emitting lasers," *Appl. Phys. Lett.*, **56**, 2496–2498, 1990.

[54] S. Adachi, "Lattice thermal resistivity of III-V compound alloys," *J. Appl. Phys.*, **54**(4), 1844–1848, 1983.

[55] G. Chen, C. L. Tien, X. Wu, and J. S. Smith, "Thermal diffusivity measurement of GaAs/AlGaAs thin-film structures," *Journal of Heat Transfer*, **116**, 325–331, 1994.

[56] K. Tai, S. L. McCall, S. N. G. Chu, and W. T. Tsang, "Chemical beam epitaxially grown InP/InGaAsP interference mirror for use near 1.55 μm wavelength," *Appl. Phys. Lett.*, **51**(11), 826–827, 1987.

[57] K. Streubel, J. Wallin, L. Zhu, G. Landgren, and I. Queisser, "High-reflective 1.5 μm GaInAsP/InP Bragg reflectors grown by metal organic vapor phase epitaxy," *Mat. Sci. Eng.*, **B28**, 285–288, 1994.

[58] P. Salet, C. Starck, A. Plais, J.-L. Lafragette, F. Gaborit, E. Derouin, F. Brillouet, and J. Jacquet, "High reflectivity semiconductor mirrors for 1.3 μm surface emitting lasers," *Proc. Conf. Lasers and Electro-Optics*, Paper CThK39, pp. 421–422, 1996.

[59] F. S. Choa, K. Tai, W. T. Tsang, and S. N. G. Chu, "High reflectivity 1.55 μm InP/InGaAsP Bragg mirror grown by chemical beam epitaxy," *Appl. Phys. Lett.*, **59**(22), 2820–2822, 1991.

[60] A. Chailertvanitkul, K. Iga, and K. Moriki, "GaInAsP/InP surface-emitting laser ($\lambda = 1.4 \mu$m) with heteromultilayer Bragg reflector," *Electron. Lett.*, **21**, 303–304, 1985.

[61] Y. Imajo, A. Kasukawa, S. Kashiwa, and H. Okamoto, "GaInAsP/InP semiconductor multilayer reflector grown by metalorganic chemical vapor deposition and its application to surface emitting laser diodes," *Jpn. J. Appl. Phys.*, **29**(7), L1130–L1132, 1990.

[62] T. Miyamoto, T. Uchida, N. Yokouchi, Y. Inaba, F. Koyama, and K. Iga, "A study on

gain-resonance matching of CBE grown $\lambda = 1.5 \, \mu$m surface-emitting lasers," *Proc. IEEE Lasers and Electro-Optics Soc. Annu. Mtg.*, Paper DLTA13.2, p. 542, 1992.

[63] M. A. Fisher, A. J. Dann, D. A. O. Davies, D. J. Elton, M. J. Harlow, C. B. Hatch, S. D. Perrin, J. Reed, I. Reid, and M. J. Adams, "High temperature photopumping of 1.55 μm vertical-cavity surface emitting lasers," *Electron. Lett.*, **29**(17), 1548–1549, 1993.

[64] K. Streubel, J. André, J. Wallin, and G. Landgren, "Fabrication of 1.5 μm optically pumped $Ga_{1-x}In_xAs_yP_{1-y}$/InP vertical-cavity surface-emitting lasers," *Mat. Sci. Eng.*, **B28**, 289–292, 1994.

[65] D. I. Babić, J. J. Dudley, K. Streubel, R. P. Mirin, E. L. Hu, and J. E. Bowers, "Optically-pumped all-epitaxial wafer-fused 1.52 μm vertical-cavity lasers," *Electron. Lett.*, **30**(9), 704–706, 1994.

[66] M. A. Fisher, Y.-Z. Huang, A. J. Dann, D. J. Elton, M. J. Harlow, S. D. Perrin, J. Reed, I. Reid, and M. J. Adams, "Pulsed electrical operation of 1.5-μm vertical-cavity surface-emitting lasers," *IEEE Photon. Technol. Lett.*, **7**(6), 608–610, 1995.

[67] H. Wada, T. Takamori, and T. Kamijoh, "Room-temperature photo-pumped operation of 1.58 μm vertical-cavity-lasers fabricated on Si substrates using wafer bonding," *IEEE Photon. Technol. Lett.*, **8**(11), 1426–1428, 1996.

[68] K. Streubel, S. Rapp, J. André, and J. Wallin, "Room-temperature pulsed operation of 1.5-μm vertical cavity lasers with an InP-based Bragg mirror," *IEEE Photon. Technol. Lett.*, **8**(9), 1121–1123, 1996.

[69] Y. Ohiso, C. Amano, Y. Itoh, K. Tateno, and T. Tadokoro, "1.55 μm Vertical-cavity surface-emitting lasers with wafer-fused InGaAsP/InP-GaAs/AlAs DBRs," *Electron. Lett.*, **32**(16), 1483–1484, 1996.

[70] A. Mircea, A. Ougazzaden, G. Primot, and C. Kazmierski, "Highly thermally stable, high-performance InGaAsP: InGaAsP multi-quantum-well structures for optical devices by atmospheric pressure MOVPE," *J. Crystal Growth*, **124**, 737–740, 1992.

[71] J. E. Schramm, D. I. Babić, E. L. Hu, J. E. Bowers, and J. L. Merz, "Anisotropy control in the reactive ion etching of InP using oxygen in methane/hydrogen/argon," *Proc. Conf. Indium Phosphide and Related Materials*, Santa Barbara, Paper WE4, p. 383, 1994.

[72] S. Lourdudos, K. Streubel, and G. Landgren, "Morphological modifications during selective growth of InP around cylindrical and paralelepiped mesas," *J. Mat. Sci. Eng.*, **B28**, 179–182, 1994.

[73] S. Lourdudoss and O. Kjebon, "Hydride vapor phase epitaxy revisited," *IEEE J. of Selected Topics in Quantum Electronics*, **3**(3), 749–767, 1997.

[74] M. J. Mondry, D. I. Babić, J. E. Bowers, and L. A. Coldren, "Refractive indexes of (Al, Ga, In)As epilayers on InP for optoelectronic applications," *IEEE Photon. Technol. Lett.*, **4**(6), 627–630, 1992.

[75] W. Kowalsky and J. Mähnss, "Monolithically integrated InGaAlAs dielectric reflectors for vertical-cavity optoelectronic devices," *Appl. Phys. Lett.*, **59**(9), 1011–1012, 1991.

[76] P. Guy, K. Woodbridge, and M. Hopkinson, "High reflectivity and low resistance 1.55 μm $Al_{0.65}In_{0.35}As$/$Ga_{0.63}In_{0.37}As$ strained quarter-wave Bragg reflector stack," *Electron. Lett.* **29**(22), 1947–1948, 1993.

[76] S.-W. Choi and H.-M. Park, "Highly reflective 1.55 μm $In_{1-x-y}Ga_xAl_y$/$AsIn_{1-z}Al_zAs$ quaternary Bragg mirrors grown by metalorganic chemical vapor deposition (MOCVD)," *Jpn. J. Appl. Phys. Part 2*, **36**(6B), L740–L742, 1997.

[77] A. J. Moseley, J. Thompson, D. J. Robbins, and M. Q. Kearley, "High-reflectivity AlGaInAs/InP multilayer mirrors grown by low-pressure MOVPE for application to long-wavelength high-contrast-ratio multi-quantum-well modulators," *Electron. Lett.*, **25**(25), 1717–1718, 1989.

[78] J. Thompson, A. K. Wood, A. J. Moseley, M. Q. Kearley, P. J. Topham, N. Maung, and N. Carr, "The use of InP-based semiconductor reflective stacks for enhanced device performance," *J. Crystal Growth*, **107**, 860–866, 1991.

[79] T. Miyamoto, K. Mori, H. Maekawa, Y. Inaba, et al., "Carrier transport in p-type GaInAsP/InP distributed Bragg reflectors," *Jpn. J. Appl. Phys. Part 1*, **33**(8), 4614–4616, 1994.

[80] P. Guy, K. Woodbridge, S. K. Haywood, and M. Hopkinson, "Highly doped 1.55 μm $Ga_x In_{1-x}$As/InP distributed Bragg reflector stack," *Electron. Lett.*, **30**, 315–317, 1990.

[81] D. G. Deppe, N. D. Gerrard, C. J. Pinzone, R. D. Dupuois, and E. F. Schubert, "Quarter-wave Bragg reflector stack of InP–$In_{0.53}Ga_{0.47}$As for 1.65 μm wavelength," *Appl. Phys. Lett.*, **56**(4), 315–317, 1990.

[82] S. S. Murtaza, R. V. Chelakara, R. D. Dupuios, J. C. Campell, and A. G. Dentai, "Resonant-cavity photodetector operating at 1.55 μm with Burstein-shifted InGaAs/InP reflector," *Appl. Phys. Lett.*, **69**(17), 2462–2464, 1996.

[83] E. Burstein, "Anomalous optical absorption limit in InSb," *Phys. Rev.*, **93**, 632–633, 1954.

[84] K. Tai, R. J. Fisher, A. Y. Cho, and K. F. Huang, "High reflectivity $AlAs_{0.52}Sb_{0.48}$/GaInAs(P) distributed Bragg mirror on InP substrate for 1.3–1.55 μm wavelengths," *Electron. Lett.*, **25**(17), 1159–1160, 1989.

[85] O. Blum, M. J. Hafich, J. F. Klem, K. Lear, et al., "Electrical and optical characterization of AlAsSb/GaAsSb distributed Bragg reflectors for surface emitting lasers," *Appl. Phys. Lett.*, **67**(22), 3233–3235, 1995.

[86] O. Blum, J. F. Klem, K. L. Lear, G. A. Vawter, and S. R. Kurtz, "Optically pumped, monlithic, all-epitaxial 1.56 μm vertical-cavity surface-emitting laser using Sb-based mirrors," *Electron. Lett.*, **33**(22), 1878–1880, 1997.

[87] O. Blum, I. J. Fritz, L. R. Dawson, and T. J. Drummond, "Digital alloy AlAsSb/AlGaAsSb distributed Bragg reflectors lattice matched to InP for 1.3–1.55 μm wavelength range," *Electron. Lett.*, **31**(15), 1247–1248, 1995.

[88] A. Kohl, J. C. Harman, J. L. Oudar, E. V. K. Rao, R. Kuszelewitz, and E. Lugagne Delpon, "AlGaAsSb/AlAsSb microcavity designed for 1.55 μm and grown by molecular beam epitaxy," *Electron. Lett.*, **33**(18), 708–710, 1997.

[89] H. Shimomura, T. Anan, K. Mori, and S. Sugou, "High-reflectance AlPSb/GaPSb distributed Bragg reflector mirrors on InP grown by gas-source molecular beam epitaxy," *Electron. Lett.*, **30**(4), 314–315, 1994.

[90] B. Lambert, Y. Toudic, Y. Rouillard, M. Baudet, B. Guenais, B. Deveaud, I. Valiente, and J. C. Simon, "High reflectivity 1.55-μm (Al)GaSb/GaSb Bragg mirror grown by molecular beam epitaxy," *Appl. Phys. Lett.*, **64**(6), 690–691, 1994.

[91] G. Tuttle, J. Kavanaugh, and S. McCalmost, "(Al,Ga)Sb long-wavelength distributed Bragg reflectors," *IEEE Photon. Technol. Lett.*, **5**(12), 1376–1378, 1993.

[92] R. P. Mirin, J. P. Ibbetson, K. Nishi, A. C. Gossard, and J. E. Bowers, "1.3 μm photoluminescence from InGaAs quantum dots on GaAs," *Appl. Phys. Lett.*, **67**(25), 3795–3797, 1995.

[93] Y. Tackeuchi, Y. Nakata, S. Muto, Y. Sugiyama, T. Inata, and N. Yokoyama, "Near-1.3-μm high-intensity photoluminescence at room temperature by InAs/GaAs multi-coupled quantum dots," *Jpn. J. Appl. Phys. Part 2*, **34**(4A), L405–L407, 1995.

[94] M. Kondow, T. Kitakani, M. C. Larson, K. Nakahara, K. Uomi, and H. Inoue, "GaInNaAs lasers," *Proc. IEEE LEOS Annu. Mtg.*, p. 325, 1997.

[95] M. C. Larson, M. Kondow, T. Kitatani, Y. Yazawa, and M. Okai, "Room-temperature continuous-wave photopumped operation of 1.22-μm GaInNAs/GaAs single quantum-well vertical-cavity surface-emitting laser," to be published in *Electron. Lett.*, **33**(11), pp. 956–960, 1997.

[96] K. Otsubo, H. Shoji, T. Fujii, M. Matsuda, and H. Ishikawa, "High-reflectivity $In_{0.29}Ga_{0.71}$As/$In_{0.28}Al_{0.72}$As ternary mirrors for 1.3 μm vertical-cavity surface-emitting lasers grown on GaAs," *Jpn. J. Appl. Phys. Part 2*, **34**(2B), L227–L229, 1995.

[97] H. Shoji, K. Otsubo, T. Kusunoki, T. Suzuki, T. Uchida, and H. Ishikawa, "$In_{0.38}Ga_{0.62}$As/InAlGaAs/InGaP strained double quantum well lasers on $In_{0.21}Ga_{0.79}$As ternary substrate," *Jpn. J. Appl. Phys. Part 2*, **35**(6B), L778–L780, 1996.

[98] I. J. Fritz, B. E. Hammons, A. J. Howard, T. M. Howard, T. M. Brennan, and J. A. Olsen, "Fabry–Perot reflectance modulator for 1.3 μm from (InAlGa)As materials grown at low temperature," *Appl. Phys. Lett.*, **62**(9), 919–921, 1993.

[99] F. E. Ejeckam, Y. H. Lo, S. Subramanian, H. Q. Hou, and B. E. Hammons, "Lattice engineered compliant substrate for defect-free heteroepitaxial growth," *Appl. Phys. Lett.*, **70**(13), 1685–1687, 1997.

[100] J. J. Dudley, D. I. Babić, R. P. Mirin, L. Yang, B. I. Miller, R. J. Ram, T. E. Reynolds, E. L. Hu, and J. E. Bowers, "Low threshold, wafer fused long-wavelength vertical-cavity lasers," *Appl. Phys. Lett.*, **64**(12), 1463–1465, 1994.

[101] Y. Qian, Z. H. Zhu, Y. H. Lo, H. Q. Hou, M. C. Wang, and W. Lin, "1.3-μm Vertical-cavity surface-emitting lasers with double-bonded GaAs-AlAs Bragg mirrors," *IEEE Photon. Technol. Lett.*, **9**(1), 8–10, 1997.

[102] J. B. Lasky, "Wafer bonding for silicon-on-insulator technologies," *Appl. Phys. Lett.*, **48**, (78), 1986.

[103] S. Bengtsson, "Semiconductor wafer bonding: A review of interfacial properties and applications," *J. Electron. Mat.*, **21**(8), 841–862, 1992.

[104] Z. L. Liau and D. E. Mull, "Wafer fusion: A novel technique for optoelectronic device fabrication and monolithic integration," *Appl. Phys. Lett.*, **56**(8), 737–739, 1990.

[105] D. I. Babić, J. E. Bowers, and E. L. Hu, "Wafer fusion for surface-normal optoelectronic device applications," *Int. J. of High-Speed Electronics and Systems*, **8**(2), 357–376, 1997.

[106] R. J. Ram, J. J. Dudley, J. E. Bowers, L. Yang, K. Carey, S. J. Rosner, and K. Nauka, "GaAs to InP wafer fusion," *J. Appl. Phys.*, **78**(6), 4227–4237, 1995.

[107] I.-H. Tan, J. J. Dudley, D. I. Babic, D. A. Cohen, B. D. Young, E. L. Hu, J. E. Bowers, B. I. Miller, U. Koren, and M. G. Young, "High quantum efficiency and narrow absorption bandwidth of the wafer-fused resonant cavity $In_{0.53}Ga_{0.47}As$ photodetectors," *IEEE Photon. Technol. Lett.*, **6**(7), 811–813, 1994.

[108] A. R. Hawkins, T. E. Reynolds, D. R. England, D. I. Babic, M. J. Mondry, K. Streubel, and J. E. Bowers, "Silicon hetero-interface photodetector," *Appl. Phys. Lett.*, **68**(26), 3692–3694, 1996.

[109] F. Jeannés, G. Patriarche, R. Azoulay, A. Ougazzaden, J. Landreau, and J. L. Oudar, "Submilliwatt optical bistability in wafer-fused vertical-cavity at 1.55-μm wavelength," *IEEE Photon. Technol. Lett.*, **8**(4), 539–341, 1996.

[110] F. A. Kish, F. M. Steranka, D. C. DeFevere, D. A. Vanderwater, K. G. Park, C. P. Kuo, T. D. Osentowski, M. J. Peanasky, J. G. Yu, R. M. Fletcher, D. A. Steigwald, M. G. Craford, and V. M. Robbins, "Very high-efficiency semiconductor wafer-bonded transparent-substrate $(Al_xGa_{1-x})_{0.5}In_{0.5}P$/GaP light-emitting diodes, *Appl. Phys. Lett.*, **64**(21), 2839–2841, 1994.

[111] G. L. Christenson, A. T. T. D. Tran, Z. H. Zhu, Y. H. Lo, M. Hong, J. P. Mannaerts, and R. Bhat, "Long-wavelength resonant vertical-cavity LED/Photodetector with a 75-nm tuning range," *IEEE Photon. Technol. Lett.*, **9**(6), 725, 1997.

[112] O. Issanchou, J. Barrau, E. Idiart-Alhor, and M. Quillec, "Theoretical comparison of GaInAs/GaAlInAs and GaInAs/GaInAsP quantum-well lasers," *J. Appl. Phys.*, **78**(6), 3925–3930, 1995.

[113] K. Moriki, H. Nakahara, T. Hattori, and K. Iga, "Single transverse mode condition of surface emitting injection lasers," *Trans. IEICE.*, **J70-C**(4), 501–509, 1987 (in Japanese); also: *Electronics and Communications in Japan, Part 2*, **71**(1), 81–90, 1988 (in English).

[114] H. Wada, D. I. Babić, M. Ishikawa, and J. E. Bowers, "Effects of nonuniform current injection in GaInAsP/InP vertical cavity lasers," *Appl. Phys. Lett.*, **60**(24), 2974–2976, 1992.

[115] J. W. Scott, R. S. Geels, S. W. Corzine, and L. A. Coldren, "Modelling temperature effects and spatial hole burning to optimize vertical-cavity surface-emitting laser performance," *IEEE J. Quantum Electron.*, **29**, 1295–1308, 1993.

[116] A. Takenouchi, T. Kagawa, Y. Ohiso, T. Tadokoro, and T. Kurokawa, "Laterally oxidised InAlAs-oxide/InP distributed Bragg reflectors," *Electron. Lett.*, **32**(18), 1671–1673, 1996.

[117] O. Blum, K. M. Geib, M. J. Hafich, J. F. Klem, et al., "Wet thermal oxidation of AlAsSb

lattice matched to InP for optoelectronic applicaitons," *Appl. Phys. Lett.*, **68**(22), 3129–3131, 1996.

[118] S. Uchiyama, N. Yokouchi, and T. Ninomiya, "Low threshold room-temperature continuous-wave operation of 1.3 μm GaInAsP/InP strained layer multiquantum well surface emitting laser," *Electron. Lett.*, **32**(11), 1011–1013, 1996.

[119] R. S. Geels, S. W. Corzine, and L. A. Coldren, "InGaAs vertical-cavity surface emitting lasers," *IEEE J. Quantum Electron.*, **27**(6), 1359–1367, 1991.

[120] D. I. Babić, J. J. Dudley, K. Streubel, R. P. Mirin, J. E. Bowers, and E. L. Hu, "Double-fused 1.52 μm vertical-cavity lasers," *Appl. Phys. Lett.*, **66**(9), 1030–1032, 1995.

[121] N. M. Margalit, D. I. Babic, K. Streubel, R. P. Mirin, R. Naone, J. E. Bowers, and E. L. Hu, "Submilliamp long-wavelength vertical-cavity lasers," *Electron. Lett.*, **32**(18), 1675–1677, 1996.

[122] P. Blixt, D. I. Babić, K. Streubel, N. M. Margalit, Thomas E. Reynolds, and J. E. Bowers, "Single-mode, 1 GB/s operation of double-fused vertical-cavity lasers at 1.54 μm," *IEEE Photon. Technol. Lett.*, **8**(5), 700–702, 1995.

[123] P. Blixt, D. I. Babić, N. M. Margalit, K. Streubel, and J. E. Bowers, "Multimode fiber transmission using room-temperature double-fused 1.54-μm vertical-cavity lasers," *IEEE Photon. Technol. Lett.*, **8**(11), 1564–1566, 1996.

[124] N. M. Margalit, J. Piprek, S. Zhang, D. I. Babić, K. Streubel, R. P. Mirin, J. R. Wesselman, J. E. Bowers, and E. L. Hu, "64°C Continuous-wave operation of 1.5 μm vertical-cavity laser," *IEEE J. Quantum Electron.*, **3**(2), 359–365, 1997.

[125] S. Z. Zhang, N. M. Margalit, T. E. Reynolds, and J. E. Bowers, "1.55 μm Vertical-cavity laser transmission over 200 km at 622 Mbit/s," *Electron. Lett.*, **32**(17), 1597–1598, 1996.

[126] E. F. Schubert, L. W. Tu, G. J. Zdyzik, R. F. Kopf, A. Benvenuti, and M. R. Pinto, "Elimination of heterojunction band discontinuities by modulation doping," *Appl. Phys. Lett.*, **60**(4), 466–468, 1992.

[127] M. G. Peters, D. B. Young, F. H. Peters, B. J. Thibeault, J. W. Scott, S. W. Corzine, R. W. Herrick, and L. A. Coldren, "High wall-plug efficiency temperature-insensitive vertical-cavity surface-emitting lasers with low-barrier p-type mirrors," *Proc. SPIE*, **2147**, 1–11, 1994.

[128] J. I. Pankove, *Optical Processes in Semiconductors*, Dover Publications, Inc., New York, 1971.

[129] W. G. Spitzer and J. M. Whelan, "Infrared absorption and electron effective mass in *n*-type Gallium Arsenide," *Phys. Rev.*, **114**(1), 59–63, 1959.

[130] F. A. Kish, D. A. Vanderwater, M. J. Peanasky, M. J. Ludowise, S. G. Hummel, and S. J. Rosner, "Low-resistance ohmic conduction across compound semiconductor wafer-bonded interfaces," *Appl. Phys. Lett.*, **67**(14), 2060–2062, 1995.

[131] Y. Okuno, "Investigation of direct bonding of III-V semiconductor wafers with lattice mismatch and orientation mismatch," *Appl. Phys. Lett.*, **68**(20), 2855–2857, 1996.

[132] H. Wada and T. Kamijoh, "Effects of heat treatment on bonding properties in InP-to-Si direct wafer bonding," *Jpn. J. Appl. Phys. Part 1*, **33**(9A), 4878–4879, 1994.

[133] M. Osinski and W. W. Chow, eds., *Physics and Simulation of Optoelectronic Devices III*, *Proc. SPIE*, **2399**, 1995.

[134] Integrated Photonics Research, *OSA Tech. Digest Ser.*, **6**, 1996.

[135] H. Shoji, K. Otsubo, T. Fujii, and H. Ishikawa, "Calculated performances of 1.3 μm vertical-cavity surface-emitting lasers on InGaAs ternary substrates," *IEEE J. Quantum Electron.*, **33**(2), 238–245, 1997.

[136] M. Shimizu, D. I. Babic, J. J. Dudley, W. B. Jiang, and J. E. Bowers, "Thermal resistance of 1.3-μm InGaAsP vertical-cavity lasers," *Microwave and Opt. Technol. Lett.*, **6**(8), 455–457, 1993.

[137] J. Piprek and S. J. B. Yoo, "Thermal comparison of long-wavelength vertical-cavity surface-emitting laser diodes," *Electron. Lett.*, **30**(11), 866–868, 1994.

[138] J. Piprek, H. Wenzel, H. Wuensche, D. Braun, and F. Henneberger, "Modeling light vs. current characteristics of long-wavelength VCSELs with various DBR materials," *Proc.*

SPIE, **2399**, *Physics and Simulation of Optoelectronic Devices III*, M. Osinski and W. W. Chow, eds., pp. 605–616, 1995.

[139] T. Baba, T. Kondoh, F. Koyama, and K. Iga, "Finite-element analysis of thermal characteristics in continuous wave long wavelength surface emitting lasers," *Optical Review*, **2**(2), 123–127 (pt. 1) and **2**(4), 323–325 (pt. 2), 1995.

[140] J. Piprek, "High-temperature lasing of long-wavelength VCSELs: Problems and prospects," in *Vertical-Cavity Surface-Emitting Lasers*, K. D. Choquette and D. G. Deppe, eds., *Proc. SPIE*, **3003**, 182–193, 1997.

[141] O. Madelung, ed., *Data in Science and Technology: Semiconductors, Group IV Elements and III-V Compounds*, Springer-Verlag, Berlin/Heidelberg, 1991.

[142] S. Adachi, *Physical Properties of III-V Semiconductor Compounds*, Wiley, New York, 1992.

9

Overview of VCSEL Applications

Richard C. Williamson

Most of the chapters in this book focus on the physics of operation and the technology of VCSELs. This chapter provides an introduction to the chapters that discuss VCSEL applications and attempts to address the question, "What potential future applications drive the advancement of VCSEL technology?" Answering this question is an exercise in prognostication because no applications exist that have generated any significant market volume. The only application for VCSELs that has progressed beyond the laboratory demonstration phase is their use in short-distance parallel fiber-optic data communications. This chapter discusses the applications areas already established for conventional in-plane semiconductor lasers, the unique features of VCSELs, those applications where the conventional lasers might be displaced by VCSELs, and those new areas in which VCSELs may be applied because of their unique features.

The total annual market for diode lasers (packaged units) is $1.6 to 1.9 billion and represents the largest segment of worldwide laser sales [1]. Applications of relatively expensive diode lasers in telecommunications yield the largest dollar volume while inexpensive lasers for compact disk players are produced in the largest unit volume. Optical storage (including CD players, CD-ROM, and higher-end storage) represents the second largest dollar volume for diode lasers and is a rapidly growing area of application. Image recording, including a range of printing applications, is the third largest dollar volume. The list of other significant markets for diode lasers includes instrumentation, bar-code scanning, medical devices, and a variety of single diode lasers and laser arrays for pumping solid-state lasers. These applications are almost exclusively served by conventional cleaved-facet edge-emitting lasers. VCSELs are potentially suitable for any of these applications, but the ability of VCSELs to displace conventional diode lasers will depend very much on their obtaining equal or superior performance at a lower cost. For conventional diode lasers, the technology, production, and cost reduction are well developed (e.g., CD lasers), and thus displacing them will not be easy. The unique geometry of VCSELs, however, promises reduced processing and packaging cost, and so this displacement may take place in some applications. The total diode laser market is so large that even a small technical niche for VCSELs could represent a significant dollar volume for the technology. More interestingly, the special features of VCSELs are likely to open up entirely new applications (e.g., laser arrays in communications and printing) not easily addressed by conventional diode lasers.

9.1 Special VCSEL Features

What are the special features of VCSELs that may allow them to displace conventional diode lasers or to open up new applications? These features include surface emission, ease

of making arrays, on-wafer testability, facet formation by means of epitaxy, single longitudinal mode, reduced temperature sensitivity, low threshold, and unusual output beam
characteristics.

9.1.1 Surface Emission

This feature is a natural consequence of the VCSEL geometry and thus makes the devices
well suited for fabrication in arrays, including two-dimensional arrays. Surface emission
allows on-wafer testability, which has been a significant aspect of VCSEL production for
parallel data communications [2–5]. It should be noted that conventional edge-emitting
diode lasers are conveniently fabricated into one-dimensional arrays and can be turned into
surface emitters (including two-dimensional arrays) by the incorporation of turning mirrors
[6] or gratings [7]. The leverage provided by the VCSEL geometry is that VCSELs take up
less wafer area and that extra processing steps are not required.

Surface emission provides added flexibility in packaging and incorporation of optical elements. The usual mode of VCSEL operation is with emission from the top of the substrate.
However, VCSELs with $\lambda > 0.85\,\mu$m on GaAs substrates and $\lambda > 1.1\,\mu$m on InP have
substrates that are transparent at the emitting wavelength, thus allowing emission through
the substrate. This feature allows such devices to be mounted with the epitaxial layers
and VCSEL contacts on the bottom side of a substrate where the VCSELs can be bonded
to other circuits, for example, to silicon drive circuits by bump bonding. Also, emission
through the substrate allows the monolithic incorporation of lenslets on the top emitting
surface [8].

9.1.2 Facet Formation by Means of Epitaxy

Because a large number of carefully controlled epitaxial layers are required to form the
active, contact, and reflecting layers in a VCSEL, the growth process is much more demanding than for a conventional diode laser. However, once the epitaxial growth is accomplished,
a number of advantages accrue. No cleaving is required. The narrow-band reflective layers
yield a narrow emission spectrum with a well-controlled longitudinal mode, thus avoiding
the demanding processing steps required for incorporating frequency-selective gratings in
conventional lasers. However, the short cavity in a VCSEL makes it difficult to achieve the
very narrow optical linewidths achieved with grating-controlled edge-emitting lasers. To
the extent that the epitaxial layers can be varied across a wafer, arrays of VCSELs with different emission wavelengths can be fabricated [3, 9]. At the shorter wavelengths (especially
in AlGaAs-based lasers), special facet coatings or treatments are required to avoid facet
degradation. VCSELs avoid the need for such coatings. Unlike cleaved facets, the wavelength of the reflection peak of an epitaxially grown mirror stack is temperature dependent.
This can be used to advantage to separately adjust the reflection peak and the gain peak so
as to yield lasers in which the threshold current is nearly independent of temperature over a
wide range [10]. Another approach to reduce temperature sensitivity is to broaden the gain
bandwidth [3]. Whereas high-reflectivity mirror stacks are readily achieved with AlGaAs
layers on GaAs, epitaxially grown mirrors with the desired characteristics are much more
difficult to achieve in the longer-wavelength alloys grown on InP [11, 12]. This means
that VCSEL technology is currently much better suited for shorter-wavelength ($<1\,\mu$m)
applications.

9.1.3 Low Threshold

This feature is important in low-power lasers for two major reasons – speed and efficiency. Response speed is increased and turn-delay is decreased by operating a diode laser with currents well above the threshold current. Thus, the current requirements consistent with a given speed are reduced by lowering thresholds. In some applications, for instance, high-density arrays, smart pixels, and cryogenic links [13, 14], it is highly desirable to minimize the drive power requirements for the lasers and also to minimize on-chip thermal dissipation. In most of these applications, the lasers need not put out many photons to provide adequately robust system performance. Usually, 1 mW of optical power is ample. For many applications, the optical power is limited to less than 1 mW by standards set for eye safety [2]. Thus the technical challenge can be phrased, "How do I achieve adequate speed and minimum power dissipation (high efficiency) for optical outputs less than a mW?" The answer is to achieve thresholds well below 1 mA and high efficiency at the operating current. These requirements appear to be better met by VCSELs than by conventional diode lasers. Achieving high efficiency at low currents has depended heavily on reductions in the series resistance of VCSELs and on maintaining low threshold current densities while shrinking VCSEL diameters [10, 15]. The recent advances in oxide-confined VCSELs [15–18] have yielded excellent performance in small-diameter VCSELs with threshold current densities well below 1 mA.

9.1.4 Output Beam Characteristics

In a conventional edge-emitting diode laser, thin epitaxial layers must be used, which has the consequence that the width of the optical mode in the direction perpendicular to the layers is always quite small (typically about two free-space wavelengths). As a result, the emission angle in that direction is large (often of the order of 30° or more). In the direction parallel to the layers, the width of the aperture is usually wider, thus yielding a noncircular beam. As the width of the aperture is widened beyond a few wavelengths, multiple transverse modes are excited and the beam becomes highly nonsymmetric and no longer diffraction limited. In contrast, a VCSEL with a circular aperture puts out a circular beam, which simplifies the coupling into subsequent optical components (e.g., lenses and fibers). Moreover, a VCSEL with a single lateral mode can have an aperture several wavelengths across, thus yielding a narrower output beam than a conventional diode laser.

As the diameter of a VCSEL is increased, a single lateral mode can no longer be maintained and the beam breaks up into multiple lateral modes. This can be very desirable for applications in which a broadened spatial and spectral output suppresses modal noise [2]. Large-diameter VCSELs have characteristics and applications very similar to LEDs except that the VCSELs are much faster and more efficient [2]. The primary example of the use of multimode VCSELs is in conjunction with multimode fibers.

In many applications, it is desirable to have a high-brightness laser. The brightness B of an optical source is given by $B = P/A\Omega$, where P is the emitted power, A is the emitting aperture, and Ω is the solid angle into which the power is emitted. For a diffraction-limited source, $A\Omega$, the etendu, is a constant, and more power directly yields higher brightness. When the VCSEL diameter is increased, the optical beam eventually breaks up into multiple lateral modes. Maximum power and A increase, but Ω remains relatively constant so that brightness remains nearly constant or may even decrease. For applications in which high brightness is important, VCSELs yield one to two orders of magnitude lower brightness than has been achieved with conventional diode lasers. Significant advances in near-diffraction-limited VCSEL power from large apertures will be needed before VCSELs can compete with conventional lasers in applications requiring high brightness. However, it should be noted that very high power and brightness have been achieved from an optically pumped VCSEL whose beam is controlled by an external cavity [19].

9.2 Parallel Fiber-Optic Data Communications

The application of VCSELs that is being mostly heavily pursued is their use in parallel fiber-optic (fiber ribbon) data communications. Two aspects are pushing this application. These are the increasing need in the digital industry for higher data rates over relatively short distances (i.e., much shorter than the distances addressed by the telephone companies) and the excellent fit between VCSELs and the source requirements in these applications. Major efforts in the development of VCSEL-based fiber–ribbon interconnections have been carried out by IBM [4], Hewlett Packard [2], Motorola [5], NEC [3], NTT, Vixel, AMP, Honeywell, Methode, Siemens, and other organizations. However, the field is not the exclusive domain of VCSELs. Many different ribbon–ribbon interconnections have been developed that employ arrays of conventional diode lasers (e.g., [20]). Much of this work, especially in Japan (e.g., [21]), has concentrated on linear arrays of 1.3 μm edge-emitting lasers with which interconnections are useful over much longer distances than can be addressed with shorter-wavelength interconnects.

As interconnection distances shrink, the links become more cost sensitive and the technical winners in the competition among VCSELs, conventional diode lasers, and metal interconnections will involve some critical cost/performance trade-offs including important issues such as speed, packaging, cross talk, and connectors. Optical interconnections have several significant features that suggest their use in high-speed interconnections. These include high bandwidth per optical path, bandwidth that is essentially independent of distance (out to hundreds of meters), low latency, and immunity from mutual interference and ground loops. Short-distance interconnections are very cost sensitive and tend to be dominated by packaging costs. The marriage of multimode VCSELs to arrays of multimode fiber is a powerful combination. There are claims that the VCSEL-based interconnections are becoming cost competitive with metal interconnections [4]. A more detailed examination of the use of VCSELs for optical interconnections is presented in other chapters in this book [2, 3].

Advances in digital technology are placing more stress on the speed of interconnections inside computers and in computer networks. Ever more powerful digital processors with constantly increasing clock rates such as the Pentium and Alpha chips have external memory-bus aggregate data rates that have now reached several tens of Gbits/s. The interconnections between processor and memory and between the multiple cabinets and boards of a computer system have become increasingly stressed. This is illustrated in Figure 9.1, where the inherent single-gate switching speed is downgraded by about two orders of magnitude by the intervening interconnection considerations before arriving at the overall system clock rate. Interconnection bottlenecks are growing in digital designs. The Semiconductor

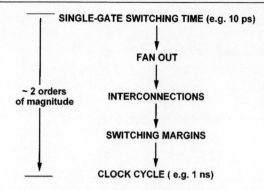

Figure 9.1. Speed of digital electronics from the gate level to the system clock.

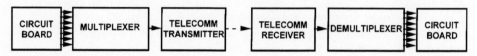

Figure 9.2. First generation of optical interconnections as developed for long-distance fiber-optic telecommunications.

Industry Association (SIA) roadmap [22, 23] for the next decades indicates increasing chip sizes, densities, output pins, and aggregate input/output rates. Clock rates of new processor chips are already at 600 MHz for the Alpha chip and are expected to be around 900 MHz for chips in production by the year 2000. Clock rates are projected to grow modestly (to around 1 GHz), thus placing greater stress on parallelism. Although metallic interconnections are relatively cheap (for short-distance interconnections) and constantly being improved, it is not at all clear that the required interconnection rates outlined by the SIA roadmap can be met by the use of metal, except for very short interconnection distances.

In parallel with the evolution toward faster and more powerful digital processors, the communications industry is undergoing a boom in the demand for increased communication capacity. This, in turn, places increasing demands on the switching equipment that must sort and route space-, wavelength-, and time-multiplexed data streams. These switches are like large digital computers and share with them a similar set of stresses on processor speeds and interconnection rates within the switch.

The first generation of intracomputer optical interconnection tests capitalized on technology adapted from the long-lines fiber-optic telecommunications industry. The basic scheme is illustrated in Figure 9.2. Parallel data streams are multiplexed to high-speed serial streams in order to take advantage of the relatively costly, but high-speed, telecommunications transmitter and receiver units. Transmission is over a single fiber, that is, there is no optical parallelism. However, the serial optical links did not prove cost effective for short-distance data communications. What was needed was a technology better suited for short distances (<500 m). This has been provided through the use of more optical parallelism and new optical technology, including VCSELs.

The economics of the push toward higher degrees of parallelism can be understood by reference to a design–trade-off graph, shown in Figure 9.3, which is typical of that used by computer developers. In this conceptual graph, the channel capacity is the aggregate data transmission rate for all the paths in parallel (i.e., the product of the number of parallel

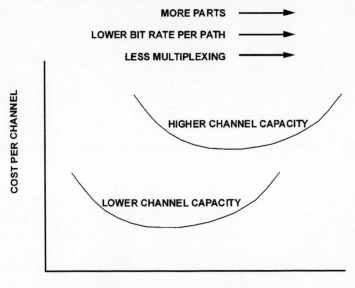

MORE PARTS ⟶

LOWER BIT RATE PER PATH ⟶

LESS MULTIPLEXING ⟶

HIGHER CHANNEL CAPACITY

LOWER CHANNEL CAPACITY

COST PER CHANNEL

NUMBER OF PARALLEL PATHS PER CHANNEL

Figure 9.3. Schematic graph of the cost of interconnections as a function of the degree of parallelism. The channel capacity is the aggregate data transmission rate for all paths operating in parallel.

paths in an interconnection and the bit rate per path). A single graph can be generated for a fixed interconnection distance in a given technology, optical or electrical. As the number of parallel paths per channel increases, more parts are required. The extra cost of packaging a multipath channel is partially offset by the fact that packaging cost can be amortized over more paths. Also, the parallelism results in a lower bit rate per path that makes the technology easier. There is great simplicity offered by operating at the clock rate of the parallel bit streams coming out of a digital component. Multiplexing (with its higher clock rates and more complex clocking schemes) is reduced or eliminated. Elimination of multiplexing also reduces interconnection latency, an important consideration in short-distance interconnections within computers. When all of these factors are put together, there is a degree of parallelism that yields a cost minimum. For long-distance communications, the cost of the installed fiber dominates, so the optimum solution is for no parallelism. However, for shorter distances the cost of the transmission medium becomes less dominant and eventually shifts the cost minimum to some degree of parallelism. A higher channel capacity will yield a cost minimum at a higher cost per channel as illustrated in Figure 9.3. It suggests that the higher channel capacity will also yield a minimum at a higher degree of parallelism, although this is not necessarily the case. The bottom line is that as interconnection distances become shorter, there is an increased impetus to utilize parallel interconnections. This is illustrated by the use of fiber–ribbon interconnections in the most high-capacity short-distance optical interconnections.

The general configuration for the second generation of optical interconnections is shown in Figure 9.4, which illustrates the transition to parallelism. The output drivers of high-speed silicon integrated circuits usually have adequate current swing to directly drive low-threshold lasers such as VCSELs. However, fiber–ribbon interconnections usually employ an interface circuit that formats and encodes the data, establishes proper communications

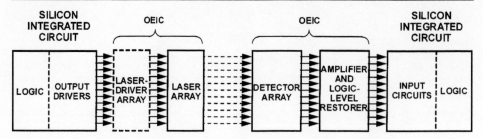

Figure 9.4. Second generation of optical interconnections employing increasing parallelism.

protocols, and includes special laser driver arrays to optimally drive the lasers. This interface circuit and laser array are packaged to yield a hybridized optoelectronic integrated circuit (OEIC). On the detector side, another OEIC implements critical functions to bring the received signal back up to levels adequate to drive the input circuits of the following silicon chips. The functions of amplification and logic-level restoration are the minimum that must be performed. In addition, clock recovery, data demodulation, and associated communication protocols are usually included. Because the circuits to be interconnected have outputs and inputs that are designed to be directly connected by wires, the optical interconnection provides no power savings. In today's silicon technology, the impedance levels, voltages, and current swings for both the chip output drivers and the chip inputs are set by the specifications of the logic family and are not optimized for use with optical interconnections. However, they yield the major advantages of high speed, high density, low spurious signals, and interconnection performance that is nearly independent of distance (except for inevitable increases in latency). Such optical interconnections are well suited for cabinet-to-cabinet connections and local-area networks (the current commercial emphasis). If the demand for increased interconnection speed continues to increase while the cost of the optical interconnections continues to decline, these parallel interconnections are likely to gravitate to shorter-distance interconnections (e.g., board-to-board).

VCSELs fit very well into this trend toward parallelism because of several factors. VCSELs are well suited for fabrication of arrays. The ability to probe the arrays is a key manufacturing feature. The standard wavelength for short-distance data communication (i.e., where fiber loss and dispersion are not critical) is around 850 nm. Thus, the optical sources can employ the AlGaAs materials system that has been the most successful VCSEL development. Also, silicon detectors can be used at this wavelength instead of the III–V detectors required at longer wavelengths. For short parallel links, cost is critical and packaging represents a significant portion of the cost. This is greatly relieved by employing the larger-diameter multimode fibers, but a multimode source is required to maintain low bit-error rates. A large-diameter VCSEL provides an ideal multimode source with a round beam whose divergence yields high-efficiency coupling into multimode fibers. Also, VCSELs with nearly temperature-stable characteristics greatly simplify the interconnections and reduce their cost by relaxing or eliminating temperature-control requirements. It should be noted that two unique features of VCSELs – surface-emission and two-dimensional arrays – are not utilized in the current generation of fiber–ribbon interconnections. In fact, surface emission can be awkward and requires miniature mirrors in order to deflect the beams parallel to the surface of the wafers.

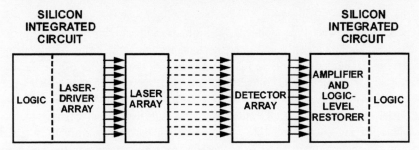

Figure 9.5. Third generation of optical interconnections in which optical and electrical elements are more intimately integrated for increased density, packaging simplicity, and power efficiency.

If fiber–ribbon optical interconnections prove to be cost effective and the speed of digital systems continues to increase, then there will be an incentive toward expanded use of optical interconnections including their use at shorter distances. These shorter-distance interconnections will push the technology toward increasing parallelism and will demand lower cost. As this occurs, the optical interconnection technology is likely to enter a third generation wherein the logic and the optical elements are more highly integrated, as seen in Figure 9.5. This integration will allow the optimization of the optical/electrical interface circuits so as to reduce the power requirements (as compared to metal connections) for high-speed interconnections [24, 25]. However, there is a major barrier to the transition from the second to the third generation. Interface circuits that are designed for use with optical interconnections must be designed into the digital chips (as distinct from designing them into special interface chips). Optical interconnections will have to become a standard part of the digital designer's parts set for this to occur. VCSELs are likely to play an important part in the third-generation interconnections because the VCSELs can operate well at very low drive levels, thus minimizing the demands on the laser-driver circuits. VCSEL drive requirements fit well with the capabilities of CMOS circuits, an attractive feature.

Over the past two decades, the use of fiber optics has become well established as a widely preferred means for high-speed communications for distances beyond a few km. The highest standard bit rates have risen to around 10 Gb/s. However, the technical and cost arguments for optical interconnections become less compelling as the bit rate drops (note that the SIA roadmap does not anticipate clock rates much beyond 1 Gb/s) and the distances become shorter. Parallel fiber–ribbon interconnections are a serious thrust into this shorter-distance regime where the current emphasis is on cabinet-to-cabinet interconnections. To the author's knowledge there is only one commercial computer system, a Cray Research super computer, which uses optics for interconnections within a computer [26]. In this case, conventional diode lasers and optical fibers provide precision fan-out of clock signals to a large number of circuit boards. Research is under way to explore the utility of highly parallel optics for shorter-distance applications. VCSELs are likely to be of importance in these areas.

9.3 Free-Space Interconnections and Smart Pixels

In all of today's optical interconnection and communication systems, optical fibers are employed to route the optical signals from the source to the receiver. A major departure from the use of fibers is the possibility of employing highly parallel free-space optical communications between logic elements. Fiber ribbons and one-dimensional arrays of lasers and detectors, along with the required connectors, become more difficult to implement as the degree of parallelism increases. Taking advantage of the strengths of two-dimensional optics offers a potential route around this bottleneck, and VCSELs are well suited for implementing two-dimensional arrays of surface emitters, which would be useful in free-space interconnections. A digital system designed with free-space optics would be very different from today's digital technology in terms of both hardware and architecture and thus there is no clear evolutionary path to such systems. A major departure from an evolutionary path may occur if the digital industry identifies technically attractive cost-effective uses for free-space optical processing. Meanwhile, there is a lot of active research exploring the technology and applications for such free-space-connected systems. VCSELs for free-space interconnections are described in two chapters of this book [3, 27].

The next step in the evolution of optics for shorter-distance interconnections is at the back-plane level. The rate at which signals can be passed electrically through a back plane with low-bit-error rate is usually less than the connection rates within a digital board, unless highly specialized and costly back planes are employed. This bottleneck places considerable emphasis on getting maximum processing capability on each board of a digital system. In a similar vein, high-speed massive interconnections are more easily achieved within a chip than within a board. These factors have driven silicon technology to ever more powerful chips along with board and back-plane hardware and architectures that best exploit the chip capabilities. High-speed massively parallel optical interconnections hold the promise of providing an alternative to this scenario. The ability of optical interconnections to alleviate the board-level interconnection problem would have a major influence on the architecture and capacity of future high-speed computers. The first use of optics for back planes will employ fiber–ribbon interconnections. Free-space interconnections may then follow if the optical technology and the digital demands adequately mesh.

As optical interconnections become more intimately integrated with electronic circuits, as interconnection distances shorten, and as parallelism increases, the packaging complexity and cost associated with parallel optical fibers will make free-space optical interconnections more attractive. The use of two-dimensional optical interconnections along the back plane [28] and point-to-point free-space interconnections between boards [29] have been explored. In these demonstrations, the optics interconnects large assemblages of conventional silicon chips. There are also more exploratory efforts aimed at assessing the utility of optics for interconnections of arrays of processing elements that contain only a few logic gates. This fine-scale optical interconnection and the more intimate integration of optics and logic are called "smart pixels," a subject covered in more detail in Chapter 12 in this book [27] and in a special journal issue [30]. The distinction between optically interconnected conventional processing elements and smart pixels is not clear cut. The finer-grained free-space optical interconnections in nonconventional architectures are usually referred to as smart-pixel systems. This fine-grained integration associated with the fourth generation of optical interconnections is illustrated in Figure 9.6. The highly integrated elements can utilize high-throughput two-dimensional optics to interconnect planes of processing circuits as shown in Figure 9.7. This integration can occur over the surface of a circuit board or in special optoelectronic smart-pixel chips. Not shown are the optical routing elements (e.g., lenses or holograms) needed to interconnect the planes.

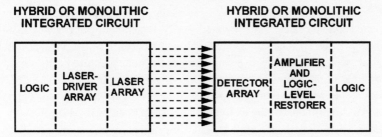

Figure 9.6. Fourth generation of optical interconnections in which optical and electrical elements are highly integrated and the capabilities of two-dimensional free-space optics are exploited.

Figure 9.7. Smart-pixel arrays for processing and interconnection in the fourth generation of optical interconnections. **L** represents a surface-emitting laser array and **D** represents a detector array. Optical elements (e.g., lenses and holograms) and additional electrical interconnections among the elements in a plane are not shown.

Extensive work has already been carried out for the development of optoelectronic smart pixels. Efforts at several laboratories have focused on the development of a smart-pixel-based switching system aimed at meeting the increasing demands for switching capacity in communications networks [31]. The long list of other applications that appear well matched to the two-dimensional format of smart-pixel arrays includes data sorting, image processing, pattern recognition, intelligent back planes, massively parallel processing systems [32], signal processing, and novel smart memory systems.

The majority of the research on smart pixels has focused on the use of optical modulators, not lasers, as the output. This choice was made because of the relative simplicity of modulators as compared to lasers and because of the thresholdless low-power operation of modulators fabricated in III–V epitaxial structures [31] and in liquid crystals on silicon [33, 34]. However, relatively complex optics is required to bring an array of optical beams to the modulators, to route an array of optical beams to the detectors, and then to bring out the modulated optical beams. The progress in VCSEL technology appears to be shifting the choice toward active sources rather than modulators because of the advances in reducing the operating power and the relative simplicity of the optics required. The ability to make large two-dimensional arrays of efficient surface-emitting sources (i.e., VCSELs) is key. Special techniques for hybrid integration of the silicon circuits and the lasers and detectors (usually in III–V compounds) are evolving (e.g., bump bonding).

At a stage when diode lasers were less developed than today, it was perceived that the power to drive the lasers would dominate any optical interconnection. However, this is not the case today. Because of the increasing efficiency and low power of the VCSELs, most of the interconnection power for optical interconnections at any level will be dissipated in the detector and receiver circuits, not in the lasers. This level of power dissipation combined with limits on power density will set a fundamental limit on the density of optical interconnections and on the ability to intimately integrate optical and electronic components.

9.4 Long-Distance Fiber-Optic Communications

Because single-mode long-distance (i.e., where loss and dispersion are significant) fiber-optic communications at 1.3 and 1.5 μm represents the largest market volume for diode lasers, this area is a very tempting target for the application of VCSELs. Several features of VCSELs make them attractive choices. Single-lateral-mode VCSELs have a circular beam with a narrower emission angle than conventional diode lasers, thus offering easier and more efficient coupling to single-mode fibers. Both fabrication and packaging costs can potentially be reduced. The increasing emphasis on wavelength-division multiplexing suggests the potential use of arrays of VCSELs with different wavelengths. Such arrays have been demonstrated at the shorter wavelengths on GaAs [3, 9, 35, 36].

There are two impediments to the use of VCSELs at the longer wavelengths. First, a highly developed industry already exists for producing 1.3 and 1.5 μm edge-emitting diode lasers. This situation has resulted in decreasing costs and considerable investment in the technology for manufacturing such lasers. For VCSELs to be cost effective, they must provide at least equivalent performance at a lower cost. Second, a major technical impediment to the use of VCSELs at 1.3 and 1.5 μm has been the difficulty of achieving high-reflectivity VCSEL mirrors. The index difference between InP and GaInAsP alloys is small, thus requiring that epitaxially grown mirror stacks be relatively thick. Performance of VCSELs employing such mirrors has been poor. Many approaches to implementing alternative types of mirrors have been demonstrated at various laboratories. A discussion of the longer-wavelength VCSELs and their implementation via wafer fusion is provided in Chapter 8 in this book [11]. The complexities of implementing mirrors and the performance achieved have yielded long-wavelength VCSELs that are not yet compelling enough to replace conventional lasers at 1.3 and 1.5 μm, but significant opportunities remain in this area.

9.5 Short-Wavelength VCSELs

The technology and applications of short-wavelength VCSELs are covered in detail in a Chapter 7 in this book [37]. As with conventional diode lasers, the technology becomes more difficult as the wavelength shortens, and yet there is a considerable drive toward developing the technology at shorter wavelengths because of significant applications including optical memories, printing, and the growth of low-cost plastic-fiber interconnections. Optical losses are minimized in plastic fibers at wavelengths near 650 nm. The advantages of VCSELs for interconnection via multimode fibers are the same at this wavelength as at 850 nm. Also, potentially lower cost may allow short-wavelength VCSELs to penetrate applications such as visible-wavelength scanners and pointers, which now employ He–Ne lasers or conventional diode lasers.

Applications of lasers to printing and data storage share some common features. In both cases, good resolution is required which means that the optical sources must be close to diffraction limited and that shorter wavelengths are desirable to increase printing resolution and storage density. For optical storage, the ultimate in diffraction-limited resolution is demanded while the requirements for many types of printing applications can use optical sources which are several times diffraction limited. Also, the physics of excitation that takes place in some print media and optical memories tends to favor more energetic shorter-wavelength photons. Increasing demands for high-speed optical memories and printing place emphasis on optical sources that deliver higher amounts of diffraction-limited optical power. There are two basic approaches: Either increase the brightness of the optical source (a difficult task for VCSELs as compared to edge-emitting diode lasers or solid-state lasers) or use parallel arrays of sources, an approach that capitalizes on the unique features of VCSELs. The excellent spectral qualities of VCSELs become a potential difficulty when used for the readout of optical memories. Conventional edge-emitting diode lasers for use in memories are usually implemented as self-pulsating lasers in order to decrease source coherence and reduce deleterious effects of reflections from the optical memory surface. Initial steps toward realizing self-pulsing VCSELs have been made [38]. Despite their limited brightness, VCSELs have the potential for significant use, especially in arrays, for optical printing and memories. Also because of their potentially low cost of production and packaging, VCSELs at 780 nm may be able to displace conventional low-power (~ 1 mW) diode lasers in compact disk optical heads [39].

9.6 Summary

The basic operation and fabrication technology and the unique features of VCSELs are described in several chapters in this book. These unique features include surface emission, novel reflectors, low thresholds, and special modal characteristics. These features suggest the use of VCSELs in a number of applications.

Commercial VCSEL-based short-distance fiber–ribbon data communications links have been developed and some are on the market. One prediction is that the incessant increases in demand for computer and communications capacity along with advances in VCSEL technology and packaging will make these VCSEL-based interconnections cost effective for use in many computer and switching systems. The SIA roadmap for digital technology predicts speeds and interconnection rates that will not be readily achievable or cost effective with metallic interconnections. The second generation of short-distance optical interconnections is a reality and is poised to be a commercial success.

The acceptance of optics as an attractive interconnection medium means that chip designs will anticipate the use of optics. This situation will change chip I/O design in anticipation and thus provide attractive technical advances that will be part of the third generation of optical interconnections.

The use of free-space optics and smart pixels in computing and switching systems implies a major departure from the course that has been taken for digital technology. This fourth generation of optical interconnections is quite speculative and there exists no clear evolutionary path to such systems. Advances in VCSEL technology, especially low thresholds, will shift the emphasis in laboratory research efforts and feasibility demonstrations toward the use of VCSELs rather than modulators.

GaAs-based VCSELs around 0.85 to 1.0 μm wavelength have become well developed and have very attractive performance. This is not yet the case at shorter and longer wavelengths. Significant markets for diode lasers exist at the fiber-optic wavelengths of 1.3 and 1.5 μm, but it is not clear that the features of VCSELs are compelling enough for them to penetrate these markets, especially given the complexity of the mirror technology. Further laboratory work may yield the required advances. Printing and memory applications are potentially significant areas for the use of VCSELs, especially where the use of arrays yields parallel memory access or high-speed multispot printing.

This overview is, by no means, comprehensive. The vast array of applications for conventional diode lasers presents opportunities for VCSEL technology and the unique features of VCSELs will open many more areas.

References

[1] Steele, R., "Review and forecast of laser markets: 1997 – Part II," *Laser Focus World*, **33**, 84–107, February 1997.

[2] Hahn, K. and Giboney, K., "VCSEL-based fiber-optic data communications," Chapter 11, this book, 1999.

[3] Kasahara, K., "Optical interconnection applications and required characteristics," Chapter 10, this book, 1999.

[4] Crow, J., "Parallel optical interconnect – a cost performance breakthrough," *Conf. Proc., LEOS'96*, pp. 3–4, IEEE, Piscataway, 1996.

[5] Lebby, M., et al., "Key challenges and results of VCSELs in data links," *Conf. Proc., LEOS'96*, pp. 167–8, IEEE, Piscataway, 1996.

[6] Williamson, R. C., et al., "Horizontal-cavity surface emitting lasers with integrated beam deflectors," Chapter 5 in *Surface Emitting Semiconductor Lasers and Arrays*, G. A. Evans and J. M. Hammer, eds., Academic Press, San Diego, 1993.

[7] Evans, G. A., et al., "Grating-outcoupled surface emitting semiconductor lasers," Chapter 4 in *Surface Emitting Semiconductor Lasers and Arrays*, G. A. Evans and J. M. Hammer, eds., Academic Press, San Diego, 1993.

[8] Coldren, L. A., et al., "Flip-chip bonded VCSELs with integrated microlenses for free-space optical interconnects," *Conf. Proc., LEOS'97*, pp. 343–4, IEEE, Piscataway, 1997.

[9] Wupen, Y., Li, G. S., and Chang-Hasnain, C. J., "Multiple-wavelength vertical-cavity surface-emitting laser arrays," *IEEE J. Select. Topics Quantum Electron.*, **3**, 422–8, 1997.

[10] Coldren, L. A. and Hegblom, E. R., "Fundamental issues in VCSEL design," Chapter 2, this book, 1999.

[11] Babic, D. I., Piprek J., and Bowers, J. E., "Long-wavelength vertical-cavity lasers," Chapter 8, this book, 1999.

[12] Debray, J.-Ph., et al., "Monolithic vertical cavity device lasing at 1.55 μm in InGaAlAs system," *Electron. Lett.*, **33**, 868–9, 1997.

[13] Akulova, Y. A., et al., "Low-temperature optimized vertical-cavity lasers with submilliamp threshold currents over the 77–370 K temperature range," *IEEE Photon. Tech. Lett.* **9**, 277–9, 1997.

[14] Lu, B., et al., "Gigabit-per-second cryogenic optical link using optimized low-temperature AlGaAs-GaAs vertical-cavity surface-emitting lasers," *IEEE J. Quantum Electron.*, **32**, 1347–59, 1996.

[15] Choquette, K. D. and Geib, K., "Fabrication and performance of vertical-cavity surface-emitting lasers," Chapter 5, this book, 1999.

[16] Huffaker, D. L., et al., "Improved mode stability in low threshold single quantum well native-oxide defined vertical-cavity lasers," *Appl. Phys. Lett.*, **65**, 2642–4, 1994.

[17] Choquette, K. D., et al., "Scalability of small-aperture selectively oxidized vertical cavity lasers," *Appl. Phys. Lett.*, **70**, 823–5, 1997.

[18] Hegblom, E. R., et al., "Vertical cavity lasers with tapered oxide apertures for low scattering loss," *Electron. Lett.*, **33**, 379–89, 1997.

[19] Kuznetsov, M., et al., "High-power (>0.5-W CW) diode-pumped vertical-external-cavity surface-emitting semiconductor lasers with circular TEM_{00} beams," *IEEE Photon. Tech. Lett.*, **9**, 1063–5, 1997.

[20] Karstensen, H., et al., "Parallel optical interconnection for uncoded data transmission with 1 Gb/s-per-channel capacity, high dynamic range, and low power consumption," *J. Lightwave Tech.*, **13**, 1017–30, 1995.

[21] Nishimura, S., et al., "Optical interconnections for the massively parallel computer," *IEEE Photon. Technol. Lett.*, **9**, 1029–31, 1997.

[22] *National Technology Roadmap for Semiconductors.* Semiconductor Industry Association, p. B2, 1994.

[23] Vallett, D. P. and Soden, J. M., "Finding fault with deep-submicron ICs," *IEEE Spectrum*, **34**, 39–50, Oct. 1997.

[24] Feldman, M. R., et al., "Comparison between optical and electrical interconnects based on power and speed considerations," *Applied Optics*, **27**, 1742–51, 1988.

[25] Krishnamoorthy, A. V. and Miller, D. A. B., "Scaling optoelectronic-VLSI circuits into the 21st century: a technology roadmap," *IEEE J. Select. Topics Quantum Electron*, **2**, 55–76, 1996.

[26] Kiefer, D. R. and Swanson, V. W., "Implementation of optical clock distribution in a supercomputer," In *Optical Computing*, Vol. 10, 1995 OSA Technical Digest Series, pp. 260–2, Optical Society of America, Washington, DC, 1995.

[27] Wilmsen, C., "VCSEL-based smart pixels for free-space optoelectronic processing," Chapter 12, this book, 1999.

[28] Liu, Y., et al., "Design and characterization of a microchannel optical interconnect for optical backplanes," *Applied Optics*, **36**, 3127–41, 1997.

[29] Tsang, D. Z., et al., "High-speed high-density parallel free-space interconnections," *Conf. Proc., LEOS'94*, pp. 217–8, IEEE, Piscataway, 1994.

[30] "Smart pixels," *IEEE J. Select. Topics Quantum Electron.*, **2**, 1996.

[31] Lentine, A. L., et al., "High-speed optoelectronic VLSI switching chip with >4000 optical I/O based on flip-chip bonding of MQW modulators and detectors to silicon CMOS," *IEEE J. Select. Topics Quantum Electron.*, **2**, 77–84, 1996.

[32] *Proceedings of Fourth International Conference on Massively Parallel Processing using Optical Interconnections*, IEEE Comput. Soc., Los Alamitos, 1997.

[33] Johnson, K. M., et al., "Smart spatial light modulators using liquid crystals on silicon," *J. Quantum Elec.*, **19**, 699–714, 1993.

[34] Perennes, F. and Crossland, W. A., "Optimization of ferroelectric liquid crystal optically addressed spatial light modulator performance," *Opt. Eng.*, **36**, 2294–301, 1997.

[35] Koyama, F., et al., "Low threshold multi-wavelength VCSEL arrays fabricated by nonplanar MOCVD," *Proc. SPIE*, **2399**, 629–35, 1995.

[36] Hu, S. Y., Hegblom, E. R., and Coldren, L. A., "Multiple-wavelength top-emitting vertical-cavity laser arrays for wavelength-division multiplexing applications," *Conf. Proc., LEOS'97*, pp. 222–3, IEEE, Piscataway, 1997.

[37] Thornton, R. L., "Visible light emitting vertical-cavity lasers," Chapter 7, this book, 1999.

[38] Lim, S. F., et al., "Self-pulsating and bistable VCSEL with controllable intracavity quantum-well saturable absorber," *Electron. Lett.*, **33**, 1708–9, 1997.

[39] Shin, H.-K., et al., "Vertical-cavity surface-emitting lasers for optical data storage," *Jpn. J. Appl. Phys.*, **35**, 506–7, 1996.

10

Optical Interconnection Applications and Required Characteristics

Kenichi Kasahara

10.1 Introduction

Vertical-cavity surface-emitting lasers have several unique advantages such as their small size, a single longitudinal mode due to their short cavity length, the ease with which they can be formed into an array, their low power consumption, and their low light beam divergence. VCSELs are expected to be used for the interconnection of data links in network computing and massive parallel computing, because the light sources used as transmitters in such systems must be inexpensive, consume little power, and allow easy array fabrication. Recent years have brought about great improvements in such basic characteristics as the slope efficiency, threshold voltage, wall-plug efficiency, and light output power. Today's VCSELs emitting at short wavelengths are characterized by a low threshold and good temperature characteristics, and they can be used with low-cost Si detectors. These advantages should lead to low-cost optical interconnections.

In optical interconnection applications, VCSELs will be used without a temperature-control circuit, and so it is important that the VCSEL be temperature insensitive. Because the cavity mode shifts at a rate of $0.07 \text{ nm/}^{\circ}\text{C}$ and the gain peak shifts at a rate of $0.3 \text{ nm/}^{\circ}\text{C}$, the cavity mode will not remain aligned with the gain peak as the temperature increases. This will reduce the light output power and increase the drive current. To obtain temperature-insensitive VCSELs, methods such as the offset-gain method have been proposed [1, 2]. However, developing a method of further improving temperature characteristics will be necessary.

Polarization control in VCSELs is also important because the devices in which VCSELs are used may use polarization-sensitive elements such as beam splitters. However, since the VCSEL polarization direction cannot be determined in advance, the polarization direction of the emitted light is random. Efforts have been made to control polarization by using differences in sidewall reflectivity [3] or stress from an elliptical window hole [4]. Also, an asymmetrically designed active layer has been proposed [5]. However, a method to effectively control polarization has not yet been developed.

In this chapter, I begin by introducing another approach to realizing temperature-insensitive and polarization-controlled VCSELs. After that, monolithically integrated multiple-wavelength VCSELs are described. An important feature of VCSELs is that single-mode operation is possible, which makes multiple-wavelength VCSEL arrays useful for wavelength-division multiplexing (WDM) applications. To fabricate monolithically integrated multiple-wavelength arrays, we have developed a new mask-molecular-beam-epitaxy (MBE) technique in which a shadow mask is placed on a wafer during epitaxial growth.

This technique allows us to change between selective and nonselective growth modes in an MBE chamber by using a movable mask. Grown layers have a high crystal quality and no contamination.

Next, a 16-channel two-dimensional VCSEL-array module for fiber-based parallel interconnections is described. This module has a plastic-based receptacle package directly joined to a push/pull fiber connector, and it does not require active alignment or intermediate components. This module showed floorless BER performance at a bit rate of 1 Gbit/s per channel without an isolator. This was the first two-dimensional receptacle-type laser diode (LD) module to be demonstrated that used a VCSEL and a plastic-based package that could be joined directly to a push/pull fiber connector.

Last, a novel network architecture for massively parallel processing (MPP) using VCSELs and free-space optical interconnections is presented. Here, a new routing concept is introduced. The results from a 64-channel free-space network prototype built to demonstrate and examine our approach are also presented. A WDM scheme using only a few different wavelengths is also shown to increase communication channel density.

10.2 Required Characteristics

10.2.1 Improved Temperature Characteristics

An increase in temperature reduces the light output power and increases the drive current of VCSELs. A rise in temperature may be caused by self-heating, thermal crosstalk due to simultaneous operation of array devices, or changes in ambient temperature. Since the cavity mode of VCSELs is not aligned with the gain-peak shifts as the temperature rises, unlike that of edge-emitting lasers, this is a particularly important consideration.

The offset-gain method has been proposed to prevent such degradation due to misalignment between the cavity mode and the gain peak. In this method, the cavity mode is set at a longer wavelength than the gain peak so that both peaks will tend to converge as the temperature increases. Thus, the change in the drive current required to reach the device threshold and 1 mW of output power is relatively flat over a broad temperature range from 15 to 90°C for a 8 μm diameter device [1].

Another approach is to extend the gain bandwidth itself. A similar idea was presented by Corzine et al. [6]. A broad gain bandwidth enables the cavity mode to move within a gain region over a wider temperature range than is conventionally possible. Although it was suspected that such a broad gain bandwidth would have disadvantages, such as an increased threshold current, it has been confirmed that the extended gain bandwidth has little adverse effect on the device characteristics [7, 8].

In the structure with the extended gain bandwidth (Figure 10.1), the p- and n-type distributed Bragg reflectors (DBRs) consist of alternating quarter-wavelength-thick GaAs and AlAs layers. There are 16 p-DBR pairs and 18.5 n-DBR pairs. The doping concentration is 3×10^{18} cm^{-3} for the p-DBR and 2×10^{18} cm^{-3} for the n-DBR. The active layers are InGaAs multiple quantum wells (MQWs) and the cladding layers are AlGaAs. The active region is surrounded by a highly resistive proton-implanted region that allows efficient current injection only in the light-emitting region [9]. The proton implantation dose is 5×10^{14} cm^{-2} and the proton implantation energy was 100 keV. Current is injected through a contact layer bypassing the p-DBR to reduce the electrical series resistance. The top mesa is 6 μm square for a single transverse mode oscillation. The n-electrodes are AuGe/Ni/Au and the

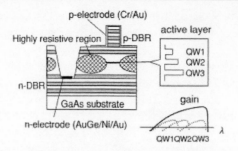

Figure 10.1. VCSEL with a broad gain bandwidth.

p-electrodes are Cr/Au. The output light is obtained from the side of the GaAs substrate. The back of the substrate is antireflection (AR) coated.

To achieve a broad gain bandwidth, the active layers consist of multiple types of InGaAs MQWs (Figure 10.1). A broad gain bandwidth may lead to disadvantages such as an increased threshold current. If the gain peaks of the individual quantum wells (QWs) are too far apart, the total gain will decrease. Therefore, a gain structure is needed that provides a broad gain bandwidth and yet has little small adverse effect on the device characteristics. To obtain an optimized active-layer structure for a broad gain bandwidth, we simulated the gain spectra, changing the gain-peak wavelength of each InGaAs well. In this simulation, the gain was calculated for a single quantum well at each bandgap energy; then the total gain was obtained by summing the individual gains obtained. Details of the calculation are given in ref. 10. In the calculation of the linear gain, only the first electron subband and the first hole subband, which are approximately described by the two parabolic hole bands, were considered; the contribution of other electron and hole subbands, which is much smaller, was ignored. The confinement factor $2d/L_c$ and the electrical field distribution were taken into consideration, where d denotes the thickness of the active layers and L_c denotes the effective cavity length. The value of L_c was assumed to be 1.5 μm. Absorption was also considered in this calculation.

In the simulation, the active layers consisted of three or four quantum wells (Figure 10.2). All active layers that consisted of more than one type of MQW had a wider gain bandwidth. The conventional device and the structure-A device had active layers with three QWs. The gain peaks were set at 980 nm for the conventional device and at 980, 970, and 960 nm for the structure-A device. The structure-B device had active layers with four QWs, and its gain was broader and flatter than that of the conventional device due to the optimization of the structure. For the optimized structure, discussed later, the gain peaks were set at 980, 980, 960, and 940 nm.

For a VCSEL, the optical standing wave is perpendicular to the quantum wells. Accordingly, the optical field intensity of each quantum well should be taken into account in a simulation. In the four-QW case, we arranged each gain peak relative to the optical field patterns as shown in Figure 10.2(b). In this simulation, the optical field intensity ratio was multiplied by the gain of each peak, considering the above optical field patterns.

In both structures, the total gain was almost as large as that of the conventional structure. This result suggests that little significant degradation in characteristics occurs due to the extended gain bandwidth.

Hole localization, which increases the threshold current, has been observed in edge-emitting lasers with a 1.5 μm lasing wavelength [11]. Two types of edge-emitting lasers

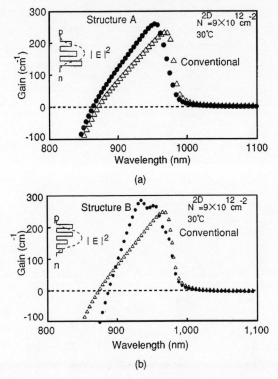

Figure 10.2. Gain spectra obtained from the simulation for extending the gain band-width. The gain peaks of the conventional device (open triangles) are set at 980 nm; that of (a) the structure-A device (closed circles) at 990, 980, and 970 nm; and that of (b) the structure-B device (closed circles) at two each of 980, 980, 960, and 940 nm.

with the following MQW active layers were fabricated. One structure consisted of five InGaAs quantum wells with a transition wavelength λ_g of 1.46 μm and four quantum wells with λ_g of 1.54 μm from the p-side. The other consisted of five InGaAs quantum wells with a transition wavelength λ_g of 1.54 μm and four quantum wells with λ_g of 1.46 μm from the p-side. The lasing wavelengths of these two lasers were 1.48 μm and 1.55 μm, respectively, which shows that the hole is localized at the p-side wells. In addition, the threshold currents were 38 mA and 20 mA, respectively. The difference is probably due to the hole localization.

Therefore, we checked whether hole localization occurred in the structure with the expected gain bandwidth. As shown in Figure 10.1, the quantum well with the lowest bandgap energy is located on the n-side. In spite of this configuration, carriers are first injected into this quantum well as shown in Figure 10.2, indicating that hole localization did not occur. This behavior is most likely due to the lower valence band discontinuity (ΔE_v) and small number of MQW layers. Carriers are injected uniformly into each well.

The temperature characteristics of VCSELs with the broad-gain-bandwidth structure-B from Figure 10.2(b) and of conventional VCSELs were measured during continuous wave (CW) operation. The measurements were made for arbitrary devices taken from 8×8 VCSEL arrays using either broad-gain bandwidth or conventional VCSELs (39 VCSELs with a broad gain bandwidth, and 24 conventional VCSELs).

Table 10.1. *Deviations in threshold current (th) and injection current at light output power of 1 mW for conventional devices and devices with a broad gain bandwidth.*

		20°C	80°C
Conventional	th	0.9 ± 0.1 (mA)	2.1 ± 0.1
	1mW	3.4 ± 0.1	6.3 ± 0.3
Broad gain	th	0.9 ± 0.1	1.7 ± 0.2
bandwidth	1mW	3.3 ± 0.2	5.1 ± 0.4

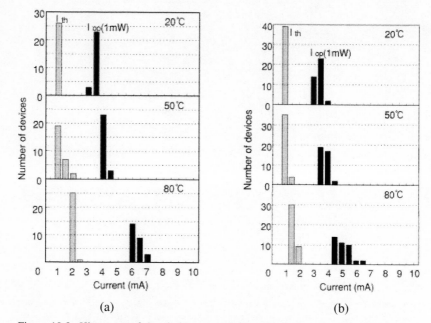

Figure 10.3. Histogram of threshold current (Ith) and injection current at light output power of 1 mW (Iop(1 mW)) for (a) conventional devices and (b) devices with a broad gain bandwidth (the structure-B).

The proportion of both VCSELs that achieves a threshold and light output power of 1 mW at a given injection current is shown in Figure 10.3 at 20, 50, and 80°C. The average and deviation of the proportions given in Figure 10.3 are shown in Table 10.1.

As shown, the temperature characteristics of the devices with a broad-gain bandwidth prove to be better than those of the conventional devices. At 80°C, the injection current at a light output power of 1 mW is 5.1 ± 0.4 mA for VCSELs with a broad-gain bandwidth and is 6.3 ± 0.3 mA for conventional VCSELs. This means the drive current needed by VCSELs with a broad-gain bandwidth was about 20% lower than that of the conventional devices. The deviation in the drive current of the devices with a broad-gain bandwidth was about 10%, whereas that of conventional devices was about 5%.

These improved temperature characteristics are mainly due to the extended bandwidth of the gain. However, some other contributions should also be taken into account. These are: (1) the unintended effect on the gain-offset caused by the uniformity of the layer thickness and (2) variations in the position of active layers caused by this uniformity relative to the optical field patterns along the vertical cavity.

The temperature characteristics of each device did not depend on the position of the device in the array, indicating that these characteristics were not affected by variations in the oscillation wavelength (the lateral-layer thickness). That is, the contributions of (1) and (2) have a negligible effect on the temperature characteristics of devices. Consequently, we can consider the improved temperature characteristics to be entirely due to the extended gain bandwidth.

The deviation in the drive current at the threshold and at a light output of 1 mW differed slightly between the devices. This slight difference is thought to be mainly due to unmatched AR coatings. A 10% deviation demonstrates that a VCSEL with a broad-gain bandwidth has sufficiently uniform temperature characteristics in an 8×8 array.

10.2.2 Polarization Control

Polarization control in VCSELs is important in the systems using polarization-sensitive elements such as beamsplitters. Additionally, the polarization needs to be well controlled to avoid polarization-induced noise caused by unstable polarization states [12]. A great advantage of VCSELs is that they can be readily used to form arrays; thus, polarization must be controlled not only for single VCSELs, but also for VCSEL arrays. However, since the VCSEL polarization direction is not limited, the polarization direction of the emitted light is random and is easily switched due to stress [4], injected current [13], or reflected light [14].

The following points must be taken into consideration when attempting to control polarization. First, the VCSEL must permit fundamental single-mode emission in both the longitudinal and transversal modes because multimode VCSEL emission does not always occur in just one polarization direction. Second, the VCSEL must be monitored during CW operation, since switching problems due to heat and spatial hole burning are not evident in pulsed operation. Third, a considerable number of the VCSELs must be monitored because the polarization direction in a small number of VCSELs may be the same owing to random effects, such as scratches or reflected light from prisms, that occur during fabrication or measuring. However, these aspects have not been considered in previous reports [3, 4, 15], probably because that research was not directly concerned with array applications.

Complete polarization control that satisfies these three criteria has been demonstrated [16]. It has also been shown that compared to conventional circular-symmetric VCSELs, such polarization-stabilized VCSELs show significantly reduced transient partition between orthogonally polarized lasing modes [17].

Yoshikawa et al. [16] have fabricated an index-guided transversal-injected VCSEL. An InGaAs 3-QW-VCSEL wafer consisting of 15.5 and 18.5 GaAs/AlAs DBR pairs was grown with good uniformity [18] by MBE on a (001)-GaAs substrate misoriented $2°$ toward $\langle 111 \rangle$A. Rectangular pattern-resist masks were made with sizes varying (in 0.25-μm increments) from $6 \times 6 \ \mu$m to $6 \times 3.5 \ \mu$m, with the shorter side in the $\langle 110 \rangle$ direction (Figure 10.4). The diffraction loss of fundamental-mode emission can be increased [19] in one direction by reducing the size of the post in one direction. This causes anisotropy of the transversal mode selectivity, leading to pinning of the polarization direction. Electron beam lithography was used to ensure accurate pattern sizes. Posts were then formed by Cl_2 reactive-ion-beam etching (RIBE) from the top layer to the 14th GaAs layer of the p-side DBR, while using in situ laser reflectometry to monitor the etching. The size

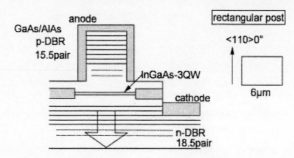

Figure 10.4. Structure of index-guided transversal injected VCSEL and upper view of rectangular post in ⟨110⟩ direction.

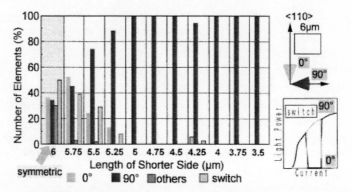

Figure 10.5. Proportion of polarization direction of 8 × 8 matrix array (64 VCSELs) versus post size from 6 × 6 μm to 6 × 3.5 μm. Polarization was completely controlled for 6 × 5 μm to 6 × 3.5 μm sizes. Probably, the post structure was broken for 6% of the 6 × 4.25 μm size array, which does not have 100% polarization pinning.

of the rectangular posts was confirmed by optical microscopic observation after etching. The active layer beside the post was inactivated by proton implantation. AuGe/Ni/Au was deposited for the cathode, and AuZn/Au was deposited for the anode. The light output side was AR coated.

The light output power emitted from the GaAs substrate side was monitored through a Grand–Thompson prism, and the prism angle at which the maximum output was obtained was classified into three polarization angles. These angles were 0°, which corresponds to 0° to the ⟨110⟩ direction, 90° corresponding to 90° to the ⟨110⟩ direction, and 45° corresponding to any intermediate angles. This type of VCSEL with a post size of less than 6.5 × 6.5 μm has a fundamental single-mode emission both in the longitudinal and transverse modes and is linearly polarized. The direction of the polarization angle, however, can switch when the operating current increases, probably because of spatial hole burning or stress from heat expansion. In such cases, the angle observed first was defined as the polarization angle. The number of polarization-switched VCSELs was also counted. The operating current was varied from 0 to 10 mA, which roughly corresponds to ten times the threshold current and a light output power of 3 mW.

The proportion of VCSELs for each polarization direction is shown in Figure 10.5 as a function of post size. The proportion of switched VCSELs is also shown. The proportion of

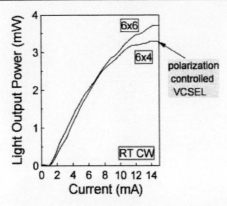

Figure 10.6. L-I characteristics of the 6×4 μm polarization-controlled VCSEL and the 6×6 μm polarization uncontrolled VCSEL. The difference in the higher current range is probably caused by heating and can be reduced when a heat sink is used.

VCSELs with 90° polarization increased, and the proportion of switched VCSELs decreased as the length of the shorter side decreased. These changes indicate that pinning of the polarization angle becomes rigid as the length of the shorter side decreases. When the length of the shorter side is less than 5 μm, the polarization of the 8×8 VCSEL matrix is 100% controlled. All of these VCSELs have a fundamental single-mode emission with linear polarization and without switching. The polarization angle distribution of the 6×5-μm VCSELs is suppressed to within 12° with a deviation of 2.9°, which is comparable to the measurement error.

As this polarization control method uses the increase in diffraction loss, light output degradation becomes a concern. The average threshold current increased as the post size decreased, but this change was small when the size exceeded 6×4 μm. Therefore, the average threshold current of the 6×5-μm polarization-controlled VCSEL array was 1.29 mA, which was only 1.17 times that of the 6×6 μm VCSEL array without polarization control. However, this higher threshold current is negligible for conventional use. The I–L characteristics of the 6×6-μm and 6×4-μm VCSELs were almost the same (Figure 10.6). The difference in the higher current injection was probably caused by heating, since these measurements were performed without a heat sink. Although thermal resistance can be high for a 6×4-μm VCSEL because of its small cross section, this can be reduced by proper bonding to a heat sink. The neat round shape of the near-field pattern of the 6×5-μm VCSEL indicates fundamental single-transverse-mode emission over the operating current range up to 10 mA. Distortion due to anisotropy of the rectangular post was not observed.

When a current injection needle is used to connect the electrode pad and current source, tension from the needle may change the polarization direction. Furthermore, when Pb-Sn is used to mount VCSEL arrays, the tension from heat expansion can change the polarization direction. Both of these effects occurred with the 6×6-μm VCSEL array that was not polarization controlled; however, the polarization-controlled VCSEL array did not experience such problems and maintained its polarization angles throughout the measuring and bonding processes. Such rigid polarization pinning is suitable for conventional use in mounted form.

The polarization controlling factor is difficult to determine accurately because the VCSEL structure is complex. It has a mesa-guided structure optically confined by post etching and electrically confined by proton implantation. Furthermore, the current injection route is not

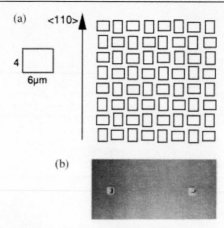

Figure 10.7. (a) Placement of 6×4 μm rectangular posts. (b) Optical microscopic photographs taken after postetching.

straight and may affect transverse mode stability and, hence, polarization stability. However, this polarization control method is simple and effective because it uses the small increase in diffraction loss that comes from reducing the post size in one direction relative to the fundamental single-mode emission. This causes anisotropy in the transverse mode selectivity. The 6-μm post size, at which the smallest threshold current is obtained for the fundamental single mode, is the optimum size for polarization control. If the post size is much larger than the fundamental mode size, the post structure does not affect mode selectivity, but if the post size is much smaller than the fundamental mode size, the diffraction loss becomes too high for stable lasing at a low injection current.

The successful polarization control of VCSELs indicated that the arbitrary placement of rectangular posts would thus result in a monolithic laser array with arbitrarily regulated polarization. This would be very useful in various applications, for example, in optical free-space routing and in a polarization-duplicated transmitter, or optical head.

To test the feasibility of arbitrarily regulated polarization, we fabricated a device where the rectangular posts (6×4 μm) were placed as shown in Figure 10.7(a). Each post was at a right-angle to its neighbor; that is, the shorter side or the longer side was alternately parallel or perpendicular to the $\langle 110 \rangle$ direction. Precise fabrication of these posts was confirmed by optical microscopy after the RIBE (Figure 10.7(b)). The VCSELs were CW-operated, one by one, at room temperature, and the polarization direction was determined according to the maximum light power obtained through a Grand–Thompson prism. We also investigated whether the polarization direction changed with the injection current, which ranged from 0 to 10 mA.

The measured polarization angle of an 8×8 matrix array is shown in Figure 10.8. The six blank elements did not emit light owing to a fabricating error. However, the expected polarization direction was obtained for almost all the other VCSELs. No VCSEL changed its polarization angle as the measured injection current was varied. This shows that the polarization can be regulated regardless of the wafer orientation. In a previous experiment, we arranged for the polarization of all the VCSELs in our 8×8 matrix array to be in one direction by reducing the length of the shorter sides of the rectangular post from 6 μm to 5 μm or less. However, setting polarization in an array in any specific direction is

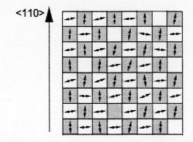

Figure 10.8. Measured polarization direction of alternately perpendicular rectangular post array. The six blank elements did not emit due to fabrication error.

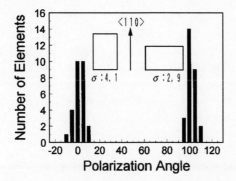

Figure 10.9. Polarization dispersion of alternately perpendicular rectangular post array. The deviations were only 4.1° and 2.9° for each perpendicular direction.

more difficult than simply arranging for it to be in one direction. In a single-directional arrangement, wafer-oriented polarization-controlling effects, such as anisotropic optical gain [20] and anisotropic gain offset [21] can simultaneously occur, turning the polarization in a particular direction. However, when the polarization is arbitrarily set, some wafer-oriented effects that are preferable for polarization determination in a certain direction will always be unfavorable for polarization determination in another direction. In addition, for arbitrarily regulated polarization, the polarization pinning must be strong enough to keep the deviation of the polarization direction small, because a wide dispersion in the polarization direction makes the regulated direction hard to determine in the array form. The polarization angle dispersion is shown in Figure 10.9. The deviation of each direction was only 4.1° and 2.9°, respectively, for 0° and 90° to the ⟨110⟩ direction. These values were almost the same as the measurement error. Under strict observation conditions, the average difference in the angle between the two polarizations was about 100°, in spite of the rectangular post disposition angle of 90°. This was probably due to the 5° mismatch between the ⟨110⟩ direction and the placement of the rectangular post. The tendency of this wafer to polarize along 90° to ⟨110⟩ was so strong that 77% of the 6×6-μm array was polarized in this direction. Therefore, owing to a 5° declination of the rectangular side in the ⟨110⟩ direction, the VCSEL tended to have a polarization angle of 5°, declined in the post direction. This makes the average angle difference between the two polarizations 10° smaller or larger depending on whether the declining angle of the post in the ⟨110⟩ direction is clockwise or counterclockwise, respectively. The fabricated VCSEL array had a counterclockwise rotation. Thus, a precise perpendicular polarization can be obtained for nonpolarized wafers and, with careful pattern placement, also for strongly polarized wafers such as the one used in this experiment. However, it is obvious that the effect of polarization control by using a rectangular post can dominate the wafer-oriented effects. An arbitrarily regulated polarization direction can be obtained regardless of the wafer orientation.

10.2.3 Multiple-Wavelength VCSELs by Mask MBE

VCSELs make useful light sources in WDM systems [22, 23] because of their two-dimensional array configuration, single-mode operation, and precisely controlled lasing wavelength. For this application, two-dimensional multiple-wavelength VCSEL arrays have been fabricated using several different growth [24–27] and postgrowth techniques [28].

In mask molecular-beam-epitaxy (mask MBE), the wavelength is controlled by controlling the growth thickness during MBE, and integration flexibility is provided by selective area growth [26]. Therefore, it is easy to obtain the designed wavelength spacing and integration area size in multiple-wavelength arrays. Also, in situ mask MBE provides a clean interface and a high-quality layer, and its single growth run enables the thickness of all the device layers to be precisely controlled. The array devices grown by mask MBE, therefore, have good characteristics comparable to those of a single device grown by conventional MBE.

Figure 10.10(a) illustrates the principle of mask MBE and shows the schematic arrangement of the mask, substrate, and molecular beams during selective growth. The mask is a

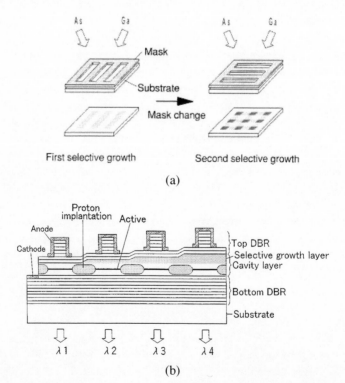

Figure 10.10. (a) The principle of mask MBE. (b) The conceptual schematic structure of a multiple-wavelength VCSEL array.

metal mask with 500 μm wide striped windows separated by a space of 500 μm. There is a clearance of 100 μm between the mask and the substrate. The mask can be moved over the substrate by using a rotary motion feedthrough, which allows switching between selective and nonselective growth in the MBE chamber. This technique allows us to make 8- and 16-wavelength arrays using 3 and 4 masks, respectively. The grown layers are expected to have a high crystal quality without contamination, and precise control of their thickness is anticipated to result from the single-run MBE growth. Also, the substrate is rotated during growth, resulting in highly uniform lateral thickness.

In comparison with two- or multistep MBE, one advantage of mask MBE is that the difficulty of making the growth rates the same can be avoided, because growth in mask MBE is performed sequentially.

We have made VCSELs that emit at four different wavelengths by using two-stripe masks. Figure 10.10(b) shows the conceptual schematic structure of a multiple-wavelength VCSEL array. The Fabry–Perot mode wavelength is varied by varying the cavity thickness. By using mask MBE for selective area growth, we can grow cavity layers of varying thicknesses, resulting in monolithic integration of lasers that emit at different wavelengths. The uniformity of the array characteristics depends on three parameters: 1) reflectivity of the distributed Bragg reflector, 2) the overlap between an active region and an electric-field standing-wave pattern within the cavity, and 3) the gain peak wavelength of the active region. The threshold gain g_{th} of the VCSEL is given by

$$g_{th} = 1/(2d\,\Gamma)\ln(1/R_1 R_2), \tag{10.1}$$

where d is the active layer thickness, Γ is the confinement factor, and R_1 and R_2 are the reflectivities of the top and bottom DBRs. Because a common DBR is used for multiple-wavelength VCSELs, the DBR reflectivity of the first parameter differs with the wavelength. The DBR has a reflectivity above 99% over a broad wavelength range, but a plot of the logarithm of the reflectivity reciprocal as a function of wavelength forms a sharp valley. The second parameter, the overlap, varies with the selective layer thickness at the upper side of the cavity layers in multiple-wavelength VCSELs. This parameter corresponds to $d\Gamma$ in Eq. (10.1). Using the first and second parameters, we can calculate the g_{th} dependence of the Fabry–Perot mode wavelength (Figure 10.11). In this calculation, the designed center wavelength of the DBR, which consists of 18 AlAs/GaAs pairs, is 980 nm. Also, the designed position of the active region, which consists of a 100 nm thick InGaAs 4-QW layer, is the

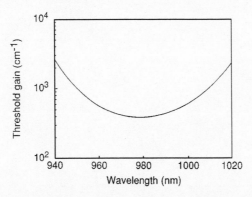

Figure 10.11. Calculated threshold gain as a function of the Fabry–Perot mode wavelength.

electric field maximum position in the cavity layer when the cavity wavelength is 980 nm. The threshold gain uniformity is ±5% within 20 nm of the wavelength bandwidth (Figure 10.11). However, the VCSEL must match the gain peak with the Fabry–Perot mode wavelength. The gain peak wavelength of the third parameter, however, is the same for each VCSEL because a common active layer is used. A broad-gain active layer consisting of multiple QWs with different bandgap energies (980, 980, 960, and 940 nm) described in Section 10.2.1 is thus useful for use in multiple-wavelength VCSELs. Such a layer has a flat gain peak over a width of about 20 nm. If this gain structure is adopted for multiple-wavelength VCSELs, a uniform threshold current is achieved in the wavelength bandwidth of 20 nm.

To fabricate four-wavelength VCSEL arrays, the selective growth was done twice, using two masks with 250 mm striped windows with 250 mm spaces, and with each mask pattern at right angles to that of the other. An InGaAs 4-QW active layer with a broad gain was used, and the designed wavelength spacing was 5 nm. The wavelength bandwidth of the four-wavelength VCSEL arrays was thus 15 nm, which is within the uniform threshold current range estimated from the calculation.

The current-versus-light-output characteristics of the four-wavelength VCSELs, whose lasing wavelengths differed by 5 nm, are shown in Figure 10.12. Uniform 1.5 mW light output was achieved under CW conditions at room temperature. For the 5×6 arrays of four-wavelength VCSEL units (10×12 VCSELs), the lasing wavelengths and the threshold currents are listed in Table 10.2 and are shown in Figures 10.13 and 10.14. The arrays show excellent uniformity in their lasing wavelengths and threshold currents, with deviations of

Table 10.2. *Lasing wavelengths and threshold currents of 5×6 four-wavelength units (10×12 VCSELs).*

Lasing wavelength, nm (Standard deviation)		Threshold current, mA (Standard deviation)	
927.4	(0.38)	0.84	(0.062)
932.4	(0.32)	0.79	(0.089)
937.9	(0.27)	0.80	(0.061)
942.9	(0.27)	0.86	(0.103)

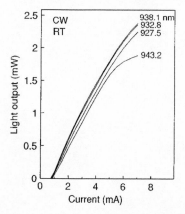

Figure 10.12. Light output versus current curves for four-wavelength VCSELs.

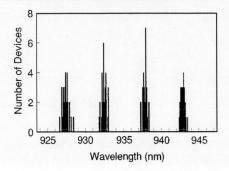

Figure 10.13. Lasing wavelength distribution of 5 × 6 four-wavelength units (10 × 12 VCSELs).

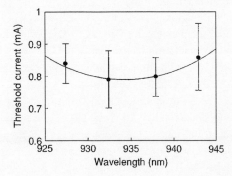

Figure 10.14. Distribution of the CW threshold currents of 5 × 6 four-wavelength units (mean ± SD, $n = 30$). The curve is the normalized curve of the calculated threshold current.

of less than 0.38 nm and 0.103 mA, respectively. The wavelength deviation, which ranges from 0.27 to 0.38 nm, is largely due to the natural nonuniformity of MBE thickness over the wafer. Low threshold currents were obtained for VCSELs at each wavelength. Figure 10.14 shows the tendency of the threshold current to increase at shorter and longer wavelengths. This increase is caused by the three parameters mentioned in the theoretical discussion: DBR reflectivity, overlap between the active region and the electric-field standing-wave pattern, and the gain peak wavelength. We calculated the threshold current of the VCSEL from the calculated threshold gain of Figure 10.11, assuming a flat gain, and inserted the normalized curve of the calculated threshold current into Figure 10.14. The measured threshold currents show the same high uniformity that the calculated curve does.

A hybrid integration technique, as an alternative to monolithic integration, has also been proposed for multiple-wavelength VCSEL arrays [29]. In this technique, a VCSEL of each wavelength is individually prepared by MBE growth and arrayed by flip-chip bonding. This is a superior technique for forming multiple-wavelength arrays that provides sufficient wavelength separation and the desired distribution. It is particularly useful for WDM systems because the lasing wavelength is sensitive to temperature variations, which may cause optical crosstalk in the detectors. In our study of monolithic arrays, however, the wavelength could be controlled by using selective growth in the mask MBE. Also, precise alignment within several microns in three dimensions is necessary for hybrid integration of VCSELs by using flip-chip bonding. For the monolithic arrays grown by mask MBE, in contrast, such alignment is not necessary.

10.3 Optical Interconnection Applications

10.3.1 Two-Dimensional Optical Fiber Interconnections

VCSEL-array modules using waveguides [29a] or 45°-polished fibers [30] between the one-dimensional VCSEL array and the fiber connector have been reported. The advantage of the VCSEL, however, is not fully utilized in a one-dimensional array, and active alignment or visual alignment to the intermediate components will spoil the cost advantage of the VCSEL. This section describes a two-dimensionally arranged 16-channel receptacle-type VCSEL module that can be joined directly to a push/pull fiber connector without active alignment or any intermediate components.

The VCSEL-array module is composed of a VCSEL-array chip, a silicon substrate, a plastic-based package, and a fiber connector. The VCSEL shown in Section 10.2.1 or 10.2.2, which shows a 3-dB down modulation bandwidth of 7 GHz [31], emits laser light with a wavelength of around 980 nm from the mechanically polished and antireflection-coated bottom surface of the 100 μm thick GaAs substrate.

Figure 10.15 illustrates the configuration of the VCSEL-array package. The VCSEL chip has an 8×2 array structure and a pitch of 250 μm. The chip is mounted on the Si substrate, and the substrate has electrical wires consisting of Ti/Pt/Au/Pt/Au, PbSn solder bumps electroplated on the wire layers with SiN as a dam, and two V-shaped holes on each side, which are chemically etched by KOH. The Si substrate is $3.2 \times 2 \times 0.625$ mm, the bumps are hemispheres about 50 μm in diameter, and the V-shaped holes are 300 μm square at the top and 120 μm deep with an angle of 54.7°.

Figure 10.16 shows the VCSEL-array module with a fiber–ribbon push/pull connector and an adapter. The package has a receptacle structure joined directly to the fiber connector. On the connector side of the package, two guide holes corresponding to the guide pins of the fiber connector were formed next to the cavity for the VCSEL chip. At each end of the package, two pairs of holes were made for the emboss pins and fixing screws. On the surface of the plastic-based package, electrical wires were formed by an electroplating technique called the MID (molded interconnection device) method, and the remaining area is covered by 30 μm thick metal to obtain efficient heat radiation. The package size is 28 mm × 10 mm × 7.5 mm.

The push/pull fiber connector has a ferrule consisting of two V-grooved silicon substrates and a silicon spacer. Sixteen optical fibers (GI-50/125) at a pitch of 250 μm are sandwiched

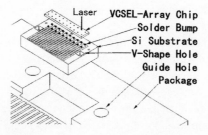

Figure 10.15. The configuration of the VCSEL-array package.

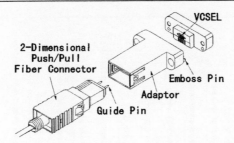

Figure 10.16. Schematic of the VCSEL-array module.

between the substrates and the spacer, and the edge facet is mechanically polished. The guide pins fixed at each side of the ferrule have the same diameter and pitch as those of a standard MT (mechanically transferable) connector. This ferrule is pushed forward by a coil spring in the connector housing.

The VCSEL-array chip was assembled on the Si substrate by using the self-alignment flip-chip bonding technique. The chip was placed in its approximate position on the substrate and then aligned precisely by reflowing at 200°C. A resin-based flux was used to obtain stable reflowing conditions. The substrate was then picked up by the master head by using a vacuum pump and was self-aligned to the head by using bumps on the head that fit into the V-shaped holes in the substrate. The bumps were made by inserting ball bearings 300 μm in diameter into the head. Two guide pins, corresponding to the guide pins on the fiber connector, were also formed on the master head to fit into the guide holes on the package. The guide pins were used to align the substrate to the package. The adapter was then roughly aligned to the package with the emboss pins and fixed in place with screws. The fiber connector was joined to the package simply by putting it into the adapter to which it was then automatically clamped.

The positioning accuracy of the VCSEL to the guide hole on the package was within 5.3 μm on average and 10 μm at most, and the positioning accuracy of the fiber to the guide pin was 1.3 μm on average and 3.1 μm at most.

Figure 10.17 shows the optical coupling loss to the fiber–ribbon connector for each of the sixteen channels without any index-matching material. The measured coupling loss was 2.1 dB with a deviation of ± 0.8 dB.

Each channel of the VCSEL-array module was operated at a bit rate of 1 Gbit/s by NRZ $2^{23} - 1$ pseudo-random bit streams (PRBS) at bias current (I_b) of 3 mA and a pulse current (I_p) of 5 mA. The receiver used for this measurement was an InGaAs-APD (NEC NDL5522P). Figure 10.18 shows the bit-error rate (BER) performance at room temperature as a function of receiver sensitivity. The optical sensitivity at a BER of 10^{-11} was -26.0 dBm on average with a power deviation of less than 0.9 dB. No significant error floor was observed for any of the sixteen channels within the measured range without an isolator. The BER difference between channels is considered to be due to the difference of output power and coupling loss. We also confirmed the operation of this module at a package temperature up to 70°C within a power deviation of 2.4 dB. Furthermore, DC measurement showed the temperature increase in the centered VCSEL to be only 20°C when the eight surrounding VCSELs were simultaneously driven at 7 mA. These results indicate that the module has sufficient temperature tolerance and heat radiation in spite of its plastic-based package.

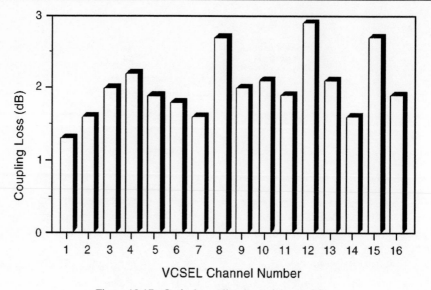

Figure 10.17. Optical coupling loss of the module.

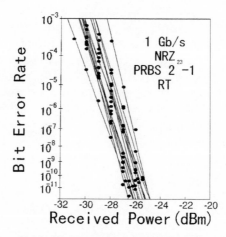

Figure 10.18. Bit-error-rate performance of the module.

10.3.2 Free-Space Optical Interconnections for Massively Parallel Processing

A prime example of a VCSEL application is in optical interconnections using optical fibers. However, there is an additional advantage offered by VCSELs that has not yet been commonly exploited. The ability of free-space optical systems to handle thousands of communication channels in combination with VCSELs could become important in the development of massively parallel processing (MPP) systems. One way to achieve this through novel network structures and routing concepts has been presented [32], and a 64-channel free-space network prototype using VCSEL arrays has been built [33].

Figure 10.19 shows the architecture of the network. The key operating principle of the network is that the functional requirements are divided between optics and electronics to make the best use of each technology's strong points. The network has two major parts. The optical part is optimized for the optical technology, to offer high connectivity and a

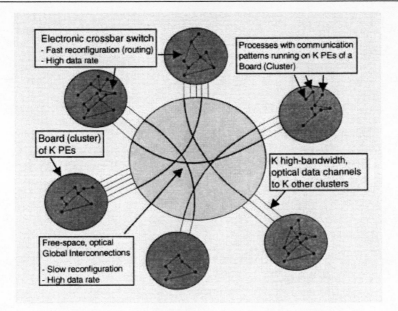

Figure 10.19. The Interconnection Cache Network (ICN).

large bandwidth. Optical interconnections are used in the circuit-switching mode. Once a point-to-point connection is established, the information flows optically, with no need for further routing or logical operations. Also, the conversion between electronic and optical signals is done only once each way. Once the optical network has been reconfigured, small electronic switches are used in a packet-switching mode. Information is routed by these switches, which we call *interconnection caches*, by selecting one out of k (e.g., $k = 16$) optical circuits that are already established. By using this combined approach, a network that can efficiently realize many MPP communication patterns is achieved.

This network architecture, named the Interconnection Cached Network (ICN) has been suggested for MPP systems [34, 35]. It benefits from the large bandwidth and high connectivity that optics has to offer and from the high processing and routing speeds that small electronic switches can achieve. Optical free-space interconnections seem to be a perfect fit with the requirements of a large, circuit-switched network. The overall effect, in many cases, is as if we had a very large network with very fast routing. More information about related aspects of our architecture and its advantages and performance benefits can be found in Lyuu et al. [36].

Figure 10.20 shows a schematic of a 1,024 channel free-space optical network that is proposed for the ICNs. This network will provide a nonblocking, reconfigurable interconnection for 512 PEs (processing elements) arranged in 32 clusters with 16 PEs each. 32×32 VCSEL arrays are used as transmitters and 8×4 fiber arrays with an optical detector/amplifier are used as receivers.

In Figure 10.20, faceted micromirror arrays have an optimum fan-in reflectance with spatially different coatings. Up to $M - 1$ sources (M is the number of boards, and in this case $M = 32$) may send to one receiver. One optical channel is allocated per receiver, and $M - 1$ VCSELs couple to this channel by means of an optimized partially reflecting micromirror. Since the various channels across the array are located at different distances from their destinations, the mirror reflectivities are set at different values as shown in

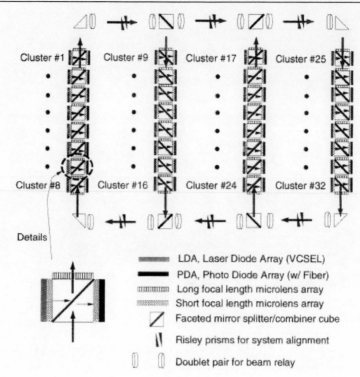

Figure 10.20. Schematic of 1,024 channel free-space optical network.

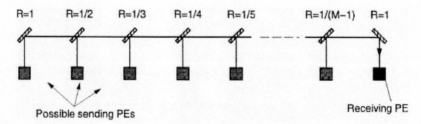

Figure 10.21. Optimum fan-in reflectivities.

Figure 10.21. Different reflectance coatings are used to optimize the power budget at each photodiode destination, so that an optimized equal optical power is received no matter from which optical beam of the VCSEL the signal was transmitted.

Arrays of micromirrors are placed between two prisms to form the beamsplitting/combining cubes shown in Figure 10.22. The individual beams within a cube remain spatially separate in the form of small microbeams. Beam diffraction limits the number of microbeam channels that can be supported within one cube.

The prototype of a 64-channel optical network [33] that interconnects 64 PEs arranged in four columns is shown in Figure 10.20. Each column consists of one board (a cluster) of 16 PEs. Four 8×8 VCSEL arrays and four 4×4 fiber arrays were used. The VCSELs in each 8×8 VCSEL array were separated by 250 μm.

Board-to-board spacing within a column, s, would be on the order of 30 mm. In the prototype of the 64-channel optical network, the output of the VCSELs was focused by

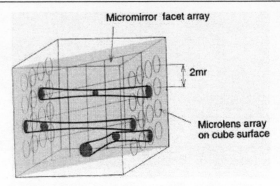

Figure 10.22. Microbeam combiner/splitter cube.

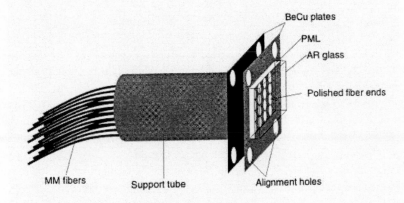

Figure 10.23. The fiber array.

the microlens arrays with a short focal length f_1 ($= 600\ \mu$m) and a very low numerical aperture NA to a waist 14 mm from the microlens array with a long focal length f_2. The short-focal-length microlens arrays were used to relay the complete array of channels from board-to-board in a column. One long-focal-length microlens array was placed between the boards. The focal length f_2 was set to be $s/2$, that is, about 15 mm. The distances and beam diameters can be adjusted to achieve the unique condition where the diffraction-limited beam waists at $\pm f_2$ are equal, enabling both transmitted and newly combined beams to propagate identically with only one relay microlens between the stages. Ray-tracing simulations of the wavefront suggest that manufacturing and alignment errors will dominate theoretical microlens aberrations.

The intercolumn distance was on the order of 300 mm, making bulk lenses more appropriate. These were used in two-lens, 4-f configurations to maintain the parallelism of the microbeams. The bulk lens focal length was about 75 mm, so individual beams had extremely small NAs. This meant that the lenses added only negligible aberration.

Metal micromirror arrays were fabricated and assembled with prisms into beam-combiner cubes under a microscope. We added the microlens arrays to the appropriate cubes for in-column relaying. Each board had one 4×4 multimode fiber array (Figure 10.23) to collect beams destined for the 16 PEs of that board.

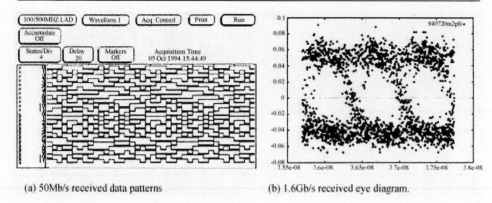

(a) 50Mb/s received data patterns   (b) 1.6Gb/s received eye diagram.

Figure 10.24. Low and high data rates results. (a) 50 Mb/s received data patterns. (b) 1.6 Gbps received eye diagram.

On each VCSEL array (i.e., each cluster board), a maximum of 12 lasers could operate in parallel at a relatively low-data rate (50 Mbit/s). These low rate channels were driven by a logic analyzer. A high-bandwidth laser (out of eight possible per board), was driven by a 3-Gbit/s bit-error tester (BERT). This allowed us to test the parallel operation (and crosstalk) and high bandwidth operation. Each fiber output was connected to a commercially available 100-Mbit/s fiber receiver module. The output of this module was connected to an input channel of the logic analyzer for all the low-bandwidth channels. For the high-bandwidth channel, the fiber output was fed to a high-bandwidth receiver connected to the bit-error tester. Figure 10.24(a) shows the results from the low-bandwidth channels taken at the various receivers by the logic analyzer. Figure 10.24(b) shows the 1.6-Gbit/s received eye diagram. All tested fast channels operated with an error rate of less than 10^{-13}, and most operated up to 1.6 Gbit/s, at which point they were limited by the driver circuitry design and the impedance mismatching of the 50-Ω signal line to the VCSEL.

The prototype of the 64-channel optical network used VCSEL arrays where all the VCSELs in an array had the same wavelength. A WDM scheme using different wavelengths, though, will increase communication channel density. A free-space optical network using four wavelengths was built, where all VCSELs in an array had the same wavelength, and the various arrays could have one out of four possible wavelengths. Although only a few wavelengths were used, the benefit in terms of higher overall network density is relatively large. Even with as few as four different wavelengths, we can quadruple the total number of channels. Also, using the same wavelength for all the VCSELs in each array will make the fabrication of such VCSEL arrays much easier.

Figure 10.25(a) shows how a fourfold communication density increase can be achieved by using a WDM technique to combine four clusters [37]. The details of the wavelength multiplexing and demultiplexing scheme are shown in Figure 10.25(b). The wavelengths of the VCSEL arrays we used in this WDM experiment were 843 nm, 950 nm, 967 nm, and 980 nm. The dielectric filter coating used for the wavelength combinations had a maximum transmittance of about 85% at each specific wavelength and an unwanted transmittance at different wavelengths of about 2%, which will increase the crosstalk associated with WDM. In a similar way, we can fabricate a 1,024-channel network for MPP using 16×16

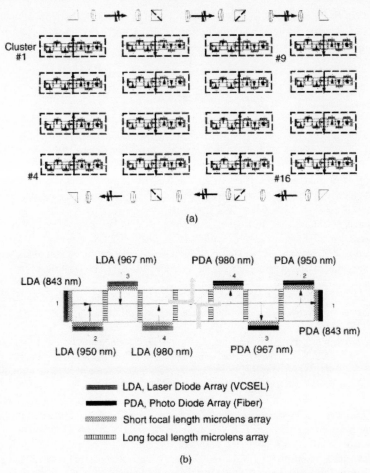

Figure 10.25. A 1,024 channel WDM system using 16×16 VCSEL arrays. (a) The overall system. (b) WDM details.

VCSEL arrays at four different wavelengths. This system uses one quarter of the space of an equivalent system using 32×32 VCSEL arrays as shown in Figure 10.20.

10.4 Conclusion

This chapter has described VCSELs that provide good temperature characteristics and controlled polarization directions. The wavelengths of these VCSELs can be changed in different areas of a substrate by using mask MBE. This technique can produce arrays that have a desired wavelength separation and that are useful in WDM applications. As examples of VCSEL applications, I also described 2-D interconnections using optical fibers and free-space optical interconnections for MPP.

In terms of optical interconnection applications, the unique features of VCSELs, such as their low power consumption and high production yield, makes them superior to conventional edge-emitting lasers, because the light sources used as transmitters in optical interconnections must consume less power and cost less to keep up with the advances in Si CMOS technology. It appears certain that VCSELs will play an important role as key components of optical interconnections–based networks.

References

[1] Young, D. B., Scott, J. W., Peters, F. H., Peters, M. G., Majewski, M. L., Thibeault, B. J., Corzine, S. W., and Coldren, L. A., "Enhanced performance of offset-gain high-barrier vertical-cavity surface-emitting lasers," *IEEE J. Quantum Electron.*, **29**, 2013–2022, 1993.

[2] Catchmark, J. M., Morgan, R. A., Kojima, K., Leibenguth, R. E., Asom, M. T., Guth, G. D., Focht, M. W., Luther, L. C., Przybylek, G. P., Mullally, T., and Christodoulides, D. N., "Extended temperature and wavelength performance of vertical cavity top surface emitting lasers," *Appl. Phys. Lett.*, **63**, 3122–3124, 1993.

[3] Shimizu, M., Mukaihara, T., Koyama, F., and Iga, K., "Polarization control for surface emitting lasers," *Electron. Lett.*, **27**, 1067–1068, 1991.

[4] Mukaihara, T., Koyama, F., and Iga, K., "Engineered polarization control of GaAs/AlGaAs surface-emitting lasers by anisotropic stress from elliptical etched substrate hole," *IEEE Photon. Technol. Lett.*, **5**, 133–135, 1993.

[5] Vakhshoori, D., "Symmetry considerations in vertical-cavity surface-emitting lasers: prediction of removal of polarization isotropicity on (001) substrates," *Appl. Phys. Lett.*, **65**, 259–261, 1994.

[6] Corzine, S. W., Scott, J. W., Geels, R. S., Young, D. B., Thibeault, B., Peters, M., and Coldren, L. A., "Reducing the effects of temperature on the threshold current of vertical-cavity surface-emitting lasers," *Proc. IEEE/LEOS Summer Topical Meet. on Integrated Optoelectronics*, Santa Barbara, CA, Aug., 51–52, 1992.

[7] Kajita, M., Kawakami, T., Nido, M., Kimura, A., Yoshikawa, T., Kurihara, K., Sugimoto, Y., and Kasahara, K., "Temperature characteristics of a vertical-cavity surface-emitting laser with a broad gain bandwidth," *IEEE J. Select. Topics Quantum Electron.*, **1**, 654–660, 1995.

[8] Kajita, M., Kurihara, K., Saito, H., Yoshikawa, T., Sugimoto, Y., and Kasahara, K., "Temperature-insensitive vertical-cavity surface-emitting laser array with a broad gain bandwidth," *Electron. Lett.*, **31**, 1925–1926, 1995.

[9] Lee, Y. H., Tell, B., Brown-Goebeler, K., and Jewell J. L., "Top-surface-emitting GaAs four-quantum-well lasers emitting at 0.85 μm," *Electron. Lett.*, **26**, 710–711, 1990.

[10] Nido, M., Naniwae, K., Shimizu, J., Murata, S., and Suzuki, A., "Analysis of differential gain in InGaAs-InGaAsP compressive and tensile strained quantum-well lasers and its application for estimation of high-speed modulation limit," *IEEE J. Quantum Electron.*, **29**, 885–895, 1993.

[11] Yamazaki, Y., Sasaki, Y., Yamaguchi, M, and Kitamura, M., "Measurement on nonuniform carrier distribution of MQW-LDs," *Ext. Abstr. (54th Autumn Meet.)*, Japan Society of Applied Physics and Related Societies, Sapporo, Sept., 28p-H-7 (in Japanese), 1993.

[12] Koyama, F., Morito, K., and Iga K., "Intensity noise and polarization stability of GaAlAs-GaAs surface emitting lasers," *IEEE J. Quantum Electron.*, **27**, 1410–1416, 1991.

[13] Chang-Hasnain, C. J., Harbison, J. P., Hasnain, G., Von Lehmen, A. C., Florez, L. T., and Stoffel, N. G., "Dynamic, polarization, and transverse mode characteristics of vertical cavity surface emitting lasers," *IEEE J. Quantum Electron.*, **27**, 1402–1409, 1991.

[14] Pan, Z. G., Jiang, S., Dagenais, M., Morgan, R. A., Kojima, K., Asom, M. T., Leibenguth, R. E., Guth, G. D., and Focht, M. W., "Optical injection induced polarization bistability in vertical-cavity surface-emitting lasers," *Appl. Phys. Lett.*, **63**, 2999–3001, 1993.

[15] Choquette, K. D. and Leibenguth, R. E., "Control of vertical-cavity laser polarization anisotropic transverse cavity geometries," *IEEE Photon. Tech. Lett.*, **6**, 40–42, 1994.

[16] Yoshikawa, T., Kosaka, H., Kurihara, K., Kajita, M., Sugimoto, Y., and Kasahara, K., "Complete polarization control of 8 × 8 vertical-cavity surface-emitting laser matrix arrays," *Appl. Phys. Lett.*, **66**, 908–910, 1995.

[17] Kuksenkov, D. V., Temkin, H., and Yoshikawa, T., "Dynamic properties of vertical-cavity surface-emitting lasers with improved polarization stability," *IEEE Photon. Technol. Lett.*, **8**, 1–3, 1996.

[18] Saito, H., Kosaka, H., Sugimoto, M., Ogura, I., Kasahara, K., and Sugimoto, Y., "Integration of vertical-cavity surface-emitting devices by molecular beam epitaxy regrowth," *J. Vac. Sci. Technol. B*, **12**, 2905–2909, 1994.

[19] Kosaka, H., Kurihara, K., Uemura, A., Yoshikawa, T., Ogura, I., Numai, T., Sugimoto, M., and Kasahara, K., "Uniform characteristics with low threshold and high efficiency for a single-transverse-mode vertical-cavity surface-emitting laser-type device array," *IEEE Photon. Technol. Lett.*, **6**, 323–325, 1994.

[20] Ohtoshi, T., Kuroda, T., Niwa, A., and Tsuji, S., "Dependence of optical gain on crystal orientation in surface-emitting lasers with strained quantum wells," *Appl. Phys. Lett.*, **65**, 1886–1887, 1994.

[21] Choquette, K. D., Richie, D. A., and Leibenguth, R. E., "Temperature dependence of gain-guided vertical-cavity surface emitting laser polarization," *Appl. Phys. Lett.*, **64**, 2062–2064, 1994.

[22] Willner, A. E., Chang-Hasnain, C. J., and Leight, J. E., "2-D WDM optical interconnections using multiple-wavelength VCSEL's for simultaneous and reconfigurable communication among many planes," *IEEE Photon. Technol. Lett.*, **5**, 838–841, 1993.

[23] Chen, Y. K., Guo, W. Y., Chi, S., and Way, W. I., "Demonstration of in-service supervisory repeatless bidirectional wavelength-division-multiplexing transmission system," *IEEE Photon. Technol. Lett.*, **7**, 1084–1086, 1995.

[24] Chang-Hasnain, C. J., Harbison, J. P., Zah, C.-E., Maeda, M. W., Florez, L. T., Stoffel, N. G., and Lee, T.-P., "Multiple wavelength tunable surface-emitting laser arrays," *IEEE J. Quantum Electron.*, **27**, 1368–1376, 1991.

[25] Koyama, F., Mukaihara, T., Hayashi, Y., Ohnoki, N., Hatori, N., and Iga, K., "Wavelength control of vertical cavity surface-emitting lasers by using nonplanar MOCVD," *IEEE Photon. Technol. Lett.*, **7**, 10–12, 1995.

[26] Saito, H., Ogura, I., Sugimoto, Y., and Kasahara, K., "Monolithic integration of multiple wavelength vertical-cavity surface-emitting lasers by musk molecular beam epitaxy," *Appl. Phys. Lett.*, **66**, 2466–2468, 1995.

[27] Yuen, W., Li, G. S., and Chang-Hasnain, C. J., "Multiple-wavelength vertical-cavity surface-emitting laser arrays with a record wavelength span," *IEEE Photon. Technol. Lett.*, **8**, 4–6, 1996.

[28] Wipiejewski, T., Peters, M. G., and Coldren, L. A., "Vertical-cavity surface-emitting laser diodes with post-growth wavelength adjustment," *Proc. LEOS'94*, **1**, 265–266, 1994.

[29] Ogura, I., Kurihara, K., Kawai, S., Kajita, M., and Kasahara, K., "A multiple wavelength vertical-cavity surface-emitting laser (VCSEL) array for optical interconnection," *IEICE Trans. Electron.*, **E78-C**, 22–27, 1995.

[29a] Hahn, K. H., Giboney, K. S., Wilson, R. E., Stranznicky, J., Wong, E. G., Tan, M. R., Kaneshiro, R. T., Dolfi, D. W., Mueller, E. H., Plotts, A. E., Murray, D. D., Marchegiano, J. E., Booth, B. L., Sano, B. J., Madhaven, B., Raghavan, G., and Levi, A. F. J., "Gigabyte/s data communications with the POLO parallel optical link," *Proc. 46th ECTC*, 301–307, 1996.

[30] Won, Y. M., Muehlner, D. J., Faudskar, C. C., Buchholz, D. B., Fishteyn, M., Brandner, J. L., Parzygnat, W. J., Morgan, R. A., Mullally, T., Leibenguth, R. E., Guth, G. D., Focht, M. W., Glogovsky, K. G., Zilko, J. L., Gates, J. V., Anthony, P. J., Tyrone, B. H., Ireland, T. J., Lewis, D. H., Smith, D. F., Nati, S. F., Lewis, D. K., Rogers, D. L, Aispain, H. A., Gowda, S. M., Walker, S. G., Kwark, Y. H., Bates, R. J. S., Kuchta, D. M., and Crow, J. D., "Technology development of a high-density 32-channel 16-Gb/s optical data link for optical interconnection applications for the optoelectronic technology consortium (OETC)," *J. Lightwave Technol.*, **13**, 995–1016, 1995.

[31] Kosaka, H., Dutta, A. K., Kurihara, K., Sugimoto, Y., and Kasahara K., "Giga-bit-rate optical signal transmission using vertical-cavity surface-emitting lasers with large-core plastic-cladding fibers," *IEEE Photon. Technol. Lett.*, **7**, 926-928, 1995.

[32] Schenfeld, E., "Massively parallel processing with optical interconnections: What can be, should be and must not be done by optics," 1995. *Proc. OC'95*, 16–18, 1995.

[33] Araki, S., Kajita, M., Kasahara, K., Kubota, K., Kurihara, K., Redmond, I., Schenfeld, E., and Suzaki, T., "Massive optical interconnections (MOI): Interconnections for massively parallel processing," *Proc. OC'95*, 8–10, 1995.

[34] Feitelson, D., Rudolph, L., and Schenfeld, E., "An optical interconnection network with 3-D layout and distributed control," *Proc. SPIE*, **1281**, 54–65, 1990.

[35] Gupta, V. and Schenfeld, E., "A heuristic classical communication topologies in the OPAM architecture," *Proc. IPPS'93*, 291–298, 1993.

[36] Lyuu, Y.-D. and Schenfeld, E., "MICA: A mapped interconnection-cached network," *Proc. 5th IEEE Symposium on the Frontiers of Massively Parallel Computation*, 80–89, 1995.

[37] Kajita M., Kasahara K., Kim T. J., Ogura I., Redmond I., and Schenfeld E., "Free-space wavelength division multiplexing optical interconnections for massively parallel processing systems," *Proc. OC'96*, 24–25, 1996.

11

VCSEL-Based Fiber-Optic Data Communications

Kenneth Hahn and Kirk Giboney

11.1 Introduction

The rapid growth in the use of optics in data communications in recent years has been brought about by several important developments:

1. continued exponential growth in bandwidth demand of computing and communication systems.
2. improvement in the performance of fiber-optic components and integrated systems, and
3. rapid decline in fiber-optic component cost.

The natural evolutionary path for optics is from long to short distances. Telecommunication systems are owned and operated by large institutions serving national and regional entities; with such large bases using relatively few telecommunication links, high per link cost is easily tolerated. Optics quickly replaced copper. At shorter distances, the number of users per link declines rapidly with distance. To the desktop, there is one user per link, and if used inside a computer, there would be but one user for a large number of optical links. Success of optics in such short-distance data communications applications will depend on extremely low cost components that are manufactured in very high volume.

Light-emitting diodes (LEDs) have been used with multimode fiber (MMF) to provide relatively low-cost solutions for data rates up to several hundred Mbit/s at distances up to 1–2 km. The large core diameters of MMF greatly ease alignment requirements and lower the cost of the fiber-optic packages and connectors. Because LEDs are considered impractical above about 622 Mbit/s, migration to Gbit/s data rates requires alternate high-speed sources with the low-cost component and packaging properties of LEDs.

The advent of VCSELs, combined with increasing demand for Gbit/s and Gbyte/s interconnections, has created a new wave of fiber-optic data communication technologies in recent years. VCSELs have desirable characteristics of both LEDs and high-speed lasers, making them ideal sources for data communications. These well-known advantages include their low drive currents, high-speed response, low numerical aperture (NA) and circular output beams, low noise, planar fabrication, wafer-level testing, and emission perpendicular to the substrate. These properties have enabled developments of low cost Gbit/s serial optical links and multi-Gbit/s parallel optical links for high-speed interconnections at distances up to several hundred meters.

Table 11.1. *Established and emerging high-performance fiber-optic data communications standards including 850 nm VSCEL-based physical specifications.*

Standard	Committee	Optical Bit Rate
Serial:		
ATM	ATM Forum	622 MBaud (OC-12)
		2488 MBaud (OC-48) (planned)
Gigabit Ethernet	IEEE 802.3	1250 MBaud
Fibre Channel	ANSI X3T11	531.25, 1062.5, 2125, 4250 MBaud
Parallel:		
SCI	ANSI/IEEE 1596	16 × 500 MBaud (+1 flag, 1clock)
HIPPI-6400	NCITS T11	8 × 1000 MBaud (+2 control, 1 frame, 1 clock)

11.2 VCSEL Applications in Fiber-Optic Data Communications

11.2.1 Data Communications Systems Requirements

Demand for interconnection bandwidth in computing and communication systems has continued to multiply for several decades. Some established and emerging high-speed serial and parallel communication standards including VCSEL-oriented specifications are listed in Table 11.1. Asynchronous Transfer Mode (ATM), Gigabit Ethernet, Fibre Channel, and HIPPI-6400 all specify line rates of at least 1 Gbaud.

Present-day desktop systems have clock rates approaching 400 MHz and high-end processors have clock speeds exceeding 500 MHz, such as DEC's latest 600 MHz Alpha processor. The sustained memory bandwidth of the Runway bus in HP 9000-series servers is 768 Mbyte/s [1]. Intel and Hewlett-Packard have been collaborating on advanced processor technologies and have announced a 64-bit processor, scheduled for production in 1999, that could run at clock speeds of around 900 MHz. This coming Intel Merced IA-64 processor is based upon a new instruction set architecture, called Explicitly Parallel Instruction Computing (EPIC), that will facilitate parallel processing. Fiber-optic links will likely serve in parallel-processor interconnections.

A desktop or network server processor operating near 1 GHz by the year 2001 will have a on-processor bus bandwidth in the vicinity of 8 Gbyte/s. When clustered together for high performance in a large multiprocessor system, for example, such processors will require tremendous interconnection performance, including bandwidth length, interconnection density, and low-cost interfaces. New Gbyte/s parallel interconnection standards, such as Scalable Coherent Interface (SCI) (ANSI/IEEE 1596-1992) and High-Performance Parallel Interface – 6,400 Mbit/s (HIPPI-6400) (NCITS 1997) are being established to meet the needs of these systems. Existing copper or LED-based fiber-optic interconnection technologies will not be able to meet all the demands of such systems, and so laser-based fiber-optic solutions are being specified.

11.2.2 Transmission Media

Simple metal wire is by far the cheapest medium within its bandwidth and distance limitations. These limitations are becoming apparent as system speeds continue to increase. Attenuation of electromagnetic waves in metal waveguides is dominated by the finite

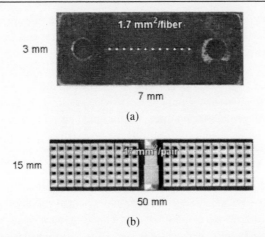

Figure 11.1. Comparison of current (a) parallel fiber-optic and (b) high-speed copper cable connector interface density. The fiber interface is denser by a factor of ten. Typical media used with such connectors would carry 1.25 Gbit/s signals 10–15 m in the copper cables and 260–650 m or farther in multimode fibers, depending on the core structure and wavelength.

resistance of the metal conductors. At high frequencies ($>$ ~ 1 MHz in typical cables) the resistance increases due to the skin effect and the loss becomes proportional to the square root of frequency [2]. (Skin effect refers to the limited penetration of electromagnetic fields into a conductor.) The high-frequency loss causes intersymbol interference in digital signals. There have been several techniques developed for reducing or compensating for the loss, but they necessarily increase the cable size and/or cost [3, 4].

The attenuation in a metallic transmission line also is inversely proportional to its cross-sectional linear dimensions, and so there are fundamental trade-offs among bandwidth, distance, and size. A problem that is increasingly vexing system designers is density limitations of the cables and connectors. This is illustrated in Figure 11.1, where connector interface densities of multichannel fiber-optic and copper-cable links are compared.

Metallic interconnects have other serious handicaps, especially with regard to electromagnetic compatibility. Crosstalk becomes significant at high frequencies. Much engineering effort is required to control high-frequency electromagnetic emissions and shield small signals. Allowing for ground differentials adds complexity in the signalling scheme. If the connected systems are separated by as little as a few meters, ground differentials can introduce a significant offset and can be a source of noise. All of these problems taken together significantly increase overall implementation cost of metallic waveguides at high frequencies.

Dielectric waveguides have much lower loss than metallic waveguides because there is no associated metal resistance. However, the minimum cross-sectional size of a dielectric waveguide to adequately confine the electromagnetic wave is on the order of a wavelength. Using light as a carrier allows very small waveguide size and very large modulation bandwidths.

Dielectric optical waveguides are free of many other problems associated with metallic waveguides. Concerns over electromagnetic interference, crosstalk, and security are relieved. Dielectric waveguides are immune to effects of voltage differentials since there is no conductive path. Low-loss optical fiber is manufactured in massive quantities, and it is easy to handle and install. Optical fiber is much smaller and lighter and has a tighter minimum bend radius than comparable metallic waveguides.

Silica fiber was investigated as a medium for guided optical communications in the 1960s and first implemented in telecommunications systems in the late 1970s. It was widely deployed in long-distance telecommunications links in the 1980s. In the 1990s, optical fiber is commonly utilized in metropolitan, campus, and building intraconnections. Present applications are bringing optical fiber into local areas, telecom switch interconnects, and multiprocessor systems.

The distance at which optical fiber is cost effective decreases as the data rate increases. A single-wavelength signal on a standard MMF can carry 1 Gbit/s data 20 times as far in 1/400th the cross-sectional area as competing copper technologies. Orders of magnitude greater bandwidth on the same fiber is available by employing wavelength-division multiplexing. The bandwidth–distance–size trade-offs and falling transmitter and receiver costs are the primary forces driving fiber optics to shorter distance links, and VCSELs are a pivotal development.

11.2.3 Serial Fiber-Optic Data Links

A number of different optical sources are used for fiber-optic communications. Several factors determine which is used for each specific application, but one general guideline is reliable: The source that allows the system to attain expected performance goals with the overall lowest cost impact is heavily preferred. Prior to the availability of VCSELs, edge-emitting lasers were used predominantly for longer distances and LEDs for shorter distances. These choices were primarily based upon cost–performance trade-offs. VCSELs change the landscape significantly by offering performance that improves upon many aspects of both edge-emitting lasers and LEDs at nearly the cost of LEDs. VCSELs are an ideal source for short- and medium-distance high-speed data communications.

Fiber-optic communications links can be divided into three broad classes that are characterized primarily by the distance between source and receiver. Each of these classes generally uses different optical sources and fiber to address unique requirements. Long-distance telecommunications can be considered to consist of links of greater than about 20 km, where dispersion and loss in single-mode fibers (SMFs) becomes significant. Links between 500 m and 20 km also transmit over SMF for metropolitan and campus area connections and long building backbones. Below 500 m multimode fiber is used for local-area interconnections between computers, to storage farms, and in building backbones.

11.2.3.1 Long-Distance Telecommunications

Very long fiber-optic links connecting cities, countries, and continents are significantly affected by loss and dispersion properties of the fiber. The primary wavelengths used for long-distance transmission are determined by the zero dispersion and minimum loss in silica glass fibers. The loss in highly pure silica is minimum around 1,550 nm and chromatic dispersion passes through zero at about 1,300 nm. Impurities in glass fiber materials in the 1970s caused high losses at these wavelengths, but left a loss minimum in the 800–900 nm range. Early installations, in the late 1970s, used 850 nm wavelength, since practical optical sources and detectors were also available in that range. Progress in fiber and semiconductor laser and photodetector manufacturing technologies allowed installation of 1,300 nm systems in the 1980s. The 1990s have seen a shift to 1,550 nm wavelength for very long links through more advanced fiber and device technologies. Erbium-doped fiber optical amplifiers are available and practical at 1,550 nm. Waveguide dispersion is used to compensate the material dispersion and shift the dispersion zero in SMF from 1,300 nm to 1,550 nm.

The great distances spanned by long-distance links determine the factors affecting overall cost. The high number of users supporting a typical long-distance communication link justifies the use of relatively expensive technologies to maximize throughput per link. While high media and installation costs motivate efficient exploitation of the fiber bandwidth, repairs can also be very expensive, such as in the case of undersea links. Fiber amplifiers have greatly extended the maximum link distance without regeneration, virtually eliminating the need for repeaters and enhancing system reliability. This reduces the demand for optoelectronic transmitters and receivers and renders the costs of optical sources and detectors negligible.

Distributed-feedback (DFB) lasers are used as sources for long-distance data and telecommunications to reduce the effects of dispersion. A DFB laser is an edge-emitting laser with a fine pitch grating integrated along its length to provide internal optical feedback. With appropriate temperature control, a DFB laser simultaneously produces a precise wavelength and a very narrow optical linewidth. These devices are not mass produced and are very expensive. While VCSELs may improve on some aspects of DFB lasers, the simultaneous control of wavelength and linewidth is difficult in a short vertical cavity. More importantly, there is little pressure in this low-volume market to drive development of less expensive optical sources.

11.2.3.2 Metropolitan

The market for transmitters and receivers in the 500 m to 20 km range is much larger than for long distance links, and it is growing. Dispersion is more easily managed over this distance range, and wavelength and linewidth controls are relaxed compared to long-distance links.

Edge-emitting Fabry–Perot lasers operating at 1,300 nm are most commonly used with SMF for this class of links. These devices are simpler and much less expensive than DFB lasers, but like virtually all high-performance edge-emitting lasers, they are difficult to mass produce. Most current edge-emitting lasers' output characteristics rely on cleaved facets that form the cavity mirrors. Cleaving has not proven to be highly reproducible, and there is significant yield loss in this operation. Additional requirements for handling cleaved parts also increase testing costs compared to wafer-scale production and testing. Packaging of edge-emitting laser-based transmitters has additional costs for heat dissipation, environmental protection of the sensitive cleaved facets, and optics for coupling to SMF. The optical output of an edge-emitting laser is generally astigmatic; so even with expensive optics, coupling efficiencies are typically around 50%.

Edge emitter development focuses increasingly on low-cost solutions. Etched facets and optical coatings can be implemented at wafer scale for lower cost volume production. Mirror coatings should improve efficiency and environmental stability. Inventions such as the integrated mode transformers that correct astigmatism may reduce packaging costs.

A 1,300 nm VCSEL-based single-mode transmitter could be much cheaper for several reasons, primarily better device yield and less expensive testing, drive electronics, and packaging. Yield and environmental robustness of VCSELs are far better than edge-emitting lasers because they consume less wafer area and they do not rely on end facets for mirrors. The small size of VCSELs affords low parasitics, and devices have been directly modulated to 10 Gbit/s. VCSELs can be fully tested in wafer form, and automated IC testing methods can be used with minor modifications. VCSELs are significantly more efficient than edge-emitting lasers, and the temperature dependence of VCSEL output characteristics can be controlled much better, making VCSELs much less demanding on the current drive circuits and power dissipation properties of the packaging.

The VCSEL output beam shape and spectrum can be tailored to application requirements. VCSELs having circular output beams can directly butt-couple to single- or multimode fiber with 80–95% efficiency. For the same input power, VCSEL-based transmitters can be expected to couple an order of magnitude higher optical power into fiber with lower cost optics than their (uncoated) edge-emitter-based counterparts.

11.2.3.3 Premises

Terminal costs compose a significant fraction of the overall link cost for shorter interconnects. A large portion of the terminal costs for a single-mode link is in precision packaging to achieve tight alignment tolerances required with SMF. Whereas SMFs have core diameters of around 9 μm, standard silica MMFs have core diameters of 62.5 or 50 μm. MMF greatly relaxes alignment tolerances and allows high-volume production methods. MMFs support thousands of modes and the bandwidth–distance product is limited by modal dispersion. Still, MMFs will transmit Gbit/s data rates up to a kilometer, depending on wavelength.

The choice of optical source has consequences for the link cost and performance. Three types of sources are used in premises links: LEDs, CD lasers, and VCSELs. LEDs are currently the least expensive, lowest performance, and most common.

Light-emitting diodes have been popular because they are inexpensive and they have met past performance demands. LEDs consume about an order of magnitude less area than edge-emitting lasers and can have high yields in planar fabrication processes. Being surface-emitting devices, their performance characteristics do not depend upon vertical (cleaved) facets as do edge-emitting lasers, and so they are fully wafer testable. LEDs couple to many fiber modes, giving a stable bandwidth–distance product and immunity to modal noise (discussed in detail in the next section). LED-based links at 850 nm or 1,300 nm are used in Ethernet (10 Mbit/s), Fast Ethernet (100 Mbit/s), FDDI (100 Mbit/s), Fibre Channel (266 and 532 Mbit/s), ATM at OC-3 and OC-12 (155 Mbit/s and 622 Mbit/s), and a number of proprietary interconnections.

LED links are reaching performance limitations that will prevent their use in higher speed links. Light from LEDs is mostly spontaneous emission, which makes them much noisier than lasers, and the long spontaneous emission lifetimes limit modulation bandwidths. The broad optical bandwidth can result in chromatic dispersion even greater than the modal dispersion in MMFs, further reducing the bandwidth–distance product. These factors are limiting LED-based links to a current best product specification of 622 Mbit/s over 500 m at 1,300 nm wavelength.

Other factors impact cost as well as performance. The current-to-light conversion in LEDs is very inefficient. The broad emission pattern limits coupling efficiency to MMF to about 10% of the emitted light. As a result, less than 0.1% of the electrical power supplied to an LED appears as guided light in a MMF. The large currents required, on the order of 100 mA, add expense to the drive electronics and packaging. The small transmitted optical power requires sensitive receivers, further adding to cost.

The performance of lasers is better, in general, than LEDs. The stimulated emission process in a laser is much quieter and faster than the spontaneous emission in an LED. The filtering properties of the laser resonator ensure relatively narrow spectral and spatial emission characteristics. Laser light usually couples to optical fiber more efficiently and disperses less as it propagates in the fiber than does LED light. Cleaved edge-emitting lasers typically have power efficiencies of a few percent.

The 1,300 nm edge-emitting lasers used in metropolitan and campus area links and lasers used in compact disk (CD) players are used in some multimode links. While high-performance edge-emitting lasers for communications are not cost competitive with LEDs, they provide a performance advantage not available with the cheaper devices. CD lasers have a cost structure based upon a large manufacturing base and they are widely available. However, they operate at 780 nm wavelength and are not optimized for high-speed modulation.

The disadvantages of edge-emitters noted for metropolitan single-mode links are even more significant for premises multimode links owing to the greater impact of the terminal costs. The power efficiency of coupled light from edge-emitters to MMF is still low, and the high drive currents result in excess heating and electromagnetic radiation. The temperature sensitivity of edge-emitters compounds the problems, resulting in expensive electronics and packaging. Furthermore, edge-emitting lasers are not well suited for multimode optical fiber communications because they couple to only a few of the available modes of MMF, which can result in large variations in the link bandwidth and puts the system at risk of modal noise.

VCSELs combine the essential advantages of edge-emitting lasers and LEDs while shunning the disadvantages. Being surface-emitting devices like LEDs, they are environmentally robust and can be manufactured and tested on the scale of integrated circuits with similar cost benefits. Like edge-emitting lasers they can be modulated at high speeds with low noise and emit lots of light, relaxing demands on the photoreceiver. Like neither, VCSELs can have spatial and spectral outputs tailored over a wide range of possibilities. This allows VCSELs to achieve the low noise and high speed of lasers with the stability and freedom from modal noise of LEDs in multimode fiber.

VCSELs owe much of their attractiveness over edge-emitters and LEDs to their very high efficiency. The efficiency of light coupled into a fiber for the amount of electrical power supplied can be at least an order of magnitude greater for VCSELs than for edge-emitting lasers and more than two orders greater than for LEDs, and VCSELs can accomplish this without coupling optics. The lower drive currents and internal temperature compensation in VCSELs greatly reduce the cost of electronics. VCSEL drive currents of a few milliamps allow the possibility of direct drive from complex CMOS application ICs. Low current densities and planar structures without exposed facets in VCSELs correlate with longer lifetimes.

These performance advantages come at the expense of more complex epitaxial structures and a slightly higher device costs than for LEDs. The overall system cost savings and performance enhancements more than offset the device expense, and as VCSEL-associated technologies mature, costs will fall further. VCSELs are clearly the source of choice for Gbit/s and faster premises links.

11.2.4   New Data Link Applications

11.2.4.1 Parallel Fiber-Optic Data Links

Parallel optical link modules are essentially packaging configurations that exploit the low power consumption, natural array form, and low cost advantages of VCSELs. The availability of VCSELs is enabling commercial development of parallel optical link modules for use in switching systems, multiprocessor computers, and premise interconnections for computer clusters and multimedia. These applications require gigabyte bandwidths over distances of hundreds of meters, low latency, and high interconnection densities on limited budgets. An example is the telephone central office switch environment, where multiple Gbit/s links up to a few hundred meters connect thousands of switches. These networks must be robust and hot reconfigurable.

With standard copper solutions, several bottlenecks exist that limit system performance, including limited backplane real estate for connectors, the sheer bulk of the cable bundles, attenuation at high data rates, and electromagnetic interference (EMI) caused by the large electrical system [5]. Serial fiber-optic solutions relieve bandwidth–distance, cable bulk, and EMI limitations of metallic interconnects. However, they still consume significant backplane space and they are somewhat more expensive than copper interconnects. A much smaller footprint is possible with parallel optical modules rather than with multiple serial modules, greatly increasing interface density. Laser arrays, multichannel driver and receiver ICs, and fiber-ribbon optical connectors are combined in the parallel optical link modules, amortizing packaging costs over several channels.

As compared to copper wire cables, ribbon-fiber cables can provide highest performance in both bandwidth–length and interconnection density. Although twisted pair and microcoaxial cables can accommodate transmission lengths of ∼10–20 m at 1 Gbit/s per line, interconnection densities are limited by the millimeter dimensions. Flexible film cables can provide high-density interconnections, but useful lengths are limited to a few meters at 1 Gbit/s per channel [6]. As illustrated in Figure 11.1, ribbon fiber maintains its high-density advantage at the connector interfaces, saving precious real estate at the card edge and in the backplane. Multifiber ribbon connectors have a pitch of 250 μm or less. High-speed copper connectors currently position pins at 2-mm pitch, and a pair of pins is required for each signal, and additional pins are used for grounds or shields. Finally, while pricing in the fiber-optics marketplace is rapidly evolving, parallel optical links are expected to become cost competitive with high-end copper cable interconnects.

Fiber ribbon is also more suitable for carrying parallel data than serial fiber or parallel copper media. Time- and space-consuming multiplexing and demultiplexing operations necessary for serial transmission can be reduced or eliminated by transmitting in parallel. The interchannel propagation delay variation, or skew, in standard fiber ribbon is around 10 ps/m, while in copper parallel cables it is typically 70–100 ps/m. By constructing the ribbon from consecutive fiber sections from the same pull, interchannel skew of less than 1 ps/m has been demonstrated, and 2 ps/m is considered manufacturable [7]. This simplifies system design by eliminating the need for skew correction, even after hundreds of meters.

11.2.4.2 Wavelength-Division Multiplexing

While parallel links reduce the costs of end components, they don't significantly cut the expense of the transmission medium. Optical fiber dominates the cost of longer links and conventional, single-wavelength links use only a small fraction of the available fiber band-width. In each of the wavelength "windows" of optical fiber, roughly 25 THz bandwidth is available. Most of the large base of installed fiber is suitable for wavelength-division multiplexing (WDM), a technique for fuller utilization of this intrinsic fiber bandwidth.

The baud rate of a given channel in a VCSEL-based MMF link is limited by modal dispersion. Dispersion limits are avoided in a WDM system by transmitting limited baud rates over multiple wavelength-specific channels. The modulated light from a set of lasers, each having distinct wavelengths, is combined and transmitted over a single optical fiber, and then separated and independently detected at the receiving end. The potential band-width utilization is limited by the precision and stability of the laser wavelengths and by multiplexing and demultiplexing component spectral characteristics. Chromatic dispersion causes each of the channels to experience a slightly different delay. This deterministic skew increases as the wavelength is increased or decreased away from the zero dispersion wave-length at about 1,300 nm in silica fiber. The modal dispersion in MMF is nearly identical at each of the wavelengths within a window and only limits the individual channel data rates.

The opportunities for VCSEL-based WDM links are in metropolitan and premises ap-plications where DFB lasers are not cost effective [8]. Because the end costs in these links comprise a larger fraction of the total link costs than in long-haul links, the VCSEL economic advantages are more significant. The higher fractional end costs motivate less expensive approaches throughout the system design. Larger wavelength spacings (coarse WDM) relax demands on component tolerances, and so source spectra have room to vary and cheaper optics can be used without thermal controls. Relatively short distances allow simpler opto-mechanical designs with greater losses than would be tolerated in a long-distance link.

Monolithic WDM VCSEL arrays would allow simpler and cheaper coarse-WDM trans-mitters. By leveraging the precision patterning of the device photolithography, optical align-ment can be accomplished more quickly and efficiently in manufacturing. However, serial and parallel links will still be cheaper alternatives for shorter distances, and the additional distance allowed by using long wavelengths is possibly critical for the cost effectiveness of VCSEL-based WDM links. Manufacturable WDM VCSEL arrays at 1,300 nm wavelength would all but ensure market application of this technology.

11.3 VCSEL-Based Link Design

While the advantages of discrete VCSELs have already placed them into serial link mod-ules, future products will exploit the multichannel advantages of VCSEL arrays in parallel or WDM link modules. Entire product lines will likely span the range from one to many channels. Basing product "families" on a common scalable platform has obvious volume manufacturing and cost advantages. As a result, single-channel VCSEL-based link module designs will incorporate features motivated by multichannel versions. These features will ap-pear in the devices, electronics, electrical and optical interfaces, packaging, and connectors.

Link rise time and power budgets specify the overall link performance. These budgets are determined by performance requirements, such as baud rate, timing margin, bit-error rate (BER), and link length, and by limitations, such as dispersion, skew, and maximum power for eye safety. The VCSEL characteristics, wavelength, performance over temperature, manufacturing tolerances, and fiber type all strongly influence the link budgets through both achievable performance and regulatory limitations.

11.3.1   Critical VCSEL Performance Parameters

The use of VCSELs for optical communications requires an understanding of many parameters in the context of the interconnection system. These include current–voltage, light–current, spectral, and modal characteristics; modulation response, noise, and reflection sensitivity; and how they interact with the transmission medium, particularly with respect to modal/chromatic dispersion. Other parameters, such as heat dissipation, temperature range, reliability, and environmental ruggedness, indirectly influence VCSEL performance and range of application. The following describes some metrics important in short-distance multimode links, where VCSELs are quickly being adopted as optical sources. VCSEL performance characteristics are covered in greater depth in Chapter 6.

11.3.1.1 Modulation Response

The multi-GHz response of VCSELs at low drive currents makes them attractive for Gbit/s operation in optical data links. With a small active layer and high photon density, VCSELs have inherent high-speed response. For example, relaxation oscillation frequencies of up to 70 GHz have been measured [9]. In practice, however, modulation response is limited by device and package parasitics as well as carrier-transport and optical cavity effects. In implanted, gain-guided VCSELs, small-signal modulation response of 14 GHz has been demonstrated [10] with 8 mA bias. With a 6-μm wide aperture oxide-confined VCSEL, a response over 16 GHz was demonstrated with only 4.5 mA bias [11].

The large-signal response is the primary metric for digital systems, and it is distinguished by a requirement to maximize the modulation depth. This allows the largest signal to noise ratio (S/N) for a given optical power level, which is generally limited by eye safety regulations. To maximize S/N, the optical power for one logic state must be large and the other must be very small. The degree to which this is successful is measured by the extinction ratio, the ratio of on to off power. This metric is specified in some standards such as Fibre Channel, where it is 9 dB$_{opt}$ minimum for the 1.06 Gbit/s baud rate and above.[1]

Figure 11.2 shows an eye diagram of a proton-implanted VCSEL driven by a $2^7 - 1$ pseudorandom bit sequence (PRBS) at 1 Gbit/s and measured with a 1 GHz bandwidth photoreceiver. An eye diagram is an overlay of all bits produced by a binary pattern, usually a pseudorandom bit sequence, displayed on an oscilloscope. Various large-signal performance metrics such as rise and fall times, overshoot, undershoot voltage levels, jitter, distortion, skew, and extinction ratio can be derived from an eye diagram.

Note the broadened rising edges in Figure 11.2 that are characteristic of pattern-dependent jitter, where the laser turn-on delay depends upon the driving signal recent history. The jitter results from the finite times required for the carrier populations to reach steady state in the device and is a function of the driving conditions, generally becoming more pronounced as the off-state current is reduced below threshold. The greater the number of on states immediately preceding a particular rising edge, the larger the carrier density will be immediately

---

[1] Decibel (dB) units: The decibel units definition is based upon a power ratio using a specified reference power, $10 \cdot \log(P/P_{ref})$. Semiconductor lasers and photodetectors used in optical communications systems are square-law devices, converting electrical current into optical power and vice versa, and so the decibel value depends upon whether the power is optical or electrical. Optical decibels are twice as large as electrical decibels. Thus a measure in optical decibels (e.g., at a photodetector input) will have one half the numerical value as in electrical decibels (e.g., at the photodetector output). A convention of denoting dB$_{opt}$ for optical decibels and dB$_{elec}$ for electrical decibels is adopted here to avoid confusion.

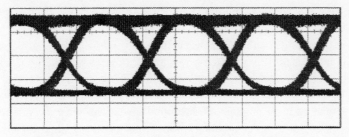

Figure 11.2. Eye diagram of a 15 μm diameter proton-implanted 850 nm VCSEL driven by a $2^7 - 1$ PRBS at 1 Gbit/s with 18 mA average current, measured with a 1 GHz bandwidth photoreceiver (horizontal scale is 400 ps/div).

prior to that rising edge. Less charge is required to reach threshold and the rising edge appears earlier. Pattern-dependent jitter is a dynamic form of duty cycle distortion that reduces the time window during which a discrete level decision can be made with confidence.

11.3.1.2 Noise Properties

The BER performance of fiber-optic links depends on the signal to noise ratio observed at the receiver [12]. Whereas the inherent receiver sensitivity is limited by a number of noise sources in the receiver subsystem, including thermal and shot noises [13, 14], BER degradation can also be caused by source- or link-induced noises. Relative intensity noise (RIN) is defined as the random fluctuations produced by the source alone. Reflection-induced intensity noise (RIIN), mode-partition noise, and modal noise result from interactions of the source and the transmission medium. Mode partition noise occurs in long single-mode links using Fabry–Perot edge-emitting lasers that have multiple longitudinal modes. Single-mode VCSELs should afford greater immunity to this phenomenon.

RIN is defined as the ratio of the source noise power density to the total power. If a frequency dependence is not specified, the noise is assumed to be constant with frequency so that the S/N in a bandwidth of Δf would be given by $S/N = (RIN \times \Delta f)^{-1}$. For example, if RIN $= -120$ dBc_{elec}/Hz and $\Delta f = 1$ GHz, then $S/N = 30$ dB$_{elec}$ (assuming no other sources of noise), which is sufficient for BER $< 10^{-18}$. In practice, standards such as Fibre Channel or ATM specify a maximum RIN of -116 dBc_{elec}/Hz for data rates approaching 1 Gbit/s. Typically, the noise performance of VCSELs operating in the fundamental mode is similar to that of edge-emitting lasers, with observed RIN below -150 dBc_{elec}/Hz [15]. However, operation with multiple and higher-order transverse modes causes degradation in noise performance.

Localized increases in RIN have been observed at kinks in the light–current curve [15, 16]. Fortunately, the RIN of multitransverse mode, fiber-coupled VCSELs operating under large-signal modulation conditions is usually less than -120 dBc_{elec}/Hz, meeting the -116 dBc_{elec}/Hz requirement of current standards [17].

RIIN is caused by external optical feedback and is a complex phenomenon that produces many effects [18]. While the higher output coupler reflectivity tends to isolate VCSELs from optical feedback, the shorter cavity length works to cancel any advantage [19]. The sensitivity to optical feedback is linked to the photon lifetime, which is roughly the same for VCSELs and edge-emitting lasers.

Modal noise can occur in a multimode system only if two conditions are present: fluctuations in the modal power distribution and mode-selective loss (MSL) [20]. Modal noise occurs when power shifts randomly between different modes and some of those modes are preferentially attenuated, leaving a net power that follows the modal fluctuations. Stated in spatial terms, modal noise occurs when variations in the spatial power distribution are filtered by a spatially dependent loss, such as a splitter or a poorly coupled connection, resulting in fluctuations in the amount of transmitted power. This gives rise to a characteristic error floor because the noise magnitude is proportional to the signal power [21, 22].

Modal noise is usually significant only in systems where a few modes are dominant so that each mode retains a sizable fraction of the total power. This implies a fairly coherent source. A bimodal system would have potential for the greatest magnitude of modal noise. The risk decreases as the power is distributed across more independent modes, which has an averaging effect.

Variations in the modal power distribution can occur in the source or through intermodal coupling in the transmission medium. Fluctuations attributed to the source generally have a much larger bandwidth. Spatial fluctuations at the optical source can lead to modal noise if there is significant coupling loss to the fiber or waveguide. Spectral variations at the source produce changing spatial patterns after propagating a distance in a multimode waveguide. MSL prior to this distance will not produce modal noise. Intermodal coupling in a multimode fiber through vibration or thermal variations will produce random shifting of power between modes and also requires propagation over a distance. There is a distance at which the potential for modal noise is maximum, however. As the distance is increased, the intermodal coupling continues to populate more modes and the susceptibility to modal noise decreases.

Low coherence sources, such as self-pulsating laser diodes [23] and LEDs [24], have been used to avoid modal noise penalties in MMF links with significant MSL elements (i.e, splitters). However, self-pulsations can be quenched with optical feedback levels as low as 1%, which are possible within typical link specifications [25]. VCSELs have also been shown to have significantly superior modal noise performance compared to edge-emitting lasers [26, 27]. Operation with a BER $< 10^{-13}$ has been demonstrated with 10 dB of MSL using large-area, multi–transverse mode VCSELs. The improved performance can be attributed to the broadened spectrum of multi–transverse mode VCSELs.

A number of standards, such as ATM and Fibre Channel, contain power penalty allocations for modal noise. An ad hoc industry group (HP, Honeywell, IBM, and VIXEL), known as the modal noise test methodology group, has compared worst-case modal noise power penalties for standard sources with a modal noise theory [28, 29].

11.3.2   VCSEL Structures

11.3.2.1   VCSEL Arrays

VCSEL arrays have the obvious advantage that the positioning of devices within an array is precise to within the tolerance of the photolithography used to make them. Furthermore, the emission angles are referenced to a single plane. The reduced degrees of freedom in an array greatly reduces the number of alignment operations necessary in die placement for efficient light coupling into an optical fiber. Handling is also simpler with an array, although arrays become more fragile as their aspect ratios increase.

The choice of VCSEL substrate type critically impacts the driving circuit topology. It is preferable to drive the lasers with electron-majority-carrier transistors because of their superior performance. An n-type transistor is referenced to the lower supply rail, and so it can act as a current sink, only. A commonly used circuit requires that the laser p-type side (anode) must be common. This can be accomplished by growing devices with the n-type side up on p-type substrates. Proton-implanted p-substrate VCSELs have proven difficult because of the greater current confinement required on the n-side to limit current spreading. Alternatively, p-up devices can be grown on semi-insulating substrates, and the epitaxial n-layers can then be mesa isolated at a distance from the active regions.

Thermal and electrical crosstalk are more significant when VCSELs are in array form. Heat dissipation in the common substrate couples the laser outputs at very low frequencies through the temperature dependence of the laser operating characteristics. Electrical crosstalk has a potentially wide bandwidth and can occur through resistive, capacitive, or inductive coupling or through direct current overlap. Resistive or inductive coupling through common occurs when current passing through series resistance or inductance in the ground path results in a significant voltage drop. This type of coupling doesn't depend strongly on device proximity. Capacitive or inductive coupling in the signal path increases as devices are brought closer to each other and the device-to-device parasitics increase relative to the device-to-common parasitics. Devices placed in very close proximity to each other may actually share injected carriers with their neighbors, depending on the degree of isolation.

11.3.2.2 Oxide-Confined VCSELs

Selective oxidation fabrication technology has recently enabled dramatic improvements in the performance of VCSELs. Threshold currents of $<100\,\mu$A [30, 31] and wall-plug efficiency $>50\%$ [32] have been demonstrated with 980 nm sources. For potential application toward 850 nm data communication standards, VCSELs with threshold current $<200\,\mu$A and wall-plug efficiencies $>30\%$ have been demonstrated [33]. Oxide confinement of n-up VCSELs may prevent excessive current spreading for p-substrate arrays.

With such enhanced performance, it is possible to operate Gbit/s links with <1 mA bias and modulation currents. Figure 11.3(a) shows an eye pattern of a 6-μm square aperture oxide-confined VCSEL in a MMF optical link at 1 Gbit/s with a $2^{31} - 1$ pseudorandom bit sequence. An average drive current of 0.5 mA (the prebias and modulation current amplitudes are $\sim$0.2 and $\sim$0.6 mA, respectively) and an extinction ratio of >9 dB are maintained under these conditions. In comparison to implanted VCSELs, which require drive currents of several milliamperes, this demonstrates more than a fourfold reduction in power required by a Gbit/s data source. Figure 11.3(b) shows the same device modulated at 2.5 Gbit/s with a 1.9 mA average current. Figure 11.4 shows a 14-μm square aperture oxide-confined VCSEL modulated at 8 Gbit/s with 12 mA average current.

Oxide-confined VCSELs also operate with a much greater number of transverse modes at low currents than implanted VCSELs, as evident in Figures 11.5(a) and 11.5(b). The benefits for modal noise are demonstrated in Figure 11.6(a), which shows the BER versus MSL of a 15 μm oxide-confined VCSEL in a $62.5/125\,\mu$m MMF link. The MSL is a fiber gap located 16 m from the source. The VCSEL is directly coupled to fiber and modulated at 622 Mbit/s with a $2^{31} - 1$ PRBS at an average drive current of 8 mA. With 15.5 dB of MSL, the oxide-confined VCSEL operated in the links without any errors for 16 hours, resulting in BER $<10^{-13}$. The corresponding performance of an implanted VCSEL, operated at an average current of 10 mA, is shown in Figure 11.6(b). A clear BER floor of 10^{-11} is seen with 13 dB of MSL.

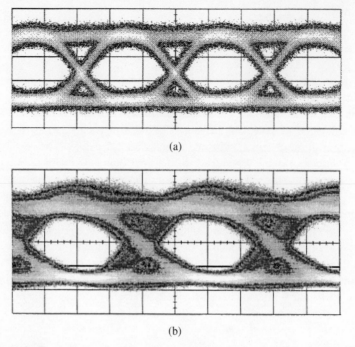

(a)

(b)

Figure 11.3. Eye patterns of a 6 μm square-aperture oxide-confined 850 nm VCSEL operating at (a) 1 Gbit/s with 0.5 mA average current (horizontal scale is 350 ps/div) and (b) 2.5 Gbit/s with 1.9 mA average current (horizontal scale is 100 ps/div). (Courtesy of R. P. Schneider, Hewlett-Packard Laboratories.)

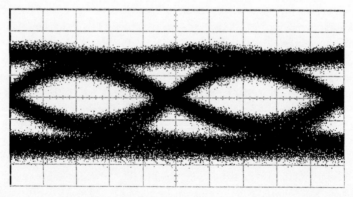

Figure 11.4. Eye diagram of a 14 μm square-aperture oxide-confined 850 nm VCSEL output at 8 Gbit/s with 12 mA average current (horizontal scale is 25 ps/div). (Courtesy of B. Zhu and I. H. White, University of Bristol, UK.)

With extremely low threshold current and enhanced noise properties, oxide-confined VCSELs will make ideal sources for advanced applications. For example, very short distance optical links (board-to-board or chip-to-chip) will require very low power consumption. The ability of these VCSELs to transmit a high signal-to-noise ratio, Gbit/s optical signal with <1 mW of electrical drive power is unmatched and offers great potential for the next generation of optical interconnections.

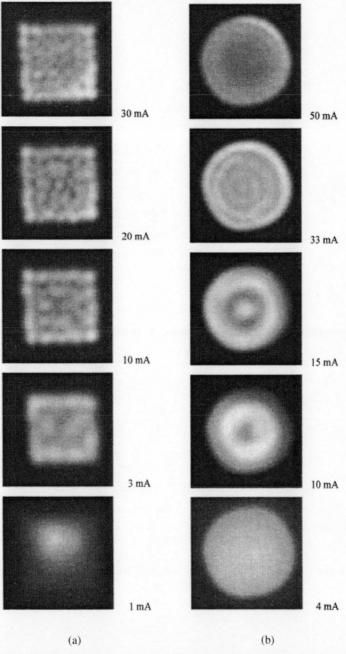

(a)   (b)

Figure 11.5. Near-field intensity images of (a) a 14 μm square-aperture oxide-confined 850 nm VCSEL and (b) a 20 μm diameter round proton-implanted 850 nm VCSEL at several DC current levels. The oxide-confined VCSEL exhibits a greater number of transverse modes at low currents. (Courtesy of R. P. Schneider, Hewlett-Packard Laboratories.)

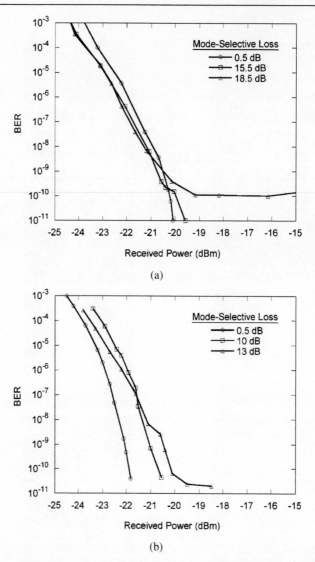

Figure 11.6. BER versus received power and varying mode selective loss (MSL) for (a) a 15 μm square-aperture oxide-confined VCSEL with 8 mA average drive current and (b) a 15 μm round proton-implanted VCSEL with 10 mA average drive current. Error floors appear above BER $= 10^{-11}$ with MSL > 18 dB$_{opt}$ for the oxide-confined device and MSL > 13 dB$_{opt}$ for the implanted device.

11.3.2.3 Long-Wavelength VCSELs

The lower loss and dispersion in glass fiber at 1,300 nm and 1,550 nm allow orders of magnitude longer transmission distances at these wavelengths than at 850 nm in SMF. The bandwidth–distance product in MMF at 1,300 nm is only up to three times that at 850 nm, depending on the manufacturing grade. FDDI-grade 62.5/125 μm MMF operating with 1,300 nm VCSELs, when they are available, will support link lengths of >800 m for Gigabit Ethernet, easily meeting the 500 m requirement of building backbones [34]. Edge-emitting lasers operating at 1,300 nm will provide an interim solution.

Although the long-wavelength transmission advantage is not as great in MMF as it is in SMF, there are also other factors that favor long wavelengths for data communications. The lower photon energy results in a lower laser forward voltage, by 1/2 V at 1,300 nm compared to 850 nm, a critical improvement for operating at lower power supply voltages. Because the VCSEL substrate is generally transparent at longer wavelengths, a substrate-emitting laser can be mounted junction-side down to improve thermal conductivity. Eye safety limits are 4 to 13 dB higher at 1,300 nm than at 850 nm and photodetector responsivity can be 50% greater. Higher optical power levels and more sensitive receivers mean relaxed link budgets and lower cost, higher performing links. Compatibility with existing long-wavelength systems would improve technology acceptance and encourage leveraged cost-reduction strategies.

11.3.3 Bandwidth–Length Limitations

The numerical aperture of graded-index fiber varies radially from the center of the core to the cladding according to $NA(r) = \sqrt{n^2(r) - n_2^2}$. The index profile, $n(r)$, is given by the power-law relation,

$$n(r) = \begin{cases} n_1\left[1 - \left(\dfrac{r}{a}\right)^{\alpha}\Delta\right], & 0 < r < a \\ n_2 = n_1(1 - \Delta), & r \geq a \end{cases} \tag{11.1}$$

where n_1 is the core maximum index, n_2 is the cladding index, r is the fiber radial vaiable, a is the raduis of the core-cladding interface, α is the index profile power parameter, and Δ is the fractional index difference between the core maximum and the cladding [12]. Optical fiber manufacturers specify MMF bandwidth with the so-called over-filled launch (OFL). The OFL is characterized by equal and uniform angular intensity up to the local NA at each point on the fiber core endface and results in equal power per fiber mode. The fiber modal delays are most closely matched when the power parameter is approximately 2. The degree to which real fibers approximate the ideal profile determines the modal dispersion characteristics. Common distortions of fiber index profile include error in the power parameter, index peaks, and dips at the core center and deviations at the core-cladding interface. The broad modal distribution produced by an OFL has an averaging effect that mitigates the deleterious impact of index distortions on modal bandwidth.

An LED source gives an approximate OFL. A direct laser launch, even a multi–transverse mode VCSEL launch, generally results in a modal distribution in the fiber that deviates significantly from that of an OFL. A laser launch can yield a larger or smaller bandwidth than that of the OFL. Determining the bandwidths for the installed MMF base with laser launches is a much needed yet daunting task, given the vast quantity of installed fiber and diversity of emission characteristics from lasers. Both the fiber index profile and the laser emission must be characterized.

Enhanced bandwidths in MMF have been observed with radial-offset launches from single-mode sources [35, 36]. Simulations with a number of index profiles approximat-ing a distribution of manufacturing distortions have revealed a 6 μm range of offset that results in a bandwidth at least equal to that of the OFL. The simulations showed that the dispersion of low-order mode groups is much greater than the mode groups above them and that the radial offset launches producing high bandwidths preferentially excite higher order modes. The tight tolerance required by the radial offset approach is not attractive for

Table 11.2. *Selected specifications for the short-wavelength (770–860 nm) 50-μm core MMF Fiber Channel (FC-0) physical links without open-fiber control.*[a]

	Link			
Data rate	100	200	400	MB/s
Bit rate, ±100 ppm	1062.5	2125	4250	MBaud
Length, min.	2	2	2	m
max.	500	300	175	
	Transmitter			
Launch optical power, min.	−10	−10	−10	dBm_{opt}
max.	−5	−5	−5	
Spectral width (RMS), max.	4	4	4	nm
Relative intensity noise, max.	−116	−118	−120	dBc_{elec}/Hz
Extinction ratio, min.	9	9	9	dB_{opt}
Optical rise and fall times, max.	0.45	0.22	0.11	ns
Timing Margin (p–p), min.	57	57	57	%
	Receiver			
Received optical power, min.	−16	−16	−16	dBm_{opt}
max.	0	0	0	
Return loss, min.	12	12	12	dB_{opt}
Rise and fall times, max.	0.6	0.3	0.15	ns

[a]Parameters are specified at BER $< 10^{-12}$ (ANSI X3YT11).

low-cost VCSEL-based transmitters. However, the same highly multimode operation shown in Figures 11.5(a) and (b) that imparts high modal-noise immunity to multi–transverse mode VCSELs may also result in preferential coupling to higher order modes in multimode fibers. Early experiments suggest that simple launches from oxide-confined VCSELs can reliably meet the OFL bandwidth.

Building wiring standards require transmission length of over 500 m using 62.5/125 μm MMF links for building backbone cabling (ISO/IEC 11801). Minimum length for campus backbones is 2 km, and hub-to-desktop requires 100 m. Whereas 50/125 μm graded-index (GI) MMF has a worst-case OFL bandwidth of 400 MHz·km at 850 nm, 62.5/125 μm GI MMF has a worst-case OFL bandwidth of only 160 MHz · km at 850 nm. Most installed MMF is of the 62.5/125 μm variety.

Table 11.1 provides a summary of current high-performance network standards with specification for a fiber-based physical layer. The first standard group to specify a VCSEL-based physical layer was the ATM Forum, which approved a long-wavelength (1,300 nm) LED solution and a short-wavelength (780–860 nm) laser solution for OC-12 MMF data links. The standard was written to accommodate both 850 nm VCSELs and 780 nm edge-emitting laser diodes, commonly known as compact disk (CD) lasers.

OFL bandwidths have been assumed in the past for laser launches. Physical specifications for high-speed, short-wavelength MMF Fibre Channel transmitters and receivers are listed in Table 11.2. ATM and Gigabit Ethernet will share similar specifications at their respective bit rates. The maximum length for 1.06 Gbaud Fibre Channel using 62.5/125 μm MMF was specified as 340 m at 850 nm assuming an OFL [34]. Gigabit Ethernet, which has a 1.25 Gbaud line rate, faced a corresponding OFL link length limit of 260 m [37]. Further examination of laser launch bandwidth measurements prompted a reduction to 220 m on FDDI grade 62.5/125 μm MMF, and a 500 m limit was specified on 50/125 μm MMF.

11.3.4 Module Design

A transmitter or receiver module generally consists of a VCSEL or photodetector chip, an electronic integrated circuit (IC), some discrete passive electronics, an optical interface, and packaging. The transmitter receives electrical signals and launches light into fiber with specified modulation rate, optical power range, and light emission pattern. The receiver converts modulated light having power within its dynamic range to standard electrical data signals. Optical interfaces in the transmitter and receiver couple light between the optoelectronics and the transmission fibers.

The packaging holds the components, protects them from the environment, dissipates heat, incorporates alignment mechanisms, and provides electrical and optical internal and external connections. Although not explicitly discussed here, packaging and alignment dominate the manufacturing costs of most fiber-optic modules. Furthermore, alignment tolerance directly affects the link performance through the required sizes of components and the fiber light coupling. The trade-off between cost and performance depends critically on packaging and manufacturing design.

11.3.4.1 VCSEL Transmitters

In a simple transmitter, the electronics receive the input data, amplify it, and provide a current to drive the VCSELs. In systems where requirements for optical power uniformity are relaxed because of a very large signal to noise ratio, VCSELs can be driven through a transmission line with a matched driver impedance. Driving a VCSEL with a low impedance source magnifies variations in the operating voltages through the nonlinear VCSEL diode current–voltage relationship.

The results are generally more uniform and predictable with current drive. Current drive is high impedance, and laser electrical impedance varies significantly with frequency and between the off and on states [38]. This makes impedance-matched transmission lines between the driver and lasers impossible. The section of line between the driver and lasers can resonate at a frequency determined by the electrical length, degrading the optical response and increasing electromagnetic radiation. These resonances can be kept out of the system frequency range by designing the propagation delay of these lines to be much shorter than the rise and fall times of the driving signal. Damping resistors can be used to control resonances [39], but they may significantly increase power consumption.

Fiber-optic transmitters and receivers typically interface to electronics via electrical interconnect standards. Electrical interconnect signalling conventions are changing to accommodate wider and faster interfaces with lower supply voltages. Differential signalling schemes cut power consumption in half compared to single-ended connections having the same nominal signal to noise ratio. Furthermore, because the power and ground currents tend to cancel in differential links, crosstalk and radiation are significantly reduced. A low-voltage differential signals (LVDS) standard (IEEE Std 1596.3-1996) has been defined for SCI (ANSI/IEEE Std 1596-1992), and a similar scheme has been adopted for HIPPI-6400 [40].

The optical launch affects fiber bandwidths, modal noise, and eye safety. Unlike an LED, which produces a nearly over-filled launch, a laser launch generally excites a subset of the available fiber modes. The associated modal bandwidth can vary widely, depending on the distribution of power among the mode groups (see Section 11.3.3). By approximating an OFL or preferentially launching higher order modes, the fiber manufacturer's specified OFL modal bandwidth can be assured with high modal noise immunity over a wide range of conditions. Furthermore, it is desirable to distribute the power throughout the fiber NA range, so that far-field intensity is minimized for a given total power. This allows the maximum light power to be transmitted within eye-safety limitations.

National and international regulations classify laser systems for eye safety.[2] In the United States, the Food and Drug Administration (FDA) has established regulations pertaining to laser products for enforcement of the Radiation Control Health and Safety Act of 1968 [41]. Manufacturers selling products subject to the performance standards under this act are required to furnish reports to the Center for Devices and Radiological Health [42]. Also in the United States, the American National Standards Institute (ANSI) has defined the Z136.2 laser safety standard. International laser safety standards are published by the International Electrotechnical Commission (IEC) [43].

FDA Class I limits apply to devices that emit light in the ultraviolet, visible, and infrared spectra. Biological hazards have not been established for exposure below these limits [41]. FDA Class I laser products are required to have identification labels and notices in user and service manuals.

IEC Class 1 lasers are considered safe under reasonably foreseeable conditions of operation [43]. IEC Class 1 products require only an identification label. IEC Class 3A lasers are regarded as safe for viewing with the unaided eye. Aversion responses including the blink reflex limit exposure in the 400 nm to 700 nm wavelength range. The hazard to the unaided eye at other wavelengths is limited to that for Class 1. Direct viewing of Class 3A laser beams with collection optics, such as binoculars, telescopes, or microscopes may be hazardous. IEC Class 3A lasers must display warning/hazard labels. A tool is required to unmate connections in IEC Class 3A fiber-optic links where access is unrestricted.

Fiber-optic communication systems are classified based on "radiation that could become accessible under reasonably foreseeable circumstances (e.g, a fiber cable break, a disconnected fiber connector, etc.)" [44]. Interlock mechanisms can allow operational power levels to exceed the eye-safety limits while preventing conditions that would expose a user to dangerous power levels. Open-fiber control (OFC) is specified in the Fibre Channel standard (ANSI X3T11) for this purpose. A low-bandwidth communication channel from the receiver back to the transmitter provides the necessary feedback.

Measurements for certifying product safety compliance are described in FDA and IEC documents [41, 43]. Figure 11.7 shows calculated accessible emission limits for FDA Class I and IEC Classes 1 and 3A at 830 nm wavelength for graded-index multifiber arrays with 0.20 NA, assuming an ideal OFL. The maximum power per fiber is plotted versus number of channels at $250\,\mu m$ pitch. The power is not linear with number of channels because the extended source cannot be focused to a single spot in the eye. Maximum power can be

---

[2] Laser safety information is provided for without guarantee of accuracy or suitability for any particular application. Manufacturers are responsible for the safety of their products and certifying compliance with applicable regulations and standards.

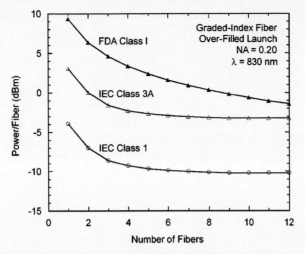

Figure 11.7. Graph of calculated FDA and IEC accessible emission limits at 830 nm wavelength in 250 μm pitch multichannel fiber ribbon. Power per fiber is plotted versus number of illuminated graded-index fibers with an overfilled launch of 0.20 NA.

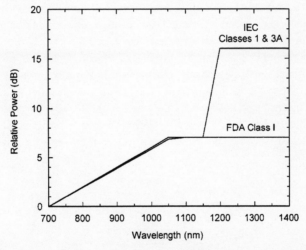

Figure 11.8. Graph of the wavelength dependence of FDA and IEC accessible emission limits. Relationships between classes are not explicit in this plot – curves are normalized within each class to the value at 700 nm wavelength.

launched if the angular spectrum is uniform and covers the available numerical aperture. Figure 11.8 shows the wavelength dependence of accessible emission limits.

Eye safety determines the upper limit of the link power budget. Array uniformity and power variations with temperature, voltage, and aging all spread the range of transmitter power output. The transmitter output power range along with the range of losses in the link must be accommodated in the receiver dynamic range. The link power budget can be tightened by using feedback in the transmitter to reduce output variations. Variations with temperature, voltage, and aging can be effectively compensated by monitoring the optical power output. However, the thermal variations usually dominate and can be limited without added complication of the optical interface by monitoring temperature.

11.3.4.2 Fiber-Optic Receivers

Semiconductor photodiodes are generally used to convert light to electricity in data communications receivers. The parameters of primary interest for photodetectors used with digital data are bandwidth, step response, efficiency, capacitance, size, and bias voltage. The step response shows how fast a photodetector responds, and it reveals ringing or other waveform anomalies that may affect receiver performance. Higher efficiency translates directly to better sensitivity.

The bandwidth, sensitivity, and power consumption of a receiver are interrelated and depend strongly on the capacitance at the receiver input [13]. Optical alignment tolerance grows with the transverse dimensions of the device, but the capacitance grows as the square of those dimensions. The available supply voltage is divided between the receiver front-end amplifier input and the photodetector. At 3.3 V supply, there may be less than 1.6 V left for the photodetector bias.

The frequency responses of GaAs metal-semiconductor-metal (MSM), GaAs p-i-n, and Si p-i-n photodetectors suitable for gigabit receivers at 850 nm wavelength are compared at various bias voltages in Figure 11.9. Actual device measurements are plotted on an absolute scale that is referred to 100% quantum efficiency.

The maximum bandwidth–efficiency product of a photodetector is proportional to the absorption coefficient and the carrier velocities [45]. While saturated carrier velocities in silicon are comparable to those in GaAs, Si has about an order of magnitude lower absorption coefficient than GaAs at 850 nm wavelength [46, 47]. The seemingly adequate bandwidth–efficiency product of around 5 GHz in silicon is compromised by absorption in the heavily doped silicon contact layers. This absorption spoils the efficiency, and long

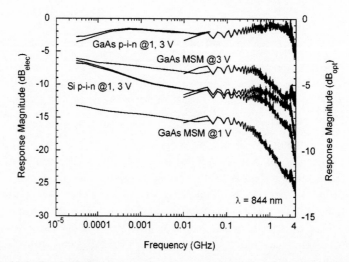

Figure 11.9. Graph of photodetector frequency responses for GaAs p-i-n, GaAs MSM, and Si p-i-n photodetectors at 1 V and 3 V biases. The plots are scaled such that 0 dB corresponds to 100% quantum efficiency or the theoretical maximum signal possible at the measurement wavelength.

carrier diffusion times in the low-field regions cause a low frequency roll-off. The Si p-i-n device response in Figure 11.9 has a 12 dB$_{elec}$ (6 dB$_{opt}$) lower effective responsivity than theoretically possible at the frequencies of interest. Not only is the responsivity in the frequency range of interest compromised, but the large relative DC response can reduce the effective dynamic range of the receiver front-end stage.

The larger absorption coefficient in GaAs means that the absorption layers can be much thinner than in silicon for a given efficiency. Furthermore, AlGaAs transparent contact layers eliminate the absorption loss and diffusion tails that plague silicon p-i-n devices. The GaAs p-i-n responsivity in Fig. 11.9 is close to the theoretical limit. The p-i-n device bandwidth and efficiency degrade little at low bias, thanks to the built-in diode junction fields.

MSM photodetectors are attractive for a number of reasons. They have low capacitance/area for large alignment tolerance. They can be fabricated in a simple two-mask process, including antireflection coating, and they can be easily integrated with field-effect transistor (FET) circuits. The GaAs MSM in Figure 11.9 performs better than the Si p-i-n at 3 V bias, with −8 dB$_{elec}$ (−4 dB$_{opt}$) response at frequencies of interest. However, because MSM photodetectors have no built-in field, responsivity and bandwidth degrade rapidly at low bias voltages. Material and structural innovations may improve the low-bias performance of MSM photodetectors.

As in the laser-transmitter case, impedance-matched transmission lines between photodetectors and the receiver IC are not normally realizable. A photodetector presents a capacitive load, and the input impedance of a high-speed transimpedance amplifier usually has some frequency dependence. The parasitic elements loading the front end directly impact the receiver performance through the bandwidth–sensitivity–power trade-off, and so the length of these lines is critical. Parasitics are minimized by integrating the photodetectors with the amplifiers.

A receiver comprises a photodetector, a low-noise analog front end, a decision circuit, digital amplifiers, and electrical transmission line drivers. The decision circuit discriminates between logic ones and zeros in the amplified analog signal. Subsequent limiting amplifiers complete the conversion to discrete binary levels. The line drivers launch the data signals into controlled impedance transmission lines according to electrical interconnect standards such as LVDS.

In serial links with clock recovery, the decision is made at the optimum time position within the bit window. The clock is usually transmitted on one of the channels in a parallel link and is used to retime all of the data channels. Interchannel skew accumulated over the link varies the clock-data relative timing across the channels, so that there is no optimum clock phase for retiming all channels. Parallel links without clock recovery or skew compensation thus require enhanced timing margin. This is equivalent to optimizing the receiver for minimum intersymbol interference rather than best noise-limited sensitivity. Such an optimization suggests a front-end bandwidth numerically equal to the NRZ (non-return to zero) baud rate rather than 0.6–0.7 times the baud rate, which is commonly used in thermal-noise-limited receivers with clock recovery [48].

The threshold level between a zero and a one can either be fixed at some DC value or it can be determined from the signal itself either by averaging or sampling. The DC-coupled receiver is simple and places no requirements on the data since the threshold level is independent. However, there are significant BER, sensitivity, and duty-cycle distortion penalties associated with using a fixed decision threshold. An AC-coupled receiver relies on the signal average value. The required capacitors consume significant area on the receiver IC or must be included as discretes in the module. In a parallel link, a hybrid DC-coupled scheme is an option that relies on uniform power levels across all channels. One AC-coupled channel then determines the threshold level for the other DC-coupled channels. Sample and hold receivers can accommodate nearly arbitrary data, although they are not commonly used at high data rates.

AC-coupled receivers depend on the average value of the data being close to midway between the zero and one levels. This means that during a specified period, there should be an equal number of zeros and ones in the data. Data of such composition are DC-balanced and AC-coupled with a low-frequency cutoff determined by the balancing period. Coding schemes can be used to shift the data spectrum up in frequency and achieve the DC balancing [49, 50]. Such schemes increase the required bandwidth and add overhead for encoding and decoding. Even with DC-balanced data, the average value fluctuates, depending on the averaging time constant and the balancing period. The BER and pattern-dependent jitter increase as the threshold level deviates from the true zero–one average. The pattern-dependent jitter results from the conversion of the threshold fluctuations to timing fluctuations through the finite signal rise and fall times.

11.3.4.3 Optical Interface

The optical interface couples light between the optoelectronics and the transmission fiber. VCSELs and planar photodetectors emit and receive light normal to their mounting plane. The optical connector may also be oriented normal to this plane or it may be in-plane. The in-plane connector case requires a turn in either the optical or electrical path of the transmitter and photoreceiver. Advantages and disadvantages of either approach depend upon the optics and packaging technologies under consideration. In all cases, the distances between the optoelectronics and the electronics critically affect speed, sensitivity, dynamic range, power consumption, and electromagnetic compatibility, as discussed above.

The optical alignment tolerance also affects all of these performance parameters. For example, launching light with a large numerical aperture into the fiber allows more power within the eye safety limit but also reduces the alignment tolerance for efficient coupling to a given fiber. A smaller core, lower NA fiber generally has a larger bandwidth–distance product but also a tighter launch alignment requirement. Inefficient VCSEL coupling calls for greater drive current, which increases power consumption and radiated emissions. Large loss ranges eat into the link power budget.

Optical interface characteristics also affect the noise performance. The resistance to modal noise is enhanced by increasing the launched modal distribution. Light aperturing in the receiver increases the likelihood of modal noise conversion to amplitude noise. Also, a lossy launch may produce noise depending on the modal stability of the laser and selectivity of the loss. For instance, a polarizing launch would convert polarization instabilities in a VCSEL to amplitude noise. A mode-selective launch also may degrade the bandwidth, since the frequency responses of some multimode VCSELs are mode dependent.

The simplest optical interface is a fiber butt-coupled (without any optics) to a VCSEL. Using 62.5 μm GI MMF and a 10 μm (multimode) VCSEL, coupling efficiency can be better than 90% with transverse alignment tolerance of $\pm 10\,\mu$m when the longitudinal separation is less than 100 μm [51]. Coupling efficiency in the far field can be estimated by assuming a point source with a given angular spectrum and the optical fiber radial NA profile. The coupling efficiency is the fractional power within the fiber radial NA limits.

Lenses can provide increased alignment tolerance by demagnifying the source, but with an attendant conversion of the numerical aperture. This is illustrated in Figure 11.10, where

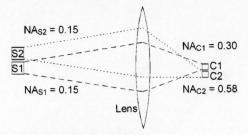

Figure 11.10. Schematic drawing of coupling from a source using a lens showing the effects of misalignment in the source plane on offset and numerical aperture in the coupling plane.

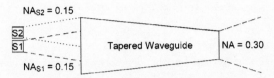

Figure 11.11. Schematic drawing of coupling from a source using a tapered waveguide. A misalignment in the source plane of a long taper does not affect the offset or numerical aperture in the coupling plane.

the optical paths for two sources are traced through a lens to a coupling plane, where they are demagnified by a factor of two. If S1 and S2 represent different alignments, it is clear that the positional alignment tolerance in the coupling plane is improved by the demagnification factor. The NA of C1, which is on the lens axis, is increased by the demagnification factor. The NA of C2, however, depends on the misalignment, the demagnification factor, and the focal length and is nearly double that of C1 in the illustration. Both the positional and NA acceptance range must be considered in coupling to the next optical element. Light from C2 would not be guided in a 0.3 NA optical fiber, but a planar photodetector with a large acceptance NA may absorb it all. Lens coupling with a laser can also lead to debilitating optical feedback [18]. When the system is aligned, any reflection at the image, C1 in Figure 11.10, is imaged back at the source, S1, producing an efficient feedback path.

Similar to a demagnifying lens system, a tapered waveguide can increase alignment tolerance and numerical aperture, as depicted in Figure 11.11. Unlike the lens system, however, a sufficiently long tapered waveguide has negligible dependence of NA on source position, and reflections are not imaged at the source. Ray angles are increased by twice the taper angle for each bounce off of the waveguide walls. The output consists of concentric rings or bands corresponding to the number of bounces. In the limit of a long or adiabatic tapered waveguide, the output is homogenized, and the overall NA increase is equal to the ratio of input to output diameters. The tapered waveguide is most effective when the input and output are in close physical proximity to the optical elements with which it interfaces. In contrast, a lens system generally works best when elements are physically separated.

Optical turns are most commonly effected by out-of-plane mirrors integrated on waveguides, as illustrated in Figure 11.12 for a parallel optical transceiver. A mirror can be formed by cutting or polishing the waveguide end at the desired angle, usually around 45°. Total internal reflection alone can be exploited, or a reflective coating is sometimes used to enhance the reflection. Mirrors can also be combined with other refractive and diffractive elements in discrete or integrated optical assemblies. Bent waveguides can also turn the

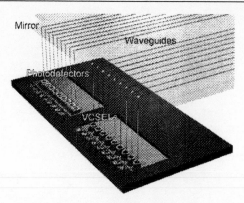

Figure 11.12. Optical turn concept using 45° out-of-plane mirrors integrated with waveguides.

light with increases in the NA and loss that depend on the bend radius. A simple waveguide turn might consist of bent optical fibers. A three-sided waveguide arc can perform the same function with some added beam spreading in the plane of the bend.

11.3.5 Multifiber Connectors

The connector is a critical enabling component, since its design ultimately determines the density, quality, and cost of fiber-optic interconnects. Overall connector size is generally determined by ergonomics. New multifiber connector designs are squeezing as many as eighty fibers into about the same size as ordinary single-fiber connectors. They trim costs for connector hardware, assembly, and cabling compared to single-fiber connectors. Not only do high-density connectors allow more efficient use of precious system board space, their smaller port area also helps reduce EMI.

The MT (mechanically transferable) ferrule, developed at NTT, is a standard multifiber interface available from several manufacturers. It is molded in one piece from an ~80% silica-filled thermosetting epoxy resin for dimensional stability [51, 52]. Stripped fibers are guided by U-shaped grooves and then tapered holes as they are inserted into the ferrule. The fibers are affixed by filling the rectangular window over the U-grooves with adhesive; then the endface is polished. Connector mating alignment is achieved through precision pins inserted into the large holes at each side of the fiber hole array. This technology can satisfy single-mode tolerances [53].

There are several versions of the MT ferrule, holding from two to eighty (5×16 array) fibers. The mini-MT ferrule, which can accommodate up to four fibers, has been selected for the emerging Gigabit Ethernet standard. Small-form-factor serial transceivers will use the two outside fiber positions. A push/pull connector and a latched connector called the MT-RJ after the familiar telephone connector have been built around the mini-MT ferrule. Two push/pull connectors have been built around the version in Figure 11.13, which positions twelve fibers linearly on 250 μm centers. The MPO (multifiber push-on, also called MTP) and the slightly smaller MPX connectors are drawn in Figures 11.14(a) and 11.14(b). The MPO is the recommended optical connector for HIPPI-6400. The latching HI-PER Link® connector, codeveloped by Alcoa-Fujikura and Motorola, is also based on the twelve-fiber MT ferrule and can be seen in Figure 11.19(a) with the Optobus I module.

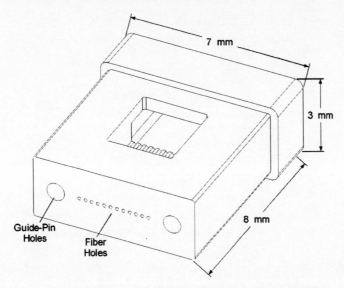

Figure 11.13. Perspective drawing of MT ferrule with 12-fiber holes on 250 μm centers.

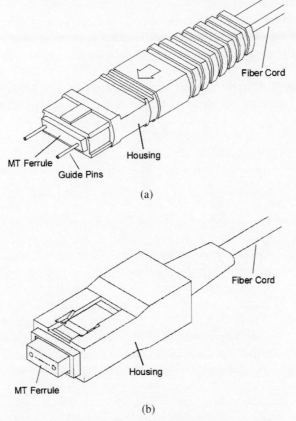

Figure 11.14. Perspective drawings of (a) MPO (MTP) (courtesy J. Keesee, US Conec.) and (b) MPX connectors based on the 12-fiber MT ferrule.

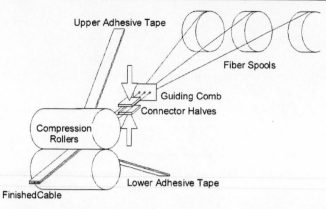

Figure 11.15. Concept drawing of the 3M INCA process for producing connectorized ribbon fiber cable assemblies in a single line. (Reproduced from Lee et al. [55].)

The INCA (IN-line Cable Assembly) process from 3M Company, which integrates automated cabling and connectorization, is illustrated in Figure 11.15 [54, 55]. As with virtually all fiber-optic connectors, the accuracy of the fiber diameter and concentricity are factors in the connection alignment tolerance. In the interest of physical robustness, the polymer coating is not stripped in the INCA process, and so it is also included in the diameter and concentricity requirements. 3M manufactures fibers that are compatible with this process. In Figure 11.15, a series of combs position the fibers in the cable assembly, with the final guiding comb setting the ultimate fiber pitch close to the compression rollers. Backing material is laminated onto the fibers with a pressure-sensitive adhesive to form the cable.

When a connector is to be installed, the combs are pulled back from the compression rollers and the connector halves are positioned where the guiding comb had been. The connector halves form face-to-face connectors that will be cut apart after they are assembled onto the cable. Fibers are aligned in V-grooves in one half of the connector and held in place by the opposite half. The connector is secured to the fibers with light-curable adhesive, and the cabling process resumes. The connector passes through the compression rollers and is laminated with the rest of the cable. The backing material is peeled back and the facing connectors are cut apart and polished. Connector bodies are installed to complete the assembly. "Type A" (Jitney) and "type B" (Argus) connectors are illustrated in Figures 11.16(a) and 16(b).

The Berg MACII (Multifiber Array Connector) positions fibers between two precision etched Si V-groove pieces, to which the alignment pins also register [56]. The standard version has eighteen fibers. The protective polyimide coating is left on the fibers to improve reliability. After the two Si pieces are epoxied together with the fibers, the end faces are polished. The alignment pins, spring clips, and housing are then installed. An alternate multimode plastic injection molded MAC can be made compatible with the MT connector [57]. The Methode MP push/pull multifiber connector accommodates up to twelve fibers in the same form factor as a standard simplex SC connector [58]. The MP is also compatible with the MT ferrule.

11.3.6 Parallel Optical Link Module Examples

The VCSEL-based high-speed parallel optical interconnect modules illustrated below not only represent the exploitation of VCSEL high efficiency and natural array form, but they

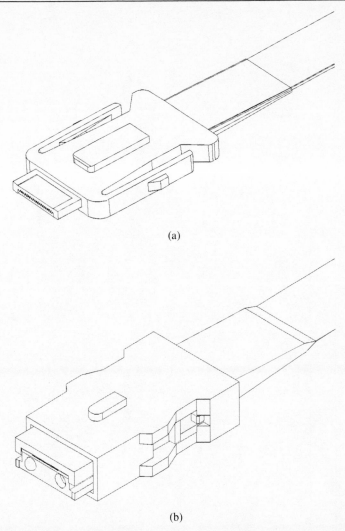

(a)

(b)

Figure 11.16. Perspective drawings of finished (a) type A (Jitney) and (b) type B (Argus) connectors designed for fabrication in the 3M INCA process. (Reproduced from Igl et al. [54].)

also employ technologies that may be used in next-generation product lines. In all cases, copper is viewed as a competing technology, and so the theme is a drive for low cost while emphasizing advantages in speed and density. The design of the optoelectronic packaging, electrical and optical interfaces, and connectors largely determine the success in these regards. Tables 11.3 and 11.4 list some specifications. Particularly important are the aggregate data rate (line rate times number of channels) and overall dimensions.

11.3.6.1 Consortia Demonstrations

The United States government and the Commission of the European Communities, citing the strategic importance of high-speed interconnects in parallel computing systems, are actively

Table 11.3. *VCSEL-based parallel optical link technology results demonstrated by various consortia.*

Consortium Name	OETC	Jitney	POLO	POINT
Participating Entities	Martin-Marietta, AT&T, Honeywell, IBM	IBM, 3M, Lexmark	Hewlett-Packard, AMP, DuPont, USC, (SDL)	General Electric, Honeywell, AMP, Allied Signal, Columbia U., UCSD
Sponsor	ARPA	NIST	DARPA	DARPA
Term	7/92–3/96	6/94–12/96	7/94–7/97	12/94–6/97
No. Channels	32 (simplex)	20 (simplex)	20 (10 Tx + 10 Rx)	up to 44 (simplex)
Max. Line Rate	500 Mb/s	500 Mb/s	1 Gb/s	1 Gb/s
Max. Distance	100 m	40 m (SI), 200 m (GI)	300 m	280 nm
Wavelength	850 nm	850 nm	980 nm & 850 nm	850 nm
VCSEL Type	Proton-implanted, top-emitting, multi-transverse mode	Proton-implanted, top-emitting, multi-transverse mode (Honeywell)	Proton-implanted, substrate- (980 nm) & top- (850 nm) emitting, multi-transverse mode	Proton-implanted, top-emitting, multi-transverse mode
Photodetector Type	GaAs MSM (integrated)	GaAs MSM (integrated)	InGaAs p-i-n, GaAs MSM	GaAs
Optical Interface	Fiber 45° mirror	Molded plastic arc	Polyguide® tapered waveguide, 45° mirror	Polyguide® waveguide, 45° mirror
Optical Connector	MACII-32	3M Jitney INCA	MPX	Super-MT ferrule
No. Fibers, Pitch	32, 140 μm	20, 500 μm	10, 250 μm	24, 100 μm
Fiber Type	62.5/125 μm GI	200/230 μm SI or 175/200 μm GI	62.5/125 μm GI	Polyguide® 50 μm SI

Tx IC	GaAs MESFET	Si CMOS	Si BJT	—
Rx IC, Coupling	GaAs MESFET, AC	GaAs MESFET, AC	Si BJT, DC & AC	GaAs, AC
Optical Coding	Manchester	Scramble optional	None	None
Electrical Levels	Diff. ECL, LVDS	LVDS	Diff. ECL	Analog (see text)
Module Technology	PQFP leadframe + polymer multilayer & Si board on alumina	Epoxy/silica QFP leadframe + Cu slug	Ceramic multilayer QFP leadframe BGA	Polymer multilayer embedded chip on alumina or plastic
Component Assembly	Conventional die attach & wire bond	Fiducial die attach & wire bond	Conventional die attach & wire bond	Thin-film lamination & planar processing
Optical Alignment, Tolerance (Tx/Rx)	Reticle/Reticle, ±15 μm/±15 μm	Passive/Passive, ±50 μm/±50 μm	Reticle/Reticle, ±15 μm/±50 μm	Passive/Passive
Supply Voltage(s)	Tx: −5.0, 5.0 V Rx: 3.3 V	3.3 V	−5.2, −2.0 V	NA
Power Dissipation	Tx: 6.2 W; Rx: 2.0 W	Tx: 2.3 W; Rx: 3.7 W	2.5 W	NA
Size (w × l × h)	33 × 45 × 5.3 mm^3	30 × 40 × 9 mm^3	25 × 61 × 10.0 mm^3	25 × 14 mm^2 (w/o connector)

Table 11.4. *Summary of some VCSEL-based parallel optical link product definitions.*

Company Name	Motorola	Vixel	Gore	Siemens
Product Name	OPTOBUS®I & II	P-VixeLink®	Parallel Seamless Migration	PAROLI
No. Channels	20 (10 Tx + 10 Rx)	4 (simplex)	4 (simplex)	12 (simplex)
Max. Line Rate	400, 800 Mbit/s (I, II)	1.25 Gbit/s	1.06 Gbit/s	1.25 Gbit/s
Max. Distance	300, 200 m (I, II)	300 m	250 m	300, 75 m (AC, DC)
Wavelength	850 nm	840 nm	850 nm	840 nm
VCSEL Type	Mesa/proton-implant, top-emitting, multi-transverse mode	Gain-guide, top-emitting, multi-transverse mode	Proton-implanted, top-emitting, multi-transverse mode	Oxide-confined, top-emitting, multi-transverse mode
Photodetector Type	GaAs p-i-n	GaAs (integrated)	GaAs p-i-n	InGaAs/InP p-i-n
Optical Interface	Molded-plastic tapered waveguide	Fiber-ribbon butt-couple	Fiber-ribbon butt-couple	Fiber 45° mirror
Optical Connector	HI-PER Link®	MTP pigtail	pigtail	SMC
No. Fibers, Pitch	10, 250 μm	4, 250 μm	4, 250 μm	12, 250 μm
Fiber Type	62.5/125 μm GI	62.5/125 μm GI or 50/125 μm GI	62.5/125 μm GI	62.5/125 μm GI
Tx IC	Si CMOS	Si	Si	Si BiCMOS
Rx IC, Coupling	Si BJT, DC	GaAs, AC	Si, AC	Si BiCMOS, AC
Optical Coding	None	None	None	None, 4b/5b (AC, DC)
Electrical Levels	Diff. CML (PECL)	Diff. PECL	Diff. PECL	LVDS
Module Technology	TAB + BT resin laminate PGA	Ceramic leadframe optical submount on FR-4 laminate PGA	BT resin laminate 2 mm connector	Organic laminate leadframe
Component Assembly	Conventional die attach & wire bond	Conventional die attach & wire bond	Conventional die attach & wire bond	Conventional die attach & wire bond
Supply Voltage(s)	5 V	5 V	5 V	3.3 V
Max. Power Dissipation	1.35, 1.65 W (I, II)	Tx: 1.25 W, Rx: 1 W	NA	Tx: 1.5 W, Rx 1 W (AC) Tx, Rx 3.8 W (DC)
Operat. Ambient	0–70 °C	0–70 °C	0–70 °C	0–80 °C
Size (w × l × h)	37 × 39 × 2mm³	25 × 66 × 11.6 mm³	21 × 57 × 12 mm³	18 × 58 × 12.7 mm³
FDA Class I	Yes, w/o OFC	Yes, w/o OFC	Yes, w/o OFC	Yes, w/o OFC

promoting the development of parallel–optical link technologies. Past European efforts have been based on edge-emitting lasers. There is little published information about present VCSEL-based efforts. Of the four U.S. government sponsored consortia that have developed VCSEL-based parallel optical links discussed below, three have targeted distances of tens to hundreds of meters. The other has focused on shorter distance backplane interconnections. The specifications of some consortia demonstration links are summarized in Table 11.3.

The Optoelectronic Technology Consortium (OETC) demonstrated several key technologies in parallel optics [57, 59]. The objective was to develop a low-latency, low-power, logic bus for 32-bit word data transfers at 500 Mbit/s. The OETC transmitter module used a 34-element monolithic VCSEL array. The two end VCSELs were used for power monitors. GaAs MESFET ICs were used for both transmitter and receiver. The receiver also had integrated MSM photodetectors. The polymer multilayer on alumina package supported the use of low parasitic thin-film resistors and low inductance decoupling capacitors. In the optical subassembly (OSA), the fiber interface 45° mirrors were visually aligned to the optoelectronics on a silicon base plate. A special 32-fiber MACII-32 connector was developed for these modules [58]. Manchester encoding allowed the use of AC-coupled receivers with minimum added delay at the expense of doubling the optical baud rate.

The Jitney team, drawing from IBM's experience with OETC, focused on low-cost manufacturability. Their design approach emphasized alignment tolerances that, as much as possible, allowed the use of current mass-production electronics, materials, processes, and assembly methods [60]. Reference edges of the optoelectronic chips were cut to within $\pm 20\,\mu m$ on standard dicing equipment so that they could be passively registered in the plastic QFP package. The optical interface used an injection-molded plastic, copper-clad, three-sided arc optical waveguide array. This one-piece optical element aligned to pins molded into the package, turned the light, and interfaced to the fiber array connector with $\pm 20\,\mu m$ net alignment tolerance. Large core step-index (SI) fiber was cheaper than graded-index (GI) fiber and allowed a greater alignment tolerance at the expense of bandwidth–distance product. A new connector, shown in Fig.11.16(a), fabricated with the 3M INCA process was developed for this project.

The ten-channel Parallel Optical Link Organization (POLO) transceiver module, shown in Figures 11.17(a) and 11.17(b), was designed to transparently convey 10 Gbit/s each direction over 10–300 m distances [61, 62]. A DC-coupled receiver used one AC-coupled channel with a very low cutoff frequency to set the threshold for all of the channels. Versions with individually AC-coupled channels were also demonstrated. Arrays of p-substrate VCSELs were developed for compatibility with the npn bipolar transistor current drive. Tapered optical waveguides were fabricated with Polyguide®, a photo-definable polymer developed by DuPont [63]. A new push pull connector based on the MT ferrule and developed under the POLO program, the AMP MPX, is shown in Figure 11.14(b). A ball-grid array (BGA) package allowed a high-density interface to the PC board with low parasitics.

There were two phases in the POLO program, primarily distinguished by wavelength. POLO-1 used 980 nm wavelength VCSELs. The system advantages of 980 nm wavelength approach those at 1,300 nm. Most importantly, the bandwidth–distance product at 980 nm is about double that at 850 nm and is large enough to reach the 500 m backbone distances at Gbit/s rates [28, 64]. The photodetectors used at 980 nm are compatible with those used at 1,300 nm, and the responsivity can be 15% greater than at 850 nm. Eye-safety limits are about double those at 850 nm. The minimum forward operating voltage at 980 nm is about

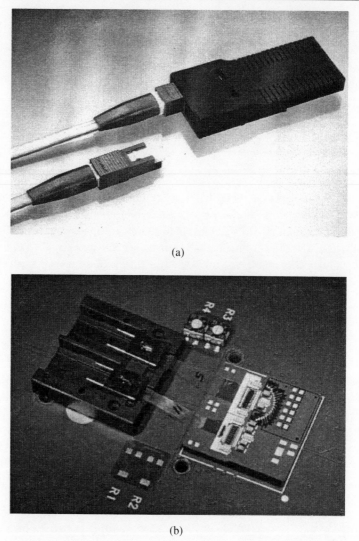

(a)

(b)

Figure 11.17. Photographs of (a) POLO parallel-optical link transceiver module with MPX connectors and (b) POLO parallel-optical link transceiver module without housing.

0.2 V lower than at 850 nm. Because the substrate is transparent, devices can be mounted junction-down for better thermal conductance. Furthermore, the reliability of 980 nm lasers is likely to be better than 850 nm lasers because of lower photon energy, lower operating temperatures, and lesser Al-content compounds near the active region.

During POLO-1, standards were defined (ATM, Fibre Channel) omitting the 980 nm wavelength in favor of 780/850 nm and 1,300 nm. There were two overt reasons: 1. a perception that the lower cost of Si p-i-n or (integrated) GaAs MSM photodetectors for 850 nm compared to InGaAs photodetectors for 980 nm offset the systems advantages of 980 nm over 850 nm and 2. The difficulties for fiber manufacturers to specify their installed products at another wavelength. POLO-2 proceeded at 850 nm wavelength.

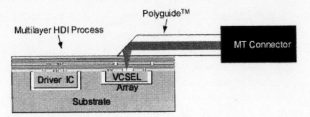

Figure 11.18. Packaging concepts of POINT module for parallel-optical backplane solution. (After Lui et al. [66].)

The primary objective of the Polymer Optical Interconnect Technology (POINT) program was to develop cost-effective optoelectronic packaging technologies for high-speed, short-distance, dense intraboard and backplane interconnections [65, 66]. Emphasis was on scalable, planar, batch fabrication processes that would be compatible with electronic board materials and assembly methods, rendering the incorporation of optical components transparent to the circuit designers. The packaging approach is illustrated in Figure 11.18. Electronic and optoelectronic chips were mounted in wells on an alumina or plastic substrate and then interconnected through overlaminated dielectric layers and metal thin films. The result was a dense circuit with well-controlled parasitics. Polyguide® waveguides were registered to precision excimer laser micromachined references using 400 μm diameter balls, yielding $\sim$1 μm passive alignment accuracy. Results for POINT cited in Table 11.3 relied on laboratory laser driver and receiver electronics.

The DARPA optical micronetworks (OMNET) program was funded for three years beginning in 1997. The primary objective was to accelerate the realization of commercially viable optical interconnect technologies for distances of one to hundreds of meters that offer orders of magnitude improvement in bandwidth per area and power compared to current technologies. The envisioned VCSEL-based technologies would enable new classes of civilian and military distributed computing and sensor systems with commercial, off-the-shelf components. Two OMNET projects, PONI, led by Hewlett-Packard [67], and ChEEtah, led by Honeywell [68], plan to achieve these goals not only through new device and packaging innovations, but also by integrating digital system functions into parallel optical link modules.

11.3.6.2 Product Developments

Parallel optical link module products are being offered or are forthcoming from a number of companies. The specifications for four of these modules are listed in Table 11.4. Note that these efforts took clearly different approaches compared to the consortium demonstrations in several ways.

Motorola announced OPTOBUS® in late 1994. This was the first introduction of a VCSEL-based parallel optical link module product and was conceived as a bidirectional ten-channel cable replacement. In order to be competitive with metallic interconnects, low-cost manufacturability dominated the design philosophy [69, 70, 71]. Motorola used a combination of a ridge waveguide and proton implants to define VCSELs with projected lifetimes above one million hours. The VCSELs are optimized for operation with maximum optical powers of around 2–3 mW. VCSELs and photodetectors used in OPTOBUS have both contacts on top so that the devices can be flip-chip bonded to the two-layer tape-automated bonding (TAB) leadframe.

Figures 11.19(a) and 11.19(b) are photos of the OPTOBUS I module. The TAB leadframe aligns to one GIUDECAST®waveguide interface, shown in detail in Figure 11.19(c), and the HI-PER Link connector mates to the other, all being aligned by the MT ferrule pins. GUIDECAST is a proprietary plastic waveguide formed by gluing two injection-molded cladding halves together with an optical compound that forms the core. In addition to providing an interface to a fiber array in a connector, this tapered waveguide spreads the beam to facilitate compliance with FDA class I eye safety. The optical path increases in size by 10 μm at each interface in the transmitter and receiver to reduce loss.

All channels in the asynchronous DC-coupled receiver of the OPTOBUS I module are identical and have a fixed decision threshold. To allow for a large dynamic range, the decision threshold for a typical link with large margin is necessarily far below the midpoint of the received logic levels. The resulting pulse-width distortion and edge skew depend on the rise and fall times and optical power levels. The distortion and skew are minimized in OPTOBUS I by designing the receiver front end to have a large bandwidth (750 MHz) compared to the baud rate (400 Mbit/s). The fixed threshold also requires a high extinction ratio, so that the lasers are biased 10% to 20% below threshold, slightly increasing turn-on delay and pattern-dependent jitter. OPTOBUS II uses a more complex receiver and a different laser biasing strategy.

W. L. Gore & Associates, Inc., has entered the premises fiber-optics business to complement its high-performance metallic interconnects and extend its range of high-speed interconnect solutions to hundreds of meters [72]. Gore's "seamless migration" strategy provides a common interface for both media, simplifying system design and implementation. Standard connectors mate to both metallic cable assemblies and fiber-optic transmitter or receiver modules. Selected ground pins are converted to power supplies for the optoelectronics, which are contained in the connector housing.

The four-channel Parallel Seamless Migration transmitter module with a 2 mm pitch electrical interface is shown with and without housing in Figures 11.20(a) and 11.20(b). A twelve-channel version is planned. An optical subassembly, consisting of a VCSEL or photodetector array mounted on a ceramic lead frame carrier, is surface-mounted perpendicular to the circuit board. The packaging of the transmitter or receiver circuits and devices in the connector shell requires particular attention to size, heat dissipation, and EMI. Size is greatly reduced by using chip-on-board technology. The metal shell acts as a heat sink and EMI shield. The temperature of the shell is limited to about 50°C for user comfort and safety. Link eye diagrams at 622 Mbit/s have been published; rates above 1 Gbit/s are targetted.

Siemens's twelve-channel 1.25 Gbit/s VCSEL-based parallel optical link family (PAROLI) transmitter modules evolved from edge-emitting laser-based modules [73, 74]. Siemens is currently advertising a parallel fiber-optic transceiver module that incorporates parallel-to-serial and serial-to-parallel conversions and optional 4B/5B coding functions. The module mates with a 12-fiber MT-based, latched connector developed by Siemens, called the Simplex MT Connector (SMC).

NEC is emphasizing two-dimensional array capabilities of VCSELs for fiber-optic links. Passive alignment techniques are employed in the 980 nm substrate-emitting VCSEL array transmitter shown in Figure 11.21 [75, 76]. The 2 × 8 VCSEL array chip is precisely aligned to the metal pattern on the Si substrate by the surface tension of molten solder bumps. The

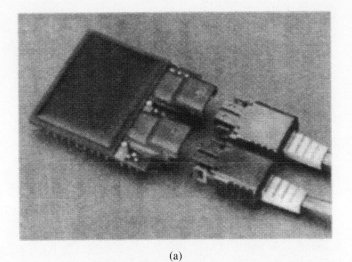

(a)

(b)

(c)

Figure 11.19. Photographs of (a) OPTOBUS®I ten-channel optical data link transceiver module with Alcoa-Fujikura HI-PER Link® connectors, (b) OPTOBUS®I module without connectors and encapsulation, and (c) close-up of GUIDECAST® and TAB electro-optic interface. (Courtesy D. Schwartz, Motorola.)

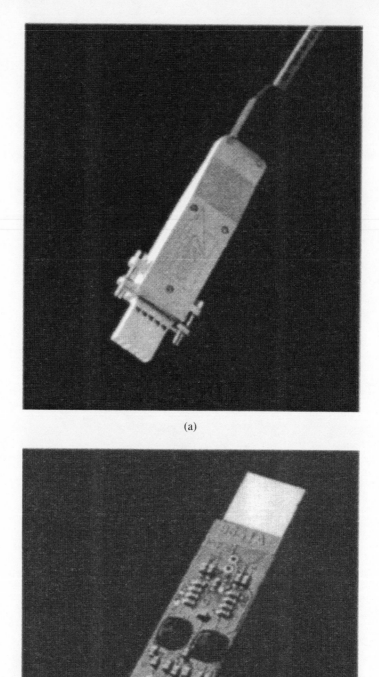

(a)

(b)

Figure 11.20. Photographs of Gore Parallel Seamless Migration Link transmitter module (a) with and (b) without housing. (Reproduced from Theorin et al. [72].)

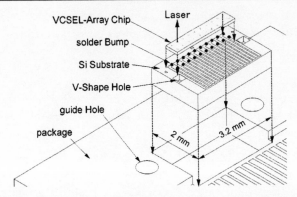

Figure 11.21. Drawing of NEC 2-D VCSEL blind-passively aligned, flip-chip optical interface assembly. (Reproduced from Kosaka et al. [76].)

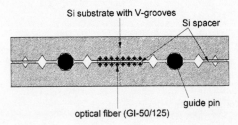

Figure 11.22. NEC 2-D MT-style ferrule design based on silicon V-grooves. (Reproduced from Kosaka et al. [76].)

V-shaped holes in the Si substrate are registered in a chuck that places the Si substrate assembly with reference to the guide holes in the metallized plastic package. The guide holes accept the alignment pins of the NEC 2D MT-style push/pull fiber connector and the lasers are butt-coupled to the fibers in the connector. The connector ferrule design, using precision-etched Si V-grooves is shown in Figure 11.22. Peak alignment tolerances of ± 5.2 μm in this assembly have been demonstrated. Other two-dimensional fiber link demonstrations at NEC using the 980 nm substrate-emitting VCSELs include a 16×16 optical crossbar connection module [77], and 6×6 optical transmission through an image fiber array [78].

Several other companies are developing VCSEL-based parallel optical link modules. Virtually all aim to produce cost-competitive cable extender/replacements that support asynchronous transmission. Hewlett-Packard and AMP are both leveraging their experience with POLO. Hewlett-Packard's n-Plex will comprise a family of products with appropriate numbers and types of channels to match the needs of various classes of applications. AMP's LIGHTRAY MPX twelve-channel transmitters and receivers will share footprints, pin-outs, electrical and optical I/O levels, electrical termination schemes, and power supply voltages with the Siemens modules, according to a multisource agreement between the two companies. Vixel announced the P-VixeLink®four-channel transmitter and receiver modules in February, 1996 [79]. NTT is developing 700 Mbit/s, 40-channel transmitter, receiver, and transceiver parallel interboard optical interconnection technology (ParaBIT) modules for distances over 100 m [80]. The modules use ten-channel p-substrate 850 nm VCSEL arrays, GaAs p-i-n arrays, and polymer waveguide interfaces with 45° mirrors. New bare fiber connectors use buckling of the fibers to provide the contact forces.

Acknowledgments

The authors thank Lewis Aronson, Brian Lemoff, Scott Corzine, Richard Schneider, Dubravko Babic, Al Yuen, David Dolfi, Steve Newton, Rick Trutna, and Waguih Ishak for helpful consulting and support.

References

[1] Bryg, W R., Chan, K. K., and Fiduccia, N. S., "A High-performance, low-cost multiprocessor bus for workstations and midrange servers," *The Hewlett-Packard Journal*, **47**, 18–24 (1996.)

[2] Pozar, D. M., *Microwave Engineering.* New York: Addison-Wesley (1990).

[3] Broomall, J. R. and Van Deusen, H., "Extending the useful range of copper interconnects for high data rate signal transmission," *1997 Proceedings of 47th Electron. Compon. & Technol. Conf.*, 196–203 (1997).

[4] Balanis, Constantine A., *Advanced Engineering Electromagnetics.* New York: Wiley (1989).

[5] Grimes, G. J., Peck, S. R., and Lee, B. H., "User perspectives on intrasystem optical interconnection in SONET/SDH transmission terminals," *1992 IEEE Global Telecommunication Conference,* New York, 201–207 (1992).

[6] Deutsch, A., Arjavalingam, G., Surovic, C. W., Lanzetta, A. P., Fogel, K. E., Doany, F., and Ritter, M. B., "Performance limits of electrical cables for intrasystem communication," *IBM J. Res. Develop.*, **38**, 659–672 (1994).

[7] Kanjamala, A. P. and Levi, A. F. J., "Subpicosecond skew in multimode fibre ribbon for synchronous data transmission," *Electron. Lett.*, **31**, 1376–1377 (1995).

[8] Aronson, L. B., Lemoff, B. E., Buckman, L. A., and Dolfi, D. W., "Low-cost multimode WDM for local area networks up to 10 Gb/s," *IEEE Photonics Tech. Lett.*, **10**, 1489–1491 (1998).

[9] Tauber, D., Wang, G., Geels, R. S., Bowers, J. E., and Coldren, L. A., "Large and small signal dynamics of vertical cavity surface emitting lasers," *Appl. Phys. Lett.*, **62**, 325–327 (1993).

[10] Shtengel, G., Temkin, H., Brusenbach, P., Uchida, T., Kim, M., Parsons, C., Quinn, W. E., and Swirhun, S. E., "High-speed vertical-cavity surface emitting laser," *IEEE Photonics Tech. Lett.*, **5**, 1359–1362 (1993).

[11] Lear, K. L., Mar, A., Choquette, K. D., Kilcoyne, S. P., Schneider, R. P., and Geib, K. M., "High-frequency modulation of oxide-confined vertical-cavity surface-emitting lasers," *Electron. Lett.*, **32**, 457–458 (1996).

[12] Keiser, G., *Optical Fiber Communications.* New York: McGraw-Hill (1991).

[13] Smith, R. G. and Personick, S. D., "Receiver design for optical fiber communication systems," In *Topics in Applied Physics*, **39**, ed.. H. Kressel, 89–160. Berlin: Springer-Verlag (1982).

[14] Muoi, T. V., "Receiver design for high-speed optical–fiber systems," *IEEE J. Lightwave Tech.*, **LT-2**, 243–267 (1984).

[15] Raddatz, L., White, I. H., Hahn, K. H., Tan, M. R., and Wang, S. Y., "Noise performance of multimode vertical cavity surface emitting lasers," *Electron. Lett.*, **30**, 1991–1992 (1994).

[16] Kuchta, D. M., Gamelin, J., Walker, J. D., Lin, J., Lau, K. Y., Smith, J. S., Hong, M., and Mannaerts, J. P., "Relative intensity noise of vertical cavity surface emitting lasers," *Appl. Phys. Lett.*, **62**, 1194–1196 (1993).

[17] Hahn, K. H., Tan, M. R., and Wang, S. Y., "Intensity noise of large area vertical cavity surface emitting lasers in multimode optical fibre links," *Electron. Lett.*, **30**, 139–140 (1994).

[18] Coldren, L. A. and Corzine, S. W., *Diode Lasers and Photonic Integrated Circuits.* New York: Wiley (1995).

[19] Langley, L. N. and Shore, K. A., "Effect of optical feedback on the noise properties of vertical cavity surface emitting lasers," *IEE Proc., Optoelectron.*, **144**, 34–38 (1997).

[20] Epworth, R. E., "The phenomenon of modal noise in analogue and digital optical fibre systems," *Proc. 4th ECOC,* Geneva, 492–501 (1976).

[21] Dandliker, R., Bertholds, A., and Maystre, F., "How modal noise in multimode fibers depends on source spectrum and fiber dispersion," *J. Lightwave Technol.,* **LT-3**, 7–12 (1985).

[22] Koonen, A., "Bit-error-rate degradation in a multimode fiber optic transmission link due to modal noise," *IEEE J. Sel. Areas Commun.,* **SAC-4**, 1415–1522 (1986).

[23] Bates, R. J. S., "Multimode waveguide computer data links with self-pulsating laser diodes," *Proc. Int. Topical Meet. Optical Computing,* Kobe, 89–90 (1990).

[24] Soderstrom, R. L., Block, T. R., Karst, D. L., and Lu, T., "Low cost high performance components for computer optical data links," *Proc. IEEE LEOS Meeting,* Orlando (1989).

[25] Raddatz, L., White, I. H., and Coles, A. N., "Reflection sensitivity of self-pulsing lasers in multimode fibre links," *Int. J. Optoelectronics,* **11**, 101–104 (1997).

[26] Hahn, K. H., Tan, M. R., Houng, Y. M., and Wang, S. Y., "Large area multi-transverse mode VCSELs for modal noise reduction in multimode fibre systems," *Electron. Lett.,* **29**, 1482–1483 (1993).

[27] Kuchta, D. M. and Mahon, C. J., "Mode selective loss penalties in VCSEL optical fiber transmission links," *IEEE Photonics Technol. Lett.,* **6**, 288–290 (1994).

[28] Cunningham, D. G., Hanson, D. C., Nowell, M. C., and Joiner, C. S., "Developing leading-edge fiber optic network link standards," *The Hewlett-Packard Journal,* **48**, 62–73 (1997).

[29] Bates, R. J. S., Kuchta, D. M., and Jackson, K. P., "Improved multimode fiber link BER calculations due to modal noise and non self-pulsating laser diodes," *Optical and Quantum Electronics,* **27**, 203–224 (1995).

[30] Huffaker, D. L., Shin, J., and Deppe, D. G., "Low threshold half-wave vertical cavity lasers," *Electron. Lett.,* **30**, 1946–1947 (1994).

[31] Hayashi, Y., Mukaihara, T., Hatori, N., Ohnoki, N., Matsutani, A., Koyama, K., and Iga, K., "Record low threshold index guided InGaAs/GaAlAs vertical-cavity surface emitting laser with a native oxide confinement structure," *Electron. Lett.,* **31**, 560–561 (1995).

[32] Lear, K. L., Choquette, K. D., Schneider, R. P., Kilcoyne, S. P., and Geib, K. M., "Selectively oxidized vertical cavity surface emitting lasers with 50% power conversion efficiency," *Electron. Lett.,* **31**, 208–209 (1995).

[33] Schneider, R. P., Tan, M. R. T., Corzine, S. W., and Wang, S. Y., "Oxide-confined 850 nm vertical-cavity lasers for multimode-fibre data communications," *Electron. Lett.,* **32**, 1300–1302 (1996).

[34] Cunningham, D. G., Nowell, M. C., and Hanson, D. C., "1.25 Gbd, 550 m links on installed 62 MMF for IEEE 802.3: Leveraging existing long wavelength specifications," *Contribution to IEEE 802.3z,* Enschede, The Netherlands (1996).

[35] Raddatz, L., White, I. H., Cunningham, D. G., and Nowell, M. C., "Increasing the bandwidth-distance product of multimode fibre using offset launch," *Electron. Lett.,* **33**, 232–233 (1997).

[36] Raddatz, L., White, I. H., Cunningham, D. G., and Nowell, M. C., "Influence of restricted mode excitation on bandwidth of multimode fiber links," *IEEE Photonics Technol. Lett.,* **10**, 534–536 (1998).

[37] Hanson, D. C., "IEEE 802.3 1.25 GBd MMF link specification development issues," *Contribution to IEEE 802.3z, Enschede,* The Netherlands (1996).

[38] Tatum, J. A., Smith, D., Guenter, J., and Johnson, R., "High speed characteristics of VCSELs," *SPIE Fabrication, Testing, and Reliability of Semiconductor Lasers II,* **3004**, 151–159 (1997).

[39] Hall, S. H., Walters, W. L., Mattson, L. F., Fokken, G. J., and Gilbert, B. K., "VCSEL electrical packaging analysis and design guidelines for multi-GHz applications," *IEEE Trans. Comp. Packag. Manufact. Technol.,* **20**, 191–201 (1997).

[40] National Committee for Information Technology Standardization (NCITS), *High-Performance Parallel Interface – 6400 Mbit/s Optical Specification (HIPPI-6400-OPT), Revision 0.2.* (1997).

[41] U.S. Code of Federal Regulations (US-CFR), *Performance Standards for Light-Emitting Products,* Title 21, Sec. 1040.10 (1997).

[42] U.S. Center for Devices and Radiological Health (US-CDRH), *Compliance Guide for Laser Products*, FDA 86–8260 (1989).

[43] International Electrotechnical Commission (IEC), *Safety of Laser Products – Part 1: Equipement Classification, Requirements, and User's Guide, Amendment 1 (1997)*, IEC 60825-1 (1993).

[44] International Electrotechnical Commission (IEC), *Safety of Laser Products – Part 2: Safety of Optical Fibre Communication Systems*, IEC 60825-2 (1993).

[45] Bowers, J. E. and Burrus, Jr., C. A., "Ultrawide-band long-wavelength p-i-n photodetectors," *J. Lightwave Technol.*, **5**, 1339–1350 (1987).

[46] Sze, S. M., *Physics of Semiconductor Devices*, 2nd ed. New York: Wiley (1981).

[47] Bean J. C., "Materials and technologies," In *High-Speed Semiconductor Devices*, ed., S. M. Sze, pp. 13–55. New York: Wiley (1990).

[48] Nuyts, R. J., Tzeng, L. D., Mizuhara, O., and Gallion, P., "Effect of transmitter speed and receiver bandwidth on the eye margin performance of a 10 Gb/s optical fiber transmission system," *IEEE Photonics Technol. Lett.*, **9**, 532–534 (1997).

[49] Widmer, A. X. and Franaszek, P. A., "A DC-balanced, partitioned-block, 8B/10B transmission code," *IBM J. Res. Develop.*, **27**, 440–451 (1983).

[50] Banwell, T. C., Von Lehmen, A. C., and Cordel, R. R., "VCSE laser transmitters for parallel data links," *IEEE J. Quantum Electron.*, **29**, 635–644 (1993).

[51] Satake, T., Arikawa, T., Blubaugh, P. W., Parsons, C., Uchida, T. K., "MT multifiber connectors and new applications," *1994 Proceedings of 44th Electron. Compon. and Technol. Conf.*, 994–999 (1994).

[52] Kevern, J., Harper, D., Knasel, D., Knight, K., and Satake, T., "Multifiber connector end face attributes for optimal connector performance," *1996 Proceedings of 46th Electron. Compon. and Technol. Conf.*, 936– 941 (1996).

[53] Satake, T., Tatsuno, T., Ouchi, Y., Knasel, D., Knight, K., Lundberg, J., and Keller, D., "Single-mode multifiber connector design and performance," *1996 Proceedings of 46th Electron. Compon. and Technol. Conf.*, 494–499 (1996).

[54] Igl, S. A., DeBaun, B. A., Lee, N. A., Smith T. L., Henson, G. D., Kuczma, A. S., and Pepeljugoski, P. K., "Automated assembly of parallel fiber optic cables," *1997 Proceedings of 47th Electron. Compon. and Technol. Conf.*, 400–409 (1997).

[55] Lee, N. A., Igl, S. A., DeBaun, B. A., Henson, G. D., and Smith T. L., "Automated method for fabrication of parallel multifiber cable assemblies with integral connector components," *SPIE Optoelectronic Interconnects and Packaging IV*, **3005**, 58–64 (1997).

[56] Million, T. P., *1995 Proceedings of 45th Electron. Compon. and Technol. Conf.*, 584–591 (1995).

[57] Wong, Y. M., Muehlner, D. J., Faudskar, C. C., Fishteyn, M., Gates, J. V., Anthony, P. J., Cyr, G. J., Choi, J., Crow, J. D., Kuchta, D. M., Pepeljugoski, P. K., Stawiasz, K., Nation, W., Engebretsen, D., Whitlock, B., Morgan, R. A., Hibbs-Brenner, M. K., Lehman, J., Walterson, R., Kalweit, E., and Marta, T., "OptoElectronic Technology Consortium (OETC) parallel optical data link: Components, system applications, and simulation tools," *1996 Proceedings of 46th Electron. Compon. and Technol. Conf.*, 269–278 (1996).

[58] Methode Electronics, Fiber Optic Products, 7444 W. Wilson Avenue, Chicago, IL 60656, 708/867-9600, Fax: 708/867-0435, http://www.methode.com. (1997).

[59] Wong, Y. M., Muehlner, D. J., Faudskar, C. C., Bucholz, B., Fishteyn, M., Brandner, J. L., Parzygnat, W. J., Morgan, R. A., Mullally, T., Leibenguth, R. E., Guth, G. D., Focht, M. W., Glogovsky, K. G., Zilko, J. L., Gates, J. V., Anthony, P. J., Tyrone, B. H., Ireland, T. J., Lewis, D. H., Smith, D. F., Nati, S., Lewis, D. K., Rogers, D. L., Aispain, H. A., Gowda, S., Walker, S. G., Kwark, Y. K., Bates, R. J. S., Kuchta, D. M., and Crow, J. D., "Technology development of a high density 32-channel 16-Gb/s optical data link for optical interconnection applications for the optoelectronic technology consortium (OETC)," *IEEE J. Lightwave Technol.*, **13**, 995–1016 (1995).

[60] Crow, J. D., Choi, J. H., Cohen, M. S., Johnson, G., Kuchta, D., Lacey, D., Ponnapalli, S., Pepeljugoski, P., Stawiasz, K., Trewhella, J., Xiao, P., Tremblay, S., Ouimet, S., Lacerte, A., Gauvin, M., Booth, D., Nation, W., Smith, T. L., DeBaun, B. A., Henson, G. D., Igl, S.A., Lee, N. A., Piekarczyk, A. J., Kuczma, A. S., and Spanoudis, S. L. "The Jitney parallel optical interconnect," *1996 Proceedings of 46th Electron. Compon. and Technol. Conf.*, 292–300 (1996).

[61] Hahn, K. H., Giboney, K. S., Wilson, R. E., Straznicky, J., Wong, E.G., Tan, M. R., Kaneshiro, R. T., Dolfi, D. W., Mueller, E. H., Plotts, A. E., Murray, D. D., Marchegiano, J. E., Booth, B. L., Sano, B. J., Madhaven, B., Raghavan, B., and Levi, A. F. J. "Gigabyte/s data communications with the POLO parallel link," *1996 Proceedings of 46th Electron. Compon. and Technol. Conf.*, 301–307 (1996).

[62] Giboney, K. S., "Parallel-optical interconnect development at HP Laboratories," *SPIE Optoelectronic Interconnects and Packaging IV*, **3005**, 193–201 (1997).

[63] Booth, B. L., "Polymers for integrated optical waveguides," In *Polymers for Lightwave and Integrated Optics*, ed., C. P. Wong, New York: Academic Press (1993).

[64] Tan, M. R., Hahn, K. H., Houng, Y. M, and Wang, S. Y., "Surface emitting laser for multimode data link applications," *The Hewlett-Packard Journal*, **46,** 67–71 (1995).

[65] Liu, Y. S., Wojnarowski, R. J., Hennessy, W. A., Bristow, J. P., Liu, Y., Peczalski, A., Rowlette, J., Plotts, A., Stack, J., Kadar-Kallen, M., Yardley, J., Eldada, L., Osgood, R. M., Scarmozzino, R., Lee, S. H., Ozgus, V., and Patra, S., "Polymer optical interconnect technology (POINT) – optoelectronic packaging and interconnect for board and backplane applications," *1996 Proceedings of 46th Electron. Compon. and Technol. Conf.*, 308–315 (1996).

[66] Liu, Y. S., Wojnarowski, R. J., Hennessy, W. A., Rowlette, J., Stack, J., Kadar-Kallen, M., Green, E. Liu, Y., Bristow, J. P., Peczalski, A., Eldada, L., Yardley, J., Osgood, R. M., Scarmozzino, R., Lee, S. H., and Patra, S. "High density optical interconnects for board and backplane applications using VCSELs and polymer waveguides," *1997 Proceedings of 47th Electron. Compon. and Technol. Conf.*, 391–398 (1997).

[67] Yuen, A., Giboney, K., Wong, E., Buckman, L., Haritos, D., Rosenberg, P., Straznicky, J., and Dolfi, D. "Parallel optical interconnections development at HP labs," *Proc. IEEE Lasers Electro-Optics Soc. 1997 Annual Meeting,* San Francisco, 191–192 (1997).

[68] Lehman, J., "An introduction to the ChEEtah project," Presented at *Hot Interconnects V*, Stanford, CA, Aug., (1997).

[69] Lebby, M., Gaw, C. A., Jiang, W., Kiely, P. A., Shieh, C. L., Claisse, P. R., Ramdani, J., Hartman, D. H., Schwartz, D. B., and Grula, J. "Characteristics of VCSEL arrays for parallel optical interconnects," *Proceedings of 46th Electron. Compon. and Technol. Conf.*, 279–291 (1996).

[70] Schwartz, D. B., Chun, K. Y., Choi, N., Diaz, D., Sylvia, P., Raskin, G., and Shook, S. G., "OPTOBUST™ I: Performance of a 4 Gb/s optical interconnect," *Proc. Third International Conf. Massively Parallel Process. Optical Interconnects,* 256–263 (1996).

[71] Raskin, G., Lebby, M. S., Carney, F., Kazakia, M., Schwartz, D. B., and Gaw, C. A., "A parallel interconnect for a novel system approach to short distance high information transfer data links," *SPIE High-Speed Semiconductor Lasers for Communication*, **3038**, 165–174 (1997).

[72] Theorin, C. R., Kilcoyne, S. P., Peters, F. H., Martin, R. D., Donhowe, M. N., "'A seamless migration' to VCSEL-based optical data links," *SPIE Vertical Cavity Surface Emitting Lasers*, **3003**, 120–130 (1997).

[73] Karstensen, H., "Parallel optical links – PAROLI, a Low cost 12-channel optical interconnection," *Proc. IEEE Lasers Electro-Optics Soc. 1995 Annual Meeting,* San Francisco, 226–227 (1995).

[74] Karstensen, H., Melchior, L., Plickert, V., Drogemuller, K., Blank, J., Wipiejewski, T., Wolf, H.-D., Wieland, J., Jeiter, G., Dal'Ara, R., and Blaser, M., "Parallel optical link (PAROLI) for multichannel gigabit rate interconnection," *1998 Proceedings of 48th Electron. Compon. and Technol. Conf.*, 747–754 (1998).

[75] Kosaka, H., Kajita, M., Yamada, M., Sugimoto, Y. Kurata, K., Tanabe, T., and Kasukawa, Y., "2D alignment free VCSEL-array module with push/pull fibre connector," *Electron. Lett.*, **32**, 1991–1992 (1996).

[76] Kosaka, H., Kajita, M., Yamada, M., Sugimoto, Y., Kurata, K., Tanabe, T., and Kasukawa, Y., "Plastic-based receptacle-type VCSEL-array modules with one and two dimensions fabricated using the self-alignment mounting technique," *1997 Proceedings of 47th Electron. Compon. and Technol. Conf.*, 382–390 (1997).

[77] Kosaka, H., Kajita, M., Yamada, M., and Sugimoto, Y., "A 16*16 optical full-cross-bar connection module with VCSEL-array push/pull module and polymer-waveguide coupler connector," *IEEE Photonics Technol. Lett.*, **9**, 244–246 (1997).

[78] Kosaka, H., Kajita, M., Li, Y., and Sugimoto, Y., "A two-dimensional optical parallel transmission using a vertical-cavity surface-emitting laser array module and an image fiber," *IEEE Photonics Technol. Lett.*, **9**, 253–255 (1997).

[79] Swirhun, S., Dudek, M., Neumann, R., Calkins, P., Brusenbach, P., Brinkmann, D., Northrop, T., Moore, A., Paananen, D., Scott, J., and White, T., "The P-VixelLink multichannel optical interconnect," *1996 Proceedings of 46th Electron. Compon. and Technol. Conf.*, 316–320 (1996).

[80] Usui, M., Matsuura, N., Sato, N., Nakamura, M., Tanaka, N., Ohki, A., Hikita, M., Yoshimura, R., Tateno, K., Katsura, K., and Ando, Y., "700-Mb/s × 40-channel parallel optical interconnection Module using VCSEL arrays and bare fiber connectors (ParaBIT: Parallel inter-Board optical Interconnection Technology)," *Proc. IEEE Lasers Electro-Optics Soc. 1997 Annual Meeting,* San Francisco, 51–52 (1997).

[81] ANSI/IEEE Std 1596–1992. *Scalable Coherent Interface (SCI).* New York: The Institute of Electrical and Electronics Engineers.

[82] ANSI X3T11. *Fibre Channel Physical and Signaling Interface.* New York: American National Standards Institute.

[83] IEEE Std 1596.3-1996. *IEEE Standard for Low-Voltage Differential Signals (LVDS) for Scalable Coherent Interface (SCI).* New York: The Institute of Electrical and Electronics Engineers.

12

VCSEL-Based Smart Pixels for Free-Space Optoelectronic Processing

Carl W. Wilmsen

12.1 Introduction

One of the major advantages of optical processing is the ability of optical beams to function as information carriers, and since multiple beams can propagate in parallel in free space with little or no interactions (crosstalk), the means for multiple data channels are readily available. VCSEL arrays make ideal light source for free-space parallel processing since they can easily be fabricated into two-dimensional arrays of individually addressed lasers that emit a low divergent column of light normal to the array surface. The VCSELs do not perform logic or function as switches but rather provide the interconnection between two or more interacting planes of electronic processors. Thus, the VCSEL arrays must be electrically connected to the processing arrays that both perform the desired processing function and provide the current drive for the VCSELs. These processors are called "smart pixels" and can be either electrically or optically addressed. As discussed in Chapters 9 and 11 VCSELs are also well suited light sources for guided wave systems (e.g., optical data links), and indeed such systems will be the first to be commercially produced because there is a large potential market and only ten to sixteen VCSELs are required for these data links. The choice of free-space interconnects over guided-wave interconnects depends on the application. Free-space interconnects become advantageous when there is a large number of channels and/or when the routing of the channels must be reconfigured or have a large fan out. Competitive cost is also a factor, but in some cases other factors must be considered, such as with large ATM switches where electronics and guided-wave solutions appear excessively cumbersome to efficiently perform cross-point switching without the aid of free-space interconnections [1].

The design and implementation of free-space optoelectronics systems requires the integration of optoelectronic components with passive optics in an optomechanical package. Therefore many compromises must be made to accomplish this task at low cost with high reliability and sufficient robustness. This chapter details the implementation of smart pixels and how to integrate the VCSELs with the electronics components. This is followed by a discussion of some of the other components of smart pixels such as photodetectors, receivers, and laser drivers. At the end of the chapter a summary offers suggestions for future directions for smart pixel design and implementation.

12.2 Smart Pixels

The heart of any optoelectronic parallel processing system (i.e., the enabling component) is the smart pixel array. Smart pixels are electronic processing cells with optical input and/or

Table 12.1. *Applications and specifications of VCSEL-based smart pixel arrays.*

Fiber data links	• Small array, 10 to 30 VCSELs • High speed, $\sim$1 GHz • Low functionality
Optical back planes	• Medium arrays, $\sim$100 VCSELs • High speed, $\sim$1 GHz • Low functionality
Free-space optical processing	• Large arrays, >1000 VCSELs • High speed, >200 M frames/sec • High functionality

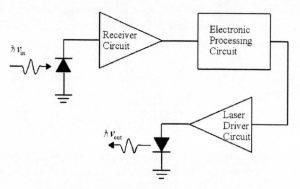

Figure 12.1. Schematic of a generic smart pixel.

optical output. In particular, this chapter is concerned with smart pixels that use VCSELs as the output device and that receive an optical input from VCSELs of the previous stage. Figure 12.1 illustrates a generalized smart pixel block diagram, which includes an input photodetector, input signal amplifier, electronic processing circuitry, a VCSEL driver, and the output VCSEL. There could, of course, be several input photodetectors and several VCSEL outputs, as well as one or more electrical inputs or outputs. If the inputs are electrical then it is preferred that these be in the form of a bus connecting all of the pixels; otherwise there would be a large number of electrical connections to the pixels, which would defeat the purpose of optical parallel processing. Such an electrical bus might also be used to set a gain or feedback level in the pixels. For the same reasons, the electrical outputs should also be a form of a bus on which data are serially outputted. All of the components of the smart pixel must be fabricated with compatible processes, occupy a limited area, have high performance, and be reliably replicated many times across the array. These are demanding requirements that can only be met with established foundry processing. Such processing foundries exist for photodetector and VCSEL arrays and, of course, for electronic circuits. However, monolithic or hybrid integration foundries that combine all three components are just now being established and it is not clear which will offer the best cost versus performance advantage. These issues are discussed in this chapter.

The three general types of applications best suited for VCSEL-based smart pixels are listed in Table 12.1 along with their general specifications. Note that all of the applications require bit rates greater than 200 Mbit/s, and thus all practical smart pixel arrays must be fast.

The data links and interconnect applications are discussed in Chapters 9 and 11 while this chapter emphasizes massively parallel optical processing, MPOP. To be competitive with electronic implementations, MPOPs must either require reconfigureable global intercon-nects or process data at 0.1 to 1 Tbit/s. The latter implies that a system composed of arrays containing 2,000 smart pixels each must operate at 50 to 500 megaframes/second. With only 200 pixels per stage, the frame rate must increase to 500–5,000 megaframes/second. Thus, there is a trade-off among problems associated with fabricating large arrays, their heat dissipation, optical alignment tolerance, and the frame rate. These and other trade-offs and design compromises are discussed at the end of this chapter.

At the present time, the development of VCSEL-based smart pixels is still in its infancy and most pixels contain only a few transistors or the array contains only a few pixels. And although there is no "typical" smart pixel, the pixels shown in Figures 12.2 and 12.3 illustrate the general range of possibilities for foundry fabricated optoelectronic circuits with a VCSEL as the output device. Figure 12.2 illustrates an example of a smart pixel [2] that contains three metal–semiconductor–metal (MSM) photodetectors, two bump bonding pads for a VCSEL, and thirty-two transistors, which form two logic gates and a four-stage VCSEL driver circuit. All of these components were fabricated by the standard Vitesse Semiconductor MESFET integrated circuit process and are monolithically integrated.

Since the VCSELs were not monolithically grown on this GaAs electronic chip, the VCSELs had to be added with a postfoundry process. The pixel shown in Figure 12.2 oc-cupies an area 250 μm $\times$ 250 μm, which translates into a pixel density of 1,600 pixels/cm^2, although the need for wire bonding pads for power supplies and test points reduces the pixel density to $\sim$1,000 pixels/cm^2. Note that a majority of the pixel area is devoted to the optical input and output components, that is, the three MSM photodetectors, the VCSEL bonding pads, and the VCSEL driver circuit. Reducing the size of these components is desirable, especially if the electronic circuit of the pixel is complex and requires a large percentage of the pixel area. The size of the photodetector depends on the size of the incident beam, which in turn is dependent on the optics and the system architecture. The number of photodetectors depends on the function to be performed, but effort should be made to reduce their number to a minimum.

The CMOS smart pixel of Figure 12.3 contains 168 transistors. This CMOS pixel [3] has two monolithically integrated pn junction photodetectors, but the VCSELs are located on a separated chip, although the VCSELs could have been bonded directly to the CMOS chip as shown in Figure 12.2 for the GaAs pixel. As can be seen, even with 168 transistors, the optoelectronic components still occupy a large part of the pixel area.

In designing a smart pixel array, the first decision to be made is the technique for inte-grating the electronic and optoelectronic components. The following section first compares the general advantages and disadvantages of monolithic and hybrid integration and then examines the details of implementing these techniques.

12.3 Hybrid Versus Monolithic Integration

The great success of electronics is primarily due to the ever-increasing monolithic integration of all of the components onto single chips. It therefore seems logical that optoelectronics should follow this example and set a goal of fabricating monolithically integrated arrays of

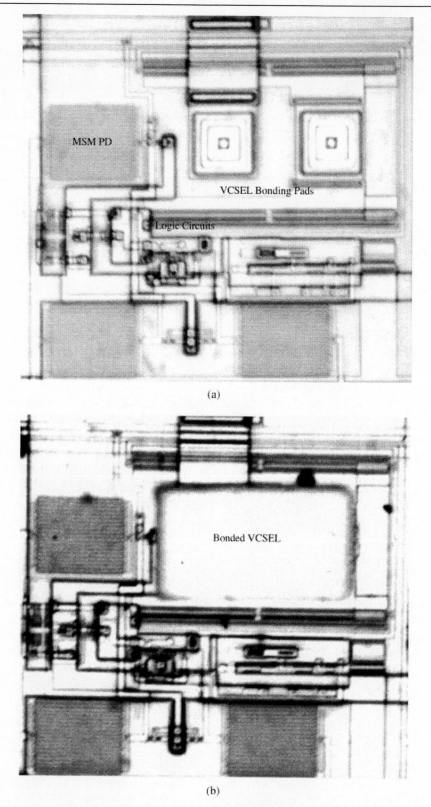

(a)

(b)

Figure 12.2. Optical photomicrographs of a GaAs MESFET smart pixel before and after flip-chip bonding of a VCSEL [2].

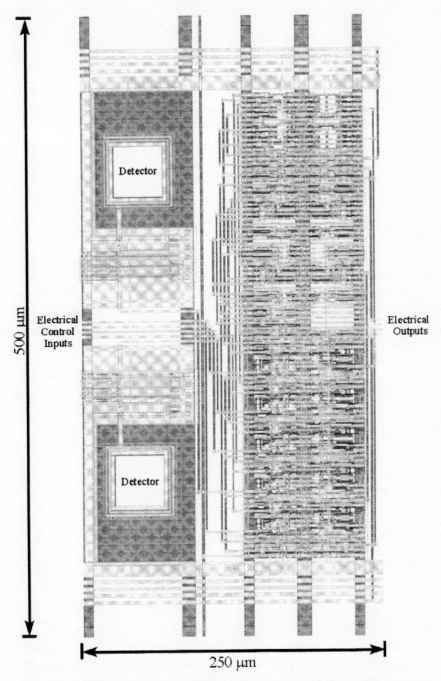

Figure 12.3. CMOS smart pixel composed of 168 transistors and 2 pn junction photo-detectors. The VCSELs were attached via bonding pads at the edge of the chip [3].

Table 12.2. *Advantages and disadvantages of hybrid and monolithic integration.*

Monolithic	Hybrid
Advantages:	**Advantages:**
• Single chip	• Lower cost
• Few packaging interfaces	• Shorter development time
	• Can use standard chip technology
	• Optimized component performance
Disadvantages:	**Disadvantages:**
• High development cost	• Many packaging interfaces
• High production/equipment cost	• Poorer mechanical reliability
• Nonplanar interconnects	• Nonplanar interconnects
• Lower performance	

processors that contain the input photodetectors, electronic processing elements, and the output VCSELs. However, the silicon integrated circuit industry has invested billions of dollars on the development of fabrication processes, fabrication/assembly equipment, and design software. The huge market for electronic integrated circuits justifies this investment. Optoelectronics based on smart pixels, however, cannot command such large markets either now or in the future and thus the development funds will be much smaller. As a result, there is no simple answer to the question of monolithic versus hybrid integration; however, it will be wise for optoelectronics to plan to use as much of the silicon and GaAs integrated circuit technology as possible and still take advantage of the characteristics of optics. Table 12.2 lists some of the advantages and disadvantages of the monolithic and hybrid approaches. The table shows that there is a trade-off among cost, performance, reliability, and development time.

The following sections discuss some of the details of monolithic and hybrid smart pixel integration and the consequences of the different design approaches. From this discussion, one can gain a better understanding of the pros and cons of the various design compromises, which should then lead to a more optimum design.

12.3.1 Monolithic Integration

Monolithic integration requires the epitaxial growth for the VCSELs, photodetectors, and electronic circuits be on the same GaAs substrate, and for the most part, these layers must also be lattice-matched to the substrate. As a result, all of these device layers are stacked on top of each other. Converting these layers into a smart pixel necessitates the removal of some of the layers in selected areas and since these layers are several microns thick, a highly nonplanar structure results. This in turn limits the minimum component size and component separation because of the difficulties with electrical interconnect and step coverage and the problems of focusing on nonplanar structures during photolithography. In addition to locally removing the layers, electrical isolation between the devices is often accomplished by etching the layers into mesas. This adds to the nonplanarity of the chip.

Several groups have successfully used monolithic integration to form simple pixels using either heterojunction bipolar transistors (HBTs)/pin diode or MESFET/MSM transistor/

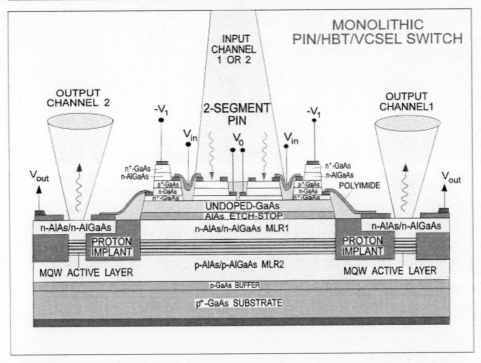

Figure 12.4. Cross section of a monolithically integrated smart pixel composed of heterojunction bipolar transistors, pin diodes, and VCSELs [4, 5].

photodetector combinations. We will see in the following examples how these different research groups provided the electrical isolation and how this affected the topography and general layer design.

Cheng's group [4, 5] fabricated 2×2 arrays of reconfigurable binary switches using HBT/pin diodes that operate up to 400 Mbit/s. A schematic cross section of this switch is shown in Figure 12.4. The HBT and pin diode layers were grown on top of the VCSEL structure and selectively removed from the active VCSEL area. Isolation was provided by an undoped GaAs layer, which acts as a nonconducting substrate for the HBTs and pin diodes. All of the top layers are then etched to form mesas, which physically separate the devices. The mesa etching requires a large amount of area, which reduces the component density and makes interconnect metalization more difficult. Note that the topography of this structure isolates the HBTs and pin diodes and results in a surface undulation of about 1.5 μm. A proton implant isolates the VCSELs and also defines the VCSEL aperture. The pin diode epitaxial layers are the same as those used for the collector of the HBT. Cheng's group has also reported other similar HPT/VCEL and pnpn/VCSEL structures [6, 7].

The smart pixels of Feld [8] used HPTs (heterojunction phototransistors) instead of HBTs since a HPT can serve both as photodetector and amplifier thereby simplifying the circuits, but the variation in topography remained about the same as that of Cheng's pixel. Feld monolithically fabricated a 4×4 array of XOR gates on top of the VCSEL layers. An undoped epitaxial layer of AlInP was used to isolate the HPTs from the proton-implanted VCSELs since the bandgap of AlInP is $\sim$150% larger than that of undoped GaAs. A cross section of this monolithic structure showing the electrical and optical feedback paths is illustrated in

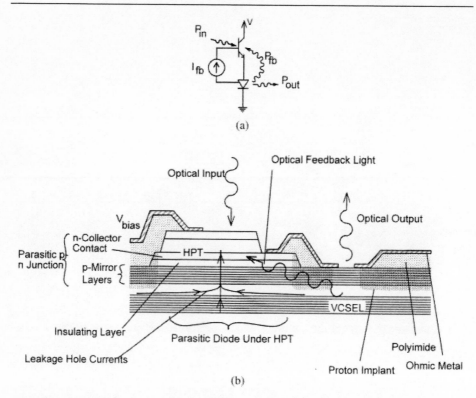

Figure 12.5. Cross section of a monolithically integrated smart pixel composed of heterojunction phototransistor and VCSEL showing the optical and electrical feedback paths [8].

Figure 12.5. The HPT and AlInP layers were regrown on top of the VCSEL in a separate MBE growth chamber, although this is not necessary if a phosphorus source is available in the VCSEL growth chamber.

Yang et al. [9] monolithically integrated a VCSEL with a MESFET as shown in Figure 12.6. The MESFET was placed on top of the undoped p-mirror, which provided the isolating platform for the electronic components. The VCSELs were isolated from the MESFET by etching a trench down to the n-mirror, which introduced $\sim 3.5\,\mu\text{m}$ steps in the surface topography. This trench also provided a via down to the n-contact. A ring of Zn around the VCSEL aperture was diffused to the undoped p-mirror in order to provide a low voltage drop conduction path from the metal contact placed on top of the VCSEL. The VCSEL had a threshold current $I_{\text{th}} = 6\,\text{mA}$ with a maximum output power of 1.1 mW. These are good characteristics considering this work was completed in 1992. The $3\,\mu\text{m}$ gate-length MESFET has a transconductance of 50 mS/mm, which produced an optical/electrical conversion factor of 0.5 W/V.

Hibbs-Brenner's group [10, 11] used a VCSEL/MESFET structure similar to Yang's [9] with the MESFETs and MSM photodetectors layers grown on top of the VCSEL. However, since their p-mirror was conducting, an undoped GaAs layer was placed between the MESFET and the VCSEL layers, as illustrated in Figure 12.7. The undoped layer electrically insulates the MSMs and MESFETs from the VCSEL layers, while an ion implant isolates the devices from each other. Tested individually, the VCSEL operated at

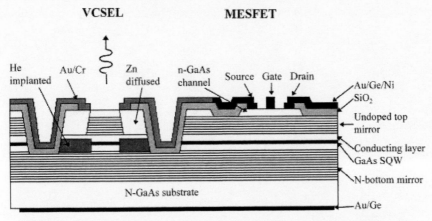

Figure 12.6. Cross section of a smart pixel monolithically integrating a VCSEL with a MESFET and defining the VCSEL by a Zn diffused pn junction. The MESFET channel is the topmost GaAs layer [9].

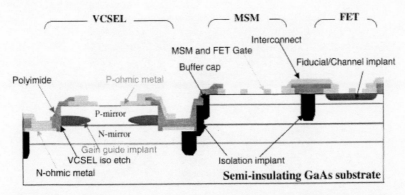

Figure 12.7. Cross section of a smart pixel fabricated by monolithically integrating a VCSEL with a MESFET and MSM photodetector. The VCSEL is defined with proton implantation and the MESFET channel is the topmost layer [10, 11].

$f > 3$ Gbit/s and the MSM detector had a rise time of 133 ps. Thus, it appears that this monolithic process is well underway and should yield high performance arrays.

Matsuo et al. [12] have taken the opposite approach from that of the others in that the VCSEL structure was grown on top of the MESFET layers as shown in Figure 12.8. This has the advantage of allowing the semi-insulating substrate to serve as the MSM optical adsorption layer as well as providing device isolation. The major disadvantage of this structure is that now the surface topography is the full height of the VCSEL plus the MESFET stack, which is $\sim$6–7 μm. Even so, they have successfully fabricated small arrays of these pixels, which shows that the problems associated with nonplanarity can be overcome. These three transistor circuits performed either as NOR or OR gates with a contrast ratio of 30 dB and an optical gain between 1.5 and 3.7, depending on the bias. A 3 dB bandwidth of 220 MHz was demonstrated for 300 μW of input power to the MSM photodetector.

A summary of the methods of device isolation and the height of the topography for the different pixels are given in Table 12.3.

Table 12.3. *Summary of the method of isolation and topographic height resulting from different methods of VCSEL-smart pixel monolithic integration.*

Research group/ Type of device	Electronic component isolation layer	Lateral electronic isolation	Lateral VCSEL isolation	Topography
Cheng (4–7) HBT/pin diode	Undoped GaAs	Etched mesas	Proton implant	1.5 μm
Feld (8) HPT	Undoped AlInGa	Etched mesas	Proton implant	1.5 μm
Yang (9) MESFET	Undoped p-mirror	Etched mesas	Etched trench	3.5 μm
Hibbs-Brenner (10–11) MESFET/MSM	Undoped GaAs	Implant	Implant	3.5 μm
Matsuo (12) MESFET/MSM	Semi-insulating GaAs substrate	Etched mesas	Etched trench	6 μm

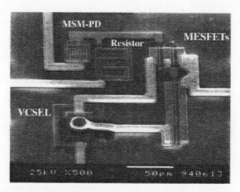

Figure 12.8. Smart pixel fabricated by monolithically integrating a VCSEL with three MESFETs, a MSM photodetector, and a diffused resistor [12].

12.3.2 Hybrid Integration

Hybrid integration usually offers more flexibility in the choice of components and the materials from which they are fabricated since the components can be fabricated separately and all of the materials do not need to be lattice-matched to the substrate. The primary problem lies in joining the various components together into large arrays such that the photodetectors and VCSELs are properly aligned and have an optical window to the next stage. In addition, the electrical interconnections from the PDs and VCSELs to the electronic circuits must be short to reduce the parasitic capacitance and inductance. The thermal conductance of the path between the VCSEL and the substrate and the substrate bonding is also an important consideration. For large arrays, there is also the difficulty of accommodating all of the interconnect "wires" (e.g., a chip with 1,000 pixels/cm^2 requires 1,000 to 2,000 electrical connections). Bryan et al. [13] discussed three methods of electrically connecting VCSELs and other optoelectronic components to electronic chips: wire bonding, bridge bonding, and flip-chip solder bonding of the whole VCSEL or PD array chip. For large, fast arrays the first two of these techniques are far from ideal and the flip-chip bonding of the whole array is suitable only when the array substrate is transparent to both the input and output wavelengths. This leads to the conclusion that it would be best if the VCSELs (and PDs) were placed in the pixels as individual devices (i.e., with the VCSEL substrate removed). To accomplish this, the VCSELs must be placed in the pixel by hybrid bonding, while the PDs can be either monolithically integrated or hybrid bonded. The following discussion concentrates on the hybrid integration of VCSELs to foundry-fabricated electronic chips, since the hybrid integration of PDs is similar, but easier than for VCSELs.

The techniques for the hybrid bonding of VCSELs to electronic chips falls into two categories: one in which the growth substrate is removed, leaving individual VCSELs, and the other that leaves the VCSELs attached to the growth substrate. The required processing and the advantages/disadvantages of both of these techniques are discussed separately.

12.3.2.1 Individually Bonded VCSELs

The techniques that result in individually bonded VCSELs can be organized into the following three processing groups: appliqué, coplanar contact, and bonded wafer.

12.3.2.1.1 Appliqué Technique

For the appliqué technique, one proceeds as follows:

- Process the wafer into VCSELs.
- Temporarily "glue" all of the VCSELs to another substrate (such as glass or Mylar).
- Remove the substrate from the VCSEL mesas.
- Place a single VCSEL onto the electronic wafer.
- Form electrical contacts between the VCSEL and the electronic chip.

This technique was first developed by Jokest and co-workers [14] for photodetectors and LEDs. They have successfully bonded an 8×8 array of thin film photodetectors onto a CMOS circuit [15]. As part of their process, the devices are attached to a Mylar membrane and then individual devices are transferred to the electronic chip (or other substrate) by applying pressure. Individual devices can be tested before final bonding and faulty devices replaced.

Maracas and coworkers [16, 17] have used a variation of this technique to place a single VCSELs onto an NMOS integrated driver circuit that was electrically driven at 200 MHz. Figure 12.9 illustrates a cross section of a VCSEL after it has been bonded to an NMOS circuit. Their present technique utilizes a mechanical probe tip to which a single VCSELs is attached with a water-soluble glue. Each VCSEL is aligned and bonded separately. Therefore this processes is not yet suitable for commercial or large-array applications; however, it does demonstrate single VCSELs, and probably VCSEL arrays can be bonded by this technique.

12.3.2.1.2 Co-Planar Contact Technique

In this technique, one must:

- Process the wafer into VCSELs.
- Deposit ohmic contacts to both the n- and p-mirrors and then electroplate the n-mirror contact to the same height as the p-mirror contact.
- Bond the VCSEL and electronic wafers together forming electrical contacts.
- Remove the VCSEL wafer substrate by etching, leaving individual VCSELs bonded directly to the smart pixels.

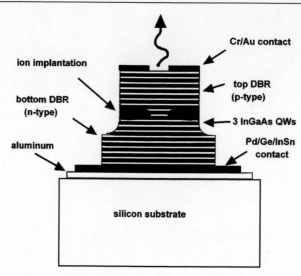

Figure 12.9. Hybrid bonding of a single VCSEL to NMOS circuitry using a mechanical probe tip [16].

Goossen et al. developed a coplanar contact process for SEED devices [18] and successfully bonded 4352 SEEDs to CMOS circuitry [19]. This technique was adapted to VCSEL arrays by Wilmsen and his students [2, 20]. The optical and SEM photomicrographs of Figure 12.10 illustrate the successful bonding of an 8×8 array of VCSELs to GaAs MESFET smart pixels fabricated by Vitesse Semiconductor Corp. VCSELs bonded by this technique have demonstrated optical output power greater than 1 mW with V_{th} as low as 1.9 V and I_{th} between 2 and 4 mA depending on the size of the oxide-defined aperture. In the present form, these chips require special handling after VCSEL bonding; however, changes in the technique could greatly increase their robustness and make this technique suitable for very large arrays.

12.3.2.1.3 Bonded-Wafer Technique

In the bonded-wafer technique, one must:

- Permanently "glue" an unprocessed optoelectronic wafer face down on the electronic circuit wafer. (The opto wafer can be composed of VCSEL and photodetector layers.)
- Remove the substrate from the opto wafer.
- Process the opto wafer into individual VCSELs and photodetectors.
- Form electrical contacts connecting the opto and electronic chips.

Yeh and Smith [21] bonded a VCSEL chip to a silicon wafer that had been coated with 300 nm of Au and 1.5 μm of In. After heating to 280°C with a weight applied, the GaAs substrate was removed, leaving the VCSEL layers bonded to the silicon. Top metal ring contacts were then deposited on the exposed p-mirror and mesas etched to form individual VCSELs.

Matsuo et al. [22] followed a similar bonding procedure except polyimide was used as the wafer bonding material to a silicon wafer that had Au bonding pads deposited in the proper locations. The GaAs substrate was then removed from the VCSEL wafer and mesas etched to form the VCSELs. After planarizing with polyimide, electroplating was used to form Au vias to connect these devices with the Au pad on the silicon wafer. This process is schematically illustrated in Figure 12.11.

(a)

(b)

Figure 12.10. Coplanar flip-chip bonded VCSELs to integrated circuits (a) a 4 × 4 array
bonded to CMOS and (b) an 8 × 8 array bonded to GaAs MESFET [2].

Matsuo et al. [23] have also proposed using this technique to simultaneously bond both
VCSELs and photodetectors to a CMOS chip. This process is illustrated in Figure 12.12 and
appears to be a powerful technique if it can be developed. The advantages and disadvantages
of three of these methods are compared in Table 12.4.

Overall, there has been significant progress in the bonding of individual VCSELs to
integrated circuits, demonstrating that the VCSELs can be bonded without either damage
or degradation of their optical performance. Therefore, it is clear that the next generation of
bonding techniques will produce smart pixel arrays that can be used in high-speed free-space
processing systems.

12.3.2.2 Bonded VCSEL Array Chips

There are several techniques for bonding or connecting VCSEL chips to electronic chips
without removing the VCSEL substrate. One of these uses a multichip module approach in
which the VCSEL and electronic chips are mounted onto a submount and then the bonding
pads of each chip is bonded to the interconnect traces. This bonding could be as simple
as wire bonding [24, 25], as illustrated in Figure 12.13 [24], or flip-chip solder bump

Table 12.4. *Comparison of the advantages and disadvantages of the VCSEL bonding processes.*

Appliqué	Coplanar contacts	Bonded wafer
Advantages: • Individual VCSELs can be tested before bonding • Good heat transfer	**Advantages:** • Planar interconnect metalization • VCSEL test structure can be pretested • Large arrays bonded simultaneously • Good heat transfer	**Advantages:** • Accurate alignment • Wafer scale fabrication • Multiple types of devices can be bonded simultaneously • Robust
Disadvantages: • One device at a time • Each device aligned separately • Nonplanar interconnect metalization	**Disadvantages:** • Breakable • Can only pretest test structures and not the array itself	**Disadvantages:** • Low heat transfer • Nonplanar interconnect metalization • Can't pretest VCSELs

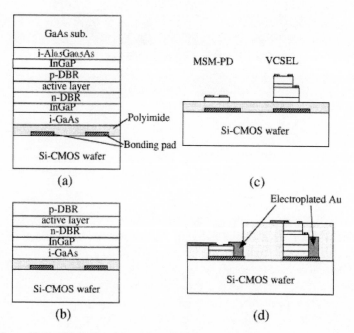

Figure 12.11. The polyimide process for bonding VCSELs to a Si wafer [22].

bonded [26–28]. These approaches becomes intractable for large arrays and also introduces excessive parasitic capacitance and inductance. Since this technique places the detector and VCSEL chips on different optical axes, the optical design is more complicated. However, the process is simple and for small, slower arrays, this approach has the advantage of low cost and ease of implementation.

A more novel approach has been reported by Jin et al. [29], as illustrated in Figure 12.14. In this process, the CMOS chip is first thinned and then large holes ($100\,\mu$m $\times\,180\,\mu$m)

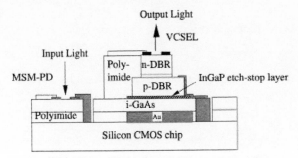

Figure 12.12. Cross section of a VCSEL and MSM photodetector bonded with poly-imide to a CMOS chip [23].

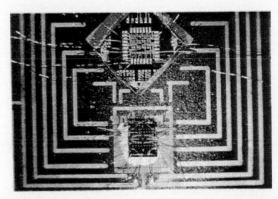

Figure 12.13. Photograph of electronic and VCSEL chips attached to a submount and wire bonded to electroplated traces [24].

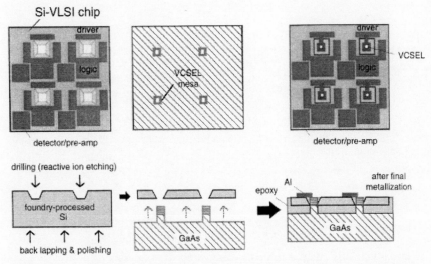

Figure 12.14. Schematic illustrations of a bonding process with plasma-etched holes [29].

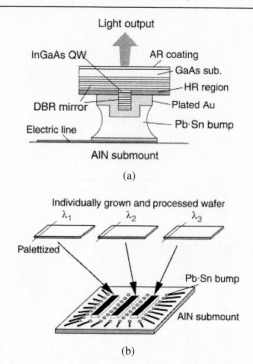

Figure 12.15. Flip-chip solder bonded linear arrays of VCSELs illustrating (a) the cross section of the bonded VCSEL and (b) arrays on a submount [30].

are etched in the appropriate locations in the chip. An array of VCSELs with 12 μm high mesas are then aligned and glued to the CMOS chip and thin film contacts are applied. This process grafts the silicon chip onto the VCSEL array, which the authors state is easier and less critical than grafting the VCSEL chip onto the CMOS. However, this technique uses a significant amount of the pixel area and does not provide a fast photodetector.

A third method of attaching VCSELs to electronic chip is used in the system described in Chapter 10. This method is a flip-chip bonding technique of VCSELs operating at a wavelength transparent to the substrate. For VCSELs grown on GaAs, λ is usually ~980 nm [30] and for VCSELs grown or wafer bonded to an AlGaAs substrate, λ is usually ~850 nm [31, 32]. An example of this technique for ~980 nm VCSELs is illustrated in Figure 12.15, which shows both a cross section of a solder bump bonded VCSEL and linear arrays of bonded VCSELs. The linear arrays have an advantage over the 2-D arrays in that they can be bonded directly to the electronic chip without covering up as much of the chip and thus avoiding the underlying photodetectors. For ~850 nm, the VCSELs are bonded top down on an AlGaAs wafer [31, 32].

12.4 Other Components of the Smart Pixel

Up to this point, this chapter has concentrated on the integration of VCSELs with the electronic chip. The following section discusses the design and integration of the other components of the smart pixel, namely the photodetector, receiver circuits, and VCSEL driver circuits.

12.4.1 Photodetectors

Even though there are many types of photodetectors, for example, pn junctions, pin diodes, metal-semiconductor-metal (MSMs), avalanche photodetectors (APDs), phototransistors, etc., almost all are based on collecting the optically generated electrons–hole pairs (EHPs) with a space charge layer (SCL) created by a pn junction. Most of these photodetectors have a responsivity in the range 0.2 to 0.6 A/W for front illumination [33]. The APD and phototransistor are the exceptions, since both have optical gain. However, the structure of the APD is more complex and phototransistors are often slow. The other photodetectors have approximately the same responsivity, and with proper design and material, they can operate at a GHz or higher. Therefore the basis for choosing a photodetector for large pixel arrays is primarily the compatibility with the integrated circuits. Monolithic integration of the PD is desirable since this is simpler, cheaper, and more reliable. However, hybrid integration may be needed to obtain the required performance and responsivity, especially for CMOS circuits. Performance, and to some extent, responsivity is determined by the optical adsorption coefficient of the photodetectors, and the geometry and width of the space charge layer.

Optical absorption in a semiconductor is characterized by an absorption coefficient, α, which is a strong function of the wavelength. The most common VCSEL wavelengths are ~670 nm, ~840 nm, and ~980 nm. The value of $1/\alpha$ (which is the $1/e$ absorption depth) for various semiconductors at these wavelengths is as shown in Table 12.5. When light is absorbed in a semiconductor, EHPs are generated both in and out of the space charge layer of the pn junction of the photodetector. The SCL has a high electric field and acts as the collector for all of the optically generated carriers. The EHPs generated in the SCL traverse this region in ~10 ps since they travel at saturated velocity. In contrast, EHPs generated outside the SCL diffuse slowly to the SCL and add long tails to the step response of the photodetector, thus limiting the frequency of the detector. For silicon, this can be a significant problem because $1/\alpha$ is usually much larger than the width of the SCL as has been shown by Beyette et al. [34], Stanko [35], and by Childers et al. [36, 37].

The doping concentration of the semiconductor, along with the applied reverse bias voltage, determines the width of the SCL. For doping concentrations of 10^{13}, 10^{14}, 10^{15}, and 10^{16} the approximate width of the SCL for a one-sided step junction with 5 V applied is ~26 μm, ~8 μm, 2.6 μm, and 0.8 μm respectively. These values are approximately independent of the semiconductor. In order to make a fast photodetector, the SCL width should be approximately equal to the absorption depth, which for Si requires the doping concentration to be of the order of 10^{14}. Childers et al. [36] have used a custom CMOS process to form a well with a doping level of 2.8×10^{14}, which provided a SCL width of ~5 μm. Unfortunately, the doping concentration of most CMOS foundry processes is ~10^{16}

Table 12.5. *The optical absorption lengths in semiconductors at common VCSEL wavelengths.*

λ	Si	GaAs	$In_{0.53}Ga_{0.47}As$	InP
670 nm	5 μm	0.2 μm	~0.2 μm	0.3 μm
850 nm	10–30	3	~0.2	1
980 nm	100	∞	~0.2	∞

or greater and cannot be altered by the foundry. As discussed below, there are circuit tricks that can be used to increase the speed and response of foundry-fabricated Si photodetectors. While the properties of InGaAs make it the best photodetector material, there are no circuit foundries for InGaAs and therefore pixels formed with these photodetectors must utilize some form of hybrid integration or be built by in-house custom fabrication. GaAs, however, has good optical absorption properties for $\lambda = 850$ and $670\,\text{nm}$ and a foundry is available to fabricate the pixel circuitry. Thus, GaAs circuits are well suited for the monolithic integration of photodetectors with electronic circuits.

For monolithically integrated GaAs MESFET circuits, MSM photodetectors are the obvious choice since both of these devices require a Schottky metal–semiconductor diode and thus can be fabricated with the same processing steps, although high performance optoelectronic circuits may require some alteration of the standard MESFET integrated circuit processing. In reality, the MSM detectors are fabricated much the same as MESFETs, as shown in Figures 12.2 and 12.8. PhotoFETs [38] are also easily fabricated with the standard MESFET process. These can have significant gain, but unfortunately, the high optical gain photoFETs are slow. The MSM detectors, in contrast, can be very fast.

For silicon CMOS circuits the two choices for a monolithically integrated photodetector using a standard foundry process are a pn junction or a lateral phototransistor. Neither of these can operate at frequencies greater than a few tens of MHz owing to the slow diffusion of carriers discussed previously. The pn junction is usually [34] formed by the n- or p-well with the substrate. For the standard CMOS process, these detectors can operate up to about 10 to 20 MHz but as slow as $\sim$1 MHz, with a responsivity of about 0.3 to 0.5 A/W for $\lambda \sim 840\,\text{nm}$. Decreasing the substrate doping to $\sim 10^{14}$ increases the maximum operating frequency by a factor of 10 [37]. This approach needs to be better developed in order that it can be easily and reliably applied to foundry-fabricated chips. It has also been suggested [39] that the formation of a surface grating can diffract the incident light causing it to propagate in the Si at an angle less than the surface normal. This causes the light to be absorbed closer to the surface and the SCL. Lateral phototransistors suffer from the same problems as the Si pn junctions; that is, they are relatively slow and have low sensitivity. Again these characteristics can be improved by lowering the substrate doping. It seems that the photodetector of choice for foundry-fabricated CMOS pixels is the p- or n-well junction. For low frequency operation, no modification is required. However, for high frequency operation, the standard process must be modified to decrease the effective doping level of the substrate and well regions. High-speed CMOS smart pixels can be achieved, even with slow photodetectors if a meta-stable receiver circuit is used, as discussed in the following section.

12.4.2 Receiver Circuits

The photodetectors provide the input to the smart pixel by converting the optical input power into the electrical current that drives the receiver circuit. There are several types of receiver circuits from which to chose, such as simple and multistage inverters, transimpedance amplifiers, and differential amplifiers, and the telecommunication industry has developed very high quality optoelectronic receivers for optical fibers. However, Woodward [40] has pointed out that the smart pixel receiver is quite different from the classical telecommunications grade receiver since the smart pixel receiver must be very compact and dissipate a minimum of power.

In inverting amplifiers (receivers), the current generated by the photodetector charges up the capacitance of the first stage; therefore the maximum frequency of the receiver can operate is limited by the input capacitance, the responsivity of the photodetector, and the optical input power. The total receiver input capacitor is the sum of the gate to source capacitance, the capacitance of the photodetector, and the parasitic capacitance. The photodetector capacitance ranges between approximately 2 and 200 fF, which corresponds to detectors with an area of $10 \times 10 \, \mu m^2$ and a space charge layer width of 5 μm and $40 \times 40 \, \mu m^2$ with a 1 μm SCL width respectively. The available input power is set by the optical power budget, but should range between 10 μW and 1,000 μW. Using these numbers, the operating frequency for the CMOS inverter type receiver is as high as 50 MHz but as low as 5 kHz for the largest photodetector and the lowest input power.

Transimpedance amplifiers (receivers) increase the maximum frequency of operation by the use of negative feedback, which reduces the output signal. This can be regained by adding one or more stages of amplification. The transimpedance receiver, therefore, occupies a larger area then the integrating receiver, dissipates more power and requires more careful circuit design but has a significantly larger bandwidth. For compactness and external control of the gain and bandwidth, both CMOS and MESFET integrated amplifiers incorporate active feedback components instead of passive resistors.

Differential amplifiers can be made very sensitive and are often used in both analog and digital input circuits. For digital circuits, very high sensitivity can be obtained by the use of meta-stable, clocked flip-flops [40–42], which utilize regenerative feedback, such as used in DRAM sense amplifiers. Although these circuits dissipate low power ($\sim$1 mW), they require synchronizing clock and careful circuit design to ensure stability over the operating temperature and in the presence of bus noise.

Until recently, most VCSEL-based smart pixels have used a simple single-stage inverting amplifier since the primary emphasis has been on the development of techniques for smart pixel fabrication and not on high performance or large arrays. Several receiver circuits specifically for smart pixels have now been reported. Some are slow (2 MHz) and simple [43, 44] while others are faster (250 MHz) [45] but are complex, dissipate a lot of power (10 mW), and occupy a large area. However, there has been significant progress in the design and analysis of receiver circuits for smart pixels. Van Blerkan et al. [46] analyzed 0.8 μm CMOS circuits to determine the number of transimpedance/voltage gain stages required as a function of the bit rate. This analysis was carried out for the conditions of either minimum optical input power or minimum total electrical + optical power. Their results are summarized in Tables 12.6 and 12.7, which indicate that minimum electric power dissipation is achieved by reducing the number of stages and increasing the optical input. This is explained by the fact that the greater number of stages provides a larger gain and therefore

Table 12.6. *Number of stages required for a minimum of input power [46].*

Bit rate	Required optical input power	Required electrical power	# of stages
250 Mb/S	1 μW	3 mW	5
500 Mb/S	20 μW	1 mW	3
1 Gb/S	200 μW	0.4 mW	1

Table 12.7. *Number of stages required for a minimum total power [46].*

Bit rate	Required optical input power	Required electrical power	# of stages
250 Mb/S	20 μW	200 μW	3
500 Mb/S	60 μW	300 μW	2
1 Gb/S	350 μW	200 μW	1

Table 12.8. *Comparison of three CMOS optoelectronic receivers [40].*

Receiver type	Area	Maximum frequency	Optical input power at this frequency	Electrical power dissipation
2-stage transimpedance (1 input beam)	350 μm^2	375 Mb/S	12 μW	3.5 mW
4-stage transimpedance (2 input beams)	1100 μm^2	1 Gb/S	200 μW	8 mW
Synchronous	1000 μm^2	320 Mb/S	12 μW	1 mW

less optical input power is required; however, each additional stage has a current path to ground, which increases the electrical power dissipation. The tables also show that 1 Gbit/s operation is possible with relatively low electrical power dissipation and optical input power, and while this is very encouraging, this analysis has not been compared with experimental circuits and therefore it serves only as a guide to what may optimally be obtained.

Woodward et al. [40, 47, 48] have reported on the design and experimental performance of three different VLSI (very large scale integration) compatible smart pixel optoelectronic receivers that were fabricated with 0.8 μm CMOS (see Figure 12.16). Table 12.8 summarizes their experimental results, which are seen to have the same trends as predicted by the analysis of Van Blerkan's analysis [46], and clearly shows that small-area receivers with low power dissipation are obtainable with CMOS for frequencies up to ~350 Mbit/s. However, to operate at frequencies approaching 1 Gbit/s with low power dissipation requires nontraditional approaches or better circuit design.

The synchronous or clock sense-amplifier type of receiver appears to be very attractive from the standpoint of low power dissipation, and several research groups have reported good results [40–42, 48, 49]. Although these circuits require more complex clocking, they have high sensitivity and thus may be the answer to the slow CMOS photodetector problem. The application of this type of receiver has only recently been investigated and much improvement is to be expected.

12.4.3 VCSEL Driver Circuits

Each VCSEL in the smart pixel requires a driver circuit whose current drive capacity matches the characteristics of the VCSEL. Over the past few years, the I_{th} and I_{max} of VCSELs have dropped significantly, as has the V_{th}. These changes in VCSEL characteristics have reduced the current drive and voltage specifications of the driver circuit, but while many receiver circuits have been reported, very few driver designs for VCSEL-based smart pixels have appeared in the literature. Probably the reason for this is the fact that every type of smart pixel (e.g., SEEDs) requires a receiver, but only the VCSEL-based smart pixel requires a driver.

For CMOS and MESFET circuits, the processing functions of the smart pixel utilize minimum size transistors in order to conserve space, increase speed, and reduce power dissipation. Unfortunately these transistors have low current sinking capability whereas the VCSELs require several milliamps of drive current. Thus, a buffer circuit must be provided to match the processing circuit to the VCSEL. In addition some method for providing a current to bias the VCSEL to I_{th} must also be provided.

It appears that the simplest method of providing the bias current is with a shunting transistor, which can be attached to any driver design. The current through this transistor is controlled by an external voltage, which sets the bias level for all of the VCSELs. This assumes that the I_{th} of the VCSELs does not vary across the array.

There are two general types of driver circuits, with the cascaded inverter amplifier being the most common. With these amplifiers, the W/L ratio of each succeeding CMOS or

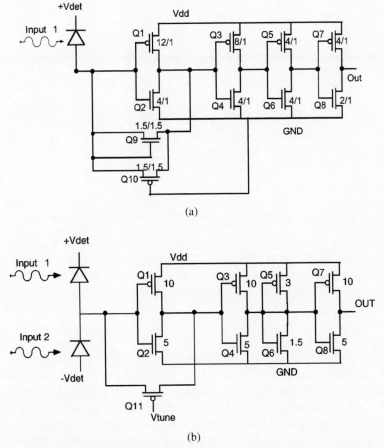

Figure 12.16. CMOS smart pixel receiver circuits: (a) single-beam transimpedance amplifier with three gain stages, (b) two-beam transimpedance amplifier, and (c) sense amp type amplifier [40].

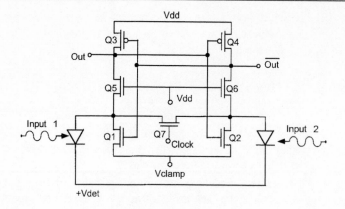

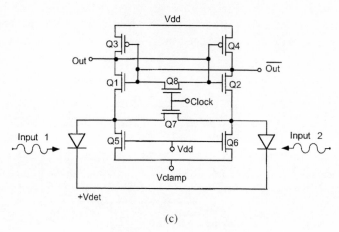

(c)

Figure 12.16. (*Continued*)

MESFET inverter increases to increase the current drive capability without unduly loading the input stage. Usually three to five stages are used, depending on the VCSEL current/signal transistor current ratio. These cascaded amplifiers are often used as bonding pad drivers on CMOS chips and in fact cell library pad driver designs can be used as VCSEL drivers. However, the current capability and impedance of these pad driver circuits usually don't match those of the VCSELs, but the same type of design methodology can be used to design the VCSEL driver.

Banwell et al. [50] have reported several driver designs such as the one shown in Figure 12.17. This circuit uses a single stage of buffering and a current mirror to drive the large output transistor. This circuit was designed to sink VCSEL currents up to 25 mA and thus the output transistor had a very large gate width of 500 μm. As a consequence, using 1 μm CMOS, the driver circuit occupied a 175 μm $\times$ 300 μm area, which is larger than the area of most pixels. The power dissipation of the circuit, ~60 mW, was also very high. Both the power and area would be significantly reduced if only ~5 mA VCSEL drive current is required and if the circuit were fabricated with 0.35 μm CMOS design rules. Even so, Banwell's circuit operated above 622 MHz. Sharma et al. [51] report a similar CMOS laser driver; unfortunately it is also large and the power dissipation is high.

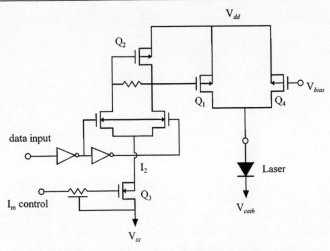

Figure 12.17. Schematic of a CMOS laser driver circuit [50].

12.5 Summary of Smart Pixel Design Concepts

Designing a smart pixel requires a number of decisions related to several interrelated factors, including:

- Choice of the type of substrate material and transistor type (e.g., Si CMOS, GaAs MESFET).
- Design of the electronic circuit that performs the desired processing function.
- Layout of this circuit and the placement of the optical components within this layout.
- Choice of photodetector type, its design, and method of integration.
- Method of integrating the VCSEL with the electronics.

Designing smart pixels, however, not only requires high performance components and techniques of integrating them, but they also require consideration of methods for the optical I/O routing, array mounting, and thermal heat sinking [52]. In addition the overall cost, size, weight, testability, and manufacturability of the pixel and the associated optoelectronic system play an important role in the design of a commercial product. At the present time there is no established methodology for smart pixel design that takes into account all of these factors and it is difficult to provide a generalized road map to the decision process. For small numbers of I/O, such as for data links, the pick and place technique using the best individual components is both workable and cost effective. However, as the size of the arrays becomes large, the optimum set of materials and component types and methods of integration become difficult and there is no clear decision path. As a result, the summary presented below does not attempt to provide a generalized approach to smart pixel design. Instead, the design choices for each of the three major electronic integrated circuit platforms are discussed separately. This approach not only simplifies the discussion, but it also fits the real world of optoelectronic design; that is, the electronic platform is usually dictated by the available process capabilities or the design experience of the designer or manager. Thus, a Si CMOS platform may be chosen even through a GaAs MESFET platform would be more optimum, or a GaAs HBT may be be chosen when a CMOS implementation would be better. The following provides a list of components and their method of integration for Si CMOS,

GaAs MESFET, and GaAs HBT implementations for large arrays of fast VCSEL-based smart pixels.

12.5.1 Silicon CMOS

- **VCSELs:** Since VCSELs cannot be grown directly onto silicon substrates, they must be attached by one of the hybrid integration techniques discussed previously. To conserve space, the VCSELs should be placed on top of the upper dielectric layer and located in or very near the pixel to reduce parasitics and reduce lead length.
- **Photodetectors:** Since silicon only poorly absorbs lights at $\sim$850 nm, the light penetrates deep into the silicon. For standard foundry-fabricated ICs the pn junction space charge layer is close to the surface and therefore the collection of the generated electron–hole pairs is primarily by diffusion, which is usually a slow process. This limits the speed of the pixels to $\sim$10–30 MHz unless special provisions are made such as reducing the substrate doping by either a custom IC process [53, 54] (this is cost effective if large batch runs are used) or by ion implantation to compensate the doping of the substrate. Differential inputs to a bistable latching amplifier (sense amplifier) can also be used but this type of circuit requires complex clocking, is more temperature sensitive, and is probably limited to a few hundred megahertz. However, this is presently the fastest CMOS monolithically integrated technique that uses standard foundry processing. This circuit also dissipates very low power, $\sim$1 mW. For hybrid integration, InGaAs or GaAs photodetectors appear to be the most suitable for realizing high-speed CMOS smart pixels. Using a GaAs detector, both the VCSELs and photodetectors can be attached at the same time. However, attaching the VCSEL and photodetector arrays independently is also a viable process. pin diodes or MSM detectors appear to be the best choices since they are fast, have reasonable responsivity, and have simple construction.
- **Isolation:** Since both the VCSELs and photodetectors are hybrid bonded on top of the passivating SiO_2 layer, isolation of the devices occurs without any additional steps. If a bonded-wafer technique is used, then mesa etching must separate the VCSEL and photodetector.
- **Summary:** Silicon CMOS circuits are and will continue to be the most advanced and densely integrated circuits available, and they are used for almost every existing electronic application. In addition, silicon is a very robust material and CMOS circuits can withstand considerable postfabrication processing. For these reasons, CMOS should be the electronic platform of choice for many smart pixel array applications. The optoelectronic properties of silicon, however, are not ideal and therefore the VCSELs (and probably also the photodetector) must be added with a hybrid bonding process. Therefore the key to success for CMOS is the development of a reliable/manufacturable hybrid process that can bond both the detectors and the VCSELs directly to the pixels.

12.5.2 GaAs MESFETs

- **VCSELs:** The VCSELs can be integrated either with a monolithic or hybrid process. The monolithic process would appear to be the most desirable since it combines the VCSELs with the electronic circuitry and photodetectors; however, this requires a specialized epitaxial growth facility and special device processing. Thus, these monolithic arrays will be available to only a select few unless ASIC-type circuits or MOSIS runs become available. The monolithic approach will probably be limited to fewer transistors and less functionality than the hybrid approach. This will also limit the applications. A hybrid bonding process similar to that required for CMOS will have to be used for many GaAs smart pixel arrays since this is a more flexible process that can accommodate a broad range of smart pixel designs. The hybrid bonding process must still be developed, but good progress is presently being made.

- **Photodetectors:** Because 850 nm light is readily adsorbed in GaAs, monolithically integrated photodetectors are both fast and have good responsivity. MSM photodetectors are the obvious choice for MESFET circuits since both devices use the same fabrication steps. Therefore, high-speed photodetectors are readily available with both standard foundry-fabricated and custom GaAs MESFET ICs with little or no change in the processing procedure. pin diodes have a somewhat higher responsivity than MSM detectors and can be fabricated with a modification of the standard MESFET processing, although for most applications, the extra effort and expense are not cost effective.

- **Isolation:** The foundry-fabrication process isolates the MSMs and MESFETs. If the VCSELs are attached with any of the hybrid bonding techniques, then VCSELs are automatically isolated from the other smart pixel components. Isolation of components of a monolithically smart pixel usually requires a semi-insulating substrate and either proton implantation and/or mesa etching.

- **Summary:** High performance GaAs MESFET circuits are readily available from standard foundries and from custom IC facilities, and while their minimum linewidth will always lag behind that of CMOS, high-density GaAs MESFET circuits are routinely fabricated. Thus, these GaAs circuits can serve as good electronic platforms for smart pixel arrays even through they are not as robust or as dense as CMOS circuits. However, the primary advantage of GaAs MESFET ICs is the availability of a monolithically integrated high-speed MSM detector that can easily be fabricated on the circuit without added expense. In addition, the VCSEL can also be monolithically integrated with the MSMs and MESFETs using custom facilities. This is a more expensive process and probably should be considered only for high performance applications where no other solution is available and cost is not the primary consideration. The hybrid integration of VCSELs onto foundry-fabricated MSM/MESFET ICs, however, appears to be viable [2]. GaAs circuits are also radiation hard, which must be considered for space and military applications. A technique for growing LEDs on a foundry-fabricated chip has also been successfully demonstrated [55] but the low speed will limit the application of this approach.

12.5.3 GaAs HBTs/HPTs

- **VCSELs:** Since VCSELs can easily be grown on the same substrate as the HBT/HPT layers and since there is no standard HBT/HPT circuit foundry, monolithic integration seems the best fabrication technique for these smart pixels. First growing a standard VCSEL on a p GaAs substrate and then growing the HBT/HPT layers on top the VCSEL appears to be the simplest growth sequence. This is an easy process since the growth of VCSELs is well established and the performance of HBTs/HPTs depends primarily on the bulk properties of the layers. These properties are easily controlled.

- **Photodetector:** The most logical photodetector for these circuits is the pin diode, since these layers are the same as those of the base-collector-subcollector of the HBT/HPT. These pin diodes are fast and have good responsivity.

- **Isolation:** Three types of isolation are required: a) the pin diodes are isolated from the HBTs/HPTs by mesa etching, b) the pin diode and HBT/HPT circuit are isolated from the substrate by a semi-insulating layer of either GaAs or AlAs, and c) the VCSELs are isolated from these components with either a proton implant or by an etched trench.

- **Summary:** This monolithic integration technique is similar to the ones used with MESFETs, MSMs, and VCSELs. Foundry-fabricated VCSEL wafers can be used as substrates with good reliability, but converting the wafer into a large operating array must be performed in specialized facilities. It also seems that these smart pixels would be limited to performing less complicated functions since the number and density of transistors would be low compared with CMOS or MESFET circuits. Isolation, leakage currents, and optical crosstalk may also be more difficult to control. Nonetheless, the HBTs/HPTs and pin diodes should be robust and reliable since they operate on bulk, not surface properties. In addition, they should be radiation hard and fast.

12.6 VCSEL-Based Free-Space Optoelectronic Systems

As pointed out in Chapter 9, the first system applications of VCSELs will be in guide-wave systems such as data links where only a few VCSELs are required. Free-space optoelectronic processing, however, will require several hundred VCSELs in order to be practical, and the techniques for fabricating these processing arrays has been the main thrust of this chapter. To close the chapter, a brief review of the free-space parallel processing systems reported in the literature is given. This includes systems that perform sorting [3, 34, 35, 56, 57], page-oriented data filtering [24], communication routing [1, 4, 5, 58–60], and interconnections [27]. The sorting and data filtering systems utilize straight pass interconnects while some of the routing and interconnection systems use global interconnections.

The sorting system was constructed using a recirculating architecture that passes data optically between two smart pixel arrays [3, 35, 56, 57]. Each smart pixel array is implemented using a hybrid combination of CMOS and VCSEL technologies in which the logic and photodetector circuitry were monolithically integrated in a CMOS chip. This chip was then connected to a separate VCSEL chip. The entire system was assembled on a slotted plate optomechanical platform, as illustrated in the photo of Figure 12.18. This system used a 4 × 4 array of VCSELs whose speed of operation was limited by the silicon photodetector to about 10 MHz.

The data filter [24] also used a similar slotted plate platform and hybrid bonding method, but the logic gates were fabricated from heterojunction phototransistors, which served to detect the light from the VCSELs, amplify the input signal, and perform the required XOR and AND logic functions. This system was demonstrated in the static mode because a multiple source input was not available.

Cheng et al. [4, 5] have demonstrated a programmable optical interconnect architecture that performs dynamic bus switching and data-path routing using monolithically integrated HBTs and VCSELs as illustrated in Figures 12.4 and 12.19. As shown in Figure 12.19, internode optical paths can be reconfigured by simply programming the connections using a two-dimensional array of smart pixels. The monolithically integrated switches have operated at close to 1 Gbit/s and transmissions experiments have been performed at 650 MHz.

Various ATM packet switching demonstrations have been reported that do not use smart pixels but rather use the VCSELs for optical interconnects. Li et al. [1, 58] have implemented an optical crossbar interconnect using beam steering from an electrically addressed array of

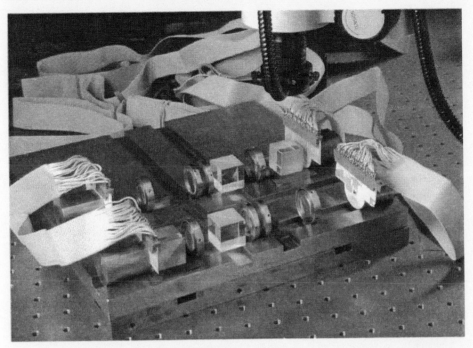

Figure 12.18. Slotted plate optomechanical platform used to assemble a CMOS/VCSEL recirculating sorting system using an array of pixels shown in Figure 12.3 [35, 54, 55].

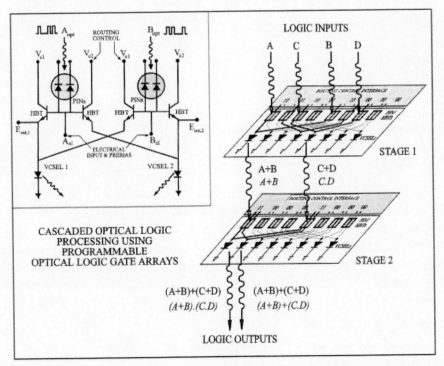

Figure 12.19. A dynamically reconfigurable, multi-access optical network using the pixels shown in Figure 12.4 [4, 5].

VCSELs. Each input channel is connected to a VCSEL array and beam steering is accomplished by turning on a VCSEL located at a different distance from the optical axis. Thus, the "output" from the channel can be focused on a different photodetector in the output detector array. A system with 240 channels has been demonstrated, with a single channel operating at up to 500 Mbit/s. A fiber bundle was used to decrease the packaging problems associated with free-space processing. Kawai and Kurita [59] used multiple wavelength VCSELs and fibers to demonstrate a space-division-multiplexing crossbar switch and Wilmsen et al. [60] used a VCSEL array and liquid crystal shutter to demonstrate a routing switch. Since these systems did not use smart pixel arrays, the integration technique is not so important. A much more complex routing system is described by Kasahara in Chapter 10.

Neff et al. [27] used an MCM method to separately bond VCSEL and CMOS arrays to a glass substrate. The input and output light passed through the substrate. By facing two of these units toward each other bidirectional processing similar to the recirculating architecture was realized. The speed of operation was limited by the silicon photodetector to about 20 MHz.

12.7 Conclusions

As discussed in the previous chapters, research on VCSEL arrays has resulted in the development of VCSELs with wall-plug efficiencies as high as 50%, threshold currents as low as a few tens of microamps, and highly controlled epitaxial growth processes that produce uniform laser characteristics across a 3 inch wafer (Chapters 1, 5 and 6). Given the complexity of the VCSEL structure, these results are truly phenomenal. Similarly, significant progress has been made on one-dimensional VCSEL arrays for fiber-optic data links (Chapter 11). In addition, Si CMOS and GaAs MESFET integrated circuits are even better established. The major issue is the merging of the optical and electronic technologies on a large scale. For small arrays (e.g., ten to sixteen VCSELs for data links), this task is relatively straightforward to implement and there are several hybrid integration techniques available. These systems must be low cost and be suitable for high-volume manufacturing. Free-space interconnect systems, in contrast, are primarily intended for applications that contain 10^2 to 10^4 optical interconnections, and perform complex tasks, and they initially would not be manufactured in large volume. There has been significant progress on both hybrid and monolithic integration techniques and, without doubt, these will produce workable arrays that will be inserted into high performance systems in the near future.

Acknowledgments

The author would like to acknowledge the many students (Fred Beyette, Jr., Steward Feld, Rui Pu, Eric Hayes, Rick Snyder, Randy Jurrat, Pat Stanko, and many others) who have contributed to the research summarized in this chapter. This work was supported in part by NSF Grant #9408371, NSF/ERC Grant #9485502, and the Colorado Advanced Technology Institute.

References

[1] Y. Li, T. Wang, and R. A. Linke, "VCSEL-Array-Based Angle-Multiplexed Optoelectronic Crossbar Interconnects," *Applied Optics*, **35**(8), 1282–1295, 1996.

[2] R. Pu, E. M. Hayes, R. Jurrat, C. W. Wilmsen, K. D. Choquette, H. Q. Hou, and K. M. Geib, "VCSELs Bonded Directly to Foundry Fabricated GaAs Smart Pixel Arrays," *IEEE Photonics Technology Letters*, **9**, 1622–1624, December 1997.

[3] F. R. Beyette, Jr., *Design and Analysis of and Optoelectronic Recirculating Sorter*, Ph. D Dissertation, Colorado State University, August 1995.

[4] J. Cheng, Y.-C. Lu, B. Lu, G. G. Oritz, A. C. Alsuino, C. P. Hains, H. Q. Hou, M. J. Hafish, G. A. Vawter, and J. C. Zolper, "High-Speed Optical Transceiver and Routing Switch Using VCSEL-Based Integrated Optoelectronics," *SPIE*, **3005**, pp. 327–333, San Jose, California, February 1997.

[5] B. Lu, Y.-C. Lu, J. Cheng, M. J. Hafich, J. Klem, and J. C. Zolper, "High-Speed, Cascaded Optical Logic Operations Using Programmable Optical Logic Gate Arrays," *IEEE Photonics Technology Letters*, **8**, 166–168, January 1996.

[6] P. Zhou, J. Cheng, C. F. Schaus, S. Z. Sun, C. Hains, K. Zheng, E. Armour, W. Hsin, D. R. Meyers, and G. A. Vawter, "Cascadable, Latching Photonic Switch with High Optical Gain by Monolithic Integration of a Vertical Cavity Surface Emitting Laser and a PNPN Photothyristor," *IEEE Photonics Technology Letters*, **3**, 1009–1012, November 1991.

[7] J. Cheng, P. Zhou, S. Z. Sun, S. Hersee, D. R. Meyers, J. Zolper, and G. A. Vawter, "Surface Emitting Laser-Based Smart Pixels for Two-Dimensional Optical Logic and Reconfigurable Optical Interconnections," *IEEE J. Quantum Electronics*, **29**, 741–756, February 1993.

[8] S. A. Feld, *Monolithically Integrated HPT/VCSEL Logic Pixels*, Ph. D Dissertation, Colorado State University, May 1996.

[9] Y. K. Yang, T. G. Dziur, T. Bardin, S. C. Wang, and R. Fernandez, "Monolithic Integration of a Vertical Cavity Surface Emitting Laser and a Metal Semiconductor Field Effect Transistor," *Applied Physics Letters*, **62**, 600–602, February 1993.

[10] J. Bristow, J. Lehman, Y. Liu, M. Hibbs-Brenner, L. Galarneau, and R. Morgan, "Recent Progress in Short Distance Optical Interconnects," *SPIE Digest*, **3005**, 112–119, February 1997.

[11] Y. Liu, M. Hibbs-Brenner, R. Morgan, J. Nohava, R. Walterson, T. Marta, S. Bounnak, E. Kalweit, J. Lehman, D. Carlson, and P. Wilson, "Integrated VCSELs, MSM Photodetectors, and GaAs MESFETs for Low Cost Optical Interconnects," *Proceedings of the Spatial Light Modulators Conference*, pp. 22–24, Lake Tahoe, Nevada, March 1997.

[12] S. Matsuo, T. Nakahara, Y. Kohama, Y. Ohiso, S. Fukushima, and T. Kurokawa, "Monolithically Integrated Photonic Switching Device Using an MSM PD, MESFETs and a VCSEL," *IEEE Photonics Technology Letters*, **7**, 1165–1167, 1995.

[13] R. P. Bryan, W. S. Fu, and G. R. Olbright, "Hybrid Integration of Bipolar Transistors and Microlasers: Current-Controlled Microlaser Smart Pixels," *Applied Physics Letters*, **62**, 1230–1232, March 1993.

[14] A. Camperi-Ginestet, M. Hargis, N. Jokerst, and M. Allen, "Alignable Epitaxial Lift-off of GaAs Materials with Selective Deposition Using Polyimide Diaphragms," *IEEE Photonics Technology Letters*, **3**, 1123–1126, 1991.

[15] S. M. Fike, B. Buchanan, N. M. Jokerst, M. A. Brooke, T. G. Morris, and S. P. DeWeerth, "8 × 8 Array of Thin-Film Photodetectors Vertically Interconnected to Silicon Circuitry," *IEEE Photonics Technology Letters*, **7**, 1168–1171, October 1995.

[16] D. L. Mathine, R. Droopad, and G. N. Maracas, "A Vertical Cavity Surface Emitting Laser Appliquéd to a 0.8 μm NMOS Driver," *IEEE Photonics Technology Letters*, **9**, 869–872, July 1997.

[17] H. Fathollahnejad, D. L. Mathine, R. Droopad, G. N. Maracas, and S. Daryanani, "Vertical Cavity Surface Emitting Lasers Integrated onto Silicon Substrates by PdGe Contacts," *Electronics Letters*, **30**, 1235–1236, 1994.

[18] K. W. Goossen, J. E. Cunningham, and W. Y. Jan, "GaAs 850 nm Modulators Solder-Bonded to Silicon," *IEEE Photonics Technology Letters*, **7**, 776–778, July 1993.

[19] L. M. F. Chirovsky, A. L. Lentine, K. W. Goossen, S. P. Hui, B. T. Tseng, L. A. D'Asaro, R. E. Leibenguth, J. A. Walker, J. E. Cunningham, G. Livercu, D. Dahringer, D. Kossives, D. D.

Bacon, R. L. Morrison, R. A. Novotny, and D. B. Bucholz, "A High Speed Optoelectronic Chip with 4352 Optical Inputs/Outputs for a 256 × 256 ATM Switching Fabric," *Optical Computing*, Conference, April 1996.

[20] E. M. Hayes, R. Jurrat, R. Pu, R. D. Snyder, S. A. Feld, P. Stanko, C. W. Wilmsen, K. D. Choquette, K. M. Geib, and H. Q. Hou, "Foundry Fabricated Array of Smart Pixels Integrating MESFETS/MSMs and VCSELs," *International J. Optoelectronics*, **11**, 229–237, 1997.

[21] H.-J. J. Yeh and J. S. Smith, "Integration of GaAs Vertical-Cavity Surface Emitting Laser on Si by Substrate Removal." *Applied Physics Letters*, **64**, 1466–1468, March 1994.

[22] S. Matsuo, K. Tateno, T. Nakahara, H. Tsuda, and T. Kurokawa, "Use of Polyimide Bonding for Hybrid Integration of a Vertical Cavity Surface Emitting Laser on a Silicon Substrate," *Electronics Letters*, **33**, 1148–1149, June 1997.

[23] S. Matsuo, T. Nakahara, K. Tateno, and T. Kurokawa, "Novel Technology for Hybrid Integration of Photonic and Electronic Circuits," *IEEE Photonic Technology Letters*, **8**, 1507–1509, November 1996.

[24] R. D. Snyder, S. A. Feld, P. J. Stanko, E. M. Hayes, G. Y. Robinson, C. W. Wilmsen, K. M. Geib, and K. D. Choquette, "Database Filter: Optoelectronic Design and Implementation," *Applied Optics*, **36**, 4881–4889, July 1997.

[25] D. N. Kabal, G. C. Boisset, D. R. Rolston, and D. V. Plant, "Packaging of Two-Dimensional Smart Pixel Arrays," *Proceedings of the Smart Pixel Conference*, pp. 53–54, Keystone, Colorado, August 1996.

[26] R. F. Carson, M. L. Lovejoy, K. L. Lear, M. E. Warren, P. K. Seigl, D. C. Craft, S. P. Kilcoyne, G. A. Patrizi, and O. Blum, "Low-Power Approaches for Parallel, Free-Space Photonic Interconnects," *Critical Reviews*, **CR62**, 35–63, January 1996.

[27] J. A. Neff, C. Chen, T. McLaren, C. C. Mao, A. Fedor, W. Berseth, Y. C. Lee, and V. Morozov, "VCSEL/CMOS Smart Pixel Arrays for Free-space Optical Interconnects," *Proceedings of the Third International Conf. on Massively Parallel Processing Using Optical Interconnections*, October 1996.

[28] K. Hatada, H. Fujimoto, T. Ochi, and Y. Ishida, "LED Array Modules by New Technology Microbump Bonding Method," *IEEE Trans. Components, Hybrids and Manufacturing Technology*, **13**, 521–527, September 1990.

[29] M. S. Jin, V. Ozguz, and S. H. Lee, "Integration of Microlaser Arrays with Thinned and Drilled CMOS Silicon Driver Arrays," *Proceedings of the Optical Computing Conference*, Sendai, Japan, pp. 68–69, April 1996.

[30] I. Ogura, M., Kajita, S. Kawai, and K. Kasahara, "Fabrication of Multiple Wavelength VCSEL Array Using Flip-Chip Bonding," *Proceedings of the Smart Pixel Conference*, Lake Tahoe, Nevada, pp. 47–48, July 1994.

[31] Y. Ohiso, K. Tateno, Y. Kohama, A. Wakasuki, H. Tsunetsugu, and T. Kurokawa, "Flip-Chip Bonded 0.85 μm Bottom-Emitting Vertical-Cavity Laser Array on an AlGaAs Substrate," *IEEE Photonics Technology Letters*, **8**, 1115–1117, September 1996.

[32] K. D. Choquette, B. Roberds, K. M. Geib, H. Q. Hou, R. D. Twesten, K. L. Lear, and B. E. Hammons, "Bottom-Emitting 850 nm Selectively Oxidized VCSELs Fabricated Using Wafer Bonding," *LEOS Conf.*, San Francisco, CA, November, 1997.

[33] P. Bhattacharya, *Semiconductor Optoelectronic Devices*, Prentice Hall, Englewood Cliffs, NJ, 1994.

[34] F. R. Beyette, Jr., P. J. Stanko, E. M. Hayes, R. D. Snyder, S. A. Feld, P. A. Mitkas, and C. W. Wilmsen, "An Optoelectronic Recirculating Sorter Based on CMOS/VCSEL Smart Pixel Arrays: Architecture and System Demonstration," *Optical Review*, **3**, 373–375, 1996.

[35] P. J. Stanko, *Demonstration and Performance of a Tomographic Recircuitating Sorter*, M.S. Thesis, Colorado State University, May 1997.

[36] J. E. Childers, J. E. Morris, and M. R. Feldman, "CMOS Compatible Photodetectors for Optical Interconnects," *SPIE*, **1849**, 292–297, 1993.

[37] J. E. Morris, M. R. Feldman, W. H. Welch, M. Nakkar, H. Yang, J. Childers, Y. Raja, I. Turlik,

G. Adrema, P. Magill, and E. Yung, "Prototype Optically Interconnected Multichip Module Based on Computer Generated Holograms," *SPIE*, **1849**, 48–53, 1993.

[38] J. Luo, A. Grot, and D. Psaltis, "High-Responsivity Optical FET's Fabricated on a FET-SEED Structure," *IEEE Photonics Technology Letters*, **7**, 760–762, July 1995.

[39] B. D. Clymer and D. Gillifillan, "Corragation Gratings for Fast Integrated Complementary MOS Photodetectors Implementation and Diffraction Analysis," *Applied Optics*, **30**, 4390–4395, October 1991.

[40] T. K. Woodward, A. V. Krishanamoothy, A. L. Lentine, and L. M. F. Chirovsky, "Optical Receivers for Optoelectronic VLSI," *IEEE J. Selected Topics Quantum Electronics*, **2**, 106–116, April 1996.

[41] J. A. B. Dines, "Smart Pixel Optoelectronic Receiver Based on a Charge Sensitive Amplifier Design," *IEEE J. Selective Topics Quantum Electronics*, **2**, 117–120, 1996.

[42] K. Ayadi, M. Kuijk, P. Heremans, G. Bickel, G. Borghs, and R. Vounckx, "A Monolithic Optoelectronic Receiver in Standard 0.7 micron CMOS Operating at 180 MHz and 176-fJ Light Input Energy," *IEEE Photonic Technology Letters*, **9**, 88–90, 1997.

[43] Y.-T. Huang, "Low-Impedance Complementary MOS Optical Receiver for Optical Interconnects," *Applied Optics*, **31**, 4623–4624, July 1992.

[44] Y.-T. Huang, "Optimized Integrated CMOS Optical Receiver for Optical Interconnects," *IEE Proceedings-J*, **140**, 107–114, April 1993.

[45] M. Lee, O. Vendier, M. A. Brooke, N. M. Jokerst, and R. P. Leavitt, "CMOS Optical Receiver with Integrated InGaAs Thin-Film Inverted MSM Detector Operating Up To 250 Mbps," *Proceedings of the Smart Pixel Conference*, Keystone, Colorado, pp. 17–19, August 1996.

[46] D. A. Van Blerkon, C. Fan, M. Blume, and S. C. Esener, "Optimization of Smart Pixel Receivers," *Proceedings of the Smart Pixel Conference*, Keystone, Colorado, pp. 68–69, August 1996.

[47] A. V. Krishnamoorthy, A. L. Lentine, K. W. Goossen, J. A. Walker, T. K. Woodward, J. E. Ford, G. F. Aplin, L. A. D'Asaro, S. P. Hui, B. Tseng, R. Leibenguth, D. Kossives, D. Dahringer, L. M. F. Chirovsky, and D. A. B. Miller, "3-D Integration of MQW Modulators over Active Submicron CMOS Circuits: 375 Mb/s Transimpedance Receiver-Transmitter Circuit," *IEEE Photonics Technology Letters*, **7**, 1288–1290, November 1995.

[48] T. K. Woodward, A. V. Krishnamoorthy, A. L. Lentine, K. W. Goossen, J. A. Walker, J. E. Cunningham, W. Y. Jan, L. A. D'Asaro, L. M. F. Chirovsky, S. P. Hui, B. Tseng, D. Dahringer, and R. E. Leibenguth, "1-Gb/s Two-Beam Transimpedance Smart-Pixel Optical Receiver Made from Hybrid GaAs MQW Modulators Bonded to 0.8 μm Silicon CMOS," *IEEE Photonics Technology Letters*, **8**, 422–424, March 1996.

[49] T. K. Woodward, A. L. Lentine, K. W. Goossen, J. A. Walker, B. T. Tseng, S. P. Hui, J. Lothian, and R. E. Leibenguth, "Demultiplexing 2.48-Gb/s Optical Signals with a CMOS Receiver Array Based on Clocked-Sense-Amplifiers," *IEEE Photonics Technology Letters*, **9**, 1146–1148, August 1997.

[50] T. C. Banwell, A. C. Von Lehmen, and R. R. Cordell, "VCSE Laser Transmitters for Parallel Data Links," *IEEE J. Quantum Electronics*, **29**, 635–644, February 1993.

[51] R. Sharma, J. E. Childers, F. E. Kiamilev, and M. R. Feldman, "High Speed CMOS Laser Drivers," *SPIE*, **3005**, San Jose, California, February 1997.

[52] F. B. McCormick, "Smart Pixel Optics and Packaging," *Proceedings of the Smart Pixel Conference*, Keystone, Colorado, pp. 45–46, August 1996.

[53] C. L. Schow, J. D. Schaub, R. Li, J. Qi, and J. C. Campbell, "A Monolithically Integrated 1-Gb/s Silicon Photoreceiver," *IEEE Photonics Technology Letters*, **11**, 120–121, January 1999.

[54] H. Zimmermann, T. Heide, and A. Ghazi, "Monolithic High-Speed CMOS-Photoreceiver," *IEEE Photonics Technology Letters*, **11**, 254–256, February 1999.

[55] A. C. Grot, D. Pasltis, K. V. Shenoy, and C. G. Fonstad, Jr., "Integration of LED's and GaAs Circuits by MBE Regrowth," *IEEE Photonics Technology Letters*, **6**, 819–821, July 1994.

[56] F. R. Beyette, Jr., P. J. Stanko, E. M. Hayes, R. D. Snyder, S. A. Feld, P. A. Mitkas, and C. W.

Wilmsen, "An Optoelectronic Recirculating Sorter Based on CMOS/VCSEL Smart Pixel Arrays," *Optical Review*, **3**, 373, 1996.

[57] F. R. Beyette, Jr., P. J. Stanko, E. M. Hayes, R. D. Snyder, and C. W. Wilmsen, "An Optoelectronic Recirculating Sorter: Architecture and System Demonstration," *Optical Engineering*, **37**, 312–319, 1998.

[58] Y. Li, T. Wang, and S. Kawai, "Distributed Crossbar Interconnects with VCSEL Angle Multiplexing and Fiber Image Guides," *Applied Optics*, **37**, 254–263, 1998.

[59] S. Kawai and H. Kurita, "Electrophotonic Computer Network with Strictly Nonblocking and Self-Routing Functions," *Applied Optics*, **35**(8), 1309–1316, 1996.

[60] C. W. Wilmsen, C. Duan, J. R. Collington, M. P. Dames, and W. A. Crossland, "VCSEL Based Optoelectronic Asynchronous Transfer Mode Switch," *Optical Engineering*, **38**, 1999.

Index